Major Financial Institutions of Continental Europe 1989/90

Managing Editor
R M Whiteside

Editors
J Forsyth
A Wilson

Published by

Graham & Trotman

A member of the Kluwer Academic Publishers Group
LONDON/DORDRECHT/BOSTON
Sterling House, 66 Wilton Road, London SW1V 1DE

Second edition published in 1989 by

Graham & Trotman Limited
Sterling House, 66 Wilton Road,
London SW1V 1DE

Graham & Trotman Ltd is a member of the
Association of British Directory Publishers

ISBN-13: 978-94-010-7011-9   e-ISBN-13: 978-94-009-1151-2
DOI: 10.1007/978-94-009-1151-2

©1989 Graham & Trotman Limited
Softcover reprint of the hardcover 2nd edition 1989

British Library Cataloguing in Publication Data
is available.

Major financial institutions of continental
Europe 1989/90.

Contents

Contents

Major Companies of
AUSTRIA

ANGLO-ELEMENTAR VERSICHERUNGS AG

Kärntner Ring 12, 1015 Wien
Tel: (0222) 501 67-0
Telex: 132355 ae wnd a
Telefax: 50 54 008

Supervisory Board: Generaldirektor Dr. Christian Beurle
(Vorsitzender), Generaldirektor Dr. Guido Schmidt-Chiari
(Stellvertreter des Vorsitzenden), Dr Jan Boetius
(Stellvertreter des Vorsitzenden), Generaldirektor Octavian
Gassauer-Fleissner, Kommerzialrat Dkfm. Dr. Theodor Hladik,
Christian Norgren, Bergrat h.c. Gewerke Dipl.-Ing. Gottfried
Pengg, Dr. Fritz Rafaseder, Generaldirektor Dr Ottheinrich
Frolich, Generaldirektor Dkfm. Dr. Erwin Bundschuh.
Arbeitnehmervertreter: Disponent Josef Geppner
(Abteilungsleiter), Sekretär Armin Hämmerle, Bezirksdirektor
Herbert Liebermann, Dr. Stefan Nemeth, Sekretär Alfred
Praier (Abteilungsleiter-Stellvertreter)
Management Board: Dipl.-Ing. Erik Skreiner (Generaldirketor
Vorsitzender), Erich Putzer (Direktor), Dr Karl Faulmann
(Direktor), Dkfm. Dr Christian Leydolt (Direktor)
Senior Executives: Bereichsleiter: Otto Dohnal (Vertrieb
Direktor), Dr. Franz Bielansky (Fach Direktor), Hanns Bartl
(Schaden Direktor), Gerhard Bockhorni (Generalsekretariat
Direktor). Leiter der Abteilung Lebensversicherung: Walter
Niederhametner (Abteilungsdirektor Mathematischer
Sachverständiger). Leiter der Landesdirektionen: Kurt Horak
(Kärnten und Osttirol Direktor), Hermann Pummer
(Niederösterreich und Burgenland, Direktor), Dr Christian Oppl
(Oberösterreich, Direktor), Othmar Horwath (Salzburg,
Direktor), Eduard Weiss (Steiermark, Direktor), Klaus
Franceschinel (Tirol, Direktor), Ing. Wolfgang Ortner (Direktor),
Dr. Peter Veith (Wien, Direktor)

PRINCIPAL ACTIVITIES: Insurance including life insurance
Parent Company: Allianz AG, Munich, Germany
Financial Information:

	31.12.87
	S'000
Premium income	2,520,000

Principal Shareholders: Allianz AG, Munich
No of Employees: 1,248

ANKER, DER, ALLGEMEINE VERSICHERUNGS-AKTIENGESELLSCHAFT

Hoher Markt 10-11, A-1010 Wien
Tel: (0222) 53 474

Management Board: Dipl Ing Dr Robert Nemling (Vorsitzender),
Walter Krejcy, Heinz Pukl

PRINCIPAL ACTIVITIES: Insurance services including life,
accident and sickness cover
Financial Information:

	1986	1987
	S'000	S'000
Premium income	1,512,000	1,636,000

No of Employees: 714

AUSTRIA VERSICHERUNGSVEREIN AUF GEGENSEITIGKEIT

Untere Donaustr 25, A-1021 Wien
Tel: (0222) 2475 11
Telex: 135308 auver a

Management Board: H Schimetschek (Chairman), Dr Franz
Korinek, H Schatzdorfer, Dr Heinz Dopplinger

PRINCIPAL ACTIVITIES: Life insurance and general insurance
Subsidiary Companies: Austria Österr. Versicherung AG; Austria
Versicherungen AG; Austria Österr. Hotelbetriebs AG; Vermi
Versicherungs-Vermittlungsges mbH; Austria Assicurazioni
SpA (90%) (Italy)

Financial Information:

	1986	1987
	S'000	S'000
Premium income	3,548,000	3,735,000

No of Employees: 1,470

BANK FÜR ARBEIT UND WIRTSCHAFT AG BAWAG

2-4 Seitzergasse, 1010 Wien
Tel: (222) 53 45 30
Cable: BAWAGBANK WIEN
Telex: 115311 bawag a
Telefax: (222) 53 4 53/2202 (L/C Department); 2840 (General)

Supervisory Board: Alfred Ströer (Chairman), Manfred Kadits
(Deputy Chairman), Fritz Klenner (Honorary Chairman), Paul
Balaz, Franz Breit, Robert Brooks, Dr Josef Cerny, Günther
Engelmayer, Dr Hermann Gerharter, Karl Hans, Hans Klingler,
Franz Pfundner, Rudolf Randus, Dr Anton Rauter, Edwin
Schuster, Karl Spörck, Anton Vitzthum, Walter Weigl, Dr Kurt
Weihs, Rudolf Weissengruber
Board of Management: Walter Flöttl (Chairman and Director
General), Dipl Ing Walter Böck, Helmut Elsner, Dr Gerhard
Partik
Senior Executives: Dr G Braunsperger (Manager, Foreign
Department), H Leichtfried (Manager, Foreign Exchange and
Deposits)

PRINCIPAL ACTIVITIES: Bank
Subsidiary Companies: Credit Institutions:- Gara Real-und
Personalkreditbank AG; Österreichische Konsumbank AG;
Venture Finanzierung GesmbH; General Motors BankgesmbH;
Sparda Bank AG; Eurocard Austria KreditkartengesmbH. Non-
Banks:- BAWAG Leasing GesmbH; BAWAG-Versicherung AG;
Steyrermühl Papierfabriks-und Verlags-AG; Reisebüro Ruefa
GesmbH. Abroad:-Bank für Gemeinwirtschaft AG (Germany)
Principal Bankers: Manufacturers Hanover Trust Co, New York
(USA); Barclays Bank, London; Bank für Gemeinwirtschaft
AG, Frankfurt (Germany); Credit Lyonnais, Paris (France);
Union Bank of Switzerland, Zurich (Switzerland); Banca
Commerciale Italiana, Milan (Italy)
Financial Information:

	31.12.86	31.12.87
	S'000	S'000
Profit before tax	693,000	899,000
Profit after tax	505,000	683,000
Retained profit	75,000	121,000
Preferred dividends	8.5%	8.5%
Common dividends	8%	8%
Earnings per share	486S	631S
Share capital	1,425,000	1,425,000
Shareholders funds	5,705,000	6,348,000
Deposits	166,029,000	170,713,000
Total assets	172,764,000	178,266,000

Principal Shareholders: Österreichischer Gewerkschaftsbund;
Konsum Österreich
No of Employees: 1,953

BUNDESLÄNDER VERSICHERUNG AG

2 Praterstrasse 1-7, PF 92, A-1020 Wien
Tel: (0222) 26 19
Telex: Bundesländervers Wien
Telefax: 134800

Management Board: Dr K Ruso, Dr H Dobin, Dr E Ramberger

PRINCIPAL ACTIVITIES: Insurance
Financial Information:

	1986	1987
	S'000	S'000
Premium income	9,598,000	9,137,000
Share capital	240,000	240,000

No of Employees: 3,831

COLLEGIALITÄT VERSICHERUNG AUF GEGENSEITIGKEIT

Lichtenfelsgasse 7, A-1016 Wien
Tel: (0222) 43 16 41
Telex: 136558 coll a

Management Board: Gen Dir Dr Theodor Detter (Vorsitzender), Mag Rudolf Gelbmann, Dr Hannes Schremser

PRINCIPAL ACTIVITIES: Life, accident and health insurance services
Financial Information:

	S'000
Premium income	900,000

No of Employees: 370

CREDITANSTALT-BANKVEREIN

Schottengasse 6, PO Box 72, A-1011 Wien
Tel: (222) 531 31-0
Cable: Credit
Telex: 133030
Telefax: (222) 330-85-54; 330-86-43

Supervisory Board: Fritz Bock (Vorsitzender), Erich Göttlicher (stv Vorsitzender), Rudolf Gruber (stv Vorsitzender), Kurt Brinner, Walter Fremuth, Carl Anton Goëss-Saurau, Anton Heschgl, Norbert Hönigsberger (Delegierter), Alfred Holoubek, Vinzenz Hübl, Herwig Hutterer, Karl Kehrer, Guido Klestil, Kurt Mészáros, Rainer Moritz, Werner Muhm, Franz Ovesny (Staff Delegierter), Robert Parzer, Karl Pfleger, Emmerich Rudavsky, Peter Schachner-Blazizek, Franz Rauch, Johannes Strohmayer, Heinrich Treichl, Edda Tremmel (Delegierter), Helmut Wasser (Delegierter), Johannes Trojan

Management Board: Dr Guido Schmidt-Chiari (Vorsitzender), Dkfm Gerhard Randa (stv Vorsitzender). Members: Dr Ottokar Finsterwalder, Dr Rudolf Schneider, Dkfm Herbert Sellner, Dr Friedrich Sequens

Senior Executives: Sekretariat: Dr Eder. Kommunikation: Dr Pellech, Stv: Hr Dreihann-Holenia. Öffentlichkeitsarbeit: Dr Kühnert. Werbe-Abteilung: Dipl-Ing Friedl. Personal-Abteilung: Dkfm Dr Milota, Stv: Hr Turba, Hr Schiechl. CA-Akademie: Hr Jankowitsch. Rechts-Abteilung: Dr Karlik, Stv: Mag Dr Potyka. Steuer-Abteilung: Dkfm Dr Kotrnoch, Stv: Dr Schönstein. Revision: Hr Kutschera, Stv: Hr Frotzbacher. Zentrale-Revision: Hr Brandl. Aussenstellen-Revision: Hr Frotzbacher. Pressesprecher des Vorstands: Dr Kühnert. Controlling und Information: Dkfm Dr Kolarik. Bankplanung: Dr Arbesser. Bankentwicklung: Mag Sigmund. Volkswirtschaftliche Abteilung: Dr Fürst. Rechnungswesen und Controlling: Dr Hollerer. Finanzbuchhaltung: Hr Nebosis. Kostenrechnung: Hr Niehsler. Budgetierung: Dr Wabitsch. Organisation: Dkfm Haberzettl, Dipl-Ing Losert (stv). International Systems Support Group: Hr Maddocks. Organisation: Dkfm Haberzettl. Organisation 1: Mag Wabnig. Organisation 2: Hr Kristinus. Organisation 3: Hr Schmölz. Systementwicklung: Dipl-Ing Losert. Systementwicklung 1: Dr Winter. Systementwicklung 2: Hr Nimführ. System- und Datentechnik: Hr Fröhlich. Verkauf Wien, Privatkunden: Hr Castellez. Verkauf Bundesländer Firmenkunden: Dr Krempler. Regionales Controlling: Hr Klamert. Verkauf/Privatkunden: Dr Fadinger. Verkauf/Firmenkunden: Dr Wursche. Privatkunden und Zahlungsverkehr: Dr Reidlinger. Privatkundenprodukte: Dr Kschwendt-Michel. Zahlungsverkehrsentwicklung: Hr Jerusalem. Primärmittel: Dr Pinz, Stv: Hr Blazek, Mag Pekarek. Finanzierungen: Dkfm Fojtl, Stv: Dr Sattler, Dr Sevelda. Einlagen und kreditorische Grosskunden: Dkfm Losch. Betriebsmittelfinanzierung: Dr Sattler. Versicherungen und Körperschaften: Hr Blazek. Kredit-Abteilung: Dr Hahn. Wechsel-Abteilung: Hr Lichtenhofer. Kreditprüfung und betriebswirtschaftliche Beratung: Hr Bogenmayr. Investitionsfinanzierung: Dkfm Fojtl. Investitionskredite: Dr Scherer. Hypothekar-Abteilung: Dr Schreiber. Projektfinanzierung: Mag Nageler. Exportfinanzierung: Dr Sevelda. Exportkredite: Hr Zboril. Auskunftei: Hr Payer. Wertpapiere/Investment Banking: Dr Musulin, Stv: Hr Petter,

Dr Hager. Wertpapier-Sekretariat: Dr Kubik. Konsortialgeschäft Inland: Dr Rauscher. Konsortialgeschäft Ausland: Dr Bienenstein. Wertpapierhandel Heijkoop: Dkfm Dr Jekl. Wertpapierverkauf Inland: Hr Beronneau. Wertpapierverkauf Ausland: Mag Dr Körner. Vermögensverwaltung und Portfolio-Management: Hr Petter. Investment Banking: Dr Hager. Betreute Tochterunternehmungen: Wertpapierverwaltung: Dr Engel. Wertpapierabwicklung: Dr Hölzl. Wertpapierverwahrung: Hr Franke. Ausland, Geld- und Devisenhandel: Mag Peschek, Stv: Hr Novy. Akkreditiv-Abteilung: Hr Stella. Devisenhandel: Hr Novy. Geldstelle: Dr Reichenwallner. Auslandszahlungsverkehr und Auslandsinkasso: Hr Müllner. Auslandszahlungsverkehr: Hr Müllner. Auslandsinkasso: Hr Steiner. Internat Geschäft: Dr Fenyves, Stv: Dr Klapper, Dr Coreth, Dkfm Dr Werner. Kreditkontrolle: Mag Steinbichler. Filialkoordination: Dr Gehberger. Bankprojekte: Projekt- und Übernahmsfinanzierung: Dr Wimmer. Innenleitung: Dr Aichhorn. Banken Ausland: Dr Coreth. Länder-Abteilung I: Hr Liska. Länder-Abteilung II: Dr Günther. Special Assets Management: Dr Gruber. Internationale Kredite und Marketing: Dkfm Dr Werner. RGW und Jugoslawien: Dkfm Dr Werner. Internationale Finanzierungen: Kreditadministration: Fr Schober. Bankbetrieb: Hr Stoschka, Stv: Hr Krenn. Datenverarbeitung und Inlandszahlungsverkehr: Hr Horak. Datenerfassung: Hr Beier. Rechenzentrum: Hr Horak. Inlandszahlungsverkehr: Hr Oujezdsky. Buchhaltung: Hr Krenn. Kontokorrentbuchhaltung: Hr Pommer. Hauptbuchhaltung: Hr Biesenbender. Verwaltung, Bau und Technik: Dipl-Ing Dr Schober. Verwaltung: Hr Schuster. Technik: Dipl-Ing Dr Schober. Bauprojekte: Hr Stradner. Wirtschaftsbetriebe: Hr Schilling. Ökonomat: Hr Schilling. Angestelltenmesse: Hr Flotzinger. Beteiligungen: Dr Pimmer. Stabsstelle Beteiligungen: Mag Fromwald. Referat Dienstleistungen und Industrie: Dkfm Dr Stalf. Referat Handel und Industrie: Dr Iro. Referat Kreditinstitute: Dr Laa. Betriebswirtschaftl techn Beratung und Sonderprojekte: Dipl-Ing Klein. Post und Zentralkasse: Hr Stöckel

PRINCIPAL ACTIVITIES: Commercial and investment banking services
Trade Names: CREDITANSTALT-BANKVEREIN
Subsidiary Companies: (100% owned unless stated): Domestic: ALAG Anlagen Leasing GmbH; AWT Internationale Handels- und Finanzierungs AG (60%); Bürohaus Vermietungs- und Verwaltungs GmbH; CA-Versicherungsdienst GmbH; CA-3 Banken Beteiligungsfonds AG (60.7%); CA-EA Investment GmbH (50%); Eurconsult Energie und Umwelt Beratungs- und PlanungsgesmbH (50%); Grundstücke- und Gebäudeverwaltungs GmbH (74.99%); Handelsbeteiligungs-Gesellschaft mbH (50%); Heid Agrartechnik GmbH (50%); Hutter & Schrantz AG (78.5%); Imperial Hotels Austria AG (71.4%); Jenbacher Werke AG (58.8%); Management Data; Datenverarbeitungs- und Unternehmensberatungsges mbH; Österreichische Armaturen GmbH; Österreichische Hotel- und Fremdemverkehrs-Treuhandges mbH (50%); Österreichische Investment Gesellschaft (50%); Österreichische Leasing GesmbH (75%); Österreichische Realitäten AG (98%); Papierindustrie-Beteiligungsges mbH; Semperit AG Holding (86.4%); Steyr-Daimler-Puch AG (82%); Steyr-Daimler-Puch Fahrzeugtechnik GmbH (50%); Universale-Bau AG (70.4%); Vereinigte Papierindustrie- u. Allgemeine Warenhandels AG (51%); Wienerberger Baustoffindustrie AG (75.5%). International: CA Asia Ltd (Hong Kong); CAMINVEST Ltd (indirect subsidiary) (Hong Kong); Vienna Holdings Corp (USA); McKenzie Walker Investment Management Inc (Indirect subsidiary) (USA); CAWT Trade Finance Co (51%) (indirect subsidiary) (USA); Creditanstalt Finance Inc (USA); CA Industrial Finance Ltd (UK); Creditanstalt Finanziaria (Italy); Banco BBA-Creditanstalt SA (50%) (Brazil); CA-BB Brokers AG (50%) (Hungary); Wirtschafts- und Privatbank (50%) (Switzerland); Banco Interfinanzas SA (20%) (Argentina); Central-European International Bank Ltd (11%) (Hungary); European American Bancorp (5.5%) (USA)

Financial Information:

	31.12.87	31.12.88
	S'000	S'000
Profit before tax	1,110,000	1,130,000
Profit after tax	992,000	1,020,000
Retained profit	587,000	602,000
Dividends	372,000	381,000
Dividends per share	60S	60S
Earnings per share	131.90S	150.50S
Share capital	3,100,000	3,400,000
Shareholders funds	11,426,000	13,263,000
Deposits(1)	364,950,000	379,120,000
Total assets(2)	384,208,000	401,944,000

(1) excluding subordinated debt

(2) excluding own acceptances outstanding and funds received on a trust basis at third party risk

Principal Shareholders: Austrian Government (60%)

No of Employees: 10,376

D SWAROVSKI & CO

See SWAROVSKI, D, & CO

DER ANKER ALLGEMEINE VERSICHERUNGS-AKTIENGESELLSCHAFT

See ANKER, DER, ALLGEMEINE VERSICHERUNGS-AKTIENGESELLSCHAFT

DIE ERSTE ÖSTERREICHISCHE SPAR-CASSE-BANK

First Austrian Bank

Graben 21, 1011 Wien

Tel: (0222) 53100

Cable: Grabenspar Wien

Telex: 114012 ESPKWA

Supervisory Board: C H Schönbichler (Chairman), Dr F Haas (Deputy Chairman), Dr K Korinek, Dr F Pfohl, H Schimetschek, Dr J Schneider, Dr W Stauffer, Dr J Taus, Dkfm J Härtel, R Jungwirth, H Kerschbaum, R Winkler

Board of Management: Dr Konrad Fuchs (Chairman), Dr Wolfgang Ulrich (Deputy Chairman), Mag Reinhard Ortner

Senior Executives: Retail Banking: Peter Aschauer (Senior Vice President), Walter Huber (Senior Vice President), Manfred Kautnik (Senior Vice President), Felix Leucht (Senior Vice President), Walter Schwimbersky (Senior Vice President), Gerhard Fasching (First Vice President), Erich Wagner (First Vice President), Johann Höbinger (Vice President). Domestic Corporate Banking: Elisabeth Bleyleben-Koren (Senior Vice President). International: Anton M Burghardt (Senior Vice President). Securities: Otto Höller (Senior Vice President). Payments and Central Administration: Georg Tucek (Senior Vice President). Marketing: Wolfgang Ronzal (Senior Vice President). Organization and Data processing: Erwin Standl (Senior Vice President). Internal Services: Kurt März (Senior Vice President). Internal Audit: Josef Schmidradner (Senior Vice President). Personnel Department: Engelbert Gruber (Senior Vice President). Legal Services: Elfriede Marek (General Counsel). Credit Analysis and Financial Advisory Services: Walter Hügel (Senior Vice President). Credit Department Retail: Anton Schmoll (First Vice President). Accounting: Helmuth Kaltenbrunner (First Vice President). Financial Planning and Budgeting: Gerhard Srb (Vice President). Advisor: Hanns Christian (Senior Vice President)

PRINCIPAL ACTIVITIES: Bank

Financial Information:

	31.12.88
	S'000
Profit after tax	440,000
Retained profit	440,000
Shareholders funds	6,322,000
Deposits	125,357,000
Total assets	143,687,000

No of Employees: 2,528

DONAU ALLGEMEINE VERSICHERUNGS-AG

Schottenring 15, PF 770, A-1011 Wien

Tel: (0222) 31 46 11

Cable: donauassekuranz, wien

Telex: 114588 donre

Telefax: (0222) 34 65 87

Management Board: Dr Gerhard Puschmann (Chairman), Dr Ing Heinz Rinklak, Dr Heinz Traxler, Dr Helene Fischer, Dr Rudolf Ertl, Dkfm Hans Raumauf

PRINCIPAL ACTIVITIES: All classes of insurance

Subsidiary/Associated Companies: Interrisk Internationale Vers. AG (16.67%); Komma Marketingberatung Ges mbh (90%); RD-Beteiligungsgesellschaft mbH (30%); MEDIAL Beteiligungs-Gesellschaft mbH (29.33%); ARITHMETICA Versicherungs- und Finanzmathematische Beratungs-Gesellschaft mbH (25%); AUDATEX Autoschaden-Begutachtungs-Gesellschaft mbH (25%); DIE FREIBURGER Allgemeine Versicherung Aktiengesellschaft (10%); Realitätenverwaltungs-und Restaurantbetriebs-Gesellschaft mbH (2%); LANDHAUS-PARKGARAGEN Gesellschaft mbH & Co KG (1.23%); VENTURE Finanzierungsgesellschaft mbH (0.77%)

Financial Information:

	31.12.86	31.12.87
	S'000	S'000
Premium income	2,846,512	3,041,621
Investment income	443,823	473,888
Profit before tax	10,870	41,989
Profit after tax	5,498	20,099
Retained profit	6,984	20,333
Dividends	6,750	20,000
Share capital	160,000	160,000
Shareholders funds	382,433	441,997

Principal Shareholders: Wiener Städtische Wechselseitige Versicherungsanstalt (73.88%); Schweizer Rück Holding AG (26.12%)

No of Employees: 1,568

DOROTHEUM AUKTIONS- VERSATZ- U BANK- GESELLSCHAFT MBH

Dorotheergasse 17, A-1010 Wien

Tel: (0222) 515 60-0

Telex: 132230 doro a

Management Board: Dkfm Alfred Karny, Mag Dr Peter Pechar, Wolfgang Kraml

Senior Executives: Kurt Hrstka (Sales Manager), Herbert Maringer (Finance)

PRINCIPAL ACTIVITIES: Auctions, pawnshops and banking services

Financial Information:

	1986	1987
	S'000	S'000
Revenues	751,000	924,000

No of Employees: 517

ERSTE ALLGEMEINE VERSICHERUNGS-AG
EA
Brandstätte 7-9, Postfach 173, A-1010 Wien
Tel: (0222) 6338
Cable: Erstallges
Telex: 11-4085 eagda
Telefax: (0222) 6338-226

Supervisory Board: Präsident KR. Karl Kornis (Wien, Vorsitzender), Präsident Dr Enrico Randone (Triest, Vorsitzender-Stellvertreter), Generaldirektor-Stellvertreter Dr Guido Schmidt-Chiari (Wien, Vorsitzender-Stellvertreter), Direktor Cavaliere Teodoro Adler (Mailand), Deleg. Verwaltungsrat Dr Eugenio Coppola di Canzano (Triest), August von Finck (München), Generaldirektor Dr Carl H Hahn (Wolfsburg), Präsident Dkfm Dr Hans Igler (Wien), Generaldirektor Dr Franz Leibenfrost (Wien), Dr Ing. Georg Prinz von und zu Liechtenstein (Wien), Direktor Dr Alberto Tiberini (Triest). vom Zentralbetriebsrat delegiert: Disponent Franz Juraczka (Wien), Bezirksdirektor Horst Detitscheg (Graz), Manfred Thalhammer (Wien), Zentralinspektor Walter Kovarik (Wien), Disponent Walter Roider (Salzburg), Obersekretär Edgar Gruber (Bregenz)
Management Board: Generaldirektor Dr Dietrich Karner (Vorsitzender des Vorstandes), Direktor Edgar Messerschmidt, Direktor DDr Günter Neumann, Direktor Dkfm Ulrich Pott, Direktor Dr Wilfried Wagner, Direktor Dr Hans Peer (stv Vorstandsmitglied)
Senior Executives: Direktoren: Erhard Graser, Dr Franz Kisielewski, Dr Robert Kriegel, Hedwig Krzyzanowski, Dr Richard Nathschläger, Ing. Stefan Novak, Dr Robert Pöltl, Herbert Smoliner, Dipl. Ing. Harald Stelzer, Dr Heinz Zidek

PRINCIPAL ACTIVITIES: Non-life insurance and reinsurance
Trade Names: Erste Allgemeine Versicherungs-Aktiengesellschaft
Subsidiary Companies: EA-Rechtsschutzversicherungs-AG (Germany); Holdux-Beteiligungsgesellschaft (Switzerland); Beleggingsmaatshcappij "De Grachten" BV (Netherlands); A.B.V. Advies en Bemiddeling in Verzekeringszaken BV (Netherlands); AV Algemene Holding en Financierungs Maatschappij (Netherlands)
Principal Bankers: Creditanstalt-Bankverein
Financial Information:

	31.12.86	31.12.87
	S'000	S'000
Premium income	10,878,240	11,430,000
Share capital	550,000	550,000

Principal Shareholders: Assicurazioni Generali, Trieste, Italy, through group companies
No of Employees: 3,840

EXPORT FINANZIERUNGSBANK GMBH
Prinz-Eugen Strasse 8-10, 1040 Wien
Tel: (0222) 65 36 92
Telex: 132484 efba a

PRINCIPAL ACTIVITIES: Financial services, banking
Financial Information:

	S'000
Turnover	13,621,000

No of Employees: 10

F M TARBUK & CO
See TARBUK, F M, & CO

FACTOR-BANK GESELLSCHAFT MBH
Naglergasse 1, A-1014 Wien
Tel: (0222) 533 17 76

Management Board: Dr L Binder-Degenschild (Geschäftsführer), Theo Hibler (Geschäftsführer)

PRINCIPAL ACTIVITIES: Financial services, factoring

Financial Information:

	S'000
Turnover	4,050,000

No of Employees: 2256

GEBRUEDER WEISS GES MBH
See WEISS, GEBRUEDER, GES MBH

GENERALI ALLGEMEINE LEBENSVERSICHERUNG AG
Bauernmarkt 2, A-1010 Wien
Tel: (0222) 63 56 81
Telex: 114534 gligd a

Management Board: Dr Dietrich Karner (Vorsitzender), Dr Alvaro Costa, DDr Günther Neumann, Dr Ladislaus Viragh, Peter Vogt
Executives: Direktoren:

PRINCIPAL ACTIVITIES: Life insurance
Financial Information:

	31.12.86	31.12.87
	S'000	S'000
Premium income	3,091,000	3,311,000

Principal Shareholders: Erste Allgemeine Versicherung AG, Wien; Assicurazioni Generali, Trieste
No of Employees: 1,240

GENOSSENSCHAFTLICHE ZENTRALBANK AG
GZB-Vienna
Herrengasse 1-3, Postfach 50, 1010 Wien
Tel: 71707-0
Cable: Zentralbank Wien
Telex: 136989

President: Dr Karl Gruber
General Manager: Dr Klaus Liebscher
Senior Executives: Dr Robert Wychera (Deputy General Manager)

PRINCIPAL ACTIVITIES: Central institution of the Austrian Raiffeisen Banking Group; all banking services
Subsidiary/Associated Companies: Raiffeisen Bausparkasse GmbH; Raiffeisen Finanzierung AG; Raiffeisen Garantiegesellschaft mbH; Raiffeisen Versicherung AG; Raffeisen Leasing Gmbh; Raiffeisen Reisebüro GmbH; Raiffeisen Kapitalanlageges.mbH; Raiffeisen Datennetz GmbH; Raiffeisen Invest GmbH; Raiffeisen-Rent Vermögensberatung und Treuhand GmbH; Kathrein & Co Bank AG; Schoeller & Co. Bank AG; F J Elsner & Co; Schoeller Holding AG; DG Bank (Schweiz) AG (Switzerland); London & Continental Bankers Ltd (UK); Canadian Cooperative Credit Society (Canada); Rakyat First Merchant Bankers Berhad (Malaysia); European Asian Trade Service Co Ltd (Japan); GZB-World Travel Co Ltd (Hong Kong); DG Bank-GZB (Asia) Ltd (Singapore); Unicbank Rt (Hungary); Centrotrade (M) SDN BHD (Malaysia); Centro Internationale Handelsbank AG (Austria)
Principal Bankers: Members of UNICO Banking Group
Financial Information:

	31.12.86	31.12.87
	S'000	S'000
Profit after tax	129,100	217,800
Dividends	128,000	200,000
Share capital	2,000,000	2,000,000
Shareholders funds	5,678,500	6,713,500
Deposits	151,897,300	158,297,000
Total assets	160,116,100	168,224,200

Principal Shareholders: Regional cooperative banks
No of Employees: 912

GIROZENTRALE U. BANK DER ÖSTERR. SPARKASSEN AG

Schubertring 5, A-1010 Wien
Tel: (0222) 711 94 0
Cable: SPARGIRO WIEN
Telex: 133006 giro a
Telefax: (0222) 713 70 32 (GZ-General); 713 70 55 (GZ-Int'l
Operations); 713 60 78 (GZ-New Issues); 713 18 06 (GZ-
Public relations)

Supervisory Board: Dr. Karl Vak (Chairman), Dr. Hans Haumer
(Deputy Chairman)
Managing Board: Dr. Karl Pale (Chairman and Managing
Director) Dr. Rudolf Fiala (Deputy Chairman and Deputy
Managing Director), Dr. Karl Ludwig (Executive Director), Dr.
Theoderich Mellich (Executive Director)
Senior Executives: Dr. Gerhard Kandler (Divisional Manager,
Corporate Customers and Public Authorities/Lending), Franz
Caspar (Manager International Division), Karl Stumwöhrer
(Divisional Manager, Private and Institutional Investors/Capital
Market Financing)

PRINCIPAL ACTIVITIES: Bank
Subsidiary Companies: Holdings of over (50%):
Fremdenverkehrsbetriebs-und Management mbH; WBG
Betriebs-Wirtschafts-Beratungsges mbH; Gesellschaft für
energie-und umwelttechnische Projekte GesmbH;
"CONSULTCO INTERNATIONAL" Projektierung u
Koordination von Bauten u Anlagen GesmbH; BTX-
Hotelpartnerring GesmbH; Real Service
Realitätenvermittlungs-ges mbH; Gesellschaft für
Unternehmenscommunication; AUG-Betriebsanlagen-und
Unternehmensvermittlungs-ges mbH; Vienna Commerz
Handelsges mbH; Spakassengebäudeverwaltungsgesellschaft
mbH; Job Creation Betriebsansiedlungsges mbH; Sparconsult
Wirtschafts-und Managementberatung für Sparkasses Ges
mbH; Garage am Beethovenplatz Ges mbH & Co KG;
IMMORENT Ges mbH; Sparinvest-Kapitalanlage-Ges mbH;
GZ-Vermögensverwaltungs-ges mbH; GZ
Beteiligungsfinanzierungs Ges mbH; Pinkafelder Spar-und
Kreditbank AG; Rechnitzer Spar-und Kreditbank AG;
Sparkassen-Datendienst GesmbH; "KOMDAT" Kommunale
Datendienst Ges mbH; Vorarlberger Rechenzentrum Ges
mbH; Sparkassenversicherungsservice GesmbH; "Leasfinanz"
Vermietung von Investitionsgütern GesmbH & Co KG;
ImmoRent Liegenschaftsanlage-und Verwaltungsges mbH;
Bankinvest AG (Switzerland)
Financial Information:

	31.12.86	31.12.87
	S'000	S'000
Surplus before tax*	926,580	473,210
Surplus after tax*	769,376	392,820
Profit	277,513	185,243
Net profit	277,717	185,433
Share capital	2,725,000	2,725,000
Total assets	265,737,000	276,429,599

*and before reserve changes

Principal Shareholders: Austrian Savings Banks
No of Employees: 1,704

GRAZER WECHSELSEITIGE VERSICHERUNG

Herrengasse 18-20, A-8010 Graz
Tel: (0316) 7037-0
Telex: 31414 grawe a
Telefax: 70 37 414

Management Board: Dr Friedrich Fall (Vorsitzender), Dr Othmar
Ederer, Siegfried Ettl, Gottfried Krainer
Senior Executives: Gerhard Kollaritsch, Dr Siegfried Grigg

PRINCIPAL ACTIVITIES: Insurance including life and accident
insurance; reinsurance

Financial Information:

	1986	1987
	S'000	S'000
Premium income	2,107,000	2,240,000

No of Employees: 1,070

INTERMARKET FACTORING GMBH

Linke Wienzeile 234, 1150 Wien
Tel: (0222) 857626-0
Telex: 135953 imfac a

Management Board: Mag Dr Harald Schranz (Geschäftsführer),
Dr Helmut Paris (Geschäftsführer)

PRINCIPAL ACTIVITIES: Credit and financial services; factoring
Financial Information:

	1986	1987
	S'000	S'000
Total assets	7,300,000	7,000,000

No of Employees: 50

INTERNATIONALE UNFALL-UND SCHADENVERSICHERUNG AG
INTERUNFALL

Tegetthoffstr 7, Pf 867, A-1010 Wien
Tel: (0222) 51403-0
Telex: 112111 iuwn a

Supervisory Board: Dr Fridolin Kristinus (Chairman)
Management Board: Dr Josef Cudlin (Chairman), Dr Otto Agler,
Josef Svoboda, Hellmuth Wandschneider

PRINCIPAL ACTIVITIES: All classes of insurance excluding life
insurance; reinsurance
Trade Names: INTERUNFALL
Parent Company: RAS-Riunione Adriatica di Sicurtà, Italy
Subsidiary Companies: Österr. Kreditversicherungs-AG;
"Continentale" Lebensversicherungs-Gesellschaft
(Switzerland); Münchener Lebensversicherung AG (Germany);
"Nordeuropa" Forsikrings A/S (Norway); Canadian Home
Assurance Company (Canada); Interunfall Allgemeine
Versicherungs-AG (Germany)
Financial Information:

	31.12.86	31.12.87
	S'000	S'000
Premium income	4,903,000	5,133,000
Share capital	200,000	200,000

No of Employees: 2,158

MERKUR WECHSELSEITIGE VERSICHERUNGSANSTALT

Neutorgasse 57, A-8011 Graz
Tel: (0316) 7034-0
Telex: 31267 merkur a

Management Board: Harald Beranek (Vorsitzender), Werner
Reimelt
Senior Executives: Mag Franz Voves, Paul Lang, Reinhard
Affenberger, Harald Kastner

PRINCIPAL ACTIVITIES: Insurance
Financial Information:

	1986	1987
	S'000	S'000
Premium income	1,609,000	1,650,000

No of Employees: 902

METRO EINKAUFSGESELLSCHAFT MBH

Ortsstr 23-27, A-2331 Vösendorf
Tel: (0222) 67 15 24

Management Board: Otto Beisheim
Senior Executives: Franz Scheichenost, Friedrich Stelzer, Dipl
Vw Erwin Stock

PRINCIPAL ACTIVITIES: Administration and services for group companies with interests in the food trade and consumer goods

Subsidiary Companies: Metro Selbstbedienungsgrosshandel Ges mbH; Möbel Grossvertriebsgesellschaft mbH

Financial Information:

	S'000
Sales turnover	8,700,000

No of Employees: 2,300

NORDSTERN VERSICHERUNGS-AKTIENGESELLSCHAFT

Uraniastrasse 2, A-1010 Wien
Tel: (0222) 73 35 61-0
Telex: 132193 nstwa

Management Board: Friedrich Steirer (Vorsitzender), Mag Dr Konstantin Klein, Peter Neudecker

PRINCIPAL ACTIVITIES: Insurance services

Financial Information:

	1986	1987
	S'000	S'000
Premium income	950,000	1,035,000

No of Employees: 540

OBERÖSTERREICHISCHE WECHSELSEITIGE VERSICHERUNGSANSTALT

Gruberstrasse 32, 4020 Linz
Tel: (0732) 2891
Telex: ʋ2-1594
Telefax: (0732) 2981-566

Supervisory Board: Landesrat Leopold Hofinger (Präsident), Kommerzialrat Oskar Winkler (Vizepräsident), Bezirkshauptmann Wirkl Hofrat Dr Gerald Glinz, Gemeinderat Hilde Kerschbaum, Bundesrat Engelbert Lengauer, Gemeinderat Karl Maier, Kammerrat Ing Wilhelm Paschinger, Landesjägermeister Hans Reisetbauer, Wirkl Hofrat i R Dr Othmar Seefeldner, Regierungsrat Walter Steiner, Kommerzialrat Gustav Tatschl, Vizebürgermeister Leopold Wippersberger, *Prokurist Rudolf Barth (Betriebsratsobmann), *Konsulent Rudolf Orthofer (Betriebsratsobmann-Stellvertreter), *Sekretär Peter Bamberger, Sekretär Herbert Plöderl, *Obersekretär Mathäus Gössweiner, *Disponent Valentin Stöckler, *vom Betriebsrat delegiert

Management Board: Reg Rat Josef Hackl (Generaldirektor), Dr Helmuth Karigl (Generaldirektor-Stv), Dr Herbert Roitinger (Vorstandsdirektor)

PRINCIPAL ACTIVITIES: Insurance

Financial Information:

	31.12.86	31.12.87
	S'000	S'000
Premium income	1,467,000	1,511,500

No of Employees: 431

ÖSTERREICHISCHE KONTROLLBANK AG

Am Hof 4, Postfach 70, A-1011 Wien
Tel: (0222) 6627-0
Cable: Kontrolbank Wien
Telex: 13-2747 OEKBS A (Board of Executive Directors); 13-2771 OEKMI A (General Secretariat)

Board of Executive Directors: Prof Dr Helmut H Haschek (Chairman), Erich Groiss, Paul Castellez,

PRINCIPAL ACTIVITIES: Export finance and control of Austrian export credit guarantees

Financial Information:

	31.12.86	31.12.87
	S'000	S'000
Total assets	184,300,000	173,700,000

No of Employees: 335

ÖSTERREICHISCHE LÄNDERBANK AKTIENGESELLSCHAFT

Am Hof 2, Postfach 271, A-1010 Wien
Tel: (222) 531 24-0
Cable: Laenderbank; Swift: OELB AT WW
Telex: 115561 LBW
Telefax: (222) 531 24 55

Supervisory Board: Alois Piperger (Chairman), Arthur Mussil (Deputy Chairman), Herbert Schmidtmeier (Deputy Chairman), Otto Binder (Deputy Chairman), Wilhelmine Goldmann, Fritz Heiss, Franz Kafka, Richard Kirchweger, Emmy Lauteren, Erich Meyringer, Kurt Mühlbacher, Kurt Neuner, Werner Obermayer, Johann Penz, Walter Petrak, Hugo Michael Sekyra, Peter Wrabetz. Representatives of the Employees: Wilhelm Franta, Josef Vogelsinger, Siegfried Hammerle, Sonja Heyny, Heinz Hufnagl, Sylvia Krenstetter, Heribert Kruschik, Thomas Schlager

Board of Managing Directors: Gerhard Wagner (Chairman), Dr Manfred Drennig (Deputy Chairman), Konrad Rumpold (Managing Director), Dr Herbert Cordt (Managing Director)

Senior Executives: Divisional Managers: Dr Ernst Löschner (Senior General Manager, International), Oskar Latzko (Senior General Manager, Corporate Customers), Peter Fischer (Senior General Manager, Treasury), Wolfgang Feuchtmüller (Senior General Manager, Securities), Wilfried Hanreich (Senior General Manager, Marketing), Leopold Röhrer (Senior General Manager, Retail Banking), Alfred Brandstetter (Senior General Manager, Credit), Werner Du-Rieux (Senior General Manager, Internal Services), Michel Ghanes (General Manager, Organisation and EDP), Fritz Pferschy (Senior General Manager, Accounting and Controlling), Ludwig Neumann (Senior General Manager, Personnel), Karl Tambornino (Senior General Manager, Legal), Peter Weber (General Manager, Banking and Industrial Holdings), August Bitschofsky (Senior General Manager, Internal Audit), Josef Fritz (Senior General Manager, Corporate Planning)

PRINCIPAL ACTIVITIES: Commercial banking

Subsidiary Companies: (100% owned unless stated): Österreichisches Creditinstitut AG; Mercur Kapitalanlageges mbH; Eisenstädter Bank AG (99%); Länderbank-Exportbank AG; LB-Leasingges mbH; Mercurbank GmbH (85%). Affiliates: Domestic:- Österreichische Kontrollbank AG; Adria Bank AG; Factor-Bank Ges mbH; Internationale Bank für Aussenhandel AG; Österreichische Investitionskredit AG; Factor-Bank Ges mbH; Eurocard Austria Kreditkarten Ges mbH. International:- B.A.I.I. (France); B.A.I.I. Holdings (Luxembourg); Eulabank (UK); The International Investment Corp for Yugoslavia (Luxembourg)

Financial Information:

	31.12.86(1)	31.12.87
	S'000	S'000
Surplus for the year(2)	454,400	751,001
Shareholders funds(3)	7,266,300	8,175,500
Deposits(4)	195,915,408	198,883,299
Total assets net	239,578,000	247,277,313

(1) restated according to new Austrian Banking Act in force since January 1987
(2) before transfers to reserves and allocation of dividend to holders of participation certificates
(3) liable capital pursuant to Austrian Banking Act
(4) including customer deposits

Principal Shareholders: Government (53%); Private (47%)

No of Employees: 3,966 (Länderbank AG); 5,200 (Länderbank Banking Group)

ÖSTERREICHISCHE POSTSPARKASSE
P.S.K.

Georg Coch-Platz 2, 1018 Wien
Tel: (0222) 514 00-0
Cable: "Postsparkasse Wien"
Telex: 111663 psk a
Telefax: (0222) 514 00 3951

Supervisory Board: Helmut Frisch (Chairman), F Eduard Demuth
(Deputy Chairman), Günther Chaloupek (Deputy Chairman).
Members: Erich Amerer, Kurt Bronold, Robert Ecker, Irene
Geldner, Ernst Massauer, Klaus Liebscher, Josef Pinter,
Anton Rainer, Josef Sindelka, Herbert Tumpel. Delegated by
the Staff Committee: Eleonore Bauer, Volkmar Harwanegg,
Hilde Hornig, Franz Satzinger, Herta Somek, Erika Texler
Management Board: Kurt Nösslinger (Governor), Viktor Wolf
(Vice Governor), Erich Hampel (Vice Governor)

PRINCIPAL ACTIVITIES: Bank
Subsidiary Companies: Bank der Österreichischen
Postsparkasse AG ("P.S.K. Bank AG"); Bankhaus Feichtner &
Co AG
Financial Information:

	31.12.86	31.12.87
	S'000	S'000
Profit before tax	633,000	985,832
Profit after tax	258,000	279,000
Share capital	-	160,000
Reserves	-	3,843,754
Shareholders funds	2,307,000	-
Deposits	149,279,000	142,389,000
Total assets	155,079,000	163,107,000

No of Employees: 1,784

ÖSTERREICHISCHE VOLKSBANKEN
AKTIENGESELLSCHAFT

Peregringasse 3, A-1090 Wien
Tel: (0222) 31 34
Telex: 114084 vbag a

Management Board: Robert Mäde (Vorsitzender), Johann Kubu

PRINCIPAL ACTIVITIES: Banking
Financial Information:

	31.12.86	31.12.87
	S'000	S'000
Total assets	55,100,000	57,800,000

No of Employees: 453

PORSCHE HOLDING GESMBH

Porschehof, Fanny v Lehnert Str 1, Postfach 164, A-5020
Salzburg
Tel: (0662) 50581-0
Telex: 633291

Management Board: Dr Hans Himmer (Managing Director)

PRINCIPAL ACTIVITIES: Import and trade in Porsche,
Volkswagen, Audi and NSU motor vehicles; repairs and
servicing; leasing; manufacture of turbines and agricultural
vehicles
Trade Names: VW, Audi, Porsche
Subsidiary Companies: Porsche-Inter Auto Ges mbH; Porsche
Austria Ges mbH & Co; Porsche Konstruktionen AG; Porsche
Produktions-und Handelsgesellschaft mbH
Financial Information:

	1986	1987
	S'000	S'000
Sales turnover	16,127,000	15,528,000

No of Employees: 3,389

QUELLE AG ÖSTERREICH

Industriezeile 47, PF 398, A-4021 Linz
Tel: (0732) 2809-0
Cable: Quelle Linz
Telex: (047) 21611 qlinz a
Telefax: (0732) 2809-254

Management Board: Werner Aldenhoff, MOB Wolfgang König,
MAG Nikolaus Maximoff
Directors: T Auer (Logistik), M Cech (Controlling), M Fühling
(Verkauf T/E), H Gerdenitsch (Vertrieb VS), O Leeb
(Organisation/DVA), J Pera (Einkauf)
Senior Executives: Dr J Aufreiter (Recht), H Schwenke
(Marketing/Werbung), MAG J Haan (Finanzen)

PRINCIPAL ACTIVITIES: Department stores chain; mail order
house
Trade Names: UNIVERSUM, PRIVILEG, SENATOR, MARS,
EUROVAL, REVUE, WOHNFORM, SILENTA, Apollo optik,
Reise-Quelle
Subsidiary/Associated Companies: Bankhaus CA Steinhäusser
AG, Wien; Unterstützungkasse der Quelle Aktiengesellschaft
GesmbH; Foto-Quelle Ges mbH; WR Neudorf
Principal Bankers: Zentralsparkasse Wien; CA-BV; Chase
Manhattan; Amex; Kathrein Wien
Financial Information:

	31.12.86	31.12.87
	S'000	S'000
Sales turnover	3,618,000	3,804,000
Share capital	120,000	120,000
Shareholders funds	392,000	454,000
Total assets	1,542,000	1,647,000

Principal Shareholders: G.u.G Schickedanz KG, Furth, Germany
No of Employees: 2,755

RAIFFEISEN-VERSICHERUNG AG

Wiedner Hauptstrasse 100, A-1050 Wien
Tel: (0222) 55 55 55
Telex: 113782 raivs a

Management Board: Dr Mag Christian Sedlnitzky (Vorsitzender),
Johann Stadler, Dr Günter Lukas

PRINCIPAL ACTIVITIES: Life and accident insurance
Financial Information:

	1986	1987
	S'000	S'000
Premium income	2,082,000	2,117,000

No of Employees: 272

RAS ÖSTERREICH ADRIATISCHE
VERSICHERUNG AG

Tegetthofstr 7, A-1010 Wien
Tel: (0222) 51 403-0
Telex: 131794 iuldw a

Supervisory Board: Dr Fridolin Kristinus (Chairman)
Management Board: Dr Josef Cudlin (Chairman), Dr Otto Agler,
Josef Svoboda, Hellmuth Wandschneider

PRINCIPAL ACTIVITIES: Insurance services
Financial Information:

	31.12.86	31.12.87
	S'000	S'000
Premium income	1,309,000	1,365,000
Share capital	50,000	50,000

No of Employees: 597

SWAROVSKI, D, & CO

A-6112 Wattens
Tel: (05224) 2411-0 (5000-0)
Cable: SWAROVSKI WATTENS
Telex: 533656 swaw a
Telefax: (05224) 2335 (52335)

The Board: Daniel Swarovski, Manfred Swarovski, Gernot Langes Swarovski, Dipl. Ing. Helmut Swarovski, Christian Schwemberger Swarovski, Gerhard Swarovski, Andreas Schiestl-Swarovski

PRINCIPAL ACTIVITIES: Crystal: Crystal Jewellery Stones (Jewellery Stones). Trimmings and Transfers (Rhinestone). Crystal Chandelier Parts "Strass" (Chandelier Crystals). Synthetic Stones (Synthetic and Natural Stones). Animals, decorative and functional gifts (Swarovski Silver Crystal/Swarovski Collectors Society). Fashion Jewellery (Swarovski, Savvy). Abrasives and Grinding Tools: "Tyrolit". Optical Products: binoculars, telescopes, telescopic sights "Habicht" (Swarovski Optik). Reflectors for traffic safety: (Swareflex). ECG-Electrodes: (Swaromed). Electronic Control Systems: (Swarotronic)

Subsidiary Companies: Swarovski has 19 manufacturing units/sales companies, 43 sales companies and 10 service companies

Principal Bankers: CA-BV Innsbruck; Sparkasse Innsbruck-Hall; Girozentrale der Österr. Sparkassen AG

Financial Information:

	1986	1987
	S'000	S'000
Sales turnover	5,186,000	5,072,000

No of Employees: 8,000

TARBUK, F M, & CO

Davidgasse 90, Postfach 283, A-1100 Wien

Tel: (0222) 62-202

Telex: 134691

Managing Partners: Dr Mario Seiller-Tarbuk, Dkfm Gert Langer, Dr Dieter Blahut

PRINCIPAL ACTIVITIES: Wholesale and retail trade in motor vehicles, diesel engines, hydraulic cranes, fork lifts, parts and accessories; import, wholesale and retail business for Nissan, Saab, Jaguar/Rover, Cummins, Atlas, Cov Climax; leasing; general trading

Trade Names: Nissan, Saab, Jaguar/Rover, Cummins, Atlas, Cov Climax

Principal Bankers: Creditanstalt-Bankverein; Schoeller & Co

Financial Information:

	S'000
Sales turnover	2,200,000

No of Employees: 541

TAUERNAUTOBAHN AG

Alpenstr 94, A-5020 Salzburg

Tel: (0662) 20511

Telex: 633718 taag a

Supervisory Board: Sektionschef i.R. Dipl.-Ing. Dr. techn. Franz Matl (Vorsitzender), Dipl. Ing. DDr. Hans Lechner (stellv. Vorsitzender), Dr. jur. Wolfgang Nolz (stellv. Vorsitzender), Dipl. Ing. Josef Zeisler (stellv. Vorsitzender), Dipl. Ing. Manfred Hessle, Walter Leitner, Dr. jur. Herbert Martinek, Dr. jur. Ludwig Schuberth, Dipl. Ing. Anton Supersperg, Dr. jur. Wilfried Trabold. Vom Betriebsrat waren delegiert: Günter Neuhofer, Theodor Deutinger, Franz Gruber, Johann Müller, Hans Prodinger

Management Board: Dkfm. Karl Just, Dipl.-Ing Arnold Tautschnig

Confidential Clerks: Dr C Heu, Dr E Wagner

PRINCIPAL ACTIVITIES: Building, finance and maintenance of highways

Principal Bankers: Salzburger Landes- Hypothekenbank

Financial Information:

	31.12.87
	S'000
Sales turnover	524,000
Share capital	700,000

Principal Shareholders: Republic of Austria
No of Employees: 200

UNIVERSALE-BAU AG

Renngasse 6, A-1010 Wien

Tel: (0222) 53461-0

Cable: Universalbau Wien

Telex: 114703 uhtwn a; 135416 ubaua

Telefax: 5334710

Supervisory Board: Dr Guido N Schmidt-Chiari (Generaldirektor und Vorsitzender des Vorstandes der Creditanstalt-Bankverein, Vorsitzender), Dkfm Gerhard Randa (Generaldirektor-Stellvertreter und Stellvertretender Vorsitzender des Vorstandes der Creditanstalt-Bankverein, Stellvertreter des Vorsitzenden), Dkfm Dr Theo Ettel (Vorsitzender des Aufsichtsrates der Österr. Philips Industrie Gesellschaft mbH), Baurat h.c. Dipl.-Ing Bruno Freibauer, Dr Rudolf Gruber (Generaldirektor und Vorsitzender des Vorstandes der EVN Energie-Versorgung Niederösterreich AG), Komm.-Rat Professor Dr Helmut H Haschek (Generaldirektor und Vorsitzender des Vorstandes der Oesterr. Kontrollbank AG), Professor Dr Walther Kastner, Abg.z.NR Kammerrat Josef Kerschbaum, Komm.-Rat Karl Kornis (Generaldirektor i.R., Vorsitzender des Aufsichtsrates der Ersten Allgemeinen Versicherungs AG), Franz Krammer, Johann Obermaier, Kammerrat Franz Perdacher, Friedrich Schödl, Julius Sommer, Dkfm Dr Helmut Stalf, Dr Heinz Traxler (Mitglied des Vorstandes der Donau Allgemeine Versicherungs-AG)

Management Board: Dr Josef Vlcek (Generaldirektor und Vorsitzender), Dr Ferdinand Birkner, Dipl.-Ing. Helmut Frey-Trauer, Ing. Alois Salzmann

PRINCIPAL ACTIVITIES: All forms of construction work; construction of industrial plants; structural engineering; civil engineering; foundation work, comprising large bridges, roads, waterways, earthmoving; the building of power plants, shafts and tunnels, subways, subterranean civil engineering including foundation piles, concrete piles, anchorage and injection work as well as demolition

Parent Company: Creditanstalt Bankverein, Vienna

Subsidiary Companies: Shopping Center, Planungs-und Entwicklungs GmbH & Co Werbeberatung KG; Tourist Center Hotelbau GmbH & Co KG; "Universale-Bau" GmbH (Germany); Universale Bau de Venezuela (Venezuela); Sonderbau GmbH; "Universale" Grundbau GmbH; Transportbeton GmbH & Co KG and other participations

Principal Bankers: Creditanstalt-Bankverein

Financial Information:

	31.12.86	31.12.87
	S'000	S'000
Sales turnover	4,556,000	4,571,000
Profit after tax	633	9,296
Dividends	-	10,000
Dividends per share	-	50S
Share capital	200,000	200,000

Principal Shareholders: Creditanstalt-Bankverein
No of Employees: 3,500

WEISS, GEBRUEDER, GES MBH

Bundestr 110, 6923 Lauterach, Vorarlberg

Tel: (05574) 38611-0

Cable: Weissexpedit

Telex: 57733 gw lt a

Telefax: (05574) 39303

Chairman: P Senger-Weiss
President: H Senger-Weiss

Senior Executives: A Neier (Finance)

PRINCIPAL ACTIVITIES: International freight forwarding, international transport services, warehousing, customs clearing, airfreight, express service, hanging garment and leasing, management service and consulting
Trade Names: Gebrueder Weiss
Parent Company: G W Verwaltungsges. m.b.H.
Subsidiary Companies: Friedrich Platzer & Co; Confracht Container Transport Ges mbH
Principal Bankers: All major Austrian Banks
Financial Information:

	30.6.87	30.6.88
	S'000	S'000
Sales turnover	5,635,000	6,322,000
Share capital	60,000	60,000

Principal Shareholders: G W Verwaltungsges.m.b.H.
No of Employees: 1,476

WIENER ALLIANZ VERSICHERUNGS-AG

Hietzinger Kai 101-105, 1131 Wien
Tel: (0222) 94 85 11
Telex: 134222 alwn a
Telefax: 94 85 11 260

Supervisory Board: Dipl. Ing. C A Goess-Saurau (Chairman), Dr W Schieren, H Birnstingl, Dr E Coppola di Canzano, H K Jannott, Dr A Tiberini, DDr F Padoa, Dr W Jeanée, K Kuefstein, Dr G N Schmidt-Chieri, Dr J Taus, D I O Voisard
Management Board: Dr E Baumgartner (Chairman), Dr F Vogler (Deputy Chairman), Dr K Frisch, Dr H-R Grüber, Dr R Huber

PRINCIPAL ACTIVITIES: All classes of insurance except life insurance
Subsidiary Companies: (100% owned unless stated): Opernringhof Bau-und Betriebsgesellschaft mbH; "Providentia" Vermögensverwaltungs-Betriebsgesellschaft mbH; Wiener Allianz Werkküchen-Betriebsgesenschaft mbH; Braun & Co Gesellschaft mbH; Wiener Allianz Leasing Gesellschaft mbH & Co KG (45%); Wiener Allianz Leasing Gesellschaft mbH; EBA Vermögensverwaltungs-und Betriebsgesellschaft mbH (32%); ESTRA Liegenschaftsvermietungsgesellschaft mbH (35%); L'Equité Compagnie d'Assurances (25%) (France)
Financial Information:

	31.12.86	31.12.87
	S'000	S'000
Premium income	4,942,000	5,221,000
Investment income	392,000	408,000
Profit before tax	36,000	59,000
Profit after tax	19,500	19,500
Dividends	19,200	12,200
Dividends per share	60S	60S
Share capital	160,000	160,000
Shareholders funds	565,000	627,000

Principal Shareholders: Assicurazioni Generali, Triest; Allianz AG, München; Münchner Rück, München
No of Employees: 2,542

WIENER STÄDTISCHE WECHSELSEITIGE VERSICHERUNGSANSTALT

Schottenring 30 (Ringturm), A-1010 Wien
Tel: (0222) 53139
Cable: Municipal
Telex: 135140
Telefax: 5353437

Management Board: Dr S Sellitsch (Chairman), Dr Helene Fischer, Dr F Lauer, Dkfm K Fink, Dr G Geyer, Mag M Kothbauer
Secretary-General: Dr M Ferstl

PRINCIPAL ACTIVITIES: Insurance
Trade Names: Wiener Städtische Wechselseitige Versicherungsanstalt

Subsidiary/Associated Companies: Donau Allgemeine Versicherungs-Aktiengesellschaft; Volksfürsorge-Jupiter Allgemeine Versicherungs-Aktiengesellschaft; Union Versicherungs-Aktiengesellschaft; Wiener Verein Lebens- und Bestattungsversicherung auf Gegenseitigkeit; Österreichische Beamtenversicherung Versicherungsverein auf Gegenseitigkeit; Wiener Rückversicherungs-Gesellschaft; Interrisk Internationale Versicherungs-Aktiengesellschaft; Österreichische Kreditversicherungs-Aktiengesellschaft; Europäische Güter- und Reisegepäck-Versicherungs-Aktiengesellschaft; Wüstenrot Versicherungs-Aktiengesellschaft; Folksamerica Holding Company; Die Freiburger Allgemeine Versicherung AG
Principal Bankers: Länderbank; Creditanstalt; Zentralsparkasse; Bank für Arbeit und Wirtschaft; Erste Österr. Sparkasse; Girozentrale
Financial Information:

	31.12.86	31.12.87
	S'000	S'000
Premium income	8,995,821	9,856,892
Investment income	1,678,150	1,800,526
Profit before tax	716,185	804,907
Profit after tax	636,700	739,396
Retained profit	119,699	180,396
Dividends*	499,046	733,457
Shareholders funds	1,376,165	1,543,606

*paid to policy holders

No of Employees: 4,122

ZENTRALSPARKASSE UND KOMMERZIALBANK, WIEN

Vordere Zollamtsstrasse 13, A-1030 Wien
Tel: (0222) 72 91 0;72 92 0
Cable: SPARZENTRALE WIEN
Telex: 133615; 133256; 133167
Telefax: 73 46 84

Supervisory Board: Dr H Zilk (Chairman), H Mayr (Vice Chairman), Prof Dr J Bandion, Dr E Busek, K Eder, Dr E Hofbauer, W Nettig, G Sallaberger, H Schmidtmeier, F Rücker, J Wimmer, S Beran, H Schrefler, P Brandmüller
Management Board: Dr K Vak (Chairman), Dr A Haiden (Deputy Chairman), P Höfinger, H Lugmayr, Dr R Riedl

PRINCIPAL ACTIVITIES: Bank
Trade Names: Z-Bank of Vienna
Subsidiary Companies: Z-Beteligungsfond AG; Z-Export und Handelsbank GmbH; Z-Invest GmbH; Girozentrale und Bank der Österreichischen Sparkassen AG; Adria Bank AG; Österreichische Investitionskredit AG; Union Versicherungs AG; Dataservice, Organisations-und Datenverarbeitungs GmbH; Swift; Z-Treuhand und Vermögensverwaltung GmbH; Z-Leasing GmbH; Kur-und Sporthotel GmbH and others
Financial Information:

	31.12.86	31.12.87
	S'000	S'000
Total assets	191,037,365	201,800,000

No of Employees: 4,592

ZÜRICH KOSMOS VERSICHERUNGEN AG

Schwarzenbergplatz 15, A-1015 Wien
Tel: (0222) 65 36 31
Telex: 133375 zueko a

Management Board: Dr Werner Faber (Vorsitzender), Dr Herbert Nowak

PRINCIPAL ACTIVITIES: Insurance
Parent Company: Zürich Versicherungs Gesellschaft, Zürich, Switzerland
Financial Information:

	1986	1987
	S'000	S'000
Premium income	1,585,000	1,707,000

AUSTRIA

No of Employees: 1,091

Major Companies of
BELGIUM

ABB ASSURANTIE VAN DE BELGISCHE BOERENBOND NV

Minderbroederstr 8, 3000 Leuven

Tel: (016) 24.33.33

Telex: 26448

Telefax: (016) 24.32.12

Board of Directors: J Hinnekens (Chairman), I Van Thielt (Managing Director), A Debruyne, S De Schaetzen, R Eeckloo, J Mertens, R Van De Keere, V Van Rompuy, G Jonckheere

Management Board: J Daniels (General Manager), C De Keyser (Manager Administration), W Duron (Manager Life and Reinsurance), G Haven (Manager Claims), A Wouters (Manager Technical), J Boon (Manager Marketing)

PRINCIPAL ACTIVITIES: Insurance

Subsidiary Companies: SA AVM (70.5%); SC Maatschappij voor Brandherverzekering (90.9%); SA Secura (68.8%); SA Concert Noble (99.9%)

Principal Bankers: Cera CV

Financial Information:

	31.12.86	31.12.87
	BF'000	BF'000
Premium income	13,014,000	14,642,000
Investment income	4,252,000	4,544,000
Profit before tax	509,000	600,704
Profit after tax	494,000	602,000
Retained profit	462,000	570,000
Dividend	32,000	32,000
Dividend per share	80BF	80BF
Earnings per share	1,235BF	1,505BF
Share capital	400,000	400,000
Shareholders funds	5,184,313	6,622,350

No of Employees: 1,426

AG, GROUPE,

BD Emile Jacqmain 53, 1000 Bruxelles

Tel: (02) 214.81.11

Cable: Assurances Générales

Telex: 22766 agbel b

Telefax: 214.81.50

Board of Directors: Maurice Lippens (Président-Administrateur-Délégué), Jacques Carlier, le Baron de Fauconval, le Comte Eric de Villegas de Clercamp, Pierre Pluys, le Baron Claude de Villenfagne de Vogelsanck, le Baron Jean van de Put, Jozef Van Riet, le Chevalier de Broux, le Vicomte Jean de Jonghe d'Ardoye, Bernard t'Serstevens, Valère Croes (Directeur Général), Comte Michel de Broqueville, Piet Van Waeyenberge

Management: M Lippens (Président Administrateur-Délégué), V Croes (Administrateur-Directeur Général), M Archambeau, M Dachy, G Valckenaere, M van Pee, R Verelst

PRINCIPAL ACTIVITIES: All types of insurance and reinsurance

Parent Company: Compagnie Financière et de Réassurance du Groupe AG

Subsidiary Companies: AG de 1824 Cie Belge d'Assurances Générales "Vie"; AG 1830 Cie Belge d'Assurances Générales

Principal Bankers: Société Générale de Banque

Financial Information: Consolidated figures

	31.12.86	31.12.87
	BF'000	BF'000
Premium income	36,035,752	35,821,580
Investment income	16,281,301	16,744,618
Profit before tax	2,155,595	2,570,655
Profit after tax	1,751,620	2,354,850
Retained profit	491,403	559,916
Dividends(1)	1,005,220	1,178,442
Dividends per share(1)(2)	105BF	115BF
Dividends per share(1)(3)	102.14BF	112.14BF
Dividends per new AFV share(1)	161.92BF	170.37BF
Earnings per share(1)	176.7BF	193.5BF
Share capital	3,750,000	4,050,479
Shareholders funds	85,360	85,360

(1) Parent Company only
(2) fully-paid share
(3) partly-paid share of BF400

Principal Shareholders: Société Générale de Belgique; Raffinerie Tirlemontoise; AN-HYP; private shareholders

No of Employees: 3,315

AN-HYP

Antwerpsche Hypotheekkas, Caisse Hypothecaire Anversoise

Grote Steenweg 214, B-2600 Antwerpen
Tel: 32 3 218.21.11
Telex: 33100

Board of Directors: Baron Jean van de Put
Management: Jozef Van Riet (General Manager)
Senior Executives: Georges De Wit (Head of International Division)

PRINCIPAL ACTIVITIES: Banking
Financial Information:

	31.12.87
	BF'000
Total assets	161,802,000

ASLK-CGER BANK

Algemene Spaar-en Lijfrentekas

Caisse Générale d'Epargne et de Retraite

PO Box 1436, 1000 Brussels
Tel: 32-2-213.61.11
Cable: Eparspark; SWIFT: CGAK BE BB
Telex: 26860; 61189 CGE ASK; 62921 ASLKFIX (dealing room); 62999 CGERFIX
Telefax: 32-2-213.67.99

Chairman: H R H Albert Prince of Liège; *President:* A Devreker
Executive Committee: L Aerts (Chairman)
International Department: P Henrion (Managing Director and Vice Chairman of the Executive Committee), M Bienstman (Managing Director), L Peeters (Managing Director), G Debouverie (General Manager, International Relations and Corporate Banking), R Cocquyt (Deputy General Manager, Financial Department-Foreign Currency), P Detournay (Deputy General Manager, Back Office Swift and Teleprocessing-Accounting and Control), J Gilman (Senior Representative, London Representative Office), R Hock (Senior Manager, International Settlements), J Lenders (Treasurer of the Bank), E Maes (Deputy General Manager, Corporate and Country Risk), M Mortier (Deputy General Manager Correspondent Banking), R Organe (Senior Manager, Bond Trading and Selling), H Poelaert (Deputy General Manager, Chief Dealer), F Serreyn (Senior Manager, Affiliates), M Vermeulen (Deputy General Manager, International Relations and Corporate Banking Dept)

PRINCIPAL ACTIVITIES: 1. ASLK-CGER Bank: originally a savings bank now acting as a fully fledged international bank.

2. Life insurance and fire, accident and miscellaneous risks insurances

Trade Names: ASLK-CGER Bank; Algemene Spaar-en Lijfrentekas; Caisse Générale d'Epargne et de Retraite

Subsidiary Companies: ASLK-CGER Finance NV (Netherlands); BUCL-Banque Universelle et Commerciale du Luxembourg (51%) (Luxembourg); ASLK-CGER American Inc (USA); ASLK-CGER Investments Ireland (Eire); ES-Bond Fund; ES-Bond Capital Fund; ES-Capital Advisory; ES-Advisory; ES-Belbond; ES-Belbond Capital Fund; ASLK-CGER Bank (Representative Office; UK)

Financial Information:

	31.12.86*	31.12.87*
	BF,000	BF,000
Profit before tax	4,907,000	4,636,000
Profit after tax	2,772,000	2,957,000
Deposits	1,059,653,000	1,148,163
Total assets	1,248,762,000	1,369,553

*Figures include the figures of the insurance company

No of Employees: 9,840

ASSUBEL, GROUPE,

Rue de Laeken 35-39, B-1000 Bruxelles

Tel: (02) 214 61 11

Telex: 22 886 asbel b

Management: Roger Selschotter (Administrateur-Directeur Général)

PRINCIPAL ACTIVITIES: Provision of all classes of insurance

Subsidiary Companies: Assubel Vie SA; Assubel Accidents et Dommages SA; Assubel Accidents de Travail; Assubel Salaire Garanti; Assubel Epargne SA; Assubel Etudes Gestion

Financial Information:

	BF'000
Premium income	12,370,000

BACOB SAVINGS BANK SC.

Trierstraat 25, 1040 Bruxelles

Tel: 32/2/237.82.11

Telex: 62199 bacob b

Telefax: 32/2/230.71.78

Board of Directors: Willy D'havé (Chairman), Guido Allegaert, Rik Branson, Paul Caeyers, Guy Cambier, Germain Capelleman, Jan Cuyvers, Josée De Bruijckere, Edouard Descampe, Anton Desloovere, Julien Dessein, Hubert Detremmerie, André Devogel, Robert D'Hondt, Karel Janssen, Kamiel Konings, Michel Manteleers, François Martou, Willy Peirens, Rik Smits, Robert Soete, Willy Thys, Robert Van den Heuvel, Walter Vanderbeken, Hendrik Van Dyck, Antoon Van Melkebeek, Jean Verjans

Management Board: Hubert Detremmerie (Président), Guido Allegaert, Paul Caeyers, Guy Cambier, Kamiel Konings, Walter Vanderbeken. Secretary Prosper Gielen

PRINCIPAL ACTIVITIES: Savings bank

Subsidiary Companies: BACOB Finance NV (The Hague), Banque UCL (Luxembourg); International Financial Markets Trading (London); Bacob Investments (Dublin); Bacob Delaware Inc (Delaware)

Financial Information:

	31.12.86	31.12.87
	BF'000	BF'000
Profit before tax	2,914,000	2,908,000
Profit after tax	967,108	1,161,012
Retained profit	547,638	705,283
Dividend gross*	419,471	455,729
Dividend per share net	8%	8%
Share capital	4,210,000	4,500,000
Shareholders funds	11,365,939	13,436,068
Deposits	262,021,017	296,656,221
Total assets	303,808,182	342,932,825

*A change in the definition of this item with respect to last year has been introduced as follows: The withholding tax on Bacob's own dividends which is payable by the Bacob Savings Bank, has hitherto been entered as a provision on the debit side of the profit and loss account at 31 December before the Annual General Meeting of shareholders at which the appropriation of the profit is approved. In 1987 for the first time the withholding tax is not to be entered into the accounts until the date on which the dividend is approved. This has the effect of increasing the annual profit by BF104,563,643. Last year's figure has been updated accordingly so as to obtain a consistent basis for comparison

Principal Shareholders: COPLUS (Co-operative Holding Company) (38.2%); Fédération Nationale des Coopératives Chrétiennes (27.4%); Christian Workers' Movement and personnel of BACOB (34.4%)

No of Employees: 2,781

BANK BRUSSELS LAMBERT
BBL

24 Avenue Marnix, B-1050 Brussels
Tel: (32-2) 517.21.11
Cable: Bruxelat
Telex: 21421; 21742; 26392; 63965 bbl b

Board of Directors: Jacques Thierry (Chairman), G Eskénazi (Vice President), D Gillet (Vice President), J Daniëls, P De Meester, Baron J De Staercke, Baron Philippe Lambert, Y Truffert, P Henderson, Mme E le Hodey, J Samain, A Tuytens, Baron Vaxelaire, Baron Bekaert, Karl-Ludwig Bresser, Théo Peeters, Rudolf Vanmoerkerke, Maurice Velge, J Schupp, J M Heymans

Directors and members of the Executive Committee: John Dils (President), H Ruhl, Yves Petit, Comte Thierry de Broqueville, Daniel Cardon de Lichtbuer, Charles Coppieters de Gibson, Bernard Peelman, E Andersen, J F de le Court

PRINCIPAL ACTIVITIES: All bank activities

Subsidiary/Associated Companies: BBL Australia Ltd (Australia); Crédit Européen SA (Luxembourg); Banque Bruxelles Lambert Trust Company Ltd (Jersey, CI); BBL International (UK) Ltd (UK); Williams de Broë Hill Chaplin (UK); Springfield Capital Management Ltd (UK); Banque Louis-Dreyfus SA (France); Banque Bruxelles Lambert (Suisse) SA (Switzerland); Finanziaria ICCRI Banque Bruxelles Lambert SpA (Italy); Fiduciaria ICCRI Bruxelles Lambert SpA (Italy);

Leasing SpA ICCRI Bruxelles Lambert (Italy); Banque Belgo-Africaine Burundi SARL (Burundi)

Financial Information:

	30.9.86	30.9.87
	BF'000	BF'000
Profit before tax	14,731,000	16,300,000
Profit after tax	12,190,000	13,354,000
Dividends	1,727,000	2,046,000
Dividends per share	140BF	153.33BF
Share capital	14,483,000	15,003,935
Shareholders funds	31,665,000	33,802,112
Deposits	1,340,427,000	1,616,132,119
Total assets	1,520,107,000	1,667,992,041

No of Employees: 11,943 units by taking account of only personnel remunerated by BBL in Belgium and by calculating number in terms of full-time employment

BANQUE NATIONALE DE BELGIQUE SA

Blvd de Berlaimont 5, 1000 Bruxelles
Tel: (02) 221.21.11
Cable: Bankionale Brussels
Telex: 21355 bknle b (Board of Directors-banking operations); 21105 bnbsg b (other operations)
Telefax: 32 2 221 31 01 (Board of Directors-banking operations); 32 2 221 31 00 (other operations)

Governor: Jean Godeaux; Vice Governor: Alfons Verplaetse
Directors: William Fraeys, Frans Junius, Jean-Pierre Pauwels, Guy Quaden, Jean-Jacques Rey
Senior Executives: Jean Poullet (Secretary), Jacques Van Droogenbroeck (Treasurer)

PRINCIPAL ACTIVITIES: Central Bank of Belgium

Financial Information:

	31.12.87	31.12.88
	BF'000	BF'000
Profit before tax	4,329,768	4,429,529
Profit after tax	924,854	958,971
Retained profit	22,291,471	25,655,335
Dividends	730,668	757,332
Dividends per share	1,826.67BF	1,893.33BF
Earnings per share	1,370.00BF	1,420.00BF
Share capital	400	400
Deposits	15,093,261	20,552,577
Total assets	652,177,303	657,861,847

Principal Shareholders: State: (50%) Registered and untransferable shares; Private sector: (50%) shares negotiable on the stock exchange

No of Employees: 3,227

CERA SPAARBANK CV

Parijsstraat 52, 3000 Leuven
Tel: (016) 24 49 99
Telex: 24166 cera
Telefax: (016) 24 49 32

Board of Directors: R Eeckloo (Chairman), M Smeers (Vice Chairman), C André, E Baert, J Baert, M Beddegenoots, W Breesch, M Broes, M Convents, R Cremers, W Danckaert, R Donckels, P Hermans, W Janssens, W Matton, G Mertens, J Palm, R Roseleth, P Tanghe, R Terrijn, P Van Biesen, E Vandermeeren, P Van Hoef, J Vanhulst, W Vantrappen
Executive Committee: W Danckaert (Chairman), E Baert (Deputy Chairman), W Breesch, J Vanhulst, P Tanghe

PRINCIPAL ACTIVITIES: General banking services
Subsidiary Companies: NV CERA LAGE LANDEN PRODUCTS; NV CERINVEST; Cooperative Guarantee Fund

Financial Information:

	31.12.86	31.12.87
	BF'000	BF'000
Profit before tax	3,200,850	2,754,991
Profit after tax	1,833,137	2,166,976
Retained profit	1,354,895	1,571,851
Dividends	478,242	595,125
Dividends per share	8%	8%
Share capital	8,405,753	10,121,624
Deposits	329,097,200	379,474,314
Total assets	367,167,779	424,126,126

Principal Shareholders: The CERA Spaarbanken are the principal shareholders in CERA Head Office. Private persons are the shareholders in the CERA Spaarbanken

No of Employees: 3,836

CH.M, GROUPE,

5A Square Emile des Grées du Loù, 1190 Bruxelles
Tel: (02) 376.00.90
Telex: 21132
Telefax: (02) 376.02.63

Board of Directors: Ch Mannerie (Président Directeur Général)
Senior Executives: Marcel Lalmand (Tourist Vehicles)

PRINCIPAL ACTIVITIES: Non-food retailing; banking and financial services; insurance; leasing and rental services; mechanical engineering; industrial motors; printing, publishing and information services; accountancy and legal services; advertising and public relations; manpower services; management services and consultants; recreational services; travel and tourism; clothing and footwear; motor vehicles and motor vehicle components

Financial Information:

	BF'000
Sales turnover	5,000,000

No of Employees: 650

COMPAGNIE GÉNÉRALE MOSANE

See MOSANE, COMPAGNIE GÉNÉRALE,

CRÉDIT COMMUNAL DE BELGIQUE SA

Blvd Pachéco 44, 1000 Bruxelles
Tel: (02) 214 41 11
Telex: 26354 cregem b
Telefax: 214.40.32; 214.40.38

Board of Directors: F Swaelen (President), J Donfut (Vice President), L Hurez, P Meyers, L Tobback, P Havelange, M Imler, A Evers, H Van Steenberge, J Michel, M Mahieu
Managing Committee: F Narmon (President), A Brouwers (Vice President), M Goethals-Muyldermans, M Mahler, L Onclin, J Asselbergh, J Labar. Government Commissioners: J Rousseaux, E Beyens
Senior Executives: F Narmon (Managing Director, President of the Managing Committee), J-P Grandgagnage (Director of General Secrtariat Division), H Coppe (Director of Legal Division), C Bourgeois (Director of Research Division), A Brouwers (First Manager, Vice-President of the Managing Committee), L Malvoz (Manager-Adviser), C Verbist (Manager-Adviser), Mrs M Decamps (Director of Financial Division), R Vanbrabant (Director of Public Customers' Market Division), W Wera (Director of Public Customers' Transactions Division), W Meulemans (Director of Enterprises' Market Division), A Halleux (Director of Loans to Individuals and Enterprises Division), Mrs M Goethals (Director), Miss N Henrion (Director-Adviser for Collective Relations), G Van Den Hende (Director of Training Division), P Hautecler (Director of Staff Division), L Blanchart (Director of General Organization Division), M Mahler (Director), R Verstockt (Director of Data Processing Division), P Delesalle (Director of Development Division), J-P Leburton (Director of National and International Issues Division), L Onclin (Director), R Dumortier (Director-Adviser for Data Processing), R Van Den Hoof (Director of

BELGIUM

Branches Division), R De Beukeleer (Director North-East
Sector), P De Meester (Director North-West Sector), G
Schifflers (Director Central Sector), J Naeye (Director South-
West Sector), J Demeffe (Director South-East Sector), M
Bouteille (Director of Individuals' Market Division), J
Asselbergh (Director), M Ernaelsteen, (Director of Purchasing
and Real Estate Division), M Panneels (Director of Securities
Division), L Vanderheyden (Director of Individuals' Accounts
Division), J Labar (General Inspection), Mrs N Jacobs
(Director-Adviser), C Caudron (Director of External Audit,
Inspection and Safety Division), J Demarbaix (Director of
Internal and Management Audit Division)

PRINCIPAL ACTIVITIES: Bank
Financial Information:

	31.12.86	31.12.87
	BF'000	BF'000
Profit after tax	1,201,881	1,277,928
Dividend	385,000	381,500
Share capital	3,500,000	3,500,000
Deposits	1,286,234,461	1,468,574,907
Total assets	1,308,920,165	1,491,606,142

Principal Shareholders: Belgian municipalities and provinces
No of Employees: 3,300

D'IETERENS, SA, NV

50 rue du Mail, 1050 Bruxelles
Tel: (02) 536.51.11
Cable: Dietefre-Bruxelles
Telex: 21410 ditren b
Telefax: 537.31.81

Board of Directors: Roland D'Ieteren (President and General
Manager), Marcel Bruyns, Jules-Maurice Anspach, Maurice
Perier, Alain Philippson
Management Committee: Roland D'Ieteren (President and
General Manager), Marcel Bruyns (Adviser), Raoul de
Brouchoven de Bergeyck (Executive Vice President), Ingmar
Jonkers (Executive Vice President), Gilbert van Marcke de
Lummen (Executive Vice President). General Secretary: Roger
Lemmens
Central Departments: Guy Tapernoux (Manager of International
Affairs) Paul Doutrepont (Manager of Central Services),
Jacques de Smet (Finance Manager), G Danneels (Personnel
Manager)

PRINCIPAL ACTIVITIES: Import and distribution of motor
vehicles and components; financial services; leasing of
vehicles; automobile engineering
Trade Names: Volkswagen, Audi, Porsche, SEAT, Yamaha,
Avis-Rent-a-Car Systems, Credimatic, M.B.K.
Subsidiary/Associated Companies: D'Ieteren Antwerpen NV;
D'Ieteren Sport SA; D'Ieteren Lease NV; FND Industries;
D'Ieteren Vehicle Trading; Immo D'Ieteren SA; Afidif SA;
Locadif SA; Computer Valley SA; Iberauto SA; D'Ieteren
Services; D'Ieteren Trading BV; Metropolitan Bank;
Metropolitan Finance SA
Principal Bankers: Banque Bruxelles Lambert; Société Générale
de Banque
Financial Information:

	31.12.86	31.12.87
	BF'000	BF'000
Sales turnover	25,592,958	28,491,720
Profit before tax	709,170	882,702
Profit after tax	365,551	485,142
Retained profit	464,239	628,529
Dividends	67,450	82,833
Dividends per share	760BF	933.33BF
Earnings per share	570BF	700BF
Share capital	300,000	300,000
Shareholders funds	2,081,623	3,641,867

No of Employees: 1,176

DINERS CLUB BENELUX SA

36 Rue Ravenstein, 1000 Bruxelles
Tel: (02) 515 9511
Cable: Dinclub-Bruxelles
Telex: 24415 dinclub b
Telefax: 32.2.5135266

Board of Directors: John H Dobson (Président), Seymour Flug
(Vice Président, USA), Ronald E Geesey (Germany), Pierre
Hautain (Administrateur délégué)
Management: Pierre Hautain (Directeur Général)

PRINCIPAL ACTIVITIES: Credit card finance
Subsidiary Companies: S A Dinshopping (99.87%)
Principal Bankers: Citibank
Financial Information:

	1986	1987
	BF'000	BF'000
Sales turnover	16,054,485	17,316,442
Profit before tax	69,885	87,807
Profit after tax	39,685	42,582
Retained profit	945	190
Dividends	38,892	43,337
Dividends per share	14.0BF	15.6BF

Principal Shareholders: Citicorp
No of Employees: 172

GB INNO BM SA

Avenue des Olympiades 20, (Boulevard Leopold III), B-1140
Bruxelles
Tel: (02) 243.21.11
Cable: Gbinbm-Bruxelles
Telex: 21768
Telefax: (02) 243.29.88

Board of Directors: Baron F Vaxelaire (Chairman and Managing
Director), J Dopchie (Deputy Chairman and Managing
Director), Count Y du Monceau de Bergendal (Vice-Chairman),
A H Bienfait, Y Boël, R Bourgeois, B De Corte, Count de
Launoit, D du Monceau de Bergendal, A Frère, S Golstein, J
J Goossens, Mrs N Hardenne, A Leysen, J P Ruquois, P
Schohier, A Van Maele, G Vaxelaire, R Vaxelaire, Th Masset
Honorary Deputy Chairmen: R Louis, E Masset
Executive Committee: F Vaxelaire (Chairman and Chief
Executive Officer), J Dopchie (Deputy-Chairman and
Managing Director), D du Monceau de Bergendal (General
Manager, Restructuring), S Golstein (General Manager and
Chief of Corporate Staff), R Bourgeois (General Manager,
Mass Retail Trade), J Pitz (General Manager, Specialized
Retail Trade), R Vaxelaire (General Manager, Department
Stores), A Van Maele (General Manager, Unic Chain), R
Vercruysse (General Manager, Catering)

PRINCIPAL ACTIVITIES: Retail and department stores,
supermarkets, hypermarkets, restaurants, autocentres,
garden centres; do-it-yourself; wholesalers
Trade Names: Maxi, Super GB, Brico GB, Supergarden, Auto 5,
Resto GB, Inno, Grand Bazar, G B Zaïre, Unic, Club, Disport,
Vision Center, FNAC, IKEA, Gecotec, Transcontinental-
Generalcar, Pizza Hut, Christiaensen, Country Grill, Ten
Carvery, Ten Weyngaert, Cafeteria GB, Mr Grill & Buffet's,
Quick, France Quick, Homebase, Scotty's, OBI, Handy Andy,
VETIR, AKI, SARMA, NOPRI, Lunch Garden, Sarma Star,
Sarma Lux, Sarma Newshop
Subsidiary Companies and Participations: Bazimo; Immobilière
Beaufays; Imodis; Financières GB-Inno-BM; Saboma; Safe
Insurance; Supertransport; Motorest; UNIC; Transcontinental;
Distirest, CLUB; SPORTLAND; SANTAL; Fresh Wash; Van
Hemelrijck; SODAL; France Quick (France); SOMABRI
(France); HOMEBASE (UK); Scotty's (USA); GECOTEC;
SOIGESTI; IKEA; SODIF Corporation; Quick International; All
Insure; GB International; COBOMA; COMIZA; Belux Tours;
Wirtzair; Reisbureau Wirtz; EUFIDIS (Netherlands); GIB
Coordination Centre; GIB Invest; MAGINVEST; Visions
Centers; PROTRAVEL; DISPORT (France); AOK of Delaware

(USA); Shopping Services SA; Pizza Belgium; SA Award
Services; Safe Reinsurance (UK); Service Investment (UK);
Sarmag; Nopri Holding; Bricobi (Spain); GB Zaïre (Zaire);
VETIR (France)

Financial Information: GIB Group figures

	31.1.86	31.1.87
	BF'000	BF'000
Total turnover	128,171,000	140,813,000
Profit before tax	1,857,001	2,530,764
Profit after tax	1,557,719	1,945,485
Group earnings	1,513,788	1,904,859
Dividends	973,037	1,055,395
Dividends per ord share	22.50BF*	24.50BF
Dividends per AFV share	34.77BF	36.43BF
Earnings per share	38.90BF	46.70BF
Share capital	3,100,000	3,104,844
Shareholders equity	10,589,345	11,543,050

*after division by 10

No of Employees: 24,285 (group: 30,629)

GÉNÉRALE DE BANQUE SA/GENERALE BANK NV

Montagne du Parc 3, B-1000 Bruxelles
Tel: (2) 518.21.11 (Domestic Activities); (2) 516.21.11
(International Activities)
Cable: "GENERALE"
Telex: 21283 Geba b (Domestic Activities); 61050 Geba b
(International Activities)
Telefax: (2) 5164222 (G3); SWIFT CODE: GEBA BE BB

Board of Directors: Jacques Groothaert (Chairman)
Board of Managing Directors: Count Eric de Villegas de
Clercamp (Communication and Personnel Relations, Centre
for Strategic Integration), Baron Paul-Emmanuel Janssen
(Corporate Planning Information, Legal Department, Internal
Audit and Security), Henri Fayt (Credits), Xavier Malou
(Corporate and Investment Banking), Armand Hoeck
(Electronic Banking and Payment-Systems, Central,
Functional Support Services, Operations and Bank
Engineering), Roland Demol (Human Resources, Organisation
and Material Resources and Data Processing), Léon Derwa
(National Network-National Network Specialist Subsidiaries),
André Dwickx (Treasury and Foreign Exchange, Corporate
and Investment Banking), Marc Y Blanpain (Budgetary Control
of the Domestic Commercial Network, Private Banking
Activities, Retail, Upscale and Small and Medium-Sized
Enterprises Departments), Michel Berges (International
Commercial Banking) Secretary: Baron Hubert Simonart

PRINCIPAL ACTIVITIES: All banking activities
Trade Names: GENERALE BANK
Affiliates and Investments: Foreign Offices: BRANCHES: Italy:
Milan, Rome; Japan: Tokyo; Portugal: Lisbon; Singapore:
Singapore; Spain: Madrid, Barcelona; UK: London; USA: New
York. REPRESENTATIVE OFFICES: Australia: Melbourne;
Brazil: Sao Paulo; China: Beijing, Shanghai; Colombia:
Bogota; India: Bombay; Iran: Tehran; Japan: Tokyo; Mexico:
Mexico City; Turkey: Istanbul; USSR: Moscow; Venezuela:
Caracas. SUBSIDIARIES: Bahamas: Générale Bank and Trust
(Bahamas); Belgium: Banque Belgo-Zaïroise (Brussels);
Banque Européene pour l'Amerique Latine (BEAL) (Brussels)
with branches in Argentina, Brazil and Uruguay, and
Representative Offices in Chile and Columbia; Antwerpse
Diamantbank (Antwerp); Genconsultoria (Sao Paulo, Brazil);
Générale de Banque Belge (France) (Paris and Lille, France);
Generale Bank & Co (Cologne, Germany); Generale Bank
Overseas (Belgium) (Hong Kong); Générale Belgian Finance
Company Ltd (Hong Kong); Banque Générale du Luxembourg
(Luxembourg); Genservicos (Madrid, Spain); Banque Belge
Ltd (London, UK); Banque Belge Ltd (Guernsey, UK); Quinn
Cape Ltd (London, UK); Generale Investment Corporation
(New York, USA). AFFILIATED AND ASSOCIATED BANKS:
Banque de Crédit de Bujumbura (Burundi); Société
Camerounaise de Banque (Yaoundé, Douala, Cameroon);

Générale de Crédit (Cyprus); OBU (Nicosia, Cyprus); Société
Générale de Banques en Guinée (Conakry, Guinea); Société
Générale de Banques en Côte d'Ivoire (Abidjan, Ivory Coast);
Société Générale Libano-Européenne de Banque (Beirut,
Lebanon); Banque Marocaine du Commerce Extérieur
(Casablanca, Morocco); Banque de Kigali (Kigali, Rwanda);
Société Générale de Banques au Sénégal (Dakar, Senegal);
Compagnie de Gestion et de Banque Gonet SA (Geneva and
Nyon, Switzerland); European American Bank (New York,
USA); Banque Commerciale Zaïroise (Zaïre)
Principal Bankers: Most major banks throughout the world
Financial Information: Consolidated using new consolidation
methods (1)

	31.12.86	31.12.87
	BF'000	BF'000
Turnover	206,432,000	206,843,000
Profit before tax	8,964,000	7,656,000
Profit after tax	5,896,000	6,409,000
Dividends	3,560,000	3,918,000
Dividends per ord share(2)	245BF	255BF
Earnings per share	590BF	599BF
Share capital	14,867,000	14,983,000
Shareholders funds	46,057,000	47,665,000
Deposits	1,924,752,000	2,005,905,000
Total assets	2,080,365,000	2,175,143,000

(1)after full integration of
subsidiaries and
proportional integration of
banking institutions more
than 10% held
(2)after tax

Principal Shareholders: SA Société Générale de Belgique; Amro
Bank via TUBA Holding International; Groupe Assurances
Générales
No of Employees: Generale Bank Belgium 15,875

GEVAERT PHOTO-PRODUCTEN NV

Septestraat 27, 2510 Mortsel, Antwerp
Tel: (3) 440.73.46
Telex: 73055 AGNV B
Telefax: (03) 440.04.78

Board of Directors: A Leysen (Président), P Francken (Vice
President and Managing Director), F Vreys (Managing
Director), P Delacroix, P Collin, Ridder S de Schaetzen van
Brienen, J Hinnekens, A Gregoir, Mme Ch Koumans-Van
Hoestenberghe, S L Salas, P Saverys, P Scohier, J Van den
Wouwer, Mme A M Schluep-Beken, Mme C Van Damme-De
Waele, L Vlerick, G Declercq, Mme C Vlerick-Sap, R Van de
Steen, J Van Wassenhove, Mme B Bonnet-Germeau, Mme M
Smith-Cappuyns, Mme N Hardenne, Y Brasseur, T Masset, F
De Beir, F Arts
Management: P Francken, F Vreys
Senior Executives: A Claes (Manager)

PRINCIPAL ACTIVITIES: Holding company
Subsidiary Companies: (100% owned unless stated): Ortelius
Algemene Verzekeringsmaatschappij NV (99.99%); "Gebema"
Gevaert Beleggingsmaatschappij NV; Mia NV; Immori NV;
Difico NV; Gevafin SAH (Luxembourg); Secural SAH
(Luxembourg). Associated Companies: Almanij NV (8.98%);
Bayer AG (3.22%) (Germany); C.P.I. Compagnie de
Participations Internationales SAH (15.14%) (Luxembourg);
Hapag-Lloyd AG (10%) (Germany); K.B.L. Kredietbank SA
Luxembourgeoise (5.9%) (Luxembourg); O.T.V. Onafhankelijke
Televisie Vlaanderen NV (21.5%); SODIN Société
d'Investissements Internationaux SA (31.27%);
Uitgeversbedrijf TIJD NV (5.72%); VUM Vlaamse
Uitgeversmaatschappij NV (12.5%). Other Investments:
Artémis SA (3.34%) (Luxembourg); Belgamar NV (2.04%);
C.M.B. NV (5%); Communication Technologies NV (3%);
Copibel SC (0.29%); G.B.L. Groupe Bruxelles Lambert SA
(1.01%); Generale Maatschappij van België NV (1.06%);
Baring European Capital Trust (3.28%) (Guernsey); Hibernia

Corporation (4.6%) (USA); IBI Holding Company NV (1.67%) (Netherlands Antilles); OPFI-Paribas SA (0.14%) (France); Pallas Group SA (2.86%, of which 0.94% for third parties) (Luxembourg); Parfinance SA (1.15%) (France); Paribas Bank Belgie NV (1.78%); Paribaven Overseas Investors NV (2.6%) (Netherlands Antilles); Petrofina NV (0.23%)

Principal Bankers: Kredietbank NV; Paribas Bank NV; Petrofina NV (0.23%)

Financial Information:

	31.12.86	31.12.87
	BF'000	BF'000
Profit before tax	1,155,388	1,602,588
Profit after tax	1,154,432	1,601,246
Retained profit	303,498	698,579
Dividend	818,400	868,000
Dividend per share net	165.00BF	175.00BF
Earnings per share	310.33BF	430.44BF
Share capital	2,000,000	2,000,000
Shareholders funds	10,171,383	10,869,962
Total assets	11,133,765	11,810,981

No of Employees: 8

GROUPE AG
See AG, GROUPE,

GROUPE ASSUBEL
See ASSUBEL, GROUPE,

GROUPE CH.M
See CH.M, GROUPE,

GROUPE ROYALE BELGE
See ROYALE BELGE, GROUPE,

KREDIETBANK NV
Arenbergstraat 7, B-1000 Brussels
Tel: (02) 517.41.11
Cable: Creditbank Brussels
Telex: 21207 KBCOMP B; 61402 KBCOMP B
Telefax: (02) 517.42.09

Executive Committee: Jan Huyghebaert (President of the Bank), Edward Wauters (Managing Director), Jozef Lambrechts (Managing Director), Marcel Cockaerts (Managing Director), Wilfried Janssens (Managing Director)

Board of Directors: André Vlerick (Chairman), Paul Francken (Vice-Chairman), Marc Santens (Vice-Chairman), Jean Blondeel, René Brück, Godfried Lannoo, Edward Thielemans, Eugène Van Dyck, Valeer Van Rompuy, J Ch Velge, Luc. V Wauters

Management: Edgar Plan (International Financial Adviser), Hugo De Moor (Economics and Statistics Department), Michiel Beerts (Secretariat/Administration-Executive and General Management/Board of Directors), Jos Aerts (Assistant General Manager), Jacques Serckx (Security Service), Raymond Van Assche (General Management-Auditing), Louis Van Den Broeck (Internal Audit), Johan Vranken (Head Office Control). Secretariat-general: Urbain Neesen (Secretary-general), Johan Deron (Secretariat of the Board of Directors and Executive Committee), Luc Van Bruyssel (Management Accounting Department), Geert Vermander (Legal Department), Guy Bollen (Fiscal Department), Piet Jaspaert (Press and Public Relations Department). Domestic Banking Directorate: Headed by Walter Martens, Herman Hillewaere (Central Management-Private Customers and Local Enterprises), Alfons Lambrechts (Credit to Private Customers and Local Enterprises Division), Napoleon Van Tilborgh (Central Management-Investment Clientele), Willy Landuyt (Central Management-Corporate Customers), Paul Matthijs (Policy and Communications Support Division), Jan Blockx (Payments and Electronic Banking Division), Paul Vranken (Organisation and Automation Support Service-Domestic Banking), Alfons Verboven (Assistant General Manager).

Domestic Credit Risks Directorate: Headed by Jan De Vis, Alois Vertongen (Deputy), Maurits Stroeykens (Domestic Risk Management and Credit Assessment Division), Frans Maes (Credit Adviser), Jan Roels (Credit Adviser), Paul Custers (Domestic Credit Division), Assistant General Managers: Fernand Devaux, Roger Vanderhoeven. Treasury, Forex and Alm Directorate: Headed by André Swings, Willy Kestens (Treasury and Forex Division). Investment Banking Directorate: Headed by Remi Vermeiren, Marcel Vanderbeck (Deputy), Gabriel H De Smet (Corporate Finance Division), Joseph Nerincx (Assistant General Manager), Jacques Vandenberghen (Assistant General Manager), Marcel Stappers (Securities Trading Division), Jan Verhaeghe (Investment Management Division). International Directorate: Headed by Paul Vergote, Eugeen Cleemput (International Banking Division), Jaak Praet (Assistant General Manager), Herman Agneessens (International Network Division), André Bauwen (Assistant General Manager), Jozef Demesmaeker (International Banking Administration Division). International Credit Risks Directorate: Headed by Rony Van Hoeck, Pieter Heering (International Credit Division), Guido Segers (International Risk Management Division). Directorate Processing and Automation: Headed by Jean-Paul Van Kiersbilck, Alex Van der Auwera (Assistant General Manager), Pierre Coffé (EDP Division), Roger Troch (Securities Administration Division). Directorate Personnel, Training and Logistics: General Management-Personnel and Logistics Headed by André Ockerman, Rudy Broeckaert (Personnel Division), Assistant General Managers: René Coppye. Raymond Calewaert (Bank Real Estate Division), Leo Vandensteen (Services Division). General Management-Training and Management Development: Headed by Herman van de Velde, Elor Voets (Education and Training Division), E Robert Vreven (Career Planning and Management Development Division), Dirk Syx (Organization Department)

PRINCIPAL ACTIVITIES: All banking activities

Subsidiary/Associated Companies: Crédit Général SA de Banque; HSA NV Spaarbank; Bemab NV; Ateka NV; Irish Intercontinental Bank Ltd; IIB Finance Ltd; SA Gestion KB Income Fund: Holding SA Eurinvest; Bankverein Bremen AG; KB Internationale Financieringsmaatschappij NV; SA Gestion KB Capital Fund; Bank Van der Hoop Offers NV; Kredietbank SA Luxembourgeoise; Kredietbank (Suisse) SA; KB International (Hong Kong) Ltd; SA ECU Conseil; SA Decarenta Conseil. Kredietfinance (UK) Ltd; Kredietbank International Finance NV; Kredietbank North American Finance Corp; KB Financial Services (Ireland) Ltd; KB Financial Services (Dublin) Ltd (Eire). Foreign Branches: New York, London

Financial Information: Consolidated figures

	31.3.87	31.3.88
	BF'000	BF'000
Turnover	105,707,846	116,011,579
Profit before tax(1)	6,227,532	6,183,175
Profit after tax(1)	3,609,532	4,270,175
Retained profit(2)	1,285,415	1,502,764
Dividends(2)	1,870,448	2,147,864
Dividends per ord share(2)	115.00BF	127.00BF
Dividends per AFV share(2)	190.64BF	200.42BF
Earnings per share	281BF	310BF
Share capital	7,112,173	7,139,822BF
Shareholders funds(3)	28,481,639	33,576,197
Deposits	1,068,518,813	1,222,448,488
Total assets	1,191,862,303	1,359,664,126

(1) minority interests in the profit of 293.169 (1987) and 344, 447 (1988) excluded
(2) KB unconsolidated
(3) without minority interest and profit of the year

Principal Shareholders: Almanij (40.0%)
No of Employees: 11,042

MOSANE, COMPAGNIE GÉNÉRALE,

Centre Opéra, 41 Place de la République Française, B-4000
Liège
Tel: (041) 23.26.26
Telex: 41967 IMOSAN B

Board of Directors: D Collinet (Président)

Management Board: T Masset (Président), B de Corte
(Administrateur Délégué), Mlle N Hardenne, D Collinet, R
Comhaire, P Scohier

PRINCIPAL ACTIVITIES: Holding company with interests in the
fields of finance, real estate and building

Subsidiary/Associated Companies: (100% owned unless stated):
Imosa (93.3%); Materne (5.69%); SAIT (12.49%); Cofilim
(36.5%); Dujardin (32%); Cosylva (99.9%); Groupement
Financier Liegois (99.8%); Groupe Bruxelles-Lambert (1.4%);
Cie de Developpement International (Luxembourg); CPI
(3.76%) (Luxembourg)

Principal Bankers: Banque Paribas Belgique SA

Principal Shareholders: Cobepa (49.8%); Fidepa (8.1%)

PETROFINA SA

Rue de l'Industrie 52, 1040 Bruxelles
Tel: (02) 233.91.11
Cable: Petrofina Bruxelles
Telex: 21556 pfina b

Board of Directors: Baron Wolters (Honorary President), Jean-
Pierre Amory (Chairman and Managing Director), Pierre de
Tillesse (Deputy Chairman and Managing Director), François
Cornelis (Executive Director), Baron Paulo de Garcia de la
Vega (Executive Director), Pierre Jungels (Executive Director),
Baron Didrik Snoy (Executive Director). Directors: Yves Boël,
Gérard Eskenazi, Roger Fauroux, Albert Frère, Dr Paul
Janssen, René Lamy, Comte Jean-Pierre de Launoit, Emile
Guevrin, Jacques van der Schueren, Luc Wauters, Baron Guy
de Wouters, Fred E Zollinger. Honorary Vice President: Baron
Lambert. Honorary Directors: Carmille Bonnami, Richard I
Galland, Pierre Vanderstappen. Secretary General: J Van
Uytvanck

Management: Jean-Pierre Amory (President and Managing
Director), Pierre de Tillesse (Deputy President and Managing
Director), François Cornelis (Executive Director, Chairman's
Office), Baron Paulo de Garcia de la Vega (Executive Director
Supplies), Pierre Jungels (Executive Director), Baron Didrik
Snoy (Executive Director, Transport, Insurance, Coal, Trade
Relations, Studies and Acquisitions), Henrique Bandeira Vieira
(General Manager, Exploration, Production), Axel de
Broqueville (General Manager, Chemicals, Research and
Development, New activities), Jean Buckens (General
Manager, Insurance, Corporate Finance, Insurance), André
Florival (General Manager, Marketing), Georges vander
Straeten (General Manager, Refining, Data Processing, Safety
and Environment), Jean Van Uytvanck (General Manager,
Legal Affairs, Personnel), Pierre Masson (Senior Adviser,
Exploration)

PRINCIPAL ACTIVITIES: All sectors of the oil industry:
exploration and production, transportation and oil refining,
petrochemicals, marketing of petrochemical products and
research. Complementary fields: paints and coatings,
oleochemicals from natural oils and fats, coal, engineering,
insurance

Subsidiary Companies: (100% owned unless stated): Fully
consolidated companies:- Belgium: Etmofina; Fina; Labofina;
Liévin; Petrochim (99.93%); Fina Marine; Eurocoal-Spiers;
Synfina-Oleofina (99.99%); Sigma Coatings; CEAI; Sigma
Résines; Fina Exploration Denmark; Petrofina International
Group. Other Countries: Petrofina Holding Luxembourg
(Luxembourg); Fina Luxembourg (Luxembourg); Deutsche
Fina (Germany); Fina Bitumenwerk (Germany); Erdöl
Raffinerie Duisburg (Germany); Sigma Coatings (Germany);
Protherm Fernwärme (Germany); Fina Exploration Norway
(USA); Norkse Fina (Norway); Fina France (99.99%) (France);
Carnot (France); Maritime Fina (France); Pétronaphte (France);
Sigma Coatings France (France); Nord-Chauffe (France);
Petrofina UK (UK); Petrofina Refineries (UK); Sigma Coatings
(UK); Hercock Simpson (UK); Fina Exploration (UK); Fina
Chemicals (UK); Charterhouse Petroleum (UK); Energy
Equipment (UK); Fina Nederland (Netherlands); Mafina
(Netherlands); Sigma Coatings (Netherlands); Fina Aviation
and Chemicals (Switzerland); Fina Iberica (Spain); Finapetro
(Liechtenstein); Fina Italiana (Italy); Raffineria di Roma (57.5%)
(Italy); Sigma Coatings (Italy); Fina Petroleos de Angola
(67.11%) (Zaïre); Zaïre Fina (60%) (Zaïre); Zaïrep (Zaïre); Fina
Exploration Gabon (Gabon); Fina Oil and Chemical (USA);
American Petrofina Holding (USA); American Petrofina
Exploration Company (USA); American Petrofina (84.53%)
(USA); Petrofina Delaware (USA); Petrodel (USA); Brittany
Holdings (Bermuda); Brittany Insurance (Bermuda); Diglame
Finance (Bermuda). Companies Valued by Equity Method:-
Belgium: Amoco Fina (50%); Belgosoute (50%); Prodira (90%);
SBGP (50%); SIBP (50%); Sogetrol Antwerpen; Autopneu
Service (97%); Montefina (50%); Citax 2000 (90.93%); RAPL
Belgium (27.78%); Lilachim (50%); Fina Exploration Tunisie;
Continental Insurance Broker (27.5%); De Vogeleer; Motorway
Service (95%); Sabra; Senec (94.5%); Sodelac (50%); City
Parking; Savex (89.1%); Parts Trading Cy (89.1%); Leasinvest
(16.67%); Offshore Europe (33.33%); Offshore Belgium
(33.33%). Other Countries: Norsea Gas (30%) (Germany);
Norsea Gas (30%) (Norway); Sopami (France); Solunor (25%)
(France); Entrepôts Pétroliers Nancy (23.45%) (France);
Entrepôts Pétroliers Régionaux (29.12%) (France); Norsea
Pipe-Line (30%) (UK); Petrogaz Grèce (50%) (Greece); Fina
Hellas (Greece); Société Franco-Portugaise (65.55%)
(Liechtenstein); Fina Portuguesa (Portugal); Finalube (Italy); De
Ba (26%) (Italy); Fina/Total Tunisie (50%) (Tunisia); Fina/BP
Burundi (50%) (Burundi); SEP Burundi (53.92%) (Burundi);
TFZ (77.45%) (Zaïre); Sep Zaïre (36.6%) (Zaïre); Sozatole
(47.33%) (Zaïre); Sofimmo (Zaïre); Fina/Total Gabon (36%)
(Gabon); BP/Fina Rwanda (50%) (Rwanda)

Financial Information:

	31.12.86	31.12.87
	BF'000	BF'000
Sales turnover	447,000,000	424,000,000
Profit before tax	29,380,000	24,618,000
Profit after tax	18,340,000	18,071,000
Retained profit	49,281,000	57,055,000
Dividends	8,707,000	8,447,000
Dividends per share	360BF	380BF
Earnings per share	1,011BF	967BF
Share capital	33,771,000	33,806,000

Principal Shareholders: Petrofina personnel (9%); Société
Générale de Belgique Group (7.2%); Imperial Continetal Gas
Association Group (6.6%); Bruxelles Lambert Group (1.1%);
Electrafina (1.6%); Siderur (2%)

No of Employees: 22,100

PSA CREDIT BELGIUM

99 Rue de l'Étoile, 1180 Bruxelles
Tel: (02) 3781200
Telex: 62491 psabel b

PRINCIPAL ACTIVITIES: Financial services including loan
facilities

Financial Information:

	BF'000
Sales turnover	11,000,000

ROYALE BELGE, GROUPE,

Boulevard du Souverain 25, 1170 Bruxelles
Tel: (02) 661.61.11
Cable: Roybelas
Telex: 23.000 rb ass
Telefax: (02) 661.93.40

BELGIUM

Honorary Directors: Chevalier Georges Martin (Président-Directeur Général Honoraire); Jean-François Delori (Vice-Président Honoraire)

Board of Directors: P Van Der Meersch (Président), J Dromer (Vice Président), A Frere (Vice Président), Y Boël, Baron H Cappuyns, Comte J-P de Launoit, J-M de Munter (Administrateur Délégué), L Deschuyteneer, J Dromer, G Eskénazi, A Frère, P Garny (Administrateur Délégué-Directeur Général), M Goblet, J-H Gougenheim, P Labadie (Administrateur Délégué), J Moulaert, D Pfeiffer, Baron F van den Bergh

Senior Executives: Pierre Garny (Administrateur Délégué-Directeur Général, Jean-Pierre Gérard (Directeur Général Adjoint), Jacques Olivier (Directeur Général Adjoint), Julien Vanderhulst (Directeur Général Adjoint), Guy Blampain (Directeur Finances), Roger Lapaille (Directeur Placements Immobiliers), Paul Gervy (Secrétaire Général), Jacques De Droog (Directeur Administratif et du Personnel), Jean Marchal (Directeur Organisation et Informatique), Michel Martin (Contrôleur Général), Robert Boigelot (Directeur IARD Particuliers), Marc Herinckx (Directeur Auto), Pierre Mercier (Directeur IARD Entreprises), Jacques Crickx (Directeur Réassurance), Christian Defrancq (Directeur Vie), Michel Meyfroidt (Directeur Commercial), Jacques Mahaux (Directeur Affaires Général), Jean Collart (Conseiller Développement IARD)

PRINCIPAL ACTIVITIES: The group is engaged in life, accident and fire insurance, reinsurance and financial services (through its subsidiary Ippa)

Parent Company: Royale Belge

Subsidiary Companies: Members of the Royale Belge Group: Royal Belge Finance; Urbaine UAP; Belgische Lloyd; L'Assurance Liégeoise; Garantie Belge; Ippa; La Belgique; Royale Belge-Utrecht; I.A.R.D.; Utrecht-Royale Belge Vie; Associated Companies: SBAI; Sogefra; Canadian Commerce Insurance Cy.

Principal Bankers: BBL; Générale de Banque; Ippa

Financial Information: Consolidated figures

| | 31.12.86 | 31.12.87 |
	BF'000	BF'000
Gross premium income	33,769,000	35,127,000
Net premium income	31,440,000	33,441,000
Gross investment income	29,835,000	30,476,000
Net investment income	18,297,000	18,846,000
Profit before tax	4,777,000	3,658,000
Profit after tax	3,381,000	3,078,000
Retained profit(1)	11,773,000(2)	7,458,000(2)
Dividends(1)	2,115,000	2,344,000
Dividends per ord share(3)(4)	100BF(5)	106BF
Earnings per share(3)	202BF(5)	165BF
Share capital	7,821,000	7,821,000
Shareholders funds	22,673,000	34,387,000

(1) Group figures

(2) Including BF6,322,000 transferred to untaxed reserves

(3) Figures for Parent Company only

(4) Net dividends after witholding tax of 25%

(5) Shares split by 8

Principal Shareholders: Royale-Vendôme; Royale Belge Finance; UAP; Société Générale de Belgique; Sofina Belgique

No of Employees: 4,993 Group

SA D'IETERENS NV

See D'IETERENS, SA, NV

SMAP
Société Mutuelles Administratif Publiques

Rue de Croisiers 24, B-4000 Liège
Tel: (041) 20 31 11
Telex: 41216 smap b

Management: Léon Lewalle (Directeur Général), Roger Winnen (Directeur Général Adjoint)

PRINCIPAL ACTIVITIES: General insurance services
Principal Bankers: CCB
Financial Information:

	BF'000
Premium income	14,650,000

No of Employees: 1,000

SOCIÉTÉ GÉNÉRALE DE BELGIQUE SA

Rue Royale 30, B-1000 Bruxelles
Tel: 02/517 15 11
Cable: "Generale" Bruxelles
Telex: 24697 GENBEL B

Governor: René Lamy; Vice Governor: Jacques van der Schueren
Directors: Philippe Cols, Vicomte Etienne Davignon, Baron de Fauconval, Baron Guy de Wouters, Marcel Goblet, Robert Novis, Pierre-Michel Oury, Pierre Pluys, Julien Van Hove, Reynald Moretus (Secretary)
Consultative Board: Valère Croes, Jozef De Jaegere, The Baron de Woot de Trixhe, Richard Gandibleux, Jean Gandois, Pierre Garny, Hartwig Geginat, André C Jacques, Herman Liebaers, Baron Paul Meyers, Roelof Nelissen, Roger Selschotter, Henri Simonet, Robert Stouthuysen, Emmanuel Tesch, Ivo Van Thielt, Paul Washer

PRINCIPAL ACTIVITIES: Holding company with interests in financial services, energy engineering, industrial contracting and technical services, electronics, telecommunications and media, non-ferrous metals, cement, diamond industry, chemicals, international trading
Subsidiary Companies: Companies in which the Societe Generale-Generale Maatschappij has a sizeable shareholding:- Financial Services:- Tanks Consolidated Investments "TCI" (UK); Generale International Finance, Luxembourg "GIF" (formerly Métalchim); Centre de Coordination "Générale" Coördinatiecentrum "Generale"; Sodecom; Générale de Banque-Generale Bank; Vlaamse Investeringsvennootschap "VIV"; Synerfi; Groupe AG; Royale Belge; Assubel-Vie; Compagnie Belge d'Assurance-Crédit "COBAC". Energy-Engineering, Industrial Contracting and Technical Services-Electronics, Telecommunications and Media: Compagnies Réunies Electrobel et Tractionel "TRACTEBEL"-Verenigde Maatschappijen Electrobel en Tractionel "TRACTEBEL"; Compagnie Européenne pour le Développement Electrique et Electronique "CEDEE"; Electrafina. Belgatel; Havas. Non-Ferrous Metals: Union Minière. Cement: Cimenteries CBR. Transport: CMB. Diamond Industry: Sibeka-Société d'Entreprise et d'Investissements. Chemicals GECHEM. International Trading: Generale Trading Cy. Financial: Sofina; Compagnie Financière Européene et d'Outre Mer "Finoutremer"; Compagnie Financière de Suez; Belgamanches. Other shareholdings: Industrial or Services: ARBED (Luxembourg); Compagnie d'Entreprises CFE; Compagnie Générale d'Electricité "CGE" (France). Fabrique Nationale Herstal "FN"; Compagnie Immobilière de Belgique; PEBELTEC Cimenteries CBR Cementbedrijven. Transport: CMB. International Trading: Generale Trading Cy (formerly Laura & Vereeniging) (Netherlands). Other Holdings:- Holdings of a Financial nature: Compagnie Financière Européenne et d'Outre-Mer "Finoutremer"; Sofina. Holdings of an Industrial or Service nature: ARBED (Luxembourg); Constructions Ferroviaires et Métalliques "BN" Spoorwegmaterieel en Metaalconstructies "BN"; Compagnie François d'Entreprises CFE-Aannemingsmaatschappij François CFE; Fabrique Nationale

Herstal "FN"; Compagnie Immobilière de Belgique-Immobiliën Vennootschap van België; PABELTEC (formerly Papeteries de Belgique-Papierfabrieken van België)

Principal Bankers: Générale de Banque; Banque Belge Limited; Générale de Banque Belge (France); Banque Générale du Luxembourg

Financial Information: Consolidated figures

	31.12.86	31.12.87
	BF'000	BF'000
Turnover	341,011,900	336,803,700
Profit before tax	6,276,000	6,492,500
Profit after tax	4,220,300	4,261,200
Dividends	3,816,389	3,980,937
Dividends per ord share net	110BF	115BF
Dividends per AFV share net	-	20.98BF
Share capital	30,100,000	35,254,300
Capital and reserves	187,391	203,587,600
Total assets	523,029,300	562,371

Principal Shareholders: Group Royale Belge (3.16%); Group Assurances Générales (3.01%); Group Assubel (1.01%)

No of Employees: 141

SOCIÉTÉ NATIONALE DE CRÉDIT À L'INDUSTRIE SA
Nationale Maatschappij Voor Krediet Aan de Nijverheid

Avenue de l'Astronomie 14, B-1030 Bruxelles
Tel: 02 214 1211
Telex: 25996

Board of Directors: Karel Dierckx (Chairman)
Management: Alfred Rampen (General Manager)

PRINCIPAL ACTIVITIES: Banking services including loans and credit

SOFINA SA

38 Rue de Naples, B-1050 Bruxelles
Tel: (02) 512.66.10
Cable: Sofina
Telex: 21361
Telefax: 513.96.45

Administrative Board: *R Lamy (Président), *Pol Boël (Vice Président), *Yves Boël (Administrateur Délégué), *Marcel Goblet (Administrateur Délégué), Baron de Fauconval, Robert Novis, Raymond Pulinckx, *Pierre-Michel Oury, Rainer E Gut, Jean Guyot, Comte Richard Goblet d'Alviella, Sir Philip de Zulueta, Vicomte Etienne Davignon, Guy de Cordes, Léo Deschuyteneer, *Membres du Comité de Direction

Management: R Lamy (Président), Yves Boël (Administrateur Délégué), Marcel Goblet (Administrateur Délégué), Benoît Van Uytvanck (Secrétaire Général)

PRINCIPAL ACTIVITIES: Holding company with interests in electricity, petroleum, gas, finance, property, manufacturing, industries and technical services

Subsidiary Companies included in the consolidated accounts: (100% owned unless stated): Energy:- Rebelco (83%); Société Centrale pour l'Industrie (53.7%) (France); Sofilec (Luxembourg); Interamerican Finance (Panama); Atlantic Energy Investments (AEI) (Luxembourg); Sobadi (Luxembourg); Trufidee (Luxembourg); Dofico; Brefigep (99.9%); Centrale d'Etudes, Placements et Services-CEPSER (99.9%); Centrale d'Investissements et de Loisirs-CIEL; Société Centrale et Financière (69.2%); Finance et Gestion (99.6%); Société Industrielle et Financière-Sever (99.9%); SIAV (99.9%); Solec Technology; Petigars et Falguières (Anciens Etablissements) (99.9%); L Prud'homme & Fils (64.9%); Sodiag (65.6%); Interealty Fin; Interinvest; Telinter

Principal Bankers: Générale de Banque

Financial Information:

	31.12.86	31.12.87
	BF'000	BF'000
Profit after tax	2,252,262	934,770
Retained profit	525,052	539,316
Dividends	631,743	892,612
Dividends per share	335BF	355BF
Earnings per share	873.0BF	371.8BF
Share capital	3,007,656	3,007,656

Principal Shareholders: Société Générale de Belgique
No of Employees: 44

TRACTEBEL SA
Compagnies Réunies Electrobel et Tractionel SA

Place du Trône 1, 1000 Bruxelles
Tel: (02) 510.71.11
Cable: TRACTEBEL-Bruxelles
Telex: 21.852 TRACT b
Telefax: (02) 511.88.61

Board of Directors: Chairman: René Lamy. Chairman of the Executive Committee and General Management: Baron Guy de Wouters. Vice-Chairmen: Yves Boel, Albert Frère. President: André Claude. Executive directors: Pierre-Alain de Smedt, Paul Haine, Manfred Loeb, Albert Rolin, Alexandre Sariban. Directors: Henri Cambier, Vicomte Etienne Davignon, Léo Deschuyteneer, Gilbert Domken, Alexandre Dumont, Jacques Glorieux, Marcel Goblet, Maurice Herzog, Eric Jolly, Jacques Moulaert, Gérard Munera, Pierre-Michel Oury, Alain Philippson, Didier Pineau-Valencienne, Pierre Scohier, Henri Simonet, René Thiebaut, Chevalier Jean-Paul van Gysel, Julien Van Hove, Vicomte Arnold van Zeeland

General Management: Chairman: Baron Guy de Wouters. Members: André Claude (Energy and Engineering), Pierre Danniau (Labour Relations), Pierre-Alain De Smedt (Energy and Engineering), Paul-Henri Denuit (Communication and Media), Paul Haine (Administrative and Financial Services), Manfred Loeb (Electronics and Telecommunications, American and Pacific subsidiaries), Albert Rolin (Industrial Construction and Technical Services, Spanish subsidiaries), Alexandre Sariban (Construction and Technical Services, French subsidiaries)

Senior Executives: M-Ch Dallemagne (Secrétairiat, Audit, Public Relations), P Danniau (Personnel), R Deltombe (Comptabilité), M-Th Ortmans (Fiscal), Ph Soumoy (Finances), H Valschaerts (Juridique), J Fierens; J-P Neirynck, J Defraigne, A Voué (Energie et Engineering), A Gossen, D Deroux, G Catteau, J Laurent, R Godesar (Electronique-Télécommunications), L Thielemans, P Vilain, R Lachapelle (Affaires Industrielles et Commerciales)

PRINCIPAL ACTIVITIES: Industrial holding company; engineering consultants; research department

Trade Names: TRACTEBEL

Subsidiary Companies: Intercom; Fimeuse; Satom; Power Subsidiary Company; Inec; Financisa; Coditel; L'Immobilière Electrobel; Fabricom; La Construction Soudée; Ebes; Unerg; Mediafin; Reva; Coyne et Bellier; Gecifer-Sodecotra; Nobema; Telfin; T.E.I-Chema; Tractebel Finance; Trasys; Distrigaz; Petrofina; Compagnie Européene de Chauffe; Compagnie Générale de Chauffe; Coditel Brabant; Coditel Liège; Coditel Vlaanderen; American Tractebel Corp; Frances Spécialités; Seinsa; Tractebel Pacific Ltd; Aquinter; Antwerpse Gasmaatschappij; Electrobel Engineering International; Groupe Fabricom; Facair; SEEE; OCA; Belgatel

Principal Bankers: Générale de Banque; Banque Bruxelles Lambert; Banque Paribas Belgique; Royal Bank of Canada (Belgium); Banque Degroof; Kredietbank

BELGIUM

Financial Information:

	31.12.86 BF'000	31.12.87 BF'000
Sales turnover	9,454,600	9,019,886
Profit before tax	5,197,166	5,891,939
Profit after tax	4,698,138	5,760,511
Retained profit	959,232	-
Dividends	3,312,110	3,543,767
Dividends per share	260BF	270BF
Earnings per share	492BF	585BF
Share capital	15,000,000*	15,510,060

*after arrangement of share capital and increase in capital of BF 212.5m reserved for the personnel, conforming to the resolutions of the Annual General Meeting of 28.05.86, 212,500,000 BF réservée au personnel, conformément aux résolutions de l'A.G.E. du 28.05.86

Principal Shareholders: Société Générale de Belgique; Groupe Bruxelles Lambert; SOFINA; ELECTRAFINA
No of Employees: 1,610

UNION MINIÈRE S.A.
54 Avenue Louise, Box 10, B-1050 Bruxelles
Tel: +32/2/517.12.11
Cable: Uniminière
Telex: 21.551 UM-B
Telefax: (02) 517.12.30

Board of Directors: *René Lamy (Président), *Gérard Munera (Administrateur délégué), Pierre De Merre (Administrateur-directeur), *Sir Philip de Zulueta (Administrateur), *le Baron Jean de Fauconval (Administrateur), *Pierre Pluys (Administrateur), Jean-Louis Blain (Administrateur), Edgard Bossuyt (Administrateur), Michel Bouchat (Administrateur), Pierre Gousseland (Administrateur), Charles Lejeune (Administrateur), Noël Masson (Administrateur), Julien Van Hove (Administrateur), *Membre du Comité de direction
Management: Bernard Cardon de Lichtbuer (Directeur), William Kirkpatrick (Directeur), Jean-Marie Mortier (Directeur), Etienne Denis (Directeur Adjoint), André de Troyer (Directeur adjoint), Alain Godefroid (Directeur Adjoint), Georges-Claude Mathonet (Directeur adjoint), Jean Moutury (Directeur Adjoint), Michel Schicks (Directeur Adjoint), Pierre Van Weyenbergh (Directeur Adjoint)

PRINCIPAL ACTIVITIES: Pivotal company of the non-ferrous metals sector of the Société Générale de Belgique group
Parent Company: Société Générale de Belgique
Subsidiary/Associated Companies (100% owned unless otherwise stated): Union Minière Canada Inc (UMICAN) (Canada); Union Mines Inc (USA); União Mineira e Metalúrgica Ltda (UNIMETA) (Brazil); Metallurgie Hoboken-Overpelt SA (61.57%); Société des Mines et Fonderies de Zinc de la Vieille-Montagne SA (50.15%); Compagnie Royale Asturienne des Mines SA (50.5%); Générale Trading Cy (34.42%); Mechim SA (98.33%); Kabelwerk Eupen AG (34.3%); Câbleries de Dour (49.19%); Cie Générale d'Electrolyse du Palais (42.45%); Chevaning Mining Co Ltd (CMC) (50.1%)
Principal Bankers: Générale de Banque; Generale Bank (New York Office); Belgolaise

Financial Information:

	31.12.86 BF'000	31.12.87 BF'000
Profit before tax	616,872	582,503
Profit after tax	616,872	582,503
Retained profit	109,968	81,713
Dividends	570,000	570,000
Dividends per share	570BF	570BF
Share capital	10,000,000	10,000,000
Shareholders funds	15,071,562	15,062,156

Principal Shareholders: Société Générale de Belgique (100%)
No of Employees: 90

Major Companies of DENMARK

AKTIVBANKEN A/S

PO Box 2350, Ladegaardsvej 3, DK-7100 Vejle
Tel: +45 5 8587100
Cable: AKTIVBANK
Telex: 61113 AKTIV DK
Telefax: +45 5 858155; SWIFT-address: AKBA DK 22

Board of Directors: Niels Hede Nielsen (Chairman), Sven
Hansen (Deputy Chairman), H Chr Juhl (Deputy Chairman), S
A Christensen, Signe Grønvall, Karsten Christoffersen, Jørgen
Lauridsen, Henning Laursen, Anton Schrøder, Knud
Søndergaard
Management: E K Larsen (Managing Director)

PRINCIPAL ACTIVITIES: International bank
Subsidiary Companies: Aktivbank International SA
(Luxembourg); National Bank of Long Beach (USA);
Finansieringsinstituttet for Sejlsport A/S; Aktiv Børs,
Børsmåeglerselskab A/S; Aktiv Leasing A/S; Aktiv
Ejendomsadministration A/S
Financial Information: Group figures

	31.12.86	31.12.87
	DKr'000	DKr'000
Profit (loss) before tax	(25,242)	5,828
Profit (loss) after tax	1,608	(15,337)
Dividends	24,000	18,000
Dividends per share	16%	12%
Share capital	150,000	150,000
Shareholders funds	837,402	808,065
Deposits	6,601,169	5,671,927
Total assets	13,225,015	13,172,285

No of Employees: 928

ANDELSBANKEN DANEBANK

1-3 Staunings Plads, 1643 København V
Tel: (451) 145114; (451) 144520 (Foreign Exchange)
Cable: Danebank; Swift: ANDA DK KK
Telex: 27086; 15333 (Foreign Exchange)
Telefax: (451) 912110

Board of Directors: Olaf Smitt (Chairman), Holger Klindt
Andersen (Deputy Chairman), Laurids C Boisen, Ole Peter
Christensen (Elected by the Employees), Kirsten Jensen
(Elected by the Employees), Sv E Jørgensen (Elected by the
Employees), Fonna Kjelin (Elected by the Employees), Erik
Larsen, Jorgen Hoeg Pedersen, Ib Nielsen, A Kirstein
Pedersen, Ib Thyregod
Management Board: A C Jacobson (Presiding), Mrs Bodil
Nyboe Andersen (Managing Director), H Lundager (Managing
Director), Asger Kastbjerg (Managing Director)
Executive Staff: International Division: F Junker Mortensen
(Executive VP, Head of International Division), S Herløv
Hansen (Senior VP, Deputy Head of International Division),
Hardy Larsen (Executive VP, Head of Foreign Exchange), Ivar
Espersen (First VP, Head of Area Management), Uffe Jensen
Hou (VP Project Finance and Syndications), Hans Christian
Drost (VP Corporate Services and Foreign Loans), Kai
Morville Schrøder (First VP, International Payments), Jørgen
Johansen (VP Administration), Christopher Gabel-Jørgensen
(VP Domestic Money Market), Ivan Poulsen (VP, Chief Dealer-
Money Market), Peter Meyer (VP, Chief Dealer-Spot). Other
Executive Staff: Ole Jacobsen (Executive VP Head of
Investment Division), E Due-Hansen (Executive VP Personnel
and Organisation), E Due-Hansen (Executive VP Personnel
and Organisation), Bent Mouritsen (Executive VP Loans and
Credits), Herluf Poulsen (Executive VP EDP), M Hvelplund
(Executive VP Law), Jørgen Houengaard (Senior VP
Accounting), Judith Larsen (Senior VP Loans and Credits),
Claus K Møller (Senior VP Deputy Head of Investment
Division), Preben Weissenborn (Senior VP Marketing), Dorthe
T Jørgensen (Assistant VP)
PRINCIPAL ACTIVITIES: Commercial Bank

Subsidiary Companies: Andelsbanken (I.O.M.) Ltd (Isle of Man).
Representative offices in London, Hong Kong, Hamburg and
New York
Financial Information:

	31.12.86	31.12.87
	DKr'000	DKr'000
Profit before tax	32,763	432,181
Profit after tax	19,307	259,198
Retained profit	326,386	381,592
Dividends	99,429	99,429
Dividends per share	15%	15%
Share capital	662,859	662,859
Shareholders funds	2,793,778	2,953,547
Deposits	24,574,598	24,689,650
Total assets	52,676,183	47,927,053

No of Employees: 4,250

ARBEJDERNES LANDSBANK A/S

Panoptikonbygningen 5, Vesterbrogade, 1502 København V
Tel: (01) 148877
Cable: Bankland
Telex: 15633; 27235 Albank kh
Telefax: (01) 32 18 73

Board of Directors: Georg Poulsen (Chairman), Agner
Christensen, Anton Johansen, Hardy Hansen, Arnold Larsen,
Bent Larsen, Peter Falk, Susan Nielsen, Poul Carsten
Hansen, Ellen Andersen
Board of Managers: Sv Nibelius, Jørgen Christensen, E
Midtgaard, P E Leth, E Castella

PRINCIPAL ACTIVITIES: International bank
Subsidiary Companies: A/S Fagbyg
Principal Bankers: Correspondents in all major international
financial centres
Financial Information:

	31.12.87	31.12.88
	DKr'000	DKr'000
(Loss) before tax	(60,575)	(27,626)
(Loss) after tax	(60,525)	(34,466)
Retained (loss)	(90,525)	(64,466)
Dividends	30,000	30,000
Dividends per share	10%	10%
Share capital	300,000	300,000
Shareholders funds	1,169,252	1,114,226
Deposits	11,416,205	9,739,027
Total assets	13,148,780	13,568,984

No of Employees: 1,315

ASSURANCE COMPAGNIET BALTICA A/S

See BALTICA, ASSURANCE COMPAGNIET, A/S

ASSURANCE COMPAGNIET BALTICA LIVSFORSIKRINGSAKTS

See BALTICA, ASSURANCE COMPAGNIET,
LIVSFORSIKRINGSAKTS

BALTICA, ASSURANCE COMPAGNIET, A/S

Bredgade 40, DK-1299 København K
Tel: (01) 122 436
Telex: 16322

Board of Directors: H Brüniche-Olsen (Chairman), Peter
Christoffersen (Vice Chairman)
Management: Steen Hemmingsen, Jørgen Søltoft, Michael Pram
Rasmussen, Holger Foged

PRINCIPAL ACTIVITIES: Insurance services
Trade Names: Baltica Fosikring (Baltica Insurance)
Parent Company: Baltica Holding A/S
Subsidiary Companies: Assurance Compagniet Baltica
Livsforsikring; Baltica Arbejisskade Forsikringsaktieselskab
Principal Bankers: Den Danske Bank; Handelsbanken

DENMARK

Financial Information:

	31.12.86	31.12.87
	DKr'000	DKr'000
Premium income	3,313,000	3,537,000
Investment income	480,000	435,000
Profit before tax	165,000	177,000
Profit after tax	161,000	178,000
Retained profit	120,000	137,000
Dividend	41,000	41,000
Dividend per share	12%	12%
Share capital	340,000	340,000
Total equity capital	3,568,000	4,034,000

Principal Shareholders: Baltica Holding A/S (82%)
No of Employees: 3,400 (Group)

BALTICA, ASSURANCE COMPAGNIET, LIVSFORSIKRINGSAKTS

Bredgade 40, 1299 København K
Tel: (01) 122 436
Telex: 16322

Board of Directors: H Brüniche-Olsen (Chairman), Peder Christoffersen (Vice Chairman)
Management: Steen Hemmingsen, Jørgen Søltoft, Michael Pram Rasmussen, Holger Foged
Senior Executives: Michael Pram Rasmussen (Area General Manager)

PRINCIPAL ACTIVITIES: Direct insurance services
Trade Names: Baltica Liv (Life)
Parent Company: Assurance Compagniet Baltica A/S
Principal Bankers: Dan Danske Bank; Handelsbanken
Financial Information:

	31.12.86	31.12.87
	DKr'000	DKr'000
Premium income	1,042,000	1,134,000
Investment income	2,127,000	1,957,000
Profit before tax	540,000	633,000
Profit after tax	100,000	90,000
Retained profit	98,000	88,000
Dividend	2,000	2,000
Dividend per share	10%	10%
Share capital	20,000	20,000
Total equity capital	948,000	869,000

Principal Shareholders: Assurance Compagniet Baltica, A/S (100%)

BERENDSEN, SOPHUS, A/S

1 Klausdalsbrovej, DK-2860 Søborg
Tel: (01)69 85 00
Cable: Berendsen
Telex: 22 285 sobeas dk
Fax: (01) 69 73 00

Chairman: Ebbe J B Christensen
Management Board: Hans Werdelin (Managing), Joergen Ajslev
Deputy Directors: Ole Balle, Nick Elsass, Mrs Kirsten Andersen, Mrs Karen Højgaard

PRINCIPAL ACTIVITIES: Parent company of a group whose subsidiaries cover light-engineering, commerce, industry and service
Subsidiary Companies: Engineering and Trading Activities:-
Sophus Berendsen A/S-Engineering Department*; Viggo Petersen, Maskin-og Ingeniørfirma A/S (Sophus Berendsen A/S)*; Islef+Hagen A/S (Sophus Berendsen A/S)*; GRI Europe A/S; E. Friis-Mikkelsen A/S; E.O. Farstrup A/S; Transmotor A/S PRC Components A/S; Sophus Berendsen Marine A/S; Landis & Gyr A/S; Atlas-Abemi A/S (Norway); Ingeniørforretningen Atlas A/S (Norway); Atlas-Spong Hydraulic A/S (Norway); AB Abemi (Sweden); AB Ingeniörsfirman Titan (Sweden); Titan Cement AB (Sweden); Abemi OY AB (Finland); Hydrowa BV (Netherlands); Sweetland Corporation (USA). Paul-Munroe Hydraulics Inc (USA). Data and Informatics:-Berendsen Data A/S (Sophus

Berendsen A/S)*; Berendsen Computer Products A/S (Sophus Berendsen A/S)*; Computer Forum A/S (Sophus Berendsen A/S)*; Berendsen Cad A/S; SC Metric A/S*; Uniware Danmark A/S (SC Metric A/S); Atlas Informatik A/S (Norway); Atlas Audio-Vision A/S (Norway); Berendsen Computer Products A/S (Norway); Berendsen Computer Products B.V. (Netherlands); Berendsen Computer Products BVBA (Belgium). Agricultural Industry:- Jesma-Matador A/S; Matador Inc (USA). Textile Rental:- Dehn-Fix Linned Service A/S. Environmental Services and Property Care:- Rentokil Group PLC (UK) subsidiaries in 36 countries. Agricultural Industry: Jesma-Matador A/S; Matador Inc (USA)
*Departments and Registered Departments
Principal Bankers: Den Danske Bank af 1871 A/S
Financial Information:

	31.12.86	31.12.87
	DKr'000	DKr'000
Sales turnover	3,039,542	3,264,619
Profit before tax	383,094	417,815
Profit after tax	238,787	251,334
Retained profit	104,971	91,728
Dividends	40,900	41,036
Dividends per share	17.5%	17.5%
Earnings per share	81DKr	58DKr
Share capital	233,717	234,491
Shareholders funds	834,766	864,592

Principal Shareholders: Mr Adam Elsass; Miss Helene Elsass
No of Employees: 10,500

COPENHAGEN HANDELSBANK A/S (AKTIESELSKABET KJØBENHAVNS HANDELSBANK)

Holmens Kanal 2, DK-1091 Copenhagen K
Tel: +45 1 12 86 00
Cable: cocobank
Telex: 12186 coco dk
Telefax: +45 2 63 07 19

Board of Directors: Bernhard Gomard (Chairman, Professor, University of Copenhagen), Niels Thorsen (Deputy Chairman, Civil Engineer and Managing Director Monberg & Thorsen A/S), Christian Arnstedt (Civil Engineer, Member of Mangement of Cement Invest A/S), Niels B Brüel (Managing Director A/S Brüel & Kjær), P Tønnesen (Managing Director A/S Korn-og Foderstofkompagniet), Jens W Werner (Managing Director GN Laur Knudsen A/S), Mads Øvlisen (Managing Director Novo Industrie A/S), John Agerholm (Elected by the Bank's staff, Deputy Manager, Copenhagen Handelsbank A/S), Tove Andersen (Elected by the Bank's staff, Deputy Manager, Copenhagen Handelsbank A/S), Gretha Højen (Elected by the Bank's staff, Assistant Manager, Copenhagen Handelsbank A/S), Poul Erik Milhøj (Appointed by the Ministry of Industry, Professor, Copenhagen School of Economics and Business Administration)
Board of Managing Directors: Bendt Hansen (Chief Executive), Hans Ejvind Hansen (Deputy Chief Executive), Egon Jensen, Søren Møller Nielsen, Bo Jagd
General Managers: Cato Baldvinsson, H Baroudy, Preben Bisgaard, Birte Roll Brandt, Ditlev Engelholm, Peter Preben Hansen, Jens Frederik Jensen, Boris Kreilgaard, Svend Kæstel, Poul Lorenzen, H Martensen, Poul Mathiesen, Per Asbjørn Nielsen, Jørgen Pernø, Jørgen Wichmann, Kjeld Øberg

PRINCIPAL ACTIVITIES: Banking
International Subsidiary: (100% owned unless stated): Copenhagen Handelsbank International SA (Luxembourg). International Affiliates: CHB Invest Ltd (Gibraltar); Copenhagen Handelsbank North America Inc (USA). Foreign Branches: London Branch, England; Los Angeles Branch, USA; New York Branch, USA; Singapore Branch, Singapore; Grand Cayman Branch. Representative Offices: Hong Kong; Tokyo, Japan; Fuengirola/Malaga, Spain; Madrid, Spain;

Stockholm, Sweden; Nice, France. Consultants Offices: Tokyo, Japan; Holger Morville Holding

Financial Information:

	31.12.86	31.12.87
	DKr'000	DKr'000
Turnover	9,321,594	8,970,601
Profit (loss) before tax	(835,646)	254,313
Profit (loss) after tax	(821,255)	267,329
Retained profit (loss)	(1,042,230)	32,276
Dividends	216,975	235,050
Dividends per share	15%	15%
Earnings (loss) per share	(56.78DKr)	15.84DKr
Share capital	1,446,500	1,687,500
Shareholders funds	4,547,806	4,784,135
Deposits	73,593,589	83,808,223
Total assets	99,465,208	110,338,570

No of Employees: 6,352 (Heads Full time staff)

DANMARKS NATIONALBANK

5 Havnegade, 1093 København K
Tel: 01 141411
Cable: Nationalbanken
Telex: 27051 Natbk dk
Telefax: 01 145902

Board of Directors: P Nørregaard Rasmussen (Chairman), M Folmer Andersen (Deputy Chairman), Svend Auken, Per V Brüel, Helge Dohrmann, Bjørn Elmquist, Jørgen Estrup, Bent le Fèvre, Andreas Graversen, Flemming Skov Jensen, Knud Koch Jensen, Kjeld Larsen, Jes Lunde, Grethe Fenger Møller, Paul la Cour Møller, B Frank Nielsen, Ivar Nørgaard, Karsten Olsen, Erik B Rasmussen, Henning Rasmussen, Kirsten Stallknecht, Jens Thomsen, Finn Thorgrimson, Erland Thrane
Governors: Erik Hoffmeyer, Ole Thomasen, Richard Mikkelsen
Administration: Assistant Governors: Jørn Kjær, Henning Dalgaard. Directors: Arne Brøndum, Erik Niepoort (Leave of absence until August 31 1990), Ib Glud Konradsen, Kirsten Mordhorst, Hans Kloch. Departments of the Head Office: Accounts Department: Erik Winding (Chief Accountant), Benny Jensen (Assistant Head of Department). Audit Department: Keld Grøndahl Jensen (Chief Auditor), Geert Egtved Petersen (Assistant Head of Department). Cash Department: Erik Herly (Head of Department), Mogens Brink (Assistant Head of Department), Arne Klemen Jensen (Assistant Head of Department). Establishment Department: Hans Kloch (Director), Aage Røndbjerg (Assistant Head of Department), Leif Petersen (Assistant Head of Department), Bente Puck (Assistant Head of Department). Exchange Regulations Department: Erik Niepoort (Director. Leave of absence until August 31 1989), Carsten Freiberg Jensen (Head of Department), Flemming Farup-Madsen (Head of Department), Jens Hjorth Hald (Adviser), Erik Sønderbek (Assistant Head of Department), Henrik Peter Mogensen (Assistant Head of Department), Finn Jürgens (Assistant Head of Department), Hanne S Sørensen (Assistant Head of Department). Foreign Department: Henning Dalgaard (Assistant Governor), Hans Denkov (Head of Department), Ejner Petersen (Head of Department), Kaj Kjær (Assistant Head of Department), Preben Poulsen (Assistant Head of Department), Erik Pedersen (Assistant Head of Department), Viggo Sørensen (Assistant Head of Department), Helge Fosgaard (Assistant Head of Department). International Department: Kai Aaen Hansen (Head of Department). Monetary Policy Department: Jørn Kjær (Assistant Governor), Kjeld Røhl (Head of Department), Ove Ingvard Nielsen (Adviser). Secretariat: Kirsten Mordhorst (Director), Bent Danielsen (Assistant Head of Department), Else Marie Nielsen Christensen (Assistant Head of Department), John Larsen (Assistant Head of Department). Securities Department: Ib Glud Konradsen (Director), Benth Jensen (Head of Department), Ole Christian Hansen (Head of Department). Staff and Planning Department: Arne Brøndum (Director), Tage Heering (Assistant Head of Department), Erik Lundgreen Hansen

(Assistant Head of Department), Jørgen Ovi (Assistant Director, has been appointed Board Member for the Nordic Countries to the International Monetary Fund until June 30 1989)

PRINCIPAL ACTIVITIES: Central bank of Denmark
Financial Information:

	31.12.86	31.12.87
	DKr'000	DKr'000
Profit before allocations(1)	1,074,766	3,119,881
Profit after allocations(2)	1,074,766	3,119,881
General capital fund	50,000	50,000
Total net capital	24,552,000	24,552,000
Deposits etc.	61,702,095	62,129,200
Total assets	124,437,451	133,826,293

(1) before allocations to reserves
(2) after allocations to reserves this profit is payable to the Government pursuant to the Danmarks Nationalbank Act

No of Employees: 593

DANMARKS SKIBSKREDITFOND
Ship Credit Fund of Denmark

Klampenborgvej 205, 2800 Lyngby
Tel: (02) 93 20 20

Management: Henning Axel Nielsen, Hakon Fjeldberg
Senior Executives: Stigkaare Lyngsie

PRINCIPAL ACTIVITIES: Financial operations including credit services
Principal Bankers: Hovedbankerne
Financial Information:

	DKr'000
Turnover	1,975,000

DANSK ALFA-LAVAL HOLDING A/S

Maskinvej 5, 2860 Søborg
Tel: (01) 67 0311
Telex: 22201 alsep dk
Telefax: (01) 674 735

Board of Directors: Paul Tomsen, Gerhard Magnusson, Lars Trane, Poul Lygum, Hans Peter Jørgensen, Ingrid Møller-Nielsen, Jens Rickelt
Management: G Magnusson (President)

PRINCIPAL ACTIVITIES: Financial and insurance activity
Parent Company: Alfa-Laval AB, Sweden
Financial Information:

	31.12.87
	DKr'000
Sales turnover	1,182,800
Profit before tax	104,000
Dividend	27,000
Dividend per share	11.1DKr
Share capital	249,000

Principal Shareholders: Alfa-Laval AB, Sweden
No of Employees: 1,215

DANSK LANDBRUGS REALKREDITFOND

Nyropsgade 21, 1503 København V
Tel: (01) 151100
Telefax: (01) 939500

Board of Directors: Bent Hansen (Chairman), Peder Elkjær, A C Jacobsen, A Holm Jensen, Lars Johansen, H O A Kjeldsen, Abel Nielsen, Arne Nielsen, Henning Kruse Petersen, Ib Rasmussen, Ole Thomasen
Management: J Fløystrup Jensen (Chief Executive), P Wegge-Olsen

DENMARK

Senior Executives: B Dyreborg-Carlsen

PRINCIPAL ACTIVITIES: Credit and financial services

No of Employees: 70

DANSKE BANK AF 1871, DEN, A/S

12 Holmens Kanal, DK-1092 Copenhagen K
Tel: 45 1 15 65 00
Cable: DENDANSKE
Telex: 27 000
Telefax: 45 1 18 58 73

Board of Directors: Poul J Svanholm (Chairman), Flemming
Count of Rosenborg (Deputy Chairman), Steen Engel, Bent
Mogens Hansen, Niels Eilschou Holm, Palle Marcus, Henning
Mikkelsen, Frederik Chr Obel, Jonna Schou, Kirsten
Sørensen, Aage Stougaard Pedersen, Thorkil Zederkof,
Anders Ølgaard

General Management: Tage Andersen (Chief Executive), Knud
Sørensen, John Rammer, Peter Straarup, Henrik Thufason

Executive Vice Presidents: Jørgen Melchior, Jakob Brogaard,
Flemming Duus, Ib Christiansen, Hugo Andersen, George
Yelken, Mogens Laursen. International Department: Henrik
Lindholm (Senior-Vice President, International Administration),
Jesper Christensen (First Vice President, Treasury and
Foreign Exchange). Credits Department: Jakob Brogaard
(Executive Vice President). Merchant Bank, Energy
Department, International Capital Markets: George Yelken
(Executive Vice President)

PRINCIPAL ACTIVITIES: Commercial banking

Subsidiary Bank: Den Danske Bank International SA,
Luxembourg. Foreign branches: London, New York,
Singapore, Hamburg, Frankfurt and Cayman Islands.
Representative Office: Tokyo

Financial Information: Consolidated figures

	31.12.86 DKr'000	31.12.87 DKr'000
Profit before tax	137,700	1,253,300
Net profit	51,800	760,800
Retained profit (loss)	(191,200)	485,400
Dividends	243,000	275,400
Dividends per share	15Dkr	16Dkr
Earnings per share	3.2Dkr	41.7Dkr
Share capital	1,620,000	1,822,500
Shareholders equity	8,039,700	9,009,600
Deposits	110,617,600	122,014,900
Total assets	132,100,200	150,557,400

No of Employees: 6,309

DEN DANSKE BANK AF 1871 A/S

See DANSKE BANK AF 1871, DEN, A/S

EAST ASIATIC COMPANY LTD A/S (THE) (A/S DET ØSTASIATISKE KOMPAGNI)

2 Holbergsgade, DK-1099 København K
Tel: (01) 11 83 00
Cable: Pyramide
Telex: 12.100 EACKH DK
Telefax: (01) 12 37 00

Board of Directors: T W Schmith (Chairman), Otto F Andreasen
(Vice-Chairman), J A Hansen, P Hjelt, H O A Kjeldsen, A L
Korsberg, E Kristiansen, H Peetz-Larsen, H H Sparsø, J
Thorsen, S Thostrup V Bentsen, E Boegh Christensen, J
Rytter

Executive Board: H H Sparsø, (Presiding Managing Director), J
A Hansen (Managing Director), F Hasle (Managing Director)

Senior Executives: General Managers:- K Stock Andresen, Ib
Henry Hansen, Hans P Holst, Bent K Kierkegaard, Sv. Aa.
Larsen, A Holst Larsen, Jan H Laursen, C Dencker Nielsen,
Flemming Marcher, Erik Petersen. Public Relations Manager:-
Claus Arboe-Rasmussen. Assistant General Managers:- G A
Engelmann, J E Lertoft, N-E Lockenwitz, Chr G Lund, O J
Mohrsen, Knud Fejerskov, Hans Christen Jensen, Peer C

Bardenfleth-Hansen, Svend Aage Mønsted, Hans Christian
Sørensen, K R Thomsen, J Selsmark, N-E Olsen

PRINCIPAL ACTIVITIES: The Group is divided into seven
divisons, each of which is responsible for a defined product
area. Graphics Division: The division distributes and markets
equipment and products for the graphics industry
internationally. Informatics Division: The division is engaged in
marketing, distribution and maintenance of equipment and
systems including computers and word processing systems -
as well as software, network and programming tools. Trading
Division: The division handles timber, timber products, wool,
commodities and machinery and is also actively engaged in
projects business. Consumer Products Division: The division
handles the consumer products business outside the
Plumrose sphere of interest. Activities include marketing and
distribution of products manufactured by companies within
the EAC Group or handled on an agency basis. In addition to
consumer products, an extensive business in chemicals and
paints falls within the province of the division. Plumrose
Division: Food products manufactured at own factories and
under licence agreements are marketed under the "Plumrose"
brand in more than 100 countries. The division is engaged in
the production, marketing and distribution of meat products,
cheese, yoghurt, seafood, marmalade, herbs, tomato
products and pasta. Transport Division: The division is
engaged in international transportation by sea, land and air.
The operations include transport in containers - with all the
activities related to a container transport network - product-
and bulk carriers as well as chartering of and trading in ships.
Energy and Property Division: The division is responsible for
the Group's involvement in the hydrocarbon sector and also
includes a number of subsidiaries within the chemical and
synthetics industry

Subsidiary Companies: (100% owned unless stated): Sub-
subsidiaries and associated companies owned by subsidiary
companies are not shown): North-and Central America: EAC
USA Inc (USA); Dan Merchant Ltd (Bahama); EAC Lines
Trans Pacific Service Ltd (51.5%) (Bahama); Forest Merchant
Ltd (49%) (Bahama); Nor Merchant Ltd (49%) (Bahama);
Forest Product Carriers (International) Ltd (50%) (Cayman
Islands); Rubicon Tankers Ltd (50.1%) (Bahama); Meonia
Insurance Brokers Ltd (Cayman Islands); FAC Investments
(Cayman Islands); Selandia Insurance Company Ltd (Cayman
Islands). South America: Este Asiatico Comércio e Indústria
Ltda (Brazil); Industria Envasadora Nacional CA (70%)
(Venezuela); Inmobiliaria Titicaca CA (70%) (Venezuela); Cia
Metalurgica Bera de Colombia (Columbia); Fundicion de
Metales Bera del Peru SA (34.7%) (Peru); Bera de Venezuela
SA (Venezuela). Europe: Allfreight I/S (63%) (Denmark); Bera
Corporation A/S (44.6%) (Denmark); Bloch & Behrens ApS
(Denmark); Cavalet ApS (Denmark); CompuDan ApS
(Denmark); Cornelder's Scheepvaart Maatschappij BV (20%)
(Netherlands); DS Industries ApS (Denmark); Danish Freeze
Drying Ltd (Denmark); EAC Data Ltd (DK) (Denmark); Det
Østasiatiske Kompagnis Ejendomsselskab ApS (Denmark);
EAC Assurance Agentur ApS (Denmark); EAC Energy A/S
(Denmark); The East Asiatic Company (Korea) Ltd ApS
(Denmark); The East Asiatic Company UK (Holdings) Ltd (UK);
EAC-P.N.S.L. Service Ltd (33.3%) (UK);
Beteiligungsgesellschaft Die Ostasiatische Kompagnie mbH
(Germany); Die Ostasiatische Kompagnie GmbH; Global
Management Equipment Ltd (30%) (UK); Erik Levison ApS
(Denmark); Fibertex ApS (Denmark); Indufor (1975) NV
(Belgium); ScanDutch BV (27.5%) (Netherlands); Nakskov
Shipyard Ltd (Denmark); Plumrose A/S (Denmark); Selandia
Finance and Investment BV (Netherlands); INDUBOIS SA
(France); The East Asiatic Company (Finance) SA
(Luxembourg); SIAMAR Srl (50%) (Italy). Africa: The Old East
African Trading Company Ltd (Kenya); The East Asiatic
Company (SA) (Pty) Ltd (South Africa); EAC Graphics (SA)
(Pty) Ltd (South Africa); Business Machines Ltd (Malawi); R T
Briscoe (Nigeria) Ltd (40%) (Nigeria); The Old East African
Trading Co (T) Ltd (Tanzania). The East Asiatic Co (Pvt) Ltd

(Zimbabwe). Asia: The East Asiatic Company (Hong Kong) Ltd (Hong Kong); Carlsberg Brewery Hong Kong Ltd (50%) (Hong Kong); ScanDutch Ltd (26.1%) (Hong Kong); The East Asiatic Company (Malaysia) Berhad (65%) (Malaysia); Nedlloyd-EAC Agencies (Malaysia) Sendirian Berhad (28%) (Malaysia); The East Asiatic Company (Singapore) Pte Ltd (Singapore); EAC Ship Management Services (S) Pte Ltd (Singapore); Bakke-Sime Darby Shipping Pte Ltd (50%) (Singapore); Nedlloyd-EAC Agencies Pte Ltd (40%) (Thailand); The East Asiatic Company (Thailand) Ltd (49%) (Thailand); Dumex Ltd (78.3%) (Thailand); Thai-Dan Corporation Ltd (49%) (Thailand); The East Asiatic Company (Brokers) Ltd (49%) (Thailand); The East Asiatic Company (Transport) Ltd (49%) (Thailand); EAC Properties (Thailand) (33%) (Thailand); EAC Export Corp (Thailand) Ltd (33%) (Thailand); P.T. Danapaints Indonesia (30%) (Indonesia); P.T. Danmotors Vespa Indonesia (49%) (Indonesia); P.T.ICCO Murni Indonesia (49%) (Indonesia); EAC Services Inc (25%) (Philippines); Luzon Mahogany Corporation (40%) (Philippines); Philippine Moulded Pulp Products Inc (79%) (Philippines); Printing Machine Trading Co Ltd (32.9%) (Japan); The East Asiatic Company (Japan) Ltd (Japan); The East Asiatic Company (B) Ltd (Negara Brunei Darussalem). Australia and New Zealand: Plumrose (Australia) Ltd) (Australia); Bloch & Behrens Pty Ltd (Australia); Kauri Timber Company Ltd (Australia); Bloch & Behrens (NZ) Ltd (New Zealand); The East Asiatic Company (New Zealand) Ltd (New Zealand); Scanservice Leasing Company (Vanuatu) (21.1%) (Vanuatu); Beaufort Shipping Agency (WA) Pty Ltd (Australia)

Principal Bankers: Den Danske Bank
Financial Information:

	31.12.86	31.12.87
	DKr'000	DKr'000
Sales turnover	14,174,263	14,471,724
Profit before tax	385,172	473,039
Profit after tax*	311,218	301,719
Retained profit	189,565	214,071
Dividends	111,516	124,182
Dividends per old share	-	8%
Dividends per new share	-	4%
Earnings per share	20.3DKr	18.6DKr
Share capital	1,536,540	1,568,029
Shareholders funds	2,225,851	2,199,227

*and minority interest

Principal Shareholders: The East Asiatic Company's Holding Company Ltd (approx 35%)
No of Employees: 14,425 (Group)

F L SMIDTH & CO A/S
See SMIDTH, F L, & CQ A/S

HAFNIA INSURANCE GROUP
Holmens Kanal 22, DK-1097 Københaven K
Tel: (01) 131415
Telex: 15546 hafnia
Telefax: (01) 126293

Board of Directors: E J B Christensen, Allan Philip, A Vinding Kruse, Haldor Topsøe, Per Villum Hansen, Viggo J Rasmussen, Henrik Møller Sørensen, Leila Falk Pedersen, Ole Borre
Management: H H Koefoed, Ralf Egede Andersen, Jette M Arnskjold, H A Kovsted

PRINCIPAL ACTIVITIES: Provision of a wide range of insurance services
Parent Company: Hafnia Invest Ltd (Hafnia Insurance Company Ltd is parent company of the Hafnia Insurance Group)
Group Companies: (100% owned by Hafnia Insurance Company Ltd unless stated): Hafnia-Haand i Haand General; Arbejdsskadeforsickringsaktieselskabet Hafnia; Försäkringsaktiebolaget Holmia; Hafnia-Haand i Haand Life; Livsforsikringsaktieselskabet Hafnia Pension Invest; Ejendomsaktieselskabet Holmens Kanal; Dania Assurance

Agentur A/S; Besigtelses Kontoret af 1914 A/S; Hafnia Reassurance Ltd; Economic Holdings Ltd (UK); Economic Insurance Company Ltd (UK); Hafnia Holdings Inc (USA); Haffina Reinsurance Private Ltd (Singapore)
Financial Information: Group figures

	31.12.86	31.12.87
	DKr'000	DKr'000
Premium income	4,231,000	4,469,514
Investment income	3,152,000	3,242,002
Profit before tax	241,064	292,032
Group profit	235,447	285,376
Dividend	29,200	35,099
Share capital	146,242	146,242
Shareholders equity	3,747,921	3,749,575

Principal Shareholders: Hafnia Invest Ltd (79.9%)
No of Employees: 1,534

HAFNIA INVEST LTD
Holbergsgade 3, Postboks 1153, 1010 København K
Tel: (01) 324511
Telex: 16193 hafinv dk
Telefax: (01) 930316

Board of Directors: E J B Christensen (Chairman), Haldor Topsøe, A Vinding Kruse, Jørgen E Tandrup, Leila Falk Pedersen, Ebbe Nielsen, Ebbe Madsen, Ulrik Lett
Management: Per Villum Hansen, Henrik H Klæbel
Senior Executives: Thomas Dahl, Jesper Hansen, John Ipsen, Peter Wendt, Michael Øxenbjerg

PRINCIPAL ACTIVITIES: Holding company for a financial services group, the principal subsidiary of which is the Hafnia Insurance Group, the second largest insurance group in Denmark
Subsidiary Companies: Hafnia Bank Ltd; Hafnia Service Ltd and The Hafnia Insurance Group which includes: Hafnia Insurance Company Ltd; Hafnia-Haand i Haand General Insurance Company Ltd; Hafnia-Haand i Haand Life Insurance Company Ltd
Financial Information: Group figures

No of Employees: 2,300

INCENTIVE A/S
Kongens Nytorv 28, DK-1050 Kjøbenhavn
Tel: (01) 111142
Telex: 27142 incent dk

Board of Directors: Nils Wilhjelm, Poul Jørgen Jensen

PRINCIPAL ACTIVITIES: Finance and investment services
Financial Information:

	DKr'000
Sales turnover	1,500,000

No of Employees: 2,221

JYSKE BANK (A/S)
Vestergade 8-16, DK-8600 Silkeborg
Tel: +4586 82 11 22 (Head Office Silkeborg); +45 31 21 22 22 (International Division Copenhagen)
Cable: Jyskebank; SWIFT: JYBADKKK
Telex: 63231 JBNK (Head Office Silkeborg); 16550 JYBKCD DK; 27340 JYBKC DK (International Division Copenhagen)
Telefax: +4586 82 41 66 (Head Office Silkeborg); +45 31 24 41 22 (International Division Copenhagen)

Board of Directors: Viggo Sørensen (Chairman)
Management Board: Poul Norup (Chairman of the Board), Kaj Steenkjaer, Erik D Bruun, Benny Jensen
Senior Executives: Hans Moeller-Christensen (Assistant General Manager, Head of International Division), Torben Budde (Deputy Head of International Division), Hans-Erik Skov Nielsen (Head of International Commercial Department)

PRINCIPAL ACTIVITIES: Bank

Subsidiary Companies: Silkeborg Datacentral A/S; Jysk Garanti A/S; Jysk Diskontering A/S; A/S Jysk Leasing; A/S Gammel Skovridergaard; Jysk Børsmæglerselskab A/S; Jyske Bank AG (Schweiz) AG (Zürich, Switzerland); Jyske Bank (Deutschland) AG (Hamburg, Germany); Jyske Bank (Gibraltar) Ltd (Gibraltar); Jyske Bank (Espana) SA (Fuengirola, Spain)

Financial Information:

	31.12.87	31.12.88
	DKr'000	DKr'000
Profit before tax	155,614	486,745
Profit after tax	151,168	307,823
Dividends	10%	10%
Share capital	385,000	462,000
Shareholders funds	3,076,000	3,387,000
Deposits	24,326,000	27,525,000
Total assets	51,702,000	59,114,000

No of Employees: 2,541

KOMMUNERNES PENSIONSFORSIKRING AS

Nyropsgada 37, DK-1602 København V

Tel: (01) 122788

PRINCIPAL ACTIVITIES: Provision of pension and life insurance

Financial Information:

	DKr'000
Premium income	960,000

No of Employees: 114

PRIVATBANKEN A/S

Torvegade 2, PO Box 1000, DK-2300 Copenhagen S

Tel: 45 1 11 11 11

Cable: Privatbank

Telex: 27196 privco dk

Telefax: +45 1 95 57 69

Supervisory Board: Hugo Schrøder (Chairman), Allan Philip (Deputy Chairman), Poul Andreassen, Morten Balling, Søren Dalmark, Mogens Kierulff, Karsten Laursen, Knud Overø, Thorkild Sejer Rasmussen, Stig Rohde

Management Board: Steen Rasborg (Chairman), Lars Johansen, Erwin Lunderskov, Bent Pedersen

Senior Executives: René Andersen (Executive Vice President Administration Department), Christian Frigast (Executive Vice President, Merchant Banking Department), E Gorm Jørgensen (Executive Vice President Credit Department), Ejgil Jørgensen (Executive Vice President Privatholding Group), Peter Lybecker (Executive Vice President, Treasury Department), Peter Schütze (Executive Vice President, Retail Banking Department)

PRINCIPAL ACTIVITIES: All banking activities

Domestic Subsidiaries: (100% owned): Privatholding A/S; Privatfinans A/S; Privatleasing A/S; A/S Kjøbenhavns Ejendomsselskab. London: Privatbanken Ltd. Luxembourg: Privatbanken International (Denmark) SA. Associated Banks: Scandinavian Pacific Ltd; Scandinavian Bank Group Ltd; Banque Transatlantique SA; Kredietbank SA, Luxembourgeoise; Kredietbank (Suisse) SA; Banque Scandinave en Suisse. Privatbanken is a member of the Scandinavian Banking Partners and Inter-Alpha Group of Banks

Financial Information:

	31.12.86	31.12.87
	DKr'000	DKr'000
Profit before tax	88,400	295,000
Profit after tax	60,900	226,000
Retained profit (loss)	(110,000)	18,000
Dividends	171,000	208,000
Dividends per share	15%	15%
Earnings per share	5.4DKr	18DKr
Share capital	1,138,000	1,388,000
Shareholders funds	4,672,000	5,183,000
Deposits	77,079,000	84,614,000
Total assets	95,517,000	103,577,000

Principal Shareholders: 81,000 private shareholders
No of Employees: 5,645

PROVINSBANKEN A/S

Kannikegade 4-6, PO Box 42, DK-8100 Aarhus C

Tel: +45 86 12 25 22

Cable: Provinsbank; SWIFT: PROV DK 22

Telex: 64403 prova dk; 64692 proxar dk; 64425 prova dk

Telefax: +45 86 12 25 22; +45 86 12 51 31; +45 86 12 15 50

Board of Directors: A Bjørn Rühne (Chairman), Niels Chr Nielsen (Deputy Chairman), Ole A Busck, Carsten Didriksen, K Karmark Hansen, Nils Skibsted Hansen, Kaj Madsen, Frits Niegel, Tove Nielsen (Ministry of Industry Representative), Ove Jensen (Employee Representative), Peter Michaelsen (Employee Representative), Edith Nielsen (Employee Representative)

Managing Directors: Bjarne Dalsgaard, Hans-Verner Larsen, Jens Otto Veile, Mogens Gantriis

Senior Executives: Jørgen Clausen, Ejvind Hansen, Steen Hjort, Allan Hjælmhof, Peter Engberg Jensen, Poul Møller-Mikkelsen, Søren Møller, A C Nygard, Niels Østeraas

PRINCIPAL ACTIVITIES: All banking activities

Subsidiary Companies: Provinsbanken International (Luxembourg) SA; Danish Finance (Asia) Ltd (Hong Kong); Provinsbanken North America Inc; Pro Børs Børsmæglerselskab A/S; R Henriques jr Børsmæglerselskab A/S; Pro Finans A/S; Pro Kredit A/S; Pro-Venture A/S; Aktieselskabet 23 august 1934; Frederiksborg Erhvervs Invest A/S; Krone Erhvervs Invest A/S; A/S ADIF; Aurora Finans A/S; Ejendomsaktieselskabet af 1 juni 1987

Financial Information:

	31.12.87	31.12.88
	DKr'000	DKr'000
Profit before tax	357,000	751,000
Profit after tax	187,000	422,000
Retained profit	49,000	270,000
Dividends	138,000	152,000
Dividends per share	0.15DKr	0.15DKr
Earnings per share	0.20DKr	0.38DKr
Share capital	920,000	920,000
Shareholders funds	3,213,000	3,995,000
Deposits etc	49,512,000	54,636,000
Total assets	61,524,000	67,829,000

No of Employees: 4,116

SMIDTH, F L, & CO A/S

77 Vigerslev Allé, DK-2500 Valby, Copenhagen

Tel: +45 1 30 11 66

Cable: Folasmidth

Telex: 27040 flsco dk

Telefax: +45 1 30 44 41

Board of Directors: Erik Münter (Chairman), Christian Kjær (Deputy Chairman), Christian Arnstedt, S H Jonsson, Merete Jørgensen, *Grethe Machholm, Jan W Folting, Jens Münter, *Karl Nielsen, *Benny E Raahauge, *Appointed by the employees

Board of Management: Birger Riisager (President and Chief Executive Officer), Ib Christensen (Senior Vice President and

Executive Officer), Ole Trolle (Senior Vice President and Executive Officer),

PRINCIPAL ACTIVITIES: Design, manufacture and sale of machinery to the cement and allied industries

Major Subsidiary Companies (100% owned unless otherwise stated): Engineering and Manufacturing: FL Smidth & Co A/S; FL Smidth SA (Brazil); FL Smidth & Co of Canada Ltd (Canada); FL Smidth & Co Limited (UK); FL Smidth & Cie France, SARL (France); FL Smidth & C Italiana Srl (Italy); FL Smidth & Co (Japan) Ltd (Japan); FL Smidth & Cia SA de CV (Mexico); FL Smidth & Cia Española SA (Spain); FL Smidth & Co (SA) (Pty) Ltd (South Africa); FL Smidth & Co GmbH (Germany); FL Smidth & Co (USA); FLS Participaçoes e Representaçoes Ltda (Brazil); Iwasa Kikai KK (Japan); Iwasa Real Estate Co Ltd (Japan); FLS-Singapore Pte Ltd (Singapore); Fabricación de Maquinaria Pesada, SA de CV (49%) (Mexico); SECIL SARL (21%) (Portugal). Concrete and Building Materials: Unicon Beton I/S (65%) Fabriksbeton A/S Næstved; Fabriksbeton Sønderborg A/S; Holbæk Færdigbeton A/S; A/S Københavns Betonfabrik; Interconcrete Corporation (USA); Tri-County Concrete Corporation (USA); Lynette Beton A/S (50%); VS Beton A/S; A/S Aarhus Betonfabrik; Danalith A/S (Denmark); Danbond-Danish Structural Bonding Company A/S; Dampa A/S; Dampa (Deutschland) GmbH (Germany); Dampa (UK) Ltd (UK); Dampa Korea Industry Ltd (50%) (Korea); Dampa (USA) Inc (USA); Dampa Sweden AB (Sweden); Dampa Canada Ltd (Canada); Matador Væggen A/S; Danogips A/S; Svenska AB Dano (Sweden); Danogips GmbH (Germany); Dainovo Danogips Business Development A/S; Hispayeso SA (Spain); Aktieselskabet Aalborg Portland-Cement Fabrik (34%); Dansk Eternit-Fabrik A/S (34%); B&C Danmark A/S (23%); Færdigbeton Aalborg A/S (34%); Lehigh White Cement Company (20%) (USA); Højslev Teglværk A/S (17%); Hasle Klinker-og Chamottestensfabrik A/S (34%); Danaske I/S (17%); Aalborg Energy Consultants NV (34%) (Netherlands); Densit A/S (34%); ES-Færdigbeton A/S (27%); Nordsjællands Færdigbeton A/S (40%). Other Manufacturing Industries: Bates Ventil Sække Co A/S; A/S Nyborg Plast; ASX 10369 A/S (Novoplastics Emballage); Nyborg Plastics Ltd (UK); SK Textil & Emballage A/S (90%); Pedershaab Maskinfabrik A/S; Pedershaab Inc (USA); Pedershaab Maschinenfabrik GmbH (Germany); Pedershaab Limited (UK); NEA-Lindberg A/S; Industrihærderiet A/S; Aktieselskabet Faxe Kalkbrud (14%); Aktieselskabet Nordiske Kabel-og Traadfabriker (27%); Hume Industries (Malaysia) Berhad (19%) (Malaysia); Anglo Nordic Holdings PLC (75%) (UK). Trade, Transport and Services: A/S Jernkontoret (75%); SA Svendsen A/S (75%); SA Svendsen OY (75%) (Finland); SA Svendsen AB (75%) (Sweden); SA Svendsen A/S (75%) (Norway); Dansk Facade-og Tagpladefabrik A/S (75%); Rørkontoret A/S (75%); Dan Transport A/S; Dan Transport Rejsebureau A/S; Dan Transport Corporation (USA); Dan Transport International SARL (France); Channel Tours A/S; Stevedore-Kompagniet Hafnia A/S; Renovadan System Transport A/S. Financial Services: FLS Industries A/S; Cement Investments A/S; Forenede Assurandører A/S (74%); Brdr O & C Thielst A/S (74%); Risk Research & Captive Services A/S; FL Smidth & Co (Holdings) Limited (UK); Internationalt Beton-Compagni A/S; Ejendomsaktieselskabet af 31/12/1982; Benchmark Group plc (14%) (UK); United Insurance Holding NV (28%) (Netherlands); ABB Lindberg A/S (49%); Airpol Inc (USA); Lovaux Aerospace (UK); Rich Müller Management (51%) (UK); FLS Energy; FLS Data; FLS Miljø

Principal Bankers: Handelsbanken

Financial Information:

	31.12.86	31.12.87
	DKr'000	DKr'000
Sales turnover	8,726,000	7,166,000
Profit before tax	231,316	153,000
Profit after tax	214,000	71,000
Net result	61,000	14,000
Dividends	53,152	53,152
Dividends per share	6%	8%
Share capital	664,408	664,408
Shareholders equity	1,700,000	1,738,000

Principal Shareholders: Potagua, Denmark
No of Employees: 9,801 Group

SOPHUS BERENDSEN A/S
See BERENDSEN, SOPHUS, A/S

SPAREKASSEN SDS
8 Kongens Nytorv, DK-1050 Copenhagen K
Tel: +45 (1) 13 13 39
Cable: SDSSAVING
Telex: 15745 SDSFD DK
Fax: +45(1) 11 63 72; Swift: SDSF DK KK

Board of Directors: Jens Chr Kirketerp Jensen (Chairman)
Board of Management: Laurits Ringgård (Chairman), Lars Eskesen, Arne Hagedorn, Preben Kendal, Erik Mortensen, H Kruse Petersen
Senior Executive Vice President: Niels-Aage Nielsen. Executive Vice Presidents: Birger Grubbe (International Division)

PRINCIPAL ACTIVITIES: Savings bank operating as a commercial bank
Trade Names: SDS
Subsidiary Companies: (100% owned unless stated): SDS Bank Ltd (UK); London Interstate Finance Ltd (UK); London Interstate Export Finance Ltd (UK); Sikastolyn BV (UK); London Interstate Nominees Ltd (UK); London Interstate Fund Managers Ltd (Guernsey); Spar Finans A/S (50%); Spar Finans Erhverv A/S; Damcas A/S; International Factors A/S (50%); Ejendomsaktieselskabet af 15/6 1935; SDS Børsservice; Børsmæglerselskab A/S; SDS North America Inc (USA); SDS Finans A/S
Financial Information:

	31.12.86	31.12.87
	DKr'000	DKr'000
Turnover	2,711,768	2,909,586
Profit (loss) before tax	(584,865)	376,380
Profit (loss) after tax	(584,865)	381,337
Deposits	52,376,991	56,672,214
Total assets	69,638,771	77,860,529

No of Employees: 4,443

STATSANSTALTEN FOR LIVSFORSIKRING
Kampmannsgade 4, DK-1645 København V
Tel: (01) 15 15 15
Telex: 15283 danlivdk
Telefax: +45 1 32 37 32

Board of Directors: Benned Hansen (Chairman), Frede Ahlgreen Eriksen, Aage Andersen, Karsten Olsen, Knud Bjørn, Keld Rahbæk Møller, Carsten B Andersen, Gitte Jensen, Lars Raakjær
Management: Erik Bonnerup (Managing Director), Troels Bjerre (Director), Kaj G Schou (Director)
Senior Executives: Business Sectors: Ulla Plesner (Large Customer Sector, Executive Manager, Actuary), Heinz Andersen (Private/Business Sector-Greater Copenhagen Area, Customer Area Manager), Jes Lillelund (Private/Business Sector-Jutland, Customer Area Manager, Executive Manager), Niels Weng (Private/Business Sector-The Islands, Customer Area Manager, Executive Manager). Investments: Troels Gunnergaard (Investments and Capital Management, Senior Executive Manager), Benny Nielsen

DENMARK

(Personnel, Executive Manager), Christian Richter-Friis (Actuary Office, Executive Manager), Arne Finsen Munck (Budget and Accounts, Executive Manager, Accounting), John Sonberg (Sector Service, Sector Manager), Steen Rasmussen (Management Secretariat, Executive Manager), Esben Høstager (Information and PR)

PRINCIPAL ACTIVITIES: Life, pensions, annuities
Subsidiary Companies: Dan-Liv A/S; Statsanstaltens Ejendomsinvest A/S
Principal Bankers: All principal banks
Financial Information:

	31.12.86 DKr'000	31.12.87 DKr'000
Premium income	2,734,592	2,653,000
Investment income	5,189,732	5,011,147
Profit before tax	226,384	-
Profit after tax	213,844	-
Retained profit	213,844	-
Equity capital	3,034,267	2,370,000

No of Employees: 631

Major Companies of EIRE

AER LINGUS PLC AND AERLINTE EIREANN PLC

Dublin Airport, Co Dublin
Tel: 370011
Telex: 31404

Chairman: B A Slowey
Directors: P Boushell, M A Lynch, P O Ferguson, M Rafferty, R McGahern, Mrs E MacManus, E Marren, C Mullan (Chief Executive), Ms E O'Mara Walsh, J Tatten, J D Traynor
Senior Executives: C Mullan (Chief Executive), M L Slater (Assistant Chief Executive-Finance), Capt R N White (Assistant Chief Executive), D Downing (Chief Operations Officer), O O Siochru (Secretary), D Hanrahan (Chief Executive-Ancillary Activities), C P McGrath (General Manager-Britain), J Bastable (Senior Vice-President North America), Capt M P Horgan (General Manager-Operations), K Acton (General Manager Passenger Marketing), J Hartnett (General Manager Cargo)

PRINCIPAL ACTIVITIES: Air services: scheduled and charter between Ireland and European and North American destinations; shareholding in hotel operations, tour companies, data processing services, aircraft maintenance, training facilities, aircraft leasing, helicopter services, personnel recruitment and project management services, hospital management, management services, insurance and reinsurance, research and development services
Trade Names: Parc, Blueskies, Sunbound, Enterprise, Cara Holidays, Aerfast, Getaway
Subsidiary Companies: Aer Lingus:- Aer Lingus Commuter Ltd; Aer Lingus Holidays Ltd; Aer Lingus Hotel Holdings Ltd; Airmotive Ireland (Dublin) Ltd; Cara Data Processing Ltd; Devtec Ltd; PARC Ltd; PARC Hospital Management Ltd; PARC Management Services Ltd; SRS Aviation (Ireland) Ltd; Blueskies Ltd; Cara Holidays Ltd; Enterprise Travel (Ireland) Ltd; Sunbound Holidays Ltd; Aer Turas Teo (75%); Timas Ltd; Hotel Commodore SA (France); Aer Lingus-Dunfey Corp (98%) (USA); Automation Tooling Systems Corporation (USA); ATS Automation Tooling Systems Inc (75%) (Canada); Aer Lingus Canadian Holdings Ltd (Canada); Cara Ireland (Tours) Ltd (UK). Aerlinte Eireann PLC:- Travel Leisure Developments Ltd (85%); Irish Helicopters Ltd (51%); Alpex Trading Ltd; Altek Automation Ltd; Montague Lloyd Ltd (UK); Caledonian Airborne Systems Ltd (75%) (UK); Parc Care Ltd; Copthorne Hotels Holdings Ltd (UK); Aer Securicor Ltd (50%) (UK); Aer Lingus UK Holdings Ltd (UK); Cara Consulting Ltd (UK); Enterprise Travel Ltd (UK); Aer Lingus Holidays (NI) Ltd (UK); Sunbound Holidays (Northern Ireland) Ltd (UK); London Tara Hotel Ltd (UK); PARC (UK) Ltd (UK); Aer Lingus Espana Tourist Developments SA (70%) (Spain); Dirnan Insurance Company (Bermuda); Las Rosas Marketing SA (Spain); Inversiones Santiago SA (35%) (Spain); Punta Tamaimo SA (28%) (Spain); Dirnan Insurance Company Ltd (Bermuda); Communicare Ltd (41%) (UK); Cara Properties Ltd (50%) (UK); International Airport Hotel Ltd (25%); Jury's Hotel Group plc
Principal Bankers: Ulster Bank; National Westminster Bank; Chase Manhattan Bank

Financial Information:

	31.3.87 IR£'000	31.3.88 IR£'000
AER LINGUS		
Operating revenues	393,438	419,114
Profit before tax	24,067	38,683
Profit after tax	21,695	33,956
Retained profit	29,832	35,502
Dividends	-	2,182
Dividends per share	-	10.2p
Share capital	36,393	36,393
Shareholders funds*	169,811	214,241
AER LINTE		
Operating revenues	141,858	140,504
Profit (loss) before tax	(3,341)	(88)
(Loss) after tax	(4,269)	(1,167)
Retained profit (loss)	(5,075)	(2,700)
Share capital	47,217	47,217
Shareholders funds (deficit)	(13,595)	1,461

*Consolidated

Principal Shareholders: Irish Goverment
No of Employees: 8,000 Group

ALLIED IRISH BANKS PLC

Bankcentre, PO Box 452, Ballsbridge, Dublin 4
Tel: (01)600311
Cable: Alibanex
Telex: 24173
Telefax: (01) 600196

Chairman: Niall Crowley; Director and Group Chief Executive: Gerald B Scanlan
Directors: Liam St J Devlin (Deputy Chairman), Maurice Abrahamson, J E Casey, Thomas Cavanagh, J Owen Cole, James P Culliton, Patrick M Dowling, Dermot Egan, P M Fallon, Sir Peter Froggatt, John F Keogh, John B McGuckian, Diarmuid Moore, Sir Douglas Morpeth, Denis J Murphy, Miriam Hederman O'Brien, Michael J O'Keeffe, S J Sheehy

PRINCIPAL ACTIVITIES: Banking services, merchant banking, industrial banking and leasing services, issuing and finance house. The Group has 500 outlets in Ireland and Britain as well offices in Brussels, Singapore, New York, Chicago, Isle of Man, Jersey, Sydney and a representative office in Frankfurt
Subsidiary/Associated Companies: (100% owned unless stated): AIB Investment Management Ltd; Allied Irish Finance Company Ltd; Allied Irish Investment Bank plc; Allied Irish Leasing Ltd; Allied Irish Banks (Holdings & Investments) Ltd; Allied Irish Banks (I.O.M.) Ltd (Isle of Man); Allied Irish Banks (CI) Ltd (Jersey); Allied Irish Venture Capital Ltd; Allied Irish Fund Management Ltd; Allied Irish Investment Managers Ltd; Grofund Investment Managers (Guernsey) Ltd; Grofund Managers Ltd; Allied Combined Leasing Ltd; Allied Equities Ltd (Isle of Man); Credit Finance Bank PLC (77%); Allied Combined Trust Ltd (51%); Allied Irish Holdings (Australia) Ltd (Australia); Allied Irish Securities Ltd; Grofund Investment Managers (Jersey) Ltd (Jersey); Allied Irish Moore & Schley Ltd (50%); Coyle Hamilton Group Ltd (39%); First Maryland Bancorp (49.9%) (USA); First Venture Fund Ltd (22%)
Financial Information:

	31.3.87 IR£'000	31.3.88 IR£'000
Profit before tax	102,130	124,910
Profit after tax	65,400	74,520
Retained profit	43,720	31,510
Dividends	21,740	26,120
Dividends per share	11.3p	12.3p
Earnings per share	34.0p	35.2p
Share capital	52,991	53,311
Total assets	8,560,718	9,605,698

No of Employees: 9,000

EIRE

AN POST

General Post Office, Dublin 1
Tel: (01) 728888
Telex: 33444 POST EI

Chairman: Feargal Quinn

Directors: Oswald Barton, John Colfer, Martin Doyle, Gerard
Harvey (Chief Executive), Terry Kelly, Peter McCarthy, Luke
Mooney, Brendan O'Brien, Con O'Driscoll, Brian Patterson,
Des Ryan, Jack Sheehan, Brian Walsh, Eamon Waters

Senior Executives: Gerard Harvey (Chief Executive), John
Dwyer (Director of Mails Services), Terry Reynolds (General
Manager Banking Services), Paddy Banks (General Manager-
Counter Services), Eamonn Ryan (General Manager-Dublin
Mails Operations), John Dempsey (General Manager-Parcels
and Express Mail), Michael Sheridan (General Manager-
Provincial Mails Operations), Ross Hinds (General Manager-
Support Services), John O'Callaghan (Group Manager-Human
Resources), Norman Newcombe (Group Marketing Manager),
Liam Daniel (Group Financial Controller), Jim Treacy
(Company Secretary)

PRINCIPAL ACTIVITIES: Postal services; savings bank
Trade Names: An Post
Subsidiary Companies: An Post National Lottery Company
Principal Bankers: Bank of Ireland
Financial Information:

	31.12.86	31.12.87
	IR£'000	IR£'000
Sales	183,575	194,835
Profit before tax	673	2,776
Profit after tax	673	1,576
Retained profit (deficit)	(1,508)	68
Earnings per share	IR£0.04	IR£0.015
Share capital	44,430	44,430
Shareholders funds	42,900	44,500

No of Employees: 11,414

ARNOTTS PLC

12 Henry Street, Dublin 1
Tel: (01) 72111
Cable: Arnotts
Telex: 32524 ARNO-EI
Telefax: (01) 730109

President: R S Nesbitt; *Chairman:* T C Toner

Directors: M G O'Connor, M R Nesbitt (Managing Director), V
Dudgeon, R L Nesbitt, S Duignan, J B Davy

Senior Executives: B Howard (General Manager Wholesale), J C
O'Sullivan (General Manager Administration and Group
Secretary), V W Dudgeon (General Manager, Retail)

PRINCIPAL ACTIVITIES: Retail department stores, boutiques,
wholesalers and manufacturers; cash in transit security etc
Trade Names: Arnotts, Robert Wallace, Skylark, Robbie,
Melanie
Subsidiary Companies: (100% owned unless stated): Arnotts
(UK) Ltd; Boyers & Co. Ltd; Wallace Shirts Ltd; Asgard
Financial Services Ltd. Associated Companies: Allied Couriers
Ltd (50%); Brinks Mat Ireland Ltd (50%); Brinks Allied Ltd
(50%)
Principal Bankers: Allied Irish Banks Plc; Midland Bank;
National Irish Ltd; Barclays Bank International (Ireland) Ltd; A
B N (Ireland) Ltd; Ulster Investment Bank Ltd; Investment
Bank of Ireland

Financial Information:

	24.1.87	31.1.88
	IR£'000	IR£'000
Sales turnover	53,707	52,024
Profit before tax	3,423	1,926
Profit after tax	1,855	1,089
Retained profit (loss)	786	(292)
Dividends	1,048	1,048
Dividends per share	12p	12p
Earnings per share	21.03p	12.48p
Share capital	8,700	8,700
Shareholders funds	38,073	37,837

Principal Shareholders: Alibank Nominees Ltd; Arnotts Staff
Pension Fund; Nesbitt & O'Connor Families
No of Employees: 1,144

BANK OF IRELAND

Lower Baggot St, Dublin 2
Tel: 615933
Cable: Bank Dublin
Telex: 93836 BKHO EI
Telefax: 682671/615101/615040

Governor: W J L Ryan

Directors: B J Breen (Deputy Governor), M Hely Hutchinson
(Chief Executive), J E Callaghan, Margaret Downes, T P
Hardiman, *M A Keane, *R A D Keatinge, D M Kennedy, *M J
Meagher, *P J A Molloy, R I Morrison, *F B O'Rourke, Sir
David Orr, T C Toner, K Wylie, *Managing
Senior Executives: T H Forsyth (Secretary)

PRINCIPAL ACTIVITIES: General banking and financial services
Principal Subsidiary Companies: (100% owned unless stated):
Bank of Ireland Management Ltd; Bank of Ireland (Jersey)
Ltd; Bank of Ireland Finance Ltd; Investment Bank of Ireland
Ltd; Investment Bank of Ireland (I.O.M.) Ltd; Property Loan &
Investment Co Ltd; British Credit Trust Ltd; N.I.I.B Ltd; BOI
Asia Ltd (Hong Kong); Bank of Ireland Home Mortgages Ltd
(UK); ICS Building Society (99.9%); Lifetime Assurance
Company Ltd
Financial Information:

	31.3.87	31.3.88
	IR£'000	IR£'000
Profit before tax	81,500	109,100
Profit after tax	50,900	62,700*
Retained profit	33,100	21,100
Dividends	17,800	23,300
Dividends per share	11.00p	11.75p
Earnings per share	29.20p	32.60p
Share capital	161,720	198,113
Shareholders funds	579,500	742,000
Deposits	7,122,000	7,716,100
Total assets	8,017,100	8,837,900

*before extraordinary item of
IR£ 18.3 million in respect
of provision for deferred
taxation in respect of prior
years

Principal Shareholders: Irish Life Assurance plc 10%-15% of
issued capital stock
No of Employees: 8,775

GPA GROUP LTD

GPA House, Shannon, Co Clare
Tel: (061) 61088

PRINCIPAL ACTIVITIES: Leasing of aircraft
Financial Information:

	IR£'000
Sales turnover	238,000

No of Employees: 152

INDUSTRIAL CREDIT CORPORATION PLC

32-34 Harcourt Road, Dublin 2

Tel: (01) 72 00 55

Telex: 93220 ISSU EI

Telefax: (01) 71 77 97

Chairman: J T Barton

Directors: F A Casey (Managing Director), V Daly, N Holland, M Honan, P C Molloy, L O'Donnell, J H B Ryan, P A White

Senior Executives: J C Ryan, P B Murphy, K O'Connell

PRINCIPAL ACTIVITIES: Financial services

Subsidiary Companies: ICC Corporate Finance Ltd; ICC Fund Management Ltd; ICC Finance Ltd; ICC International Financial Services Ltd; Shipping Finance Corporation Ltd

Principal Bankers: Bank of Ireland

Financial Information:

| | 31.10.87 | 31.10.88 |
	IR£'000	IR£'000
Profit before tax	3,713	3,270
Profit after tax	2,648	2,974
Retained profit	1,210	1,536
Dividend	1,438	1,438
Dividend per share	12p	12p
Earnings per share	22.1p	24.8p
Share capital	11,984	11,984
Shareholders funds	35,235	36,771
Deposits	398,835	474,920
Total assets	543,622	659,977

Principal Shareholders: Minister for Finance (99%)

No of Employees: 350

IRISH INTERCONTINENTAL BANK LTD

91, Merrion Square, Dublin 2

Tel: (01) 61 97 44, (01) 76 02 91

Cable: INCONBANK DUBLIN

Telex: 33322 IIBK EI

Telefax: (01) 78 50 34

Chairman: Donald S McAleese; Managing Director: Patrick C McEvoy

Directors: Edward A Marah (Deputy Chief Executive), Patrick A Larragy (Executive Director/Secretary), John Kelly (Executive Director), Brian MacManus (Executive Director), Marcel Cockaerts (Belgian), Rony Van Hoeck (Belgian), Paul Vergote (Belgian), T David Kingston, Liam Donlon (Executive Director), Ronnie Fitzell (Executive Director)

Senior Executives: Edward A Marah (Director, Banking Department), Patrick Bourke (Manager, Banking Department), Liam Donlon (Director Banking Department), Brian MacManus (Director, Banking Department), Gerard Quigley (Associate Director, Banking Department), John Kelly (Director Treasury/Foreign Exchange Department), Terry Sullivan (Associate Director/Chief Dealer, Treasury/Foreign Exchange Department), Ronald W Fitzell (Director/Financial Controller), Adrian B Henderson (Operations Manager), Keith O'Byrne (Manager, Treasury/Foreign Exchange), Tom Foley (Associate Director, Banking Department), Michael O'Hara (Associate Director, Operations Department), Paul Devenney (Manager, Treasury/Foreign Exchange Department), Frank Cafferty (Manager, Banking Services), Declan Treacy (Associate Director, Corporate Finance Department), Patrick A Larragy (Director/Company Secretary, Corporate Finance/Operations)

PRINCIPAL ACTIVITIES: Merchant bank

Parent Company: Kredietbank N.V. Belgium

Financial Information:

| | 31.12.86 | 31.12.87 |
	IR£'000	IR£'000
Profit after tax	3,559	4,238
Retained profit	10,003	12,376
Dividends	1,000	1,339
Share capital	7,000	7,000
Shareholders funds	17,003	19,376
Deposits	384,877	423,370
Total assets	404,880	445,747

Principal Shareholders: Kredietbank NV, Brussels (75%); Irish Life Assurance Co Ltd (25%)

No of Employees: 85

JEFFERSON SMURFIT GROUP PLC

Beech Hill, Clonskeagh, Dublin 4

Tel: (0001) 696622

Cable: Smurfit Dublin

Telex: 93411 JSGP EI

Telefax: (0001) 694481

Chairman: M W J Smurfit (and Chief Executive Officer)

Directors: P A Smurfit (Joint Deputy Chairman), D F Smurfit (Joint Deputy Chairman), H E Kilroy (President and Chief Operations Director), Senator E Ryan, D A Kilroy, J M Lynch, K C McCourt, W J MacDonald, P J P Gleeson, J B Malloy (President and Chief Operating Officer Jefferson Smurfit Corporation), D F T Austin (Executive Vice President Marketing), J O'Dwyer

Senior Executives: M R J Pettigrew (Secretary), R D Holmes (Chief Financial Officer), E P Grace (Group Treasurer), G A Fagan (Group Financial Controller)

PRINCIPAL ACTIVITIES: The company is a holding company. Its subsidiary companies are engaged principally in the manufacture of corrugated cases, paper and board, newsprint, print and packaging and in the provision of financial services and distribution of consumer durables

Principal Operating Subsidiary and Associated Companies: (100% owned unless stated): Subsidiary Companies: CartoEspana SA (Spain); Carton de Venezuela SA (78%) (Venezuela); Iona Print Ltd; Jefferson Smurfit Corporation (78%) (USA); Mercurius Verpakking BV (Netherlands); Smurfit Corrugated Ireland Ltd (51%); Smurfit Ireland Ltd; Smurfit Ltd (UK); Smurfit Newsprint Corporation (62%) (USA); VosaCec SpA (Italy). Associated Companies: Container Corporation of America (39%) (USA); MacMillan Smurfit SCA Ltd (25.5%) (UK); Smurfit Paribas Bank Ltd (50%)

Financial Information:

| | 31.1.87 | 31.1.88 |
	IR£'000	IR£'000
Sales turnover	1,067,114	1,159,837
Profit before tax	60,102	153,858
Profit after tax	43,698	91,129
Retained profit	19,215	52,213
Dividends	8,210	9,196
Dividends per share	3.61328p	3.97461p
Earnings per share	15.2p	31.6p
Share capital	50,185	51,900
Shareholders funds*	261,450	310,590
*including minorities		

Principal Shareholders: M W J Smurfit (5%); Irish Life Assurance Company plc (10%-15%); The National Coal Board Pension Fund (5%); Standard Life Assurance Company (5%-10%)

No of Employees: 15,240

SILVERMINES PLC

43 Ailesbury Rd, Dublin 4

Tel: 0001-698 155

Telex: 31522

Telefax: 0001-695 636

Chairman: J G O'Neill (Non-Executive)

EIRE

Directors: K M Baker (Deputy Chairman), F J Traynor (Managing Director), J D McFarlane (Finance Director), J H Gunn (Non-Executive), P C Kilroy (Non-Executive), T P Mahoney (Non-Executive)

PRINCIPAL ACTIVITIES: Investment in natural resources, property, industry and venture capital

Subsidiary Companies: (100% owned unless stated): Greenhills Mining Co Ltd; Silvermines Investments Ltd; Silvermines International Ltd (Bermunda); Keeper Investments Ltd; Silvermines Exploration PLC; Whitverton Ltd (Bermuda); PGM Holdings Ltd (Bermuda); PGM Ballscrews Ltd; KME International Ltd (97%); W E Sykes (1986) Ltd; PGM Ballscrews Inc (USA); P R Dalney (Gears) Ltd; Vaughan Gear Machines & Services Ltd; Solid State Induction Ltd (75%); Elequip Ltd

Financial Information:

	31.12.86 IR£'000	31.12.87 IR£'000
Sales turnover	900	15,400
Profit (loss) before tax	(2,330)	3,130
Retained profit (loss)	(2,710)	1,570
Dividend per share	2.00p	4.00p
Earnings (loss) per share	(16.2p)	36.10p

No of Employees: 407

ULSTER INVESTMENT BANK LTD

2 Hume St, Dublin 2
Tel: 613444
SWIFT: UIBL IE 2D
Telex: 93980
Telefax: 763507

Chairman: Martin Rafferty

Directors: Brian McConnell (Chief Executive Director), Leo Conway (Deputy Chief Executive), Sir Ewart Bell, Gavin Caldwell, Victor Chambers, Michael Griffin, J W Matthews, H Mackeown, Kevin Menton, Patrick McMahon, Richard Robinson, Noel Ryan, Martin Wilson, David Went

PRINCIPAL ACTIVITIES: Bank
Parent Company: Ulster Bank Ltd, Northern Ireland
Principal Bankers: Ulster Bank Ltd
Financial Information:

	30.9.86 IR£'000	30.9.87 IR£'000
Profit after tax	5,012	5,035
Share capital	3,000	3,000
Shareholders funds	31,359	36,801
Deposits	651,332	684,447
Total assets	729,603	767,813

No of Employees: 174

VHI

VHI House, Lower Abbey Street, Dublin 1
Tel: (01) 724499
Telex: 32991
Telefax: (01) 734004

Directors: T R Ryan (Chief Executive)

PRINCIPAL ACTIVITIES: Health insurance services
Financial Information:

	IR£'000
Turnover	112,000

No of Employees: 300

WOODCHESTER INVESTMENTS PLC

Woodchester House, Golden Lane, Dublin 8
Tel: 0001-784299

Chairman: C McKinney

Directors: J Gunn (Deputy Chairman), C Allan, Mary Broughan, D Dilger, W Kane, P L Goldie

PRINCIPAL ACTIVITIES: Leasing of equipment and motor vehicles; financial services including instalment credit; contract hire services for motor vehicles
Financial Information:

	31.12.86 IR£'000	31.12.87 IR£'000
Sales turnover	21,900	58,700
Profit before tax	3,830	7,800
Retained profit (loss)	2,630	(9,900)
Dividend per share	1.25p	1.13p
Earnings per share	5.55p	6.75p

Major Companies of
FINLAND

AMER-YHTYMÄ OY

Amer Group Ltd

PO Box 130, SF-00601 Helsinki
Tel: +358-0-757 71
Cable: Amercon Helsinki
Telex: 121360 amerc sf
Telecopier: +358-0-7577200

Board of Directors: Heikki O Salonen (Group Chairman and Chief Executive Officer)
President: Leif Ekström (Group President and Chief Operating Officer)
Senior Executives: Seppo Salminen (Group Executive Vice President, Finance and Administration)

PRINCIPAL ACTIVITIES: Import and distribution of cars and vans; paper merchanting and converting; tobacco industry; communications industry; investment properties; golf-business; plastics industry; design and manufacture of interior fabrics and women's clothing, licensing; import and marketing of branded goods; metal industry
Trade Names: Amer-Tupakka, Weilin + Göös, Kiviranta, Amer Institute, Finnreklama, Amerpap, Golden Leaf, Korpivaara, Marimekko, Mecamer, Rias, Time/System, Hobart McIntosh, Amer Trading, MacGregor Golf
Subsidiary Companies: Korpivaara Oy; Marimekko Oy; Time/System A/S; Rias A/S; Hobart McIntosh Paper Company; MacGregor Golf Company
Principal Bankers: Kansallis-Osake-Pankki; The Union Bank of Finland Ltd; The Bank of Helsinki Ltd
Financial Information:

	31.8.87 (12 mths) FMk'000	29.2.88 (6 mths) FMk'000
Sales turnover	3,594,800	2,537,300
Profit before tax	243,200	161,200
Profit after tax	214,000	149,900
Retained profit	152,000	123,000
Dividends	45,700	29,900
Dividends per share	4.00FMk	2.00FMk
Earnings per share	15.30FMk	17.30FMk
Share capital	299,200	299,200
Shareholders funds	816,200	806,700

Principal Shareholders: The Engineering Society in Finland STS; The Association of Graduates of the Schools of Economics SEFE; The Student Union of the Helsinki School of Economics and Business Administration KY; The Land & Water Technology Foundation; Pohjola Group
No of Employees: 3,755

ILMARINEN, ELÄKEVAKUUTUSOSAKEYHTIÖ

Ilmarinen Pension Insurance Company Ltd

Eerikinkatu 41, 00180 Helsinki
Tel: 90-1841
Telex: 122932 ilmar sf
Telefax: 90-1843445

Board of Directors: Antero Siljola (Chairman), Pertti Niemistö (Deputy Chairman), Juhani Ahava, Klaus Ahlstedt, Juhani Aho, Tuomas Anhava, Pertti Berg, George Berner, Pertti Hagren, Niilo Hakkarainen, Jaakko Heinonen, Pekka Herlin, Harri Holkeri, Kauno Honkanen, Raimo Huovinen, Maunu Ihalainen, Erkki Inkinen, Liisa Joronen, Veikko Kauranen, Orvo Kiesi, Pentti Kivinen, Pentti Koivikko, Jorma Kontio, Risto Kuisma, Juha-Veikko Kurki, Fritz Kuula, Raimo Kärkkäinen, Martta Laurén, Juhani Leppänen, Mauri Matikainen, Tauno Matomäki, Uljas Mäkelä, Johan Norrmén, Heikki Peltonen, Matti Peltonen, Raimo Pohjaväre, Asko Puumalainen, Jarmo Rantanen, Jarmo Raulo, Jaakko Riikonen, Ragni Rissanen, Ylermi Runko, Timo Setälä, Asko Tarkka, Teuvo Tuominen, Eero Utter, Erik Volmari
Board of Directors: Pentti Talonen (Chairman of the Board), Juhani Salminen (Managing Director and Deputy Chairman of the Board), Jyrki Hägglund, Lauri Ihalainen, Tapani Kahri,

Aimo Korhonen, Peter Küttner, Per-Erik Lundh, Yrjö Niskanen, Jarmo Pellikka, Riitta Prusti, Pentti Seppälä, Jorma Varis, Raimo Vuorinen, Lauri Koivusalo (Deputy Managing Director), Heikki Varho (Deputy Managing Director). Deputy Members of the Board of Directors: Kaarlo Honkasalo, Hannu Marttila, Ingvar S Melin, A Y Näräkkä, Kari Pellikka, Seppo Siitonen, Jarmo Tammenmaa
Senior Executives: Kyösti Poutiainen (Director), Maire Rantanen (Director), Antii Uski (Director), Aira Bristow (Assistant Director), Pirkko Järvela (Assistant Director), Sakari Kuikka (Assistant Director), Ilkka Mikkola (Assistant Director), Tellervo Noppa (Assistant Director), Vesa Pohjankoski (Assistant Director)

PRINCIPAL ACTIVITIES: The mandatory employment pension insurance and voluntary additional pension insurance
Parent Company: Pohjola Insurance Company Ltd
Financial Information:

	31.12.86 FMk'000	31.12.87 FMk'000
Turnover	4,503,716	4,969,368
Premium income	3,255,172	3,610,858
Investment income	1,273,778	1,377,563
Profit before tax	10,470	14,939
Profit after tax	2,113	2,622
Retained profit	16,766	19,062
Dividends	126	126
Dividends per share	1FMk	1FMk
Earnings per share	10FMk	10FMk
Share capital	2,100	2,100
Shareholders funds	28,866	31,162

Principal Shareholders: Pohjola Insurance Company Ltd (62.4%); Suomi-Salama Mutual Life Assurance Company (29.5%)
No of Employees: 450

KALERVO, KESKINÄINEN ELÄKEVAKUUTUSYHTIÖ,

PL 205, SF-00181 Helsinki
Tel: 90-69521

PRINCIPAL ACTIVITIES: Insurance
Financial Information:

	FMk'000
Premium income	930,000

No of Employees: 190

KANSALLIS-OSAKE-PANKKI

PL 10, Aleksanterinkatu 42, 00101 Helsinki 10
Tel: 90-1631
Cable: Kansallis
Telex: 124412 kopi sf
Telefax: 163 35 95 general

Supervisory Board: Aatos Erkko (Chairman), Markku Mannerkoski (Vice Chairman), Juhani Ahava, Niilo Hakkarainen, Pekka Herlin, Harri Hintikka, Jaakko Ihamuotila, Risto Ihamuotila, Matti Kankaanpää, Mikko Kivimäki, Pentti Kivinen, Timo Laatunen, Olli Lehto, Tauno Matomäki, Kurt Nordman, E O Mansukoski, Kalevi Numminen, Raili Nuortila, Yrjö Pessi, Antti Potila, Jaakko Pöyry, Urho Ruola, Pentti Salmi, Juhani Salminen, Heikki O Salonen, Reino Salonen, Pentti Talonen, Asko Tarkka, Heikki Tavela, Pertti Voutilainen
Management Board: Jaakko Lassila (Chairman and Chief Executive), Erkki Karmila (Executive Vice President International Banking), Aarto Heinonen (Executive Vice President Corporate Banking), Jarmo Raulo (Senior Vice President Internal Services), Martti Korhonen (Senior Vice President International Banking: Area Management), Matti Korhonen (Senior Vice President Corporate Communications), Heikki Koivisto (Senior Vice President), Jukka Härmälä (Senior Vice President Investment Banking), Teppo Taberman (Senior Vice President Corporate Banking), Antti Hannula

(Senior Vice President Corporate Banking), Veikko Ylitalo (Senior Vice President Retail and Private Banking)

PRINCIPAL ACTIVITIES: Bank
Subsidiary Banks: Kansallis Mortgage Bank Ltd, Helsinki; Kansallis International Bank SA, Luxembourg; Kansallis International Bank (Asia-Pacific) Ltd, Singapore; Kansallis Overseas Bank Ltd, Nassau; Nordfinanz-Bank Zürich, Zürich
Financial Information:

	31.12.86 FMk'000	31.12.87 FMk'000
Profit before tax	615,800	860,400
Profit after tax	519,900	753,600
Retained profit	52,693	115,745
Dividends	261,600	271,200
Dividends per share	2.40FMk	2.40FMk
Earnings per share	5.22FMk	5.86FMk
Share capital	2,260,000	2,260,000
Shareholders funds	5,849,000	6,928,000
Deposits	80,920,000	96,372,000
Total assets	101,569,000	117,313,000

Principal Shareholders: Pohjola Insurance Company; Eläkevakuutusosakeyhtiö Ilmarinen; Kansallis-Osake-Pankki Pension Fund; Finnish Cultural Foundation; Apteekkien eläkekassa; Rauma-Repola Oy:n toimihenkilöiden eläkesäätiö s.r.; Emil Aaltosen säätiö; Jenny ja Antti Wihurin rahasto; Tukkukauppojen Oy; Suomi Keskinäinen Henkivakuutusyhtiö
No of Employees: 8,318 (full and part-time staff)

KESKINÄINEN ELÄKEVAKUUTUSYHTIÖ KALERVO

See KALERVO, KESKINÄINEN ELÄKEVAKUUTUSYHTIÖ,

OKOBANK

Osuuspankkien Keskuspankki Oy

Arkadiankatu 23, SF-00100 Helsinki 10
Tel: +358 0 4041
Cable: OKOBANK Helsinki
Telex: 124714 okohe sf
Telefax: +358 0 404 2624

Board of Management: Pauli Komi (Chairman and Chief Executive), Matti Liukkonen (Managing Director), Arhi Palosuo (International), Kari Kukka (Assistant Director International), Risto Fredriksson (Foreign Exchange), Matti Koski (International Payments), Kylliki Pankakoski (Correspondent Banking), Ossi Hynninen (Investment Banking), Virpi Kumlander (International Capital Markets)
Senior Executives: Virpi Kumlander (Treasury Department)

PRINCIPAL ACTIVITIES: Commercial bank; central bank for co-op banks
Subsidiary Companies: OKOBANK (Sweden); OKOBANK (Cayman Islands) Ltd (Cayman Islands); OP-Finance Centre Ltd; OKO Mortgage Bank Ltd; OP-Kortti Oy
Principal Bankers: All principal banks
Financial Information: Consolidated Balance Sheet

	31.12.87 FMk'000
Share capital	695,000
Deposits	21,738,100
Total assets	30,803,300

Principal Shareholders: The Finnish Cooperative Banks
No of Employees: 13,000 (Worldwide)

PENSION-VARMA, MUTUAL INSURANCE COMPANY

PO Box 175, SF-00121 Helsinki
Tel: +358 0 61 651
Telex: 125415 elvar sf
Telefax: 6165714

Board: Matti Pekkanen (Chairman), Gay Ehrnrooth (Vice Chairman), Pekka Merenheimo, Antero Mäki, Tenho Olin,

Pekka Paasikivi, Pentti Somerto, Pentti Uusivirta, Juhani Kolehmainen, Asko Tanskanen. Deputy Members: Lasse Laatunen, Tuulikki Väliniemi, Voitto Isomäki, Ralf Lehtonen
Directors: Juhani Kolehmainen (Managing Director), Asko Tanskanen (Deputy Managing Director), Ralf Lehtonen (Deputy Managing Director), Voitto Isomäki (Director), Reijo Kaivo-oja (Director), Tapani Koski (Director), Jorma Leinonen (Actuary, Director), Jorma Luostarinen (Assistant Director), Risto Hermunen (Director), Pekka Hemmo (Chief Legal Counsel, Secretary to the Board), Sakari Tola (Senior Physician)
Senior Executives: Osmo Rinne (Pension Department), Markku Parviainen (Contact Department), Helvi Leinonen (Insurance Department), Ulla Kangas (Account Department), Taisto Lassila (Actuarial Department), Heikki Malin (Information Department), Timo Hukka (Investment Department), Juhani Salmia (Development Department), Antti Vuorinen (Administration Department), Jyrki Jussila (Financial Manager)

PRINCIPAL ACTIVITIES: Pension insurance
Trade Names: PENSION-VARMA (in English), ELÄKE-VARMA (in Finnish), PENSIONS-VARMA (in Swedish)
Jointly Owned Companies: Eläkepalvelu Oy-Pensionservice Ab; Pensionservice Ltd; Kiinteistövarma Oy; Oy Eläkesysteemi Ab; Henkivakuutusosakeyntio Nova; Livförsäkringsaktiebolaget Nova; Nova Life Insurance Co Ltd; Suomen Sijoitusosake Oy; Finska Investmentaktie Ab; Finnish Equity Investment Ltd; Nordben Life and Pension Insurance Co Ltd
Principal Bankers: Union Bank of Finland
Financial Information:

	FMk'000
Premium income	2,990,000

No of Employees: 335

POHJOLA INSURANCE COMPANY LTD

Lapinmäentie 1, 00300 Helsinki 30
Tel: (358-0) 5591
Telex: 124556

Supervisory Board: Jaakko Lassila (Chairman), E O Mansukoski (Deputy Chairman), Antti Aarnio-Wihuri, Juhani Ahava, Klaus Cederberg, Peter Fazer, Iikka Haka, Niilo Hakkarainen, Pertti Hellemaa, Pekka Herlin, Gustav von Hertzen, Harri Hintikka, Ahti Hirvonen, Jaakko Ihamuotila, Raimo Ilaskivi, Leo Jouhki, Jorma Järvi, Kari Kairamo, Matti Kankaanpää, Matti Kavetvuo, Simo Kärävä, Kauko Maijala, Markku Mannerkoski, Tauno Matomäki, Arvi Paloheimo, Antti Potila, Pentti Salmi, Heikki O Salonen, Reino Salonen, Paavo Siurua, Asko Tarkka, Heikki Tavela, A E Vehmas, Timo Vikström
Board of Directors: Pentti Talonen (Chairman), Pentti Seppälä (Deputy Chairman), Jyrki Hägglund, Lauri Koivusalo (Investments), Yrjö Niskanen, Aimo Y Näräkkä, Kari Pellikka (Reinsurance), Juhani Salminen, Jorma Varis
Directors: Pentti Seppälä (Managing Director), Jyrki Hägglund (Field), Olavi Kauppila (Corporate Group), Hannu Ketola, (Private Households Group), Kari Pellikka (Managing Director of Pohjola International), Asko Sasi (Investments), Jorma Varis (Major Clients Group), Lauri Koivusalo (Investment Group), Esa Säyrinen (Chief Actuary)

PRINCIPAL ACTIVITIES: All classes of non-life insurance, reinsurance (parent company), pension insurance, life insurance (subsidiary/associated companies)
Trade Names: The Pohjola Group
Subsidiary/Associated Companies: (100% owned unless stated): Subsidiaries:- Eurooppalainen Insurance Company Limited (84%); Ilmarinen Pension Insurance Company Ltd (67%); Northclaims Ltd (93.33%); Finnish Insurance Services Ltd (85.71%); Pohjola Insurance Company (U.K.) Ltd (UK); Pohjola America Reinsurance Co (USA); Suomi Mutual Life Assurance Co (100% of guarantee capital); Autovahinkokeskus Oy (19.2%); Huoneistomarkkinointi Oy (27%); Metsäpalvelu Oy (8.23%); Pohjolan Energianvälitys Oy (50%); Safeplan Multicontrol Ltd; Suomen Teollisuuden Vartiointi Oy (24.72%); Vahinkopalvelu Oy (46.67%); Otso Loss of Profits Insurance

Co Ltd (25%); The First New York Syndicate Corp (12.5%); ICS Reinsurance (Private) Ltd (16.75%) (Singapore); United Reinsurance Corporation of New York (50%) (USA); Kansallis International Bank SA (15%) (Luxembourg); Nord-Deutsche Versicherungs AG (5%) (Germany); Postipankki (UK) Ltd (8%) (UK); Scandinavian Insurance Services (15%) (Luxembourg); Scanrisk EC (33.33%) (Bahrain); Sinser (Bermuda) Ltd (15%) (Bermuda)

Principal Bankers: Kansallis-Osake-Pankki; Union Bank of Finland Ltd

Financial Information: Group figures

	31.12.86(1)	31.12.87(1)
	FMk'000	FMk'000
Premium income	2,302,057	2,431,446
Investment income(2)	475,937	624,497
Profit before tax(3)	44,810	54,472
Profit after tax(3)	27,442	27,982
Retained profit(3)	63,179	64,134
Dividends	29,311	29,311
Dividends per share	0.90FMk	0.90FMk
Income per share(4)	2.41FMk	3.24FMk
Share capital	162,840	203,550
Shareholders equity(5)	3,429,977	4,534,243

(1) The subsidiary Ilmarinen may not, under the Insurance Companies Act, be included in the consolidated accounts. Ownership in Ilmarinen's share capital 53% in 1983 and 67% in 1984-1987

(2) including realised gains

(3) statutory

(4) before changes in untaxed reserves per share, adjusted by the 1986 bonus issue and the 1987 share issue

(5) untaxed reserves and unrealised appreciation, excluding minority interest

Principal Shareholders: Suomi Mutual Life Assurance Company (30.3%); Kansallis-Osake-Pankki and Kansallis-Osake-Pankki's Pension Foundation (10.8%); Union Bank of Finland Ltd, Union Bank of Finland's Pension Foundation and Union Bank of Finland's Pension Fund (3.7%); Tukkukauppojen Oy and Tuko's Officers' Pension Fund Foundation (3.4%)

No of Employees: 2,441

POSTIPANKKI LTD

Unioninkatu 20, SF-00007 Helsinki 7
Tel: 358 0 1641
Cable: POSTIPANKKI, HELSINKI
Telex: General: 121698; Foreign Exchange: 121701; New Issues & Securities: 121079; Foreign Secretariat: 123687
Telefax: Foreign Secretariat: 358 0 1643766

Supervisory Board: Matti Jaatinen (Chairman), Pertti Paasio (Deputy Chairman), Arvo Aalto, Gustav Björkstrand, Teemu Hiltunen, Kauko Juhantalo, Matti Pekkanen, Pirjo Rusanen, Eila Salminen, Pekka Tarjanne, Jaakko Vihmola, Pertti Viinanen

Management: Seppo Lindblom (Chairman and Chief General Manager), Ilpo Santala (Managing Director), Reijo Pukonen (General Manager), Erkki Linturi (General Manager), Matti Niemi (General Manager), Annikki Saarela (General Manager, International Banking), Juhani Paloheimo (General Manager)

PRINCIPAL ACTIVITIES: Bank

Subsidiary Companies: PSP-Finance Ltd; PSP-Municipality Bank Ltd; Finnish Equity Investment Ltd; Postipankki (U.K.) Ltd; Finnish Credit Card Company Ltd; Mortgage Bank of Finland Ltd; Osakeyhtiö Helsingin Panttilainakontori; PSP-Sampo Fund Management Co Ltd

Financial Information:

	31.12.86	31.12.87
	FMk'000	FMk'000
Profit before tax	394,600	275,100
Profit after tax	335,100	224,700
Reserve fund	384,600	419,800
Own funds and reserves	2,250,700	2,445,000
Deposits	44,593,000	50,106,300
Total assets	56,418,400	63,548,400

Principal Shareholders: The State of Finland (100%)
No of Employees: 7,508

SAMPO INSURANCE COMPANY LTD

Yliopistonkatu 27, PO Box 216, SF-20101 Turku
Tel: +358 (21) 663311
Telex: 62242 sampo sf
Telefax: +358 (21) 665811

Board of Directors: Antti Kataja (Chairman), Heikki Löyttyniemi; Kauko Pihlava, Juhani Kangas
Management: Kauko Pihlava (Managing Director)
Senior Executives: Juhani Kangas, Hannu Kokkonen, Aarne Rossi, Ernst Wendt

PRINCIPAL ACTIVITIES: Insurance of all classes except life; reinsurance
Subsidiary Companies: (100% owned unless stated): Risk Management Ltd; Sampo Finlandia Ltd; Sampo Insurance Co (UK) Ltd (UK); Sampo International Insurance Co Ltd (99.98%); Tarmo Life Assurance Co Ltd (99.98%)
Financial Information:

	31.12.86	31.12.87
	FMk'000	FMk'000
Premium income	1,858,000	1,921,000
Investment income	307,000	327,000
Profit before tax	30,000	35,000
Profit after tax	10,000	15,000
Earnings per share	-	15.00FMk
Guarantee capital	10,000	66,000
Shareholders funds	122,000	183,000

Principal Shareholders: Nearly 710,000 shareholders at 16.5.1988 (the policyholders became shareholders at that date)
No of Employees: 2,241

SKOPBANK

PO Box 400, SF-0010 Helsinki
Tel: +358 0-13341
Cable: Skop; SWIFT: SKOP FI HH
Telex: 122284 skop sf
Telefax: +358 0 133 4896

Board of Managers: Matti Ali-Melkkilä (Chairman, Chief General Manager), Christopher Wegelius (Vice Chairman, Managing Director), Jukka Haapanen (General Manager), Eljas Sukselainen (General Manager, Corporate Banking), Juhani Riikonen (General Manager, Investment Banking), Tarmo Ruosteenoja (General Manager), Kari Wuorisalo (General Manager), Timo Santalainen (General Manager), Kalevi Kauniskangas (Managing Director, The Finnish Savings Banks' Association), Lauri Siltala (Deputy Managing Director, The Finnish Savings Banks' Association)

PRINCIPAL ACTIVITIES: General banking services; central banking and foreign trade banking services for the savings banks in Finland
Trade Names: Skopbank
Subsidiary Companies: The Finnish Real Estate Bank Ltd; Interpolator Ltd; Skop Finance Ltd; SP-Invest Ltd; Skofaks Ltd; FennoScandia Bank Ltd (UK); Banque Nordeurope SA (Luxembourg); Skopbank (Cayman) Ltd; Industrialization Fund of Finland

FINLAND

Financial Information:

	31.12.86 FMk'000	31.12.87 FMk'000
Profit before tax	255,592	562,058
Profit after tax	227,155	480,587
Retained profit	11,714	52,574
Dividends	83,908	111,764
Dividends per share	1.23FMk	1.29FMk
Earnings per share	3.65FMk	5.06FMk
Share capital	682,060	1,000,000
Shareholders funds	1,716,269	2,694,313
Deposits	11,210,876	16,174,186
Total assets	29,168,741	44,587,012

Principal Shareholders: 230 Savings Banks
No of Employees: 1,557

SUOMEN PANKKI-FINLANDS BANK
Bank of Finland
PL 160, 00101 Helsinki
Tel: 0-1831
Cable: Suomenpankki
Telex: 121224; 124446 (Foreign Exchange)
Telefax: 174872

Board of Management: Rolf Kullberg (Chairman), Pentti
Uusivirta, Kalevi Sorsa (Speaker of Parliament, on leave of
absence), Ele Alenius, Markku Puntila (Acting Member), Harri
Holkeri, Seppo Lindblom, Esko Ollila (Prime Minister, on leave
of absence)

PRINCIPAL ACTIVITIES: Central bank of Finland; banking
services; security printing
Financial Information:

	31.12.87 FMk'000	31.12.88 FMk'000
Total assets	43,489,000	53,291,000

No of Employees: 907 (worldwide)

TAPIOLA INSURANCE GROUP (THE)
Revontulentie 7, 02100 Espoo
Tel: 90-4531
Telex: 121073 tapva sf
Telefax: +358 0 4532146

Board of Directors: A Kalpala (Chairman)
Management: A Kalpala (Managing Director)

PRINCIPAL ACTIVITIES: Mutual insurance group
Principal Bankers: All banks in Finland
Financial Information:

	31.12.86 FMk'000	31.12.87 FMk'000
Premium income	2,488,600	2,886,300
Investment income	801,700	856,000
Profit before tax	17,100	19,000
Profit after tax	2,600	2,900
Equity capital	137,800	139,800

Principal Shareholders: Policyholders
No of Employees: 1,565

UNION BANK OF FINLAND LTD
PO Box 868, SF-00101 Helsinki 10
Tel: + 358 0 1651
Cable: UNITAS; SWIFT: UNITFIHH
Telex: 124407 unit sf (general); 124525 unex sf (foreign
exchange); 122161 unep sf (eurobonds); 123794 unicr sf
(international credits)
Telefax: +358 0 1652648 (General)

Board of Management: Mika Tiivola (Chairman and Chief
Executive), Ahti Hirvonen (Chief General Manager and Chief
Operating Officer), Paavo Laitinen (Deputy Chief General
Manager), Kurt Stenvall (Deputy Chief General Manager) Erik
Stadigh (Deputy Chief General Manager), Niilo Pellonmaa

(General Manager), Jaakko Iloniemi (General Manager),
Markku Mikkola (General Manager), Sakari Haukka (General
Manager), Sven-Erik Rindell (General Manager), Björn
Wahlroos (General Manager), Risto Wartiovaara (General
Manager), Eero Pynnönen (Deputy General Manager),
Christer Ekman (Deputy General Manager), Kalevi Kontinen
(Deputy General Manager), Bo Harald (Deputy General
Manager), Ari Laakso (Deputy General Manager), Markku
Pohjola (Deputy General Manager)

International Division Management: Henrik von Fieandt
(Assistant General Manager, International Network and
Administration), Björn Ekman (Assistant General Manager,
Foreign Exchange Division), Harri Hollmén (Assistant General
Manager, International Credits and Capital Markets), Risto
Hynönen (Assistant General Manager, International Trade
Finance and Payments), Kari Kuoppala (Assistant General
Manager, Corporate Division), Jussi Pohls (Assistant General
Manager, Auditing), Fanny Borgström (Assistant General
Manager, International Treasury), Margaretha Einola-Rajamäki
(Assistant General Manager, Capital Markets)

Departments: Area Management: Area 1: Ingmar Silfvast
(Manager), Lena Leskinen (Area Manager USA, Canada,
Great Britain and Ireland), John Lindholm (Area Manager
Scandinavia). Area 2: Jarmo Bergman (Manager Federal
Republic of Germany, Switzerland, Austria), Martin Björkbom
(Area Manager France, Benelux countries, Turkey, Greece,
Italy, Malta, Cyprus and Israel). Area 3: Esa Linna (Manager
Latin America, Spain, Portugal), John Krogius (Area Manager
Africa), Martin Ward (Area Manager, Asia and Australasia).
Area 4: Vesa Turtiainen (Manager), Josef Stulik (Area
Manager, CMEA Countries), Gustav Porko (Manager, Middle
East). Documentary Collections: Svante Alm (Manager).
Documentary Credits: Pekka Wäinölä (Manager). Foreign
Correspondence and Communications: Valerie Vainonen
(Manager), Pirjo Rönkä (Deputy Manager). Foreign Currency
Cashier: Marjatta Rantala. Foreign Exchange Regulations and
Travellers Foreign Exchange: Christian Björkstén (Manager).
International Capital Markets: Petri Siponen (Manager). Cash
Management Systems: Anne-Mari Tyrkkö (Manager).
International Corporate Trading: Maarit Näkyvä (Manager).
International Credits: Kari Kangas (Manager). International
Network Administration: Lars Ehrnrooth. International Sales:
Thomas Begley (Manager). International Securities: Ilkka
Siltainsuu (Assistant Manager). International Trading: Ingmar
Ehrman (Foreign Exchange Manager). Short-Term
International Credits: Tuomo Ropponen (Manager), Hannu
Reilin (Assistant Manager). Short-Term International Funding:
Arn-Johan Grotenfelt (Deputy Manager)

PRINCIPAL ACTIVITIES: All banking activities
Subsidiary Companies: Finnish Corporate Finance Ltd; Union
Bank of Finland International S.A. (Luxembourg); Lease-
Finance Luxembourg S.A. (Luxembourg); Union Bank of
Finland (Singapore) Ltd (Singapore); American Scandinavian
Banking Corporation (New York); Union Bank of Finland
(Bahamas) Ltd; Industrial Bank of Finland Ltd; Finnish Credit
Card Ltd; Customer Finance Ltd; Union Bank of Finland
(France) SA; Unitas Ltd
Financial Information:

	31.12.86 FMk'000	31.12.87 FMk'000
Profit before tax	849,000	1,200,000
Profit after tax	427,000	400,000
Retained profit	61,000	62,000
Dividends	354,000	353,000
Dividends per A share	1.70FMk	1.40FMk
Dividends per B share	1.25FMk	1.10FMk
Dividends per C share	1.70FMk	1.40FMk
Earnings per share	2.78FMk	4.18FMk*
Share capital	2,175,000	2,610,000
Shareholders funds	4,927,000	4,897,000
Deposits	84,835,000	92,766,000
Total assets	104,671,000	114,460,000

*diluted by warrants 4.12FMk

Principal Shareholders: Skandinaviska Enskilda Banken; Suomi
  Salama Mutual Life Insurance Co; Pensions-Varma Mutual
  Insurance Co; UBF Ltd Pension Fund
No of Employees: 10,165

FINLAND

Major Companies of FRANCE

AMERICAN EXPRESS CARTE FRANCE

1 avenue de Chatou, F-92508 Rueil Malmaison Cédex
Tel: (1) 47.32.92.62
Telex: 203412

Management: Claude Jouven (Président Directeur Général)
Senior Executives: Colin Reeve (Informatique et Télématique), Alain Janssen (Marketing et Ventes), Guy Bouclet (Finances), François Cerisay (Ressources Humaines)

PRINCIPAL ACTIVITIES: Financial services including credit facilities
Parent Company: American Express, USA
Financial Information:

	FF'000
Sales turnover	10,000,000

ARBEL, ETS,

46 Rue Cardinet, 75017 Paris
Tel: (1) 46 229339
Telex: 650441 etarbel

Board of Directors: Jean-Marc Blanc (Président-Directeur Général), Lucien Arbel (Directeur Général), Jean-Louis Collin, Henri Corson, Jacques Henry, François Perreau-Saussine, Pascal Porteu de la Morandière
Senior Executives: Jean-Marc Blanc (Président-Directeur Général), Lucien Arbel (Directeur Général), Claude Dollé (Directeur Général)

PRINCIPAL ACTIVITIES: Holding company whose subsidiaries are engaged in the manufacture and repair of railway freight cars of all types; containers; shunting locomotives; pressed steel and moulded plastic products; specialised automotive body making; rental of railway freight cars and trailers
Affiliated Companies: ARBEL-Industrie; ARBEL-FAUVET-RAIL; ARBEL-RAIL; LORMAFER; COFIMETA; SORENOR; AUBRY; ECRIM
Principal Bankers: Société Générale
Financial Information: Consolidated figures

	31.12.86 FF'000	31.12.87 FF'000
Sales turnover	1,157,028	1,297,220
Profit before tax	9,618	24,982
Profit after tax	7,331	18,177
Dividends*	1,183	1,373
Dividends per share*	2.50FF	2.90FF
Earnings per share*	8.04FF	8.99FF
Share capital*	47,339	47,339
Shareholders funds*	85,805	88,760

*Non-consolidated

No of Employees: 2,255

ASSURANCES DU GROUPE PARIS SA
AGP

78 rue de l'Université, F-75007 Paris
Tel: 45.49.18.78
Telex: AGEPE 280638 F
Telefax: 42.82.86.45

Board of Directors: Bernard Pagezy (Président), Etienne Bénézech (Administrateur Directeur Général), Jean-Noël Pébarthe, Jean Lamson, Compagnie du Midi (représentée par Paul Plazonich), Jean-Paul Arnal

PRINCIPAL ACTIVITIES: All classes of insurance
Trade Names: Groupe de Paris
Parent Company: Compagnie du Midi
Subsidiary Companies: La Paternelle Risques Divers; La Paternelle Vie; Seine et Rhône/Océanide Réunies; Prévoyance Mutuelle MACL; Trans Expansion Vie; La Paternal S.I.C.A. (Spain); La Prévoyance SA (Belgium and Luxembourg); Compagnie Générale de Réassurance de Monte-Carlo (CGRM) (Monaco); London & Hull Maritime Insurance Co Ltd (UK); Compagnie Riunite di Assicurazione (Italy)
Principal Bankers: Crédit Parisien
Financial Information:

	FF'000
Premium income	7,400,000

Principal Shareholders: Compagnie du Midi
No of Employees: (In France) 2,560 plus 1,300 agents; (Abroad) 1,000

ASSURANCES GÉNÉRALES DE FRANCE
AGF

87 Rue de Richelieu, 75002 Paris Cedex 02
Tel: 42.44.04.44
Cable: Asgeniart Assuragevie
Telex: 210697 AGF Paris

Board of Directors: Michel Albert (Président), M Bernasconi, M de Carmoy, M Credeville, M Dumon, M Focard, M Gaubert, M Gaymard, M Lobjeois, M Lynen, M Mercereau, M Moisan, M Moreaux, M de Mourgues, M Planet, M Rouanet, Mme Scrivener, M Thévenot
Management: R Papaz (Directeur Général), P Beraud (Directeur Général Adjoint), P Maynie (Directeur Général Adjoint), F Simonnet (Directeur Général Adjoint), L Leuret (Directeur Général Adjoint), J D Le Franc (Directeur Général Adjoint)

PRINCIPAL ACTIVITIES: Insurance
Trade Names: Assurances Générales; Assurances Le Phénix
Subsidiary Companies: (100% owned unless stated): Assurances Générales de France Vie; Assurances Générales de France IART; La Métropole IARD (93.93%); La Célérité (98.37%); AGF Réassurances (99.99%); Générale Informatique et d'Exploitation; Prévention Sécurité Automobile AGF; AGF Reinsurance Corporation of the US (USA); AGF North America Corporation (USA); AGF Re (Luxembourg); SAVAG Saarbrücker Versicherungs AG (62.77%) (Germany); SAVAG Saarbrücker Kranken Versicherungs AG (85.67%) (Germany); City General Insurance CY Ltd (UK); La Première Belge. De Eerste Belgische (Belgium); AGF Seguros (Spain); Fenix Peninsular SA de Seguros (Spain); Omnia SA de Seguros y Reaseguros (Spain); Sociedade Portuguesa de Seguros (88.04%) (Portugal); AGF Hellas (80%) (Greece); Cosmos AG (80%) (Greece); Phenix Compagnie d'Assurances sur la Vie (Switzerland); Phenix Compagnie d'Assurances (Switzerland); Assurances Générales Sénégalaises (51.38%) (Senegal); Assurances Générales de Côte-d'Ivoire (79.95%) (Ivory Coast); Atlantic and Continental Assurance CY of South Africa Ltd (95.59%) (South Africa); Brasil Seguros (49.73%) (Brazil); AGF Hong-Kong Ltd (85%) (Hong Kong); Société Française d'Assurance pour Favoriser le Crédit (44.83%); Plus Allgemeine Versicherungs AG (50%) (Germany); Phenix Soleil Vita (50%) (Italy); Phenix Soleil SPA (25%) (Italy); La Protectrice AIAR (20.52%); Sacnas (Sté d'Assistances Internationales) (24.91%); Société Française de Protection Juridique (35.13%); Schiedam (NV Verzekering Maatschappij) (38.43%) (Netherlands); Church and General Insurance CY Ltd (27.28%) (Eire); Union Générale des Assurances du Niger (39.99%) (Niger); Compagnie Africaine d'Assurances (49%) (Morocco); Astrée (26%) (Tunisia); Seguros La Territorial SA (32.46%) (Mexico); La Fourmi Immobilière (61.5%); Phenix Pierre Sicomi (97.5%); SA du Château Larose Trintaudon (99.96%); Groupement pour le Financement de la Construction (33.33%); Socim (27.07%); Compagnie Lyonnaise Immobilière (38.42%); Simnor (41.71%); Imminvest (44.55%); Banque Générale du Phénix(88.75%); Caisse d'Epargne AGF (Belgium); Compagnie Générale des Voitures (40.17%); The Insurance Group of Canada West Ltd (35%) (Canada); Agiart do Brasil (99.99%) (Brazil)
Principal Bankers: Banque Générale du Phenix; Société Générale

FRANCE

Financial Information:

	FF'000
Premium income	25,000,000

Principal Shareholders: French State (75.01%)
No of Employees: 12,248 France

ASSURANCES MUTUELLES AGRICOLES, CAISSE CENTRALE DES,

8-10 rue d'Astorg, 75413 Paris Cedex 08
Tel: (1) 42.96.77.77
Cable: AGRIMUTEL PARIS 123
Telex: AGMUT B 650 915 F
Telefax: (1) 42.96.72.74

Board of Directors: André de Bretteville (Président)
Management: André Mouly (Directeur Général), Jacques Lasne
(Directeur Général Adjoint), Régis Belhomme (Directeur
Général Adjoint), Gilles Laporte (Directeur)
Senior Executives: Régis Belhomme

PRINCIPAL ACTIVITIES: Insurance
Trade Names: GROUPAMA
Parent Company: Caisse Centrales des Mutuelles Agricoles
Subsidiary Companies: SAMDA; SORAVIE; SOREMA
Financial Information:

	1986	1987
	FF'000	FF'000
Premium income	11,314,888	11,902,133

No of Employees: 7,600

AVIS LOCATION DE VOITURES

Tour Franklin, F-92081 La Défense cédex 11
Tel: (1) 47.76.42.21
Telex: 612340

Board of Directors: Rolf Ruegg (Président Directeur Général)
Senior Executives: Jean-Paul Danon (Sogenavis), Pierre Jaeger
(Opérations), Gérard Maitre (Marketing et Commercial), Harry
Staiger (Finances)

PRINCIPAL ACTIVITIES: Car hire and rental services
Parent Company: Avis Inc, USA
Financial Information:

	FF'000
Sales turnover	1,300,000

AXA

40 Rue du Colisée, 75008 Paris
Tel: (1) 40.75.59.00
Telex: 642639 F

President: Claude Bébéar
Directors: P Garnier, J-C Aaron, H Hottinguer
Senior Executives: C Huot, L Brossier, P Gardes, V-C Rosset, C
Tendil

PRINCIPAL ACTIVITIES: All classes of insurance
Parent Company: Mutuelles Unies
Subsidiary Companies: Axiva; Drouot Assurances; La Nouvelle
Mutuelle; Mutuelle Parisienne de Guarantie; Mutuelles Unies
Assurances; Mutuelle Phocéenne; Présence Assurances;
Mutuelle de l'Quest; Mutuelle Saint Christophe. AXA
International: AXA Belgium (Belgium); Assurlux (Luxembourg),
Mare Nostrum (Spain); AXA Canada (Canada); Provinces
Unies (Canada); Anglo Gibraltar Insurance Group (Canada);
Gamma Ré (USA); Gamma Reinsurance Company (USA);
Westgate Insurance (UK); Amré; Star; Tene Reinsurance
Company (Luxembourg)
Principal Bankers: AXA Banque

Financial Information:

	31.12.86	31.12.87
	FF'000	FF'000
Premium income	19,774,000	22,339,000
Investment income	4,920,000	4,040,000
Profit before tax	1,584,000	1,530,000
Profit after tax	1,267,000	1,083,000
Shareholders' equity*	5,716,000	7,054,000

*Group and Minority interests

Principal Shareholders: Mutuelles Unies
No of Employees: 8,284

BAIL EQUIPMENT SA

22 Place Vendôme, 75001 Paris
Tel: 40.20.45.45
Telex: 210980 baleki

Board of Directors: François de Lamothe Dreuzy (Président),
Crédit Industriel et Commercial de Paris (représentée par F
Arsac), Louis Decoster, Claude Feron, Amaury Daniel de
Seze, Claude Delamotte, L'Union des Assurances de Paris-
Vie (représentée par Jean-François Gouzon), Compagnie
Financière de Suez (représentee par Michel de la Brosse),
L'Abeille Paix-Vie (représentée par Henri Katz), Les
Assurances Générales de France-Vie (représentées par Henri
Katz) Soparcic (représentée par B Madinier)
Management: François de Lamothe Dreuzy (Président-Directeur
Général), Daniel Voillereau (Directeur Général), Marc Picard
(Directeur-Général Adjoint), Gilles Stehelin (Directeur Général
Adjoint), Guilhem d'Elissagaray (Directeur Général Adjoint),
Thierry Trebert (Directeur Général Adjoint)

PRINCIPAL ACTIVITIES: Finance house leasing capital
equipment
Trade Names: B.E.
Parent Company: C.I.C. de Paris
Subsidiary/Associated Companies: (100% owned unless stated):
Bail Gestion SA; Bail-Auto SNC (99%); Technicredit SA (99%);
Factobail (51%); FAC Finance Assurance Conseil SARL
(51%); INNOBAIL SA and its subsidiaries:- Technibail SA;
Innofit SA (20%) (Italy); Locmabi SA (49%); Kienzle Bail SA
(25%); J.C.B. Flexi-Bail SA (60%); Ericsson Location SA
(33%); V.I.V. Location SA (45%)
Principal Bankers: Crédit Industriel et Commercial; Banque
Nationale de Paris; Société Lyonnaise de Banque
Financial Information:

	31.12.86	31.12.87
	FF'000	FF'000
Sales turnover	2,051,214	2,427,101
Profit before tax	48,040	51,084
Profit after tax	27,229	27,554
Dividends	25,867	26,177
Dividends per share	17.50FF	17.50FF
Earnings per share	18.48FF	18.42
Share capital	147,307	149,583

Principal Shareholders: Crédit Industriel et Commercial; Volvo
France; A.G.F.; G.A.N.; U.A.P.
No of Employees: 435

BANQUE COMMERCIALE POUR L'EUROPE DU NORD

Eurobank

79/81 bld Haussmann, F-75382 Paris 08
Tel: 42.66.92.80
Cable: Eurobank Paris
Telex: 280200 eurob x

Board of Directors: B Dupuy (Président Directeur Général), Iouri
Poletaev (Vice President Directeur Général), Banque d'Etat de
l'U.R.S.S.-Moscow, Banque Pour le Commerce Extérieur de
l'U.R.S.S.-Moscow, M Boter

Management: Vladislav Soudakov (Directeur Général), Pierre Lacoste (Directeur Général Adjoint), Henri Roche (Directeur Général), Paul Rey (Directeur Général Adjoint)

PRINCIPAL ACTIVITIES: Banking activities

No of Employees: 330

BANQUE DE FRANCE
39 Rue Croix-des-Petits-Champs, BP 140-01, 75049 Paris Cedex 01
Tel: 42.92.42.92
Cable: Banfra Paris
Telex: 220932

Governor: J de Larosière
General Council: J Waitzenegger (Deputy Governor), P Lagayette (Deputy Governor), R Lion, J M Floc'hlay, G Joulin, G Bonin, H Prevot, Y Roland-Billecart, F Morin, M Estager
Senior Executives: D Ferman (Secrétaire Général)

PRINCIPAL ACTIVITIES: Central Bank
Financial Information:

	31.12.87 FF'000	31.12.88 FF'000
Share capital and reserves	3,130,000	3,332,000
Total assets	367,857,000	787,366,000

Principal Shareholders: French State
No of Employees: 16,987

BANQUE DE L'UNION EUROPÉENNE SA
4 Rue Gaillon, 75107 Paris Cedex 02
Tel: (1) 42 66 70 00
Cable: Norebank Paris
Telex: 210-942 BUE
Telefax: (1) 42 66 78 90

Board of Directors: Paul Alibert (Président Directeur Général), Jean-Paul Dessertine (Directeur International)
Management: Paul Alibert (Président Directeur Général), Paul d'Abzac (Directeur Général Adjoint), Gilles Barnaud (Secrétaire Général), Christian Perrin (Inspecteur Général), Jean-Pierre Antoine (Directeur de l'Exploitation, de l'Organisation et de l'Informatique), Gilles Delapalme (Directeur de la Trésorerie et des Changes), Jean-Paul Dessertine (Directeur International), René Ehrmann (Directeur Financier), Jean Lacroix (Directeur des Participations), Jacques Laurioz (Directeur du Personnel), Jean-Claude Wagner (Directeur Bancaire)
Senior Executives: Attaches a la Présidence: Christian de Longevialle (Vice-Président Directeur Général de l'Union Industrielle de Credit), Jean-Marie Guehenno (Conseiller du Président), Pierre de Lauzun (Conseiller du Président)

PRINCIPAL ACTIVITIES: Bank
Financial Information:

	31.12.86 FF'000	31.12.87 FF'000
Deposits	31,556,614	39,678,285
Total assets	38,483,160	53,727,448

Principal Shareholders: The French State through Crédit Industriel et Commercial Group
No of Employees: 1,404

BANQUE FEDERATIVE DU CREDIT MUTUEL
See CREDIT MUTUEL, BANQUE FEDERATIVE DU,

BANQUE FRANÇAISE DU COMMERCE EXTERIEUR
BFCE
21 Blvd Haussmann, 75009 Paris
Tel: 42.47.47.47
Cable: Extecomexe
Telex: 660370 f

Board of Directors: Michel Freyche (Président), Rémi Baudin, François Blamont, Claude Bondonneau, Daniel Cheyroux, Jacques Delmas-Marsalet, Bernard Duhamel, Laurence Galbe, François Guillaume, Gilles Guitton, François Lagrange, Robert Raymond, Bernard Thiolon, Marc Tequi. Censors: Jacques Wahl, Marc Vienot. Government Representative: Denis Samuel-Lajeunesse (Chef de Service-Affaires Internationales Direction du Tresor, Ministère de l'Economie, des Finances et de la Privatisation). Government Co-Representative: Serge Allain. Auditors: B.D.A. Paris, Bernard Boiton, Philippe Vassor (Substitut)
General Management: Michel Freyche (Président), Pierre Antoni (Directeur Général), Francis Gavois (Directeur Général Adjoint), Patrick Haizet (Directeur Central), Louis Flaive (Directeur Central), Patrick Mesmin des Vaux (Directeur), Pierre Nadal (Directeur), Jean-Louis de Roux (Directeur), Erik Lescar (Directeur Adjoint International)

PRINCIPAL ACTIVITIES: Banking activities
Principal Bankers: All principal banks
Financial Information:

	31.12.86 FF'000	31.12.87 FF'000
Deposits	14,833,173	241,194,375
Total assets	227,832,403	350,379,110

Principal Shareholders: Banque de France (24.55%); CDC-Participations (Caisse des Dépôts et Consignations) (24.55%); Crédit National (10%); Caisse Nationale de Crédit agricole (10%); Banque nationale de Paris (10%); Crédit Lyonnais (10%); Scoiété Générale (10%); Caisse centrale de coopération économique (0.9%)
No of Employees: 2,947

BANQUE INDOSUEZ
96 Blvd Haussmann, 75008 Paris
Tel: 45.61.20.20
Cable: INDOSUEZ PARIS
Telex: 650409 INSU
Telefax: 45.61.29.56

Board of Directors: Antoine Jeancourt-Galignani (Chairman and Chief Executive Officer), Renaud de la Genière (Vice Chairman), Max Benhamou, Gérard Billaud, Bernard Bréard, Philippe Chareyre, Jacques-Henri David, Bernard Egloff, Philippe Hustache, Alfonzo Jozzo, Georges Pebereau, André Ross, Guy Touche
Executive Board: Jean-François Lepetit (President and Chief Operating Officer); Philippe Geslin (President and Chief Operating Officer), Georges Mazaud, Bernard Simon-Barboux, Pierre Strub, Emmanuel Vasseur, Alain de Korsak, Patrice Mignon

PRINCIPAL ACTIVITIES: Banking
Parent Company: Compagnie Financière de Suez
Subsidiary/Associated Companies: (100% owned unless stated): Consolidated companies: American Transit and Finance Corp Inc (Panama); Banque Indosuez Belgique (60.02%) (Belgium); BFC Holding (98.64%); BFC Antilles Guyane; BFC Océan Indien; Banque Indosuez Finlande (80%) (Finland); Banque Indosuez Italia (Italy); Banque Indosuez Luxembourg (Luxembourg); Banque Indosuez Mer Rouge (99.99%) (Djibouti); Banque Indosuez Sverige (80%) (Sweden); Banque Indosuez Vanuatu (79.99%) (Vanuatu); Banque Libano-Française (95%); Banque de Suez Nederland NV (90%) (The Netherlands); Brizay Realty Private Ltd (Singapore); Caisse de Gestion Mobilière (70.59%); Cie Française de l'Asie "CFA" (99.97%); Cie Générale de Crédit "COGENEC" (99.88%) (Monaco); Cie Parisienne de Gestion Financière "COPAGEFI" (99.94%); COPAGEFI SNC (99%); Crédit Calédonien et Tahitien (82.02%) (New Caledonia); Crédit Foncier Immobilier de la Nouvelle-Calédonie et Polynésie (93.32%) (New Caledonia); Crédit Foncier de Monaco (71.84%) (Monaco); Crédit de Monaco pour le Commerce (91.87%) (Monaco); Financière Indosuez Genève (98.27%) (Switzerland); Financière Indosuez (ex-Fireg); Finanziaria Indosuez (71.99%)

(Italy); Fininvest (95.01%); French Bank of Southern Africa (54.60%) (South Africa); French South African Investment Trust (South Africa); Immobanque (38.43%); Indoniu Finance Co Ltd (51%) (Papua-New Guinea); Indosuez Asia Investment Services Ltd (Hong Kong); Indosuez Asia Ltd (Hong Kong); Indosuez Asia Singapore Ltd (Singapore); Indosuez Asia Vanuatu Merchant Bank Ltd (Vanuatu); Indosuez Australia Ltd (Australia); Indosuez Bank International (USA); Indosuez Beteiligungs und Verwaltungs GmbH (Germany); Indosuez Handels GmbH and Co KG (Germany); Indosuez Finance Hong Kong Ltd (Hong Kong); Indosuez Finance UK Ltd (99.99%) (UK); Indosuez Italia Holding (73.77%) (Italy); Indosuez Luxembourg Holding GmbH (Germany); Indosuez New Zealand Ltd (New Zealand); Indosuez Niugini Bank Ltd (49%) (Papua-New Guinea); Indosuez Norge (80%) (Norway); Indosuez Overseas Finance Ltd (Panama); Locafrance (49.18%); Fiat Bail Industries (ex-Cefimat) (49.99%); Credimo (68.08%); Ergifrance (52.50%); Financière de Transport et de l'Industrie (59.99%); Locafrance International Leasing (Switzerland); Localease Mietfinanzierung (Germany); Locamur Sogigros (51.43%); Locasuez Asia (Hong Kong); Ordinabail; Marcard, Stein und Co (83.28%) (Germany); Maderox (99.72%); Mescas (99.72%); Société Etude et Participations Minières de la France d'Outre-Mer "SEPAMIFOM" (99.74%); Société Foncière et Financière de la Ville l'Evêque (99.99%); Suez Asia Development Capital (50%) (Netherlands); Nepal Indosuez Bank (50%) (Nepal); Groupe Union Financière de France (52.19%); W.I. Carr (Overseas) Holding (Hong Kong); W.I. Carr (UK) Holding (UK). Companies accounted for by the equity method: Al Bank Al Saudi Al Fransi BSF (40%) (Saudi Arabia); Bapindo Loka Sentra Leasing (30%) (Indonesia); Burofinance (34%); Cie Parisienne de Réescompte (18.91%); Paresco Gestion et Cie (99.95%); C.P.R. Maubeuge; Fimagest (22.5%); Malaysian French Bank Berhad (30%) (Malaysia)

Financial Information:

	31.12.86 FF'000	31.12.87 FF'000
Profit before tax	1,443,032	1,351,231
Profit after tax	827,481	867,907
Retained profit	625,053	491,735
Dividends	167,760	198,120
Dividends per share	14.94FF	14.94FF
Earnings per share	70.51FF	53.07FF
Share capital	1,275,000	2,275,000
Reserves	3,888,397	4,123,206
Shareholders funds	6,463,397	8,198,206
Deposits	210,251,873	198,583,510
Total assets	260,604,000*	255,120,962

*Securities and bills sold firum under repurchase agreements are included in these figures but are not deposits according to Banque Indosuez

Principal Shareholders: Compagnie Financière de Suez
No of Employees: 12,745

BANQUE INTERNATIONALE POUR L'AFRIQUE OCCIDENTALE

BIAO
9 Ave de Messine, BP 8908, F-75360 Paris Cedex 8
Tel: (1) 42 99 80 20
Telex: 650339 afrbk

Board of Directors: Antoine d'Arjuzon (Président Directeur Général)
Management: Antoine d'Arjuzon (Président Directeur Général), Jacques Bayle (Directeur Général Adjoint), Gérard Larrera de Morel (Directeur Général Adjoint), Jacques Chanard (Secrétaire Général)

PRINCIPAL ACTIVITIES: Banking
Parent Company: COFIFA

Subsidiary Companies: BIAO (Niger); Banque Internationale pour l'Afrique au Tchad (Chad); Banque Internationale pour l'Afrique au Zaïre (Zaire)

BANQUE NATIONALE DE PARIS

BNP
16 Blvd Des Italiens, 75009 Paris
Tel: (1) 40.14.45.46
Cable: Natiopar
Telex: 280605; 280775

Board of Directors: Henry Bizot (Honorary Chairman), Pierre Ledoux (Honorary Chairman), Pierre Calvet (Honorary Vice Chairman), René Thomas (Chairman), Jean-Louis Beffa, Gilles Bellec, Robert Bézias, Jean-Luc de Boissieu, Jean-Marie Chevalier, Léon Cligman, Jacques Dermagne, Jean-Paul Elkann, Daniel Lebègue, Alain Limouzin, Jean Louis, Pierre Marcolini, M Margulio, Jean-Pierre Moussy, Michel Prada, Philippe Remond, Jean-René Sautier
General Management: René Thomas (Chairman and Chief Executive Officer), Jacques Masson (President and Chief Operating Officer), Jacques Henri Wahl (President and Chief Operating Officer), Pierre-Yves Cossé (Senior Executive Vice President Corporate Business and Development), Jean Gagné (Senior Executive Vice President National Network), Emmanuel Philippon (Senior Executive Vice President International Affairs), Claude Reinhart (Senior Executive Vice President Investment Banking, General Accounting and Budgeting), Bernard Benoist-Lucy (Executive Vice President for General Supervision), Raymond Lambert (Executive Vice President Human Resources and General Administration)

PRINCIPAL ACTIVITIES: All banking activities
Subsidiary Companies: (100% owned unless stated): Banks and Financial Institutions: French Companies: Banexi; Natiolocation; Natioéquipement; Natiocrédibail; Natiocrédimurs; Banque Natiotrésorerie; Société Fincière Auxiliare "SFA"; Compagnie d'Investissements de Paris "CIP"; Compagnie du Crédit Universel (Group) (69.18%); Natioénergie; Scintex; Codexi; Banque Nationale de Paris Intercontinentale "BNPI" (65.27%); BNP Guyane (79.96%); BNP Nouvelle Calédonie. Foreign Companies: Europe: BNP (Luxembourg) SA; BNP plc; BNP Ireland; BNP Bank NV; BNP Norge A/S (75%); BNP Sverige; BNP SA & Co Deutschland OHG; BNP Suisse (94.35%); BNP España (78.61%); Banque Multicommerciale (78%); Banacomi; Giveco (60%); BNP Leasing SpA Italie. Africa: BICI du Niger. America: FABC; Bank of the West; BNP Canada; BNP US Finance Corporation; Banco Panama SA (91.31%); Natioleasing Overseas Corporation. Asia-Far East: BNP IFS Hong-Kong; BNP IFS Singapore; BNP Finance Hong-Kong; BNP Vila Ltd; BNP Pacific Australia. BNPI Subsidiaries: BCI Mer Rouge (51%); BMCI (50%); UBCI (50%). Other Subsidiaries: French Companies: Logexi; Société Auxiliaire de Participation et de Gestion "SAPEG"; Natioinformatique. Foreign Companies: Asia-Far East: 12 Castlereagh Street Pty; 90 William Street Pty. Affiliates carried on an Equity Basis: French Companies: Natiobail (33.5%); UFCA (22.77%); Natio Innovation (23.94%); Finovectron (24.66%); Investexport; Compagnie Française d'Équipement Automobile "CFEA"; Cie Financière France-Afrique "COFIFA"; (and, by sub-consolidation, BIAO) (41.89%); Intercomi; SOREFI (22.5%); FIREM (22.5%). Foreign Companies: Europe: UOB Holding (and, by sub-consolidation UOB Genève) (40.77%); Eurofip (40%); Crédifimo Madrid; BNP AK Bankasi AS (49%); Eurofinanceira (45%); Leasinvest Portugal (22.5%). Africa: BICI Côte d'Ivoire (and, by sub-consolidation, Cofinci) (34.44%); BICI Bail Côte d'Ivoire (42.35%); BICI Sénégal (35.59%); BICI Cameroun (23.36%); BICI A du Burkina (33.04%); UBA Nigeria (30.7%); SPGC Congo (71.33%); BICI du Gabon (37.47%); BTCI Togo (40.41%); SFOM (and, by sub-consolidation, Banque Commerciale du Burundi) (48%). Middle East: Banque du Caire et de Paris (49%); Arab European Finance (Aref) (Kuwait) (22%). America: BNP International Corp; BNP IFS

New York; BNP Venture; Banco Cidade SA Brésil (45%); Francofin Costa Rica (55.56%); Natioleasing Cayman Ltd; Francofin Panama. Asia-Far East: Oriental Bank Berhad (22.5%); Sofidema Macao (33.33%);Pt BNP Lippo Utama Leasing (70%); China International Non Ferrous Métal Leasing Cy (20%). Other Subsidiaries: French Companies: Natio-Vie; Assu-Vie (50%); Cimoxi; Meunier Promotion (Group); Orgepro; Compagnie Immobilière de France "CIF"; Société Française du Chèque de Voyage (21%). Foreign Companies: Europe: Société de Réassurance de la BNP BNP RE Luxembourg; BNP Capital Markets

Financial Information: Consolidated

| | 31.12.86 | 31.12.87 |
	FF'000	FF'000
Profit after tax	3,169,456	2,834,745
Dividends per share	11.00FF	11.00FF
Share capital	2,672,000	2,939,590
Total assets	915,775,587	975,483,870

Principal Shareholders: The French Government
No of Employees: 58,623 BNP Group

BANQUE SOFINCO
7 Rue du Bois de Boulogne, 75116 Paris
Tel: (1) 60 76 36 36
Telex: 692616 F SOFEVRY
Telefax: 60 76 37 94

Administrative Board: Thierry de La Bouillerie (Président), Michel Ancelle, Jean Arvis, Gérard Billaud, Pierre Delage, Bernard Egloff, Michel Hemar, Mme Christine Morin-Postel, Philippe Pontet, Mme Catherine Thiriet
Management: Thierry de La Bouillerie (Président, Directeur Général), Christian Maurin (Directeur Général), Jean-Paul Beti (Directeur Général Adjoint), Jean Michel Theron (Directeur Général Adjoint), Georges Juras (Délégué Général pour les Affaires Internationales), Henri Jaxel (Directeur Central), Dominique Landreau (Directeur Central), Hubert Le Pelet (Directeur Central)

PRINCIPAL ACTIVITIES: Banking; consumer credit; personal loans. Subsidiaries have interests in the leasing of cars and also joint venture societies have been created with Fiat France and BMW France for hire purchase of vehicles
Parent Company: Compagnie Financière de Suez; Credisuez
Subsidiary Companies: SOFINROUTE; SOFINLOC; EDA; SOFINREC; SOFINCAR; SOFINCO LOCATION; FIAT Crédit France; Fiat Lease Auto; BMW Credit; BMW Lease
Financial Information:

| | 31.12.86 | 31.12.87 |
	FF'000	FF'000
Profit before tax*	314,453	247,135
Profit after tax	74,722	86,697
Retained profit	30,161	45,657
Dividends and rents	34,561	41,040
Dividends per share	15.60FF	18.00FF
Earnings per share	33.69FF	38.02FF
Share capital	285,000	369,300
Deposits	691,175	695,000
Balance sheet total	11,257,233	14,240,790
*and before provisions and amortisements

Principal Shareholders: Credisuez (99.9%)
No of Employees: 1,478

BANQUE SUDAMERIS
4 rue Meyerbeer, BP 251-09, F-75009 Paris
Tel: (0) 45 23 70 90
Telex: 641669

Board of Directors: Gustave Rambaud (Président Directeur Général)
Management: Gustave Rambaud (Président Directeur Général), Tommaso Taddonio (Directeur Général), Robert Génin

(Directeur Central), Alberto Abelli (Directeur Central), Claude Jean Panel (Directeur Central)
Senior Executives: Alberto Abelli (International et Marchés), Laurent Dufau (Action Commerciale), Pierangelo Piperno (Euro Emissions, Valeurs Mob), Natale Arculeo (Affaires Internationales), Costante Penacchio (Administration), Patrick Barsot (Change et Trésorerie), Fausto Dall'o (Personnel), Alberto Abelli (Relations Bancaires)

PRINCIPAL ACTIVITIES: Banking

BANQUE WORMS
Le Voltaire Cedex 58, 92059-Paris La Defense
Tel: (16-1) 490750 50

Executive Committee: Jean-Michel Bloch-Lainé (Chairman and Chief Executive Officer), Didier Renaudin (Deputy General Manager), Marc Vuillermet (Deputy General Manager), François Blanchard, Patrice Borghetti, Christian Haas, Claude Hernandez, Charles Sposito, Henri Debourdeaux (General Management Adviser)

PRINCIPAL ACTIVITIES: Commercial and merchant bank
Subsidiary/Associated Banks and Financial Institutions: (100% owned unless stated): Banque de l'Union Occidentale (BUO) (78.28%); Banque de la Mutuelle Industrielle (BMI) (86.58%); Société pour favoriser l'accession à la Propriété Immobilière (SOFAPI) (81.08%); Union pour le Financement du Cinéma et de l'Audiovisuel (UFCA) (27.34%); Concorde Location (20%); Robert Lefevre SA (29.96%); Banque de Financement et d'Investissement SA (BFI SA) (85%) (Switzerland); BW Capital Corp (USA); Worms Finanziaria (80%) (Italy); Losango SA-Crédito Financiamento e Investimentos (10%) (Brazil); Immobilière Saint-Martin (99.98%); Société Financière de Développement (SOFINAD) (99.99%); West Constructions (99.68%); ACMER (94.74%); Société d'Analyse et de Gestion (SAG); Société Parisienne Raveau Cartier (SPCR) (21.9%); Union Financière d'Investissements (UFI) (79.95%); SOPROMEC (51.44%); Financière de Développement et de Conseil (FIDECO) (22.86%)
Financial Information:

| | 31.12.86 | 31.12.87 |
	FF'000	FF'000
Profit before tax	430,000	308,000
Profit after tax	51,000	58,000
Profit per share	11.35FF	12.91FF
Share capital	450,000	450,000
Customer deposits	16,570,000	15,253,000
Group total assets	46,172,000	49,430,000

Principal Shareholders: Union des Assurances de Paris (UAP)
No of Employees: 1,748

BANQUES POPULAIRES, GROUPE DES,
131 Ave de Wagram, 75017 Paris
Tel: (1) 47.54.54.00
Telex: 290457 CSBPCP
Telefax: (1) 47.54.54.02

Board of Directors: Jean Martineau (Président), Jean Valentin (Vice Président), Jean Desvergnes (Vice Président), Jean Steiner (Vice Président), Guy Buniet (Secrétaire), André Bouyssié, Pierre Dap, Georges Debant, Bruno Desjonquères, Georges Doittau, Dominique Dussart, Paul Paclot, Jacques Peyre, Gilbert Pouillen, Pierre Prenaud, Bertrand Quatre

PRINCIPAL ACTIVITIES: Bank
Subsidiary Companies: (100% owned unless stated): FRUCTIBAIL (98.5%); FRUCTICOMI (50.66%); La Prospérité (99.99%); SOFINGEST (Société Financière d'Investissement et de Gestion) (57.98%); SPAFIN (Société pour la promotion des Activités Financières) (99.99%); BMF (Banque Monétaire et Financière) (10%); BPFD (Banque Populaire Fédérale de Développement) (11.95%); COFIMAGE (17.46%); FRUCTIVIE (22.1%); SFF (Société Française de Factoring) (10%)

Financial Information:

	31.12.86 FF'000	31.12.87 FF'000
Profit after tax	832,000	992,000
Share capital	4,905,000	6,251,000
Shareholders equity	9,225,000	11,012,000
Deposits	138,035,000	156,638,000
Total assets	252,808,000	287,063,000

No of Employees: 28,445

BIS, SOCIÉTÉ HOLDING, SA

26-28 Rue de Madrid, F-75008 Paris
Tel: (1) 42.93.50.44
Telex: BISSIEG 641725F

Board of Directors: Laurent Négro (Président)
Management: Norbert Thevenot (Directeur Général Adjoint), Jean-Louis Chaussat (Directeur Général Adjoint), Marie Gestin (Secrétaire Général)
Senior Executives: Francis Jacob (Finances), Danielle Kappler (Information)

PRINCIPAL ACTIVITIES: Employment agencies offering temporary posts, training services, information services, debt collection; banking services
Trade Names: BIS, PIGIER, TMG, CTR, L'Appel Medical, Wells
Subsidiary/Associated Companies: BIS France; TMG Sarl; Pigier SA; SCRL; L'Appel Medical; Banque Pommier; BIS Service; Wells Management Corp
Principal Bankers: Crédit Lyonnais; BNP; CCF; BRED; Société Générale; Banque Hervet
Financial Information:

	FF'000
Group sales turnover	3,000,000

Principal Shareholders: Laurent Negro (31%); Société Civile de Gestion de Valeurs Mobilières (39%)
No of Employees: 2,653 Group

CAISSE CENTRALE DES ASSURANCES MUTUELLES AGRICOLES

See ASSURANCES MUTUELLES AGRICOLES, CAISSE CENTRALE DES,

CAISSE NATIONALE DE CRÉDIT AGRICOLE

See CRÉDIT AGRICOLE, CAISSE NATIONALE DE,

CCBP

Caisse Centrale des Banques Populaires

115 Rue Montmartre, F-75002 Paris
Tel: (1) 42 96.15.15
Cable: Caicentrac Paris
Telex: 210993

Administrative Board: Jean Martineau (President), Francis Allix, Jacques Bagard, Louis Boinot, Georges Debant, Bruno Desjonquères, Jean Desvergnes, Georges Doittau, Philippe Neuhoff, Gilbert Pouillien, Jean Steiner, Jean Valentin
Management: Paul Loriot (General Manager); J-L Martinolle (Deputy General Manager), J C Boyer (Deputy General Manager), J Fourcail (Deputy General Manager), O Haertig (Secretary General)

PRINCIPAL ACTIVITIES: Banking activities
Subsidiary Companies: (100% owned unless stated): CMCI (Caisse Mobilière Commerciale et Industrielle (78.63%); FRUCTICOMI (SICOMI) (50.66%); SCIM (Société Civile Immobilière Montmartre) (82.72%); SOFINGEST (57.98%); SPAFIN (Société pour la Promotion des Activités Financières). Associated Companies: Banque AUDI (France) (25%); BMF (Banque Monétaire et Financière (10%); BIGT (Banque Internationale de Gestion et de Trésorerie) (24.97%); BPFD (Banque Populaire Fédérale de Développement) (10.003%); SFF (Société Française de Factoring) (18.936%)
No of Employees: 1,400

COMPAGNIE BANCAIRE

5 Ave Kléber, 75116 Paris
Tel: (1) 45.25.25.25
Telex: 610 750

Supervisory Board: Jacques de Fouchier (Honorary Chairman), Jean Yves Haberer (Chairman), Jacques Brunet (Vice Chairman), Pierre Bazy, Philippe Dulac, Michel François-Poncet, Roger Goetze, Jean-Maxime Lévêque, Bruno de Maulde, Claude Pierre-Brossolette, René Thomas, Marc Viénot
Management Board: André Lévy-Lang (Chairman of the Management Board), Bertrand de Bardies (Chairman, Cardif and Cortal), Jean-Marie Bossuat (Chairman, UFB and Locabail), Pierre Boucher (Chairman, Cetelem and Cofica), Henri Charrière (Chairman, UCB and CFEC), Roland Rainaut (Company Secretary), Philippe Rousselle (Chairman, Sinvim and Segece)
Executive Officers: Division Heads:- Jacques Lesigne (Financial Operations), Gérard Senouillet (International), Alain Roussel (Personnel), Jean Fouasse (Legal and Taxation), Maurice Alpert (Finance and Accounting), Guy Laureau (Treasury), François Henrot (Planning and Development), Philippe Bioud (Data Processing and Systems), Alain Turland (Controller), Pierre Maumy (Technical and Security), Jean-Louis Tonnerre (Administration), Véronique Guillot-Pelpel (Communications). Secretary to the Management Board and Supervisory Board: Jacques Lesigne (Financial Information)

PRINCIPAL ACTIVITIES: Housing and property finance; consumer finance; small and medium sized business equipment finance, real estate development, real estate finance, data processing services
Subsidiary/Associated Companies: (100% owned unless stated): UFB Locabail (66%); Cetelem (68%); Cofica; Cofibail; UCB (33%); CFEC; Locabail Immoblier (33%); Sinvim (30%); Sinvim Construction; Segece Cardif (56%); SIS; Foncière Compagnie Bancaire; Financière Kléber; Kléber portefeuille
Financial Information:

	31.12.86 FF'000	31.12.87 FF'000
Turnover-new loans	48,050,000	58,500,000
Net operating income(1)	670,100	708,000
Retained profit	552,800	567,500
Dividends	117,300	140,900
Dividends per share(2)	8.3FF	10.0FF
Earnings per share(2)	47.3FF	49.6FF
Share capital	1,173,000	1,409,000
Shareholders equity	10,082,000	10,866,000
Total assets	139,198,000	160,300,000

(1) attributable to shareholders of Compagnie Bancaire
(2) after adjustment

No of Employees: 7,447

COMPAGNIE DE NAVIGATION MIXTE

See NAVIGATION MIXTE, COMPAGNIE DE,

COMPAGNIE FINANCIÈRE DE PARIBAS

See PARIBAS, COMPAGNIE FINANCIÈRE DE,

COMPAGNIE FINANCIÈRE DE SUEZ

See SUEZ, COMPAGNIE FINANCIÈRE DE,

COMPAGNIE FINANCIÈRE DU GROUPE VICTOIRE

52 rue de la Victoire, 75009 Paris
Tel: (1) 42.80.75.75
Telex: 650400; 280093
Telefax: 45.26.63.54

Board of Directors: Robert Gachet (Président d'Honneur), Philippe Chareyre (Président Directeur Général), Michel Marchal (Vice-Président)

Senior Executives: Philippe Chareyre (Président Directeur Général), Jean Arvis (Administrateur Directeur Général), Henri Katz (Directeur Général Adjoint)

PRINCIPAL ACTIVITIES: Insurance

Parent Companies: Compagnie Industrielle, Compagnie Financière de Suez

Subsidiary Companies: Abeille Assurances; Abeille Vie; Abeille Reassurances

Principal Bankers: Banque Indosuez; B.I.C.M.

Financial Information:

| | 1987 | 1988 |
	FF'000	FF'000
Premium income	13,267,989	15,159,677
Investment income	3,284,872	2,145,287
Profit before tax	559,209	273,194
Profit after tax	457,058	210,204
Retained profit	241,011	242,900
Dividend	115,502	144,549
Dividend per share	34.0FF	8.5FF
Earnings per share	134.38FF	12.36FF
Share capital	340,116	851,291
Shareholders funds	3,265,616	3,331,408

Principal Shareholders: Compagnie Industrielle; Groupe de la Compagnie Financière de Suez

No of Employees: 3,000

COMPAGNIE GÉNÉRALE D'ELECTRICITÉ CGE

54 rue La Boétie, 75382 Paris Cedex 08
Tel: (1) 45.63.14.14
Cable: Electricite Paris 8
Telex: 280 953
Telefax: (1) 45.63.85.45; (1) 45.63.75.10

Board of Directors: Pierre Suard (Chairman and Chief Executive Officer), René Bourdin, Guy Dejouany, Frank Drendel, Jacques Guerendel, René Lamy, Helmut Lohr, Jean Marmignon, Jean-Paul Parayre, Ambroise Roux, Pierre Suard, Guy Verdeil, Marc Vienot

Management: Pierre Suard (Chairman and Chief Executive Officer), François de Laage de Meux (President and Chief Operating Officer), Phillipe Dargenton (Senior Executive Vice President), André Wettstein (Executive Vice President, Chief Financial Officer), Bernard Pierre (Executive Vice President, Industrial and International Affairs), Claude Marcilhacy (Corporate Secretary), Philippe Bissara (Vice President, Legal and Fiscal Affairs), Pierre Bollache (Vice President, Human Resources) Françoise Sampermans (Vice President, Corporate Relations and Advertising).

PRINCIPAL ACTIVITIES: Manufacturing and service activities in: Energy and Transportation: Power generation and distribution equipment; thermal and nuclear powerplants; industrial equipment; railway traction equipment; shipbuilding. Nuclear: Nuclear powerplant engineering, manufacturing and services. Batteries: Alkaline and special-purpose batteries; lead acid batteries. Communications and cables: Public telecommunications: public network switching, transmission equipment for cable, satellite and microwave-radio links. Business communications: PBX, terminals office automation systems. Electronic systems. Cables: power and telecommunications cables. Electrical contracting and industrial process control: Equipment and systems; industrial services; technical assistance and maintenance. Others: distribution of electrical equipment and components; general engineering and consultancy; factory automation engineering and systems.

Subsidiary Companies: (over 50% owned unless stated): Energy and Transportation: Alsthom; Neyrpic; Mecanica Pesada (Brazil); Delas Weir; Stein Industrie; Alsthom Creusot Rail; Carel Fouche Industrie; Sprecher Energie (Switzerland).

Nuclear: Framatome (40%). Batteries: Saft; Saft America (USA); Compagnie Européene d'Accumulateurs. Communications and Cables: Alcatel NV (Netherlands); Alcatel CIT; Alcatel Thomson Faisceaux Hertziens; Telspace; Alcatel Espace; Telic Alcatel; Générale de Services Téléphoniques Alcatel; Thomson CSF Téléphone. Communications: Alcatel Thomson Radiotéléphone; S M H Alcatel; Friden Alcatel (USA); Roneo Alcatel; Compagnie Générale D'Automatisme - CGA-HBS; Câbles de Lyon; Câbleries de Lens; Kabelmetal Electro (West Germany); S N Câbleries de Charleroi (Belgium); Thomson Cuivre; Standard Electrik Lorenz (West Germany); Bell Telephone Manufacturing (Belgium); Alcatel FACE (Italy); Alcatel Standard Electrica SA (Spain); Standard Telephon OG Kabelfabrik (Norway); IKO Kabel (Sweden); Standard Electric Kirk A/S (Denmark); Alcatel Nederland (Netherlands); Standard Telephon and Radio AG (Switzerland); Alcatel Austria (Austria); Standard Electrica (Portugal); Standard Telephones and Cables Ltd (Australia); Alcatel NA Inc (USA); Alcatel Business Systems (USA); Alcatel Information Systems (USA). Electrical Contracting and Industrial Process Control: CGEE Alsthom; ENTRELEC; BG Checo (Canada). Financial and Banking Activities: Compagnie Electro Financière; Compagnie Financière Alcatel; Electro Banque. Distribution: CGE Distribution. Rental: Locatel. Engineering: Sogelgerg. Publishing and Media: Générale Occidentale (41%). Others: CGE Maroc

Principal Bankers: Société Générale; CCF

Financial Information:

| | 31.12.86 | 31.12.87 |
	FF'000	FF'000
Sales net of taxes	80,902,800	127,461,000
Profit before tax	2,476,300	5,288,000
Profit after tax(1)	1,721,400	3,388,000
Retained profit	7,485,900	13,730,000
Dividends	188,277,700	529,015
Dividends per share	20.00FF(2)	7.50FF
Earnings per share	123.20FF(2)	27.80FF
Share capital	941,400	2,821,000
Shareholders funds	8,427,300	16,551,000
Number of shares(3)	9,413,885	70,535,391

(1) including minority interest

(2) 1986 proforma figures, adjusted to reflect the changes in par value and conversion of non-voting shares: earnings per share: FF22.2; dividend per share: FF6.67

(3) FF100 par value, excluding non-voting shares (titres participatifs) issued in 1983 and 1985. In 1987 the par value was increased to FF120, prior to a three to one stock-split. Then, as a result of capital increases in cash and conversion of non-voting shares, share capital was raised to FF2,821, 415,640 divided into 70,535,391 shares

Principal Shareholders: (Dec 1, 1987): Société Générale (5.8%); Fonds Commun de Placement des salariés (Employees' Mututal Fund) (3.4%); UAP (2.6%); Compagnie Générale des Eaux (2.6%); Société de Banque Suisse (2.6%); Société Générale de Belgique (1.8%); Dumez (0.9%); Companies controlled by the Group (Compagnie Immobilière Meridionale, Société Financière et Immobilière de Contrôle et de Gestion; Groupe Alcatel NV) (6.8%)

No of Employees: 219,500

COMPTOIR DES ENTREPRENEURS
CDE
3 rue de la Paix, 75086 Paris Cedex 02
Tel: (1) 40.20.30.40
Telex: Centre 213610 F
Telefax: (1) 40.20.35.14

Board of Directors: Mrs P Dufour (Président Directeur Général),
J Brunier, A Chazel, Y Comolli, J M Dore, M Gontard, P
Guillem, P Jurgensen, P Maynie, J Nicolau-Guillaumet, P
Paclot, J H Wahl
Management: Mrs Paule Dufour (Président), Philippe Jurgensen
(Administrateur Directeur Général), Jean Jacques Piette
(Directeur), Alain Rannou (Secrétaire Général), François Tanto
(Directeur Adjoint)

PRINCIPAL ACTIVITIES: Finance for industry, building and
public works
Trade Names: Comptoir des Entrepreneurs (C.D.E.)
Subsidiary/Associated Companies: ICDE (99.9%); SIPARI
(99.9%); CDE BANQUE (99.9%)
Financial Information:

	31.12.86	31.12.87
	FF'000	FF'000
Sales turnover	8,895,666	9,431,232
Profit before tax	158,446	204,550
Profit after tax	34,129	32,920
Retained profit	-	4,032
Dividends	-	12,242
Dividends per share	-	6.00FF
Earnings per share	16.735FF	16.135FF
Share capital	203,940	203,033

Principal Shareholders: Sté la Mondiale; Assurances Générales
de France; Union des Assurances de Paris; Groupe des
Mutuelles du Bâtiment et des Travaux Publics; Bancoval;
Groupe de la Caisse des Dépôts et Consignations; Groupe
des Assurance Nationales; Groupe Banque de France
No of Employees: 1,950

CONCORDE, LA,
Compagnie d'Assurances Contre les Risques de Toute Nature
5, rue de Londres, F-75009 Paris Cedex
Tel: (1) 42.80.66.00
Cable: Concord-Paris
Telex: 650734

Management Board: G Alexandre (Chairman), G Cau, Giorgio
Gerolimich, C Rey, D Zerah

PRINCIPAL ACTIVITIES: Insurance and reinsurance; interests in
property
Subsidiary Companies: Compagnie Continentale d'Assurance;
La Fédération Continentale; Equité I.A.R.D.; Compagnie Dakar
St Louis; Société Holding Economique et Financière; Société
Immobiliére du Parc de la Noue; Société Immobiliére des
154/156 Boulevard de la Gare 75013 Paris. Associated
Companies: La Lutéce; The Northern Star Insurance
Company (UK); Europ-Assistance; Société Foncière de la
Place de la Trinité; Euralux (Luxembourg); Europ Assistance
Limited (UK)
Financial Information:

	FF'000
Premium income	3,400,000

No of Employees: 1,072

CONFEDERATION NATIONALE DU CRÉDIT MUTUEL
88-90 Rue Cardinet, 75017 Paris
Tel: (1) 47.66.01.51
Telex: 640373 cccmu f

President: Etienne Pflimlin; Vice President: Georges Coudray;
Managing Director: Michel Delahousse

PRINCIPAL ACTIVITIES: Cooperative bank
Financial Information:

	31.12.86	31.12.87
	FF'000	FF'000
Profit before tax	1,614,000	1,681,000
Profit after tax	1,028,000	1,039,000
Shareholders equity	9,791,000	12,948,000
Deposits	150,412,000	168,354,000
Total assets	198,840,000	231,476,000

No of Employees: 20,500

CRÉDIT AGRICOLE, CAISSE NATIONALE DE,
91-93 Bld Pasteur, 75015 Paris
Tel: (1) 43.23.52.02
Cable: Agricrédi Paris
Telex: CAGRI X 250971 F

Board of Directors: Yves Barsalou (Chairman), Alain Delaunoy
(Vice Chairman), Lucien Douroux (Vice Chairman), Jean
Batard, René-Pierre Berseget, Paul Cabarat, Marcel Deneux,
André Fontan, Joël Icard, Pierre Perromat, Jean Simon,
Jacques Stefani, Raymond Lacombe
Management: Bernard Auberger (Chief Executive and Managing
Director), Jean Fontourcy (Deputy Chief Executive), Gilles
Gutton (Deputy Chief Executive), Jean-Pierre de Longevialle
(Secretary-General), Pierre d'Haultfoeuille (Deputy Secretary-
General), Jean Bayle-Lespitau (Accounting Official), Philippe
Guérin (Secretary)
Senior Executives: Department Heads-Markets: Philippe Collet
(Agriculture and Regional Collectives), Jacques Lenormand
(Private Individuals and Professionals), Monique Bourven
(Capital Markets), Gilles Guitton (International), Francois
Jouven (Deputy, International), Jean-Louis Brunet (Real Estate
and Leisure Developments). Department Heads-
Administration: Bernard de Pasquale (Telecommunications
Network, Data Processing and Money Processing), Maurice
Lepesant (Regional Bank Relations), Jacques Valay (Human
Resources), Martial Stambouli (Finance Department),
Raymond Trébuchon (Control and Audit), Mina Darbon
(Communications and Marketing)

PRINCIPAL ACTIVITIES: Central agricultural credit association
for member institutions
Subsidiary Companies: (100% owned unless stated): CREDIT
AND FINANCIAL INSTITUTIONS: Fransabank (34%); GFFR
(99.9%); GPF (99%); Segespar Titres (50%); Unibanque;
Unicredit (96%). INSURANCE: Predica (42.5%). INVESTMENT
COMPANIES: Agrinova (31.4%); Idia (27.2%); Segespar
Holding; Sofipar (52.5%); Sopagri (61.2%); U.I. (95.7%).
LEASING COMPANIES: Agritel (83.9%); Comser (97.2%);
Unicar (97.2%); Unicomi (71.5%); Unifergie (86.1%); Unimat
(97.2%). FOREIGN SUBSIDIARIES: Agrifin Factor; Agrifin
Finanziaria; CNCA Delaware; CNCA Hong Kong Dtc; CNCA
Securities; Credit Agricole Deutschland AG; UEP.
TOURIST/PROJECTS: Sté Dev. Agay (52.6%); Sté Dev.
Bourgenay (51.8%); Dramont Amenagement (47.3%); Sa
Montchavin Les Coches (49.4%); Unimmo France (96.2%);
Unirel (70.1%); Valmer (95.7%). MISCELLANEOUS: Cedicam
(50%); Eurocard France (25%); SCI Quentyvel (95%); SCI
Raspail (97.8%); Voyage Conseil (56.9%)
Financial Information:

	31.12.86	31.12.87
	FF'000	FF'000
Profit after tax	1,551,000	2,158,000
Deposits	689,600,000	761,000,000
Total assets	996,700,000	1,144,800

No of Employees: 73,939

CRÉDIT COMMERCIAL DE FRANCE
103 Ave des Champs-Elysées, 75008 Paris
Tel: 40.70.70.40
Cable: Francial
Telex: C.C.F. X 630300 F
Telefax: 42.25.10.77

Board of Directors: Michel Pébereau (Président), Jean Blondel, Jean-René Fourtou, Alain Gomez, Jean-Claude Jolain, René de la Serre, Olivier Lecerf, Gabriel Pallez, Albert Parment, Daniel Petit, Pierre Suard, Y Bachelot (Secretary)

Management: Michel Pébereau (Président Directeur Général), René de la Serre (Directeur Général), Gérard Engel (Directeur Général Adjoint), Yves Bachelot (Secrétair Général)

Senior Executives: Département Bancaire:- Activités Bancaires en France: Jacques Petitjean (Directeur Délégué), Joseph Perez (Directeur Central). Affaires Immobilières: Francis Picard (Directeur Délégué). Implantations à l'étranger: Olivier Lavédrine (Directeur Central). Commerce Extérieur: Noël Bazoche (Directeur à l'Administration Centrale). Agence Internationale: François de la Baume (Directeur à l'Administration Centrale). Département Financier:- René de la Serre (Directeur Général). Marchés: Gérard Engel (Directeur Général Adjoint). Affaires Financières des Entreprises: Gilles Gobin (Directeur Central). Gestion des Valeurs Mobilières: Pierre Jolain (Directeur Central). Titres et Coupons: Robert Beunet (Directeur à l'Administration Centrale). Études Économiques et Financières: Bernard Petit (Directeur à l'Administration Centrale). Trésorerie et Changes: Paul Passieux (Directeur Central). Marcel Goldberg (Directeur à l'Administration Centrale)

PRINCIPAL ACTIVITIES: Banking and financial operations

Subsidiary Companies (100% owned unless stated): Banque Privée de Dépôts et de Crédit (BPDC); Banque Dérobert (90%); Banque d'Escompte et de Dépôts (80%); Banque de l'Union Maritime et Financière (80%); Banque des Pyrénées; Union de Banques pour l'Équipement (59%); Banque Kolb (51%); Crédit Commercial de France & Co-OHG (Germany); Crédit Commercial de France (Panama); Crédit Commercial de France (Canada); Crédit Commercial de France (Suisse) SA (Switzerland); Banca di Credito Commerciale e Mobiliare SA (61%) (Switzerland); Crédit Commercial de France (Moyen Orient) SAL (51%) (Lebanon); Irish Bank of Commerce (80%) (Eire); Crédit Commercial de France Australia Ltd (Australia); Banque Chaix; L'Européenne de Banque; Union de Banques à Paris; Banque de Picardie; Société Anonyme Professionnelle de Crédit (62%); Compagnie Commerciale et Financière (98%); Européenne de Monaco (Monaco); Crédit Commercial de France Ltd (Hong Kong); Société Financière et Mobilière; Compagnie Foncière Elysées; Elysées Gestion; Vernet Gestion; Financia-Elysées Crédits (99%); Société de Financement International du CCF SA (Switzerland); Crédit Commercial de France International; Nord Financier; Laurence Prust (85%) (UK); Union pour le Financement Professionnel et Privé (97%); Société Financière pour l'Amérique Latine (Sofial) (Luxembourg); Sofial Panama (Panama); Laffitte Investissement; Autobail-Equipbail (59%); Multibail (59%); Locamic (59%); Elysées Bail (67%); Sofileasing (84%) (Italy); Crédit Commercial de France Leasing (Spain); Sté Auxiliaire d'Investissement Mobilier & de Gestion "Auxim"; Finimmo (53%); Compagnie d'Investissements et de Valeurs; Sefipar (65%); SA Neuilly-Saint-Paul; Compagnie Suisse et Française; Omnifinance; Mobigest; Elyfinance Corporation BV (Netherlands); Investissements Mobiliers et Immobiliers SA (Switzerland); Nobel (53%); Serdac SA (53%); Société Parisienne de participations; Société Immobilière Malesherbes-Anjou; Société Immobilière de l'Elysée-Palace; SCI Laffitte; SCI Pillet Will; Elysées Financière Immobilière "Elyfim"; Realimo (99%); Société Française de gestion et de construction; Auxila; Elysées Fonds

Financial Information:

	31.12.86 FF'000	31.12.87 FF'000
Profit before tax	650,419	658,639
Group profit after tax(1)	360,458	432,597
Dividends(2)	52,065	103,304
Earnings per share(3)	8.72FF	10.47FF
Share capital(4)	1,033,039	1,033,039
Shareholders funds	5,304,843	5,877,712
Deposits	60,885,487	62,212,545
Total assets	206,807,322	204,569,155

(1) minority interests excluded
(2) rents distributed to la Caisse Nationale des Banques
(3) following the reduction in the par value of the shares in 1987 from FF100 to FF 25, the number of issued shares has been multiplied by four
(4) including participation shares (1987 = 1,894,754) and excluding minority interests

Principal Shareholders: The French Government
No of Employees: 12,300

CRÉDIT DU NORD
6/8 Boulevard Haussmann, BP 253.09, F-75427 Paris Cedex 09
Tel: (1) 40 22 40 22
Cable: CREDNOR
Telex: 641 379 CDN
Telefax: 47 70 71 70

Board of Directors: Bruno de Maulde (Président Directeur Général), Pierre Simon (Administrateur-Directeur Général). Administrateurs: Christian Babusiaux, René Barberye, Philippe Dulac, Georges Fourmentin, Michel François-Poncet, André Levy-Lang, Pierre Martinaud, Eliane Mosse, Christian Noyer, Alex Peytavin, Patrice Ract-Madoux, Hubert de Saint-Amand, Jean Sembely. Censeurs: Yves de Courlon, Christian Dubois, François Essig, Gilbert Granchet, Gérard Tiebot

Management: Bruno de Maulde (Président Directeur Général), Pierre Simon (Directeur Général), Jacques Bailet (Directeur Général Adjoint, Chargé de l'Exploitation Bancaire), Philippe Toussaint (Directeur Général Adjoint, Chargé des Affaires Financières et des Affaires Internationales), Bernard Pidoux (Secrétaire Général)

PRINCIPAL ACTIVITIES: Banking
Trade Names: C.N.
Subsidiary Companies: (100% owned unless stated): Banque Tarneaud; Banque Nicolet Lefanechère et de l'Isère (97.21%); Union Bancaire du Nord (55.85%); Banque Lenoir et Bernard (99.89%); Banque Arnaud Gaidan (99.98%); Norbail Paris (62.26%); Caisse Générale de Dépôts et d'Avances (98.99%); Société Immobilière et Financière d'Investissements et de Coordination (99.86%); Société d'Exploitation d'Ensembles Electroniques (96%); Société d'Etudes et de Gestion Informatique Nouvelle (73.84%); Groupe Crédit du Nord (99.99%); Crédit du Nord Belge (97.30%)

Principal Shareholders: Union Financière et Bancaire UFIBA (40%); Compagnie Financière de Paribas (10.3%); Etat Française (49.7%)
No of Employees: 11,287

CRÉDIT FONCIER DE FRANCE
19 Rue des Capucines, 75001 Paris
Tel: (1) 42.44.80.00
Telex: 230079 crefonc f
Telefax: 42.44.86.99

Governor: Georges Bonin; *Deputy Governors:* Jean-Claude Colli, Michel Euvrard

Directors: Robert Blot, Roger Burnel, Mme Yvette Chassagne, Michel Albert, Jean Deflassieux, Mme Paule Dufour, Roger Goetze, Jean-Yves Haberer, Alain Devals, Max Laxan, Robert Lion, Roland Pignol

Senior Executives: Marcel Gontard (General Secretary), Robert Fages (Assistant General Secretary), Andrée Marlin (General Inspector)

PRINCIPAL ACTIVITIES: Credit and loans for land and property

Subsidiary Companies: Compagnie Foncière de Construction; Compagnie Foncière de France; Société des Immeubles de France; Société Immobilière de Paris; Société Immobilière Marbeuf; Compagnie Foncière Rhodanienne; Foncier Habitat Participations, L'Auxiliaire du Crédit Foncier de France A.C.F.F.; Foncier Rendement (SICAV); Société anonyme Paix Daunou; Société d'Etudes Immobilières et d'Expertises Foncières; Compagnie La Hénin; Crédit-Logement; Crédit Naval C.M.A.F.; S.C.I.C.; Union de Crédit pour le Bâtiment (U.C.B.); Compagnie d'Aménagement et de Promotion Immobilière (C.A.P.R.I.); L'Auxiliaire de la Construction Immobilière (S.A.C.I.); Sinvim; Batinnova; Centre de Réalisations et d'Études pour la Planification l'Aménagement et l'Habitat (C.R.E.P.A.H.); Société Centrale pour l'Equipment du Territoire (S.C.E.T.); Cofimeg; Groupement pour le Financement de la Construction (G.F.C.); Sefimeg; Union Immobilière de France (U.I.F.); Union pour l'Habitation; Union Générale d'Investissements Immobiliers (U.G.I.M.O.); Foncier Investissement; Codétour; Immobail-B.T.P.; Immobanque; Locabail-Immobilier; Locabanque; Bail Energie; Ecobail; Finergie; Cinergie; Mutuabail; Soficomi; Sovabail; Société Nale pour L'Application de la Geothermie (GEO-CHALEUR); Banque Internationale de Gestion Trésorerie (B.I.G.T); S.C.I. Capucines; S.C.I. du Faubourg St Honoré; SOCLIM; SOFIPAR-logement; SOFONEG; Crédit-logement Développement; Société d'Etudes Immobiliers et d'Expertise Foncière (S.E.I.E.F.); Banque du Bâtiment et Travaux Publics (BTP); Ferri Ferri Germe; Eurotunnel; Sociedad Iberica de Credito Inmobilario

Financial Information:

	31.12.86 FF'000	31.12.87 FF'000
Turnover before tax	28,084,903	30,365,226
Profit before tax	1,393,185	1,565,706
Profit after tax	214,324	248,251
Dividends	203,217	210,224
Dividends per share	29FF	30FF
Earnings per share	30.59FF	35.43FF
Share capital	1,997,129	1,997,129
Total assets	289,982,500	309,054,871

Principal Shareholders: The personnel of the Society (4.02%); Assurances Générales de France-Vie; Union des Assurances de Paris-Vie; Caisse des Dépôts et Consignations

No of Employees: 3,648

CRÉDIT INDUSTRIEL ET COMMERCIAL DE PARIS
CIC
66 Rue de la Victoire, PO Box 207, F-75452 Paris Cedex 09
Tel: (1) 42.80.80.80
Cable: Crédint Paris 22
Telex: 290692; 650723; 650929
Telefax: (1) 42.80.97.17

Senior Management: Jean René Bernard (Chairman), François Arsac (Managing Director), Jean Varda (Managing Director)
Senior Executives: Jean-Jacques Desbons (Secretary)

PRINCIPAL ACTIVITIES: Bank
Parent Company: Compagnie Financière de Crédit Industrie et Commercial

Principal Shareholders: French State
No of Employees: 5,630

CRÉDIT LYONNAIS
19 Blvd des Italiens, 75002 Paris
Tel: (1) 42.95.70.00
Cable: Credionais
Telex: 230 951

Board of Directors: Jean Deflassieux (Honorary Chairman), Jean-Maxime Levêque (Chairman), Loïc Briand, Patrice Cahart, Bernard Darty, Paul Dubrule, Denise Gerlat, Pierre Gisserot, Philippe Hanrot, Philippe Jaffré, Jacques Journoud, Philippe Jurgensen, Marc de Lacharrière, Jacques Maisonrouge, Jean-Claude Quivrin, Lucien Rebuffel, Louis Sicard, André Tourniaire

Executive Committee: Jean-Maxime Levêque (Chairman), Bernard Thiolon, Michael Gallot, Jean-Paul Amiel, Alexis Wolkenstein, Etienne Bouruet-Aubertot
Senior Executives: J-F Verny (Secretary)

PRINCIPAL ACTIVITIES: Banking activities
Subsidiary Companies: (100% owned unless stated): Crédit Lyonnais Bank Nederland NV (90.85%) (Netherlands); Banco Tornquist SA (99.99%) (Argentina); Cogefo (99.96%); Crédit Lyonnais Canada; BIGT (58.66%); Banque Laydernier; Interétud (99.98%); Société Rhodanienne Mobilière et Immobilière (99.99%); Banque Chalus (91.91%); BDEI (99.99%); Société de Banque Occidentale (99.99%); Slicomi (99.99%); Slibail (99.99%); Slibailautos (99.98%); Slibail International (99.97%); Slifequip (99.99%); Crédit Lyonnais Bank Sverige (Sweden); Crédit Lyonnais Australia (99.99%); Banque Franco Hellénique pour le Commerce International Maritime (50%) (Greece)

Financial Information:

	31.12.86 FF'000	31.12.87 FF'000
Profit after tax	1,931,348	2,223,000
Dividends	393,360	498,000
Dividends per share	15.00FF	19.00FF
Share capital	3,278,000	3,278,000
Shareholders funds	18,691,202	23,699,000
Customer deposits	204,843,538	218,517,000
Savings deposits	72,306,351	83,946,000
Total assets	852,552,031	898,956,000

Principal Shareholders: The French Government (99.99%)
No of Employees: 45,176

CREDIT MUTUEL, BANQUE FEDERATIVE DU, BFCM
34 rue du Wacken, F-6700 Strasbourg
Tel: 88 35 90 35
Telex: 880034

Supervisory Board: Etienne P Flimlin (Président)
Management Board: Raymond Crombecque (Président Directeur Général), Emile Zagar (Directeur Général Adjoint), Joseph Lehmann (Directeur Général Adjoint)

PRINCIPAL ACTIVITIES: Banking

CRÉDIT NATIONAL
45 rue Saint-Dominique, F-Paris 75007
Tel: (1) 45.50.90.00
Telex: 250005

Board of Directors: Paul Mentré (Président Directeur Général)
Management: François Lagrange (Directeur), Didier Floquet (Directeur), Jean Régis (Délégué Gal au Développement)

PRINCIPAL ACTIVITIES: Medium and long term credit services
Subsidiary Companies: Caisse Française de Développement Industriel (CFDI); DOMIBAIL (95.99%); Sté de Garantie Financière (SOGAFI) (79.97%); Sté de Financement de l'Electronique et de l'Information (SOFINETI) (51%)

Principal Shareholders: The French Government; Assurances Générales de France-Vie (10.38%)
No of Employees: 1,145

DINERS CLUB DE FRANCE
18 R François 1er, F-75380 Paris Cedex 08
Tel: (1) 47 62 75 00
Cable: DINCLUB PARIS
Telex: 630 666 DINCLUB
Telefax: 47 62 76 10

Board of Directors: John Roche (Président Directeur Général)
Directors: John Roche (Président Directeur Général), J-P Debette, (Directeur Général Adjoint), Mile D Miermont (Directeur Général Adjoint)

PRINCIPAL ACTIVITIES: Credit card finance and operations
Parent Company: Citibank, USA
Subsidiary Companies: Diners Voyages; ATGG; Auximar; Diners Assurances; Getco; Mémo Service; Diners Sélection; Signature
Principal Bankers: CGB; Worms
Financial Information:

	FF'000
Sales turnover	3,500,000

Principal Shareholders: Citicorp
No of Employees: 426

DUMEZ
345 Ave Georges Clémenceau, 92022 Nanterre Cedex
Tel: 47.76.42.43
Cable: Zemud 92000 Nanterre
Telex: 620844 zemud nantr f
Telefax: 47.78.86.69

Supervisory Board: André Chaufour (President), André Duflocq (Vice-President), Jacques Fournier, Yves Le Guay, Jean-Louis Napp, Jean-Pierre Chaufour, Jean Tissier, P Fouillade, Pierre Muron
Directorate: Jean-Paul Parayre (President), J J Chaufour (General Manager), André Kamel (General Manager),

PRINCIPAL ACTIVITIES: Civil engineering and public works construction; general contractor-project management
Trade Names: Dumez
Subsidiary Companies: Dumez Travaux Publics; Dumez France; Dumez International; Dumez Gestion; Sté Immobilière des Vingts Arpents; Dumez Construction (Canada); Dumez North America; United Westburne (Canada)
Principal Bankers: Banque de France; Crédit Lyonnais; Société Générale; Banque Française du Commerce Extérieur; Banque Internationale pour l'Afrique Occidentale; Crédit Commercial de France; Banque Nationale de Paris; Banque Indosuez
Financial Information:

	1986 FF'000	1987 FF'000
Sales turnover	9,600,000	17,000,000
Profit after tax	428,000	354,000
Retained profit	350,000	270,000
Dividends	78,000	84,000
Dividends per share	26.00FF	14.00FF
Earnings per share	142.00FF	59.00FF
Share capital	300,000	300,000
Shareholders funds	2,426,000	2,793,000

No of Employees: 30,000

ETS ARBEL
See ARBEL, ETS,

EUROPE COMPUTER SYSTÈMES
ECS
16 Rue Washington, 75008 Paris
Tel: (1) 49 53 33 33
Telex: 650805 F
Telefax: (1) 42 25 86 86

Board of Directors: Gilles Tugendhat (Président Directeur Général), DCL2, GENEFITEC, Philippe Haemmerle, Société Générale, SOGEPARTS

Management: Gilles Tugendhat (Président Directeur Général), Hervé Saint-Sauveur, Louis Lacombe, Philippe Lefebvre, Jean-Claude Leconte, Claude Dimont-Mellac

PRINCIPAL ACTIVITIES: Computer rentals and DP services
Parent Company: Société Générale
Subsidiary Companies: ECS Diffusion; ECS Maintenance; ECS Bureautique; ECS Technologies; ECSIS; Proland; ECS United Kingdom (UK); ECS Deutschland (Germany); ECS Italia (Italy); ECS Japan; ECS Belgium
Financial Information:

	31.12.86 FF'000	31.12.87 FF'000
Sales turnover	3,934,062	5,101,691
Profit before tax	90,409	50,780
Profit after tax	48,239	21,765
Retained profit	34,839	5,685
Dividend	13,400	16,080
Dividend per share	6.7FF	16.7FF
Earnings per share	24.1FF	9.1FF
Share capital	50,000	180,000

No of Employees: 767

FRANCETEL
Francaise de Financement des Télécommunications SA (Société)
20 Avenue Rapp, 75007 Paris 7E
Tel: (1) 47.05.29.72
Telex: 202.520 F
Telefax: (1) 47.05.85.91

Board of Directors: Philippe Dumas (Président), Caisse Nationale des Télécommunications, Caisse de Dépôts et Consignations
Management: Jean Vidal (Directeur Général)
Senior Executives: Jean Mourlass (Secrétaire Général)

PRINCIPAL ACTIVITIES: Leasing of equipment and machinery for telecommunications
Parent Company: Caisse Nationale de Télécommunications
Financial Information:

	31.12.86 FF'000	31.12.87 F'000
Gross operating income	3,452,973	3,2 4,207
Profit before tax	4,829	4,311
Profit after tax	3,143	2,951
Retained profit	293	101
Dividend	2,850	2,850
Dividend per share	19.80FF	19.00FF
Earnings per share	21.96	19.67
Share capital	15,000	15,000

Principal Shareholders: Caisse Nationale des Télécommunications (51%); Caisse des Dépôts et Consignations (49%)
No of Employees: 7

GARANTIE MUTUELLE DES FONCTIONNAIRES, LES,
76 rue de Prony, 75017 Paris Cedex 17
Tel: (1) 47.54.10.10
Telex: 640377 F

Board of Directors: Jean-Louis Pétriat (Président Directeur Général), Marc Michelin (Vice-Président), Pierre Tougeron (Vice-Président), Daniel Soula (Secrétaire), Jean-Pierre Boucharel (Administrateur), Jean-Pierre Castets (Administrateur), Robert Jacob (Administrateur), Maurice Maigre (Administrateur), Gilbert Magal (Président d'honneur)
Management: Jean-Louis Pétriat (Président du Groupe GMF), Jean-Jacques Vial (Directeur de l'Unité Assurance et Réassurance du Groupe), Bernard Allien (Conseiller Stratégique auprès du Président), Olivier Maumus (Directeur du Département des Finances du Groupe), Guillaume de Saulieu (Conseiller Technique pour les Systèmes

d'Information du Groupe), Jacques Boittiaux (Directeur Général des Opérations du Groupe), René Jaouanet (Directeur de la Direction de la Comptabilité et du Contrôle de Gestion du Groupe)

PRINCIPAL ACTIVITIES: Insurance
Subsidiary Companies: GMF Vie; La Sauvegarde; GMF Solidarité; AMI; BCCM; UCCM; FNAC; La Maison de Valérie; Barrière Frères; Garages Multiservices Français; CSE
Financial Information:

	31.12.86 FF'000	31.12.87 FF'000
Premium income	5,000,000	5,300,000

No of Employees: 4,100

GROUPE DES ASSURANCES NATIONALES GAN

2 rue Pillet Will, 75448 Paris Cedex 09
Tel: 42.47.50.00
Telex: 280006 ganvie f

Chairman: F Heilbronner
Directors: M Bentz, M Bracalé, M Chapas, M Delacour, M Dumas, M Gayer, M Hugot, M Malinvaud, M Meadel, M Reininger, Mme Ruellan, M Bremond, M de Boissieu, M Clarou, Mme L'Allinec, M Lepatre, M Pelisson
Senior Executives: J J Bonnaud (General Manager)

PRINCIPAL ACTIVITIES: Insurance and reinsurance operations in France and abroad; financial and real estate operations; assistance
Subsidiary Companies abroad: Minster Insurance Co Ltd; Westminster Insurance Agencies Ltd; ADEA; Plus Leben; Plus Allgemeine; Bamafi; Previsa; Assurex; Al Wataniya; AGS; Nieuw Rotterdam Beheer NV; Nieuw Rotterdam Leven; SAVAG-Saarbrücker Versicherungs AG; NIT; GAT; R Bradford du Canada Ltée; GAN Anglo-American Ins Co; A.E.R.C.; Hawkeye National Life Ins. Co.; GRINITAL; Phenix Soleil SpA; Phenix Soleil Vita; Zenith Vie; GAN Canada; GAN Belgium; Hadar. Numerous branches and agencies throughout the world
Principal Bankers: Banque pour l'Industrie Française
Financial Information:

	31.12.86 FF'000	31.12.87 FF'000
Premium income	19,568,491	21,001,984
Investment income	4,095,596	4,672,325
Profit before tax	1,329,963	1,230,429
Profit after tax	1,379,162	1,330,104
Retained profit	1,273,762	1,183,904
Dividends	105,400	146,200
Dividends per share	18.60FF*	25.80FF
Earnings per share	162.25FF*	156.48FF
Share capital	314,500	314,500
Shareholders funds	6,328,832	8,165,250

*On July 17th 1986 the number of shares was increased from 850,000 to 8,500,000

Principal Shareholders: French State (85%)
No of Employees: 8,695

GROUPE DES BANQUES POPULAIRES
See BANQUES POPULAIRES, GROUPE DES,

GROUPE MAISON FAMILIALE SA
Av du Cateau, 59432 Cambrai Cedex
Tel: 27.73.30.00
Telex: 120704 Maison Cambr
Telecopier: 27.73.31.99

Board of Directors: R Leroy (Président du Conseil de Surveillance), Y Cren, R Heline, Alain Treppoz, P Deceukeleire

Management: P Leroy (Président Directeur Général)

PRINCIPAL ACTIVITIES: Holding company with interests in house building, banks and insurance companies, hotels
Trade Names: Marignan Immobilier, Residences et Loisirs, CEGITRA, Hotel et Residences du Roy
Subsidiary Companies: Europe Maison SA; CARPI (SA D'HLM); ILFA (SARL); Ficofrance; Sociafrance; AGAMI (SARL); Immobilière Berri Ponthieu (SA); Sud Construction Promotion; Rocamar (SA)
Principal Bankers: Crédit Lyonnais; Paribas
Financial Information:

	31.12.86* FF'000	31.12.87* FF'000
Sales turnover	2,873,000	2,969,000
Profit before tax	180,000	150,000
Profit after tax	126,000	118,000
Dividends	92,400	92,000

*The important modifications to significant values are due to changes in the sphere of the holding company effected in 1986

No of Employees: 2,500

LA CONCORDE
See CONCORDE, LA,

LES GARANTIE MUTUELLE DES FONCTIONNAIRES
See GARANTIE MUTUELLE DES FONCTIONNAIRES, LES,

LES MUTUELLES DU MANS I.A.R.D.
19/21 rue Chanzy, 72030 Le Mans Cedex
Tel: 43.41.72.72
Telex: 720 764
Telefax: 43.84.42.12

Board of Directors: Jean Claude Jolain (Président et Directeur Général), Jacques Dorise (Vice-Président), Patrice Adam, Jean Claude Beaufils, Robert Brunet, Joël Icard, Georges Doitteau, André Lallouet, Michel Lefebvre, Gilbert Lescop, Michel Lombardini, Marcel Elias
Management: Jean Claude Jolain (Président et Directeur Général), Michel Cosson (Directeur Général), Jean Courbe (Directeur Général Adjoint), Benoît Castel (Directeur Général Adjoint), Jean Perroud (Directeur Général Adjoint)

PRINCIPAL ACTIVITIES: All classes of non-life insurance
Parent Company: A member of The Mutuelles du Mans Group
Companies of The Mutuelles du Mans Group: In France: Les Mutuelles du Mans I.A.R.D.; Les Mutuelles du Mans Vie; Défense Automobile et Sportive D.A.S.; Société d'Assurance Crédit des Entreprises S.A.CR.EN. Foreign Subsidiaries: Norman Insurance Co Ltd (UK); Le Mans Assurances (Belgique) (Belgium); Les Assurés Réunis (Belgium); Le Mans Seguros España (Spain); Lloyd Hellénique (Greece); Norman (Hong Kong) Insurance Co Ltd (Hong Kong); MGF Vita Italia (Italy)
Principal Bankers: Société Générale; BNP; Crédit Lyonnais; Crédit Commercial de France
Financial Information:

	31.12.86 FF'000	31.12.87 FF'000
Premium income	7,270,874	8,661,886
Investment income	959,513	1,044,739
Profit before tax	124,290	313,316
Profit after tax	108,760	289,800
Retained profit	108,760	289,800
Establishment funds	1,029,217	1,230,630
Company funds*	1,695,145	2,182,153

*and reserves

No of Employees: 4,760

LES MUTUELLES DU MANS VIE

20 rue Saint Bertrand, 72006 Le Mans Cedex
Tel: 43.41.72.72
Telex: 720 764
Telefax: 43.84.42.12

Board of Directors: Jean Claude Jolain (Président et Directeur Général), Jacques Dorise (Vice Président), Patrice Adam, Jean Claude Beaufils, Robert Brunet, Joël Icard, Georges Doitteau, André Lallouet, Michel Lefebvre, Gilbert Lescop, Michel Lombardini, Marcel Elias
Management: Jean Claude Jolain (Président et Directeur Général), Michel Cosson (Directeur Général), Jean Courbe (Directeur Général Adjoint), Benoît Castel (Directeur Général Adjoint), Jean Perroud (Directeur Général Adjoint)

PRINCIPAL ACTIVITIES: All classes of life insurance (individual and group)
Parent Company: A member of The Mutuelles du Mans Group
Companies of The Mutuelles du Mans Group: In France: Les Mutuelles du Mans I.A.R.D.; Les Mutuelles du Mans Vie; Défense Automobile et Sportive D.A.S.; Société d'Assurance Crédit des Entreprises S.A.CR.EN. Foreign subsidiaries: Norman Insurance Co Ltd (UK); Le Mans Assurances (Belgique) (Belgium); Les Assurés Réunis (Belgium); Le Mans Seguros España (Spain); Lloyd Hellénique (Greece); Norman (Hong Kong) Insurance Co Ltd (Hong Kong); MGF Vita Italia (Italy)
Principal Bankers: Société Générale; BNP; Crédit Lyonnais; Crédit Commercial de France
Financial Information:

	31.12.86 FF'000	31.12.87 FF'000
Premium income	2,111,677	3,240,292
Investment income	655,328	766,330
Profit before tax	198,650	300,236
Profit after tax	186,507	290,468
Retained profit	186,507	290,468
Establishment funds	400,000	400,000
Company funds*	942,836	1,138,405

*and reserves

No of Employees: 644

LOCAFRANCE

43/47 Avenue de la Grande Armée, 75116 Paris
Tel: 45.01.54.13
Telex: 630473 locafra paris
Telefax: 45.01.97.37

Board of Directors: Gérard Billaud (Président Directeur Général), Robert Lafon (Administrateur-Directeur Général), Banque Indosuez (Antoine Jeancourt-Galignani), Banque Française du Commerce Exterieur (Jean-Louis de Roux), Banque Nationale de Paris (Roland Achard), Société Française d'Assurances pour Favoriser le Crédit (Gérard Barthélémy), La Paternelle-Risques Divers (Jacques d'Espine), Banque de l'Union Européene (Jean-Claude Wagner), Compagnie des Caoutchoucs de Padang (Bertrand de Buffevent), Union des Assurance de Paris I.A.R.D. (Henri Daru), Philippe Geslin
Executive Committee: Gerard Billaud (Président Directeur Général), Robert Lafon (Administrateur-Directeur Général), Bernard Durand (Directeur Général)

PRINCIPAL ACTIVITIES: Leasing of industrial plant and equipment
Parent Company: Banque Indosuez
Subsidiary Companies and Participations: (100% owned unless stated): Ordinabail; Immobanque (39%); Géfiroute; Promodata; Locafrance-Export (87%); Locamur-Sofigros (51%); Localease (49%); Locafrance International Leasing (82%); Locafrance International Nederland
Principal Bankers: Banque Indosuez; Banque Nationale de Paris; Banque Française du Commerce Extérieur; Crédit Industriel & Commercial; Banque de l'Union Européenne

Financial Information:

	31.12.86 FF'000	31.12.87 FF'000
Sales turnover	3,048,936	2,969,969
Profit before tax	65,510	78,417
Profit after tax	52,222	55,656
Dividends	38,118	51,158
Dividends per share	21.00FF	19.50FF
Earnings per share	28.77FF	21.21FF
Share capital	181,515	262,346
Total assets	7,848,928	8,236,608

Principal Shareholders: Groupe Indosuez (49.18%); Union des Assurances de Paris (5.11%)
No of Employees: 1,038

LOCINDUS

Compagnie Financière pour la Location d'Immeubles Industriels et Commerciaux

11 bis rue Scribe, F-75009 Paris 9 E
Tel: (1) 47.42.83.63
Telex: Locindu 680036 f

Board of Directors: Paul Alibert (Président), Banque Nationale de Paris (Jacques Masson), Banque de l'Union Européenne (Jean-Pierre Antoine), Assurances Générales de France-Vie (Dominique Eugene), Banque de Neuflize, Schlumberger, Mallet (Bernard Desforges), Banque Veuve Morin Pons (Mme Ralph P Odendall), Prévoir-Vie (René de Labrusse), Caisse Nationale de Prévoyance (Jacques Delmas-Marsalet), Cie Financière Delmas-Vieljeux (Christian Delmas-Vieljeux), La Paternelle Risques Divers (Jacques d'Espine)
Management: Paul Alibert (President-Directeur Général), Claude Desandre (Secrétaire Général), Jean-Pierre Ducrocq (Directeur Général Adjoint), Jerôme Collin (Directeur Adjoint)

PRINCIPAL ACTIVITIES: Finance house specialising in credit facilities for commercial and industrial properties, offices, factories, shops, hotels etc.
Trade Names: LOCINDUS
Subsidiary Companies: SCI Sainte Anne (99.9%); SCI San'Ambroggio (92%); SCI Nancy Ouest (99%); SCI Seibu Barcarès (99.9%); SCI Pampre d'Or (55.58%); SCI Paris Hôtel H 10 (50%); SCI Lafitte Loire (50%)
Financial Information:

	FF'000
Turnover	400,000

No of Employees: 17

LYONNAISE DE BANQUE, SOCIÉTE, SA

8 rue de la République, 69001 Lyon
Tel: 78.92.02.12
Cable: SOLYB
Telex: 330532
Telefax: 78.92.03.00

Board of Directors: Henri Moulard (Président-Directeur Général)
Management: Henri Moulard (Président Directeur Général), Michel Angé (Directeur Général)
Senior Executives: Michel Fraisse (Directeur Général Délégué), Pierre Berthier (Directeur Général Adjoint), Paul Filhol (Directeur Général Adjoint), Bernard Versini (Directeur Général Adjoint)

PRINCIPAL ACTIVITIES: Bank
Trade Names: Société Lyonnaise de Banque
Parent Company: CIC
Subsidiary Companies: (100% owned unless stated): Figepar Fribourg (Switzerland); Banque Marin et Gianola (99.99%); Financière Lyonnaise (99.98%), Banque de Vizille (99.97%); Financière du Bât d'Argent (99.83%); Cigel (98.6%); Sofisud (89.19%); Cogelex (80%) (Switzerland); Banque Pasche (72.06%); Banque Saint-Olive (49.99%); Banque Transatlantique de Monaco (34%); Sofica Créations SA (25%); Soparic (15.08%); UBR (10.27%); Axamur SA (10%)

Financial Information:

	31.12.86 FF'000	31.12.87 FF'000
Profit before tax	285,959	211,932
Profit after tax	144,201	120,976
Retained profit	92,625	78,476
Dividend	51,576	42,500
Dividend per ordinary share	6.17FF	17.00FF
Earnings per share	27.89FF	32.01FF
Share capital	400,000	400,000
Shareholders funds	1,029,908	923,907
Total assets	43,732,405	49,820,350

Principal Shareholders: CIC (100%)
No of Employees: 4,421

MUTUELLE ASSURANCE ARTISANALE DE FRANCE

MAAF

Chaban de Chauray, F-79036 Niort Cédex
Tel: 49.24.71.71
Telex: 790622

Board of Directors: Henri Sagnial (Président)
Management: Yves Thiré (Directeur Général), Jean Boudy (Directeur Général Adjoint)

PRINCIPAL ACTIVITIES: Insurance
Trade Names: MAAF
Subsidiary Companies: Banque Hypothécaire Européenne
Principal Bankers: Société Générale; Crédit Agricole; Crédit Lyonnais
Financial Information:

	FF'000
Premium income	5,500,000

No of Employees: 4,000

MUTUELLE ASSURANCE DES COMMERCANTS ET INDUSTRIELS DE FRANCE

MACIF

2 & 4 rue de Pied de Fond, F-79037 Niort Cédex
Tel: 49.09.43.21

Board of Directors: Jacques Vandier (Président)
Management: Jean Simonnet (Directeur Général)

PRINCIPAL ACTIVITIES: Insurance
Principal Bankers: BNP
Financial Information:

	FF'000
Premium income	6,300,000

No of Employees: 4,600

MUTUELLE ASSURANCE DES INSTITUTEURS DE FRANCE

MAIF

200 av Salvador Allende, 79038 Niort Cédex
Tel: 49.73.74.75
Telex: 790607
Telefax: 49.73.78.05

Board of Directors: Jean Germain (Président Directeur Général), Jean-Marc Boudières (Vice Président), Jacques Fournier (Vice Président), André Sizes (Vice Président), Hubert Arnault, Robert Ayrault, Roger Belot, Patrick Daraud, Gilles Déchamp, Raymond Facon, Jean-Marie Faivre, Charles Lapergue, Gervais Lescure, Jean-Pierre Marquet, Gérard Panisa, Serge Priou, Hervé Quemeneur, Jean-Claude Rivier, Jean Schmitt, Pierre Senac, Charles Vallon
Management: Jean Germain (Président Directeur Général), Robert Ayrault (Administrateur Délégué à la Direction Générale), Paul Marcus (Directeur), Bernard Bellec (Directeur Délégué), Dominique Thiry (Directeur Adjoint), Jean Ducos

(Directeur du Développement et de l'Organisation), Claude Chevalier (Directeur des Relations Humaines et Sociales), Guy Laugier (Directeur de la Coordination et de la Qualité), Jean-Dominique Antoni (Directeur de FILIA-Maif)

PRINCIPAL ACTIVITIES: Insurance
Subsidiary Companies: FILIA Maif; Parnasse-Vie; Inter Mutuelles Assistance
Financial Information:

	31.12.87 FF'000
Premium income	4,497,000
Profit after tax	630,990

No of Employees: 3,177

NAVIGATION MIXTE, COMPAGNIE DE, CNM

10 rue Volney, F-75002 Paris
Tel: (1) 42.61.50.27
Telex: 440079
Telefax: (1) 42.61.48.43

Board of Directors: Marc Fournier (Président Directeur Général), Max de Boysson, Alexandre Dumont, Jack Francès, Robert Gachet, Michel Gallot, Xavier Gardinier, Gérard Louis Dreyfus, Raymond Lusinchi, Pierre Muron, Henri Ortet, Jacques Soufflet, Jean-Claude Leny

PRINCIPAL ACTIVITIES: Compagnie de Navigation Mixte is a financial institution. Its major participations are in the agro-industry, insurance, banking, transportation and other services
Subsidiary Companies: (100% owned unless stated): BANKING AND FINANCE: Via Banque (53.7%); Via Finance; Via Bail (Sicomi); Société de Banque et de Financement SBF (66%). INDUSTRIES AND SERVICES: Générale de Transport et d'Industire (89%); Fichet-Bauche (93.8%); Hurel-Dubois (51.8%); Ateliers et Chentiers de Fécamp; Vivat Holdings (24.2%) (UK); France Média International (36.1%). AGRO-FOOD: Cie Française de Sucrerie (67.8%); Saupiquet (58.9%). INSURANCE: Vie Assurances Vie (96.7%); Via Assurances IARD (62.8%); La Rurale (99.6%); Rhin et Moselle Via (50%); Languedoc (20%); Cie Générale de Prévoyance (19.9%); Rhin et Moselle Vie (40.5%)
Financial Information: Consolidated

	31.12.86 FF'000	31.12.87 FF'000
Sales turnover	9,952,248	13,368,384
Profit before tax	1,261,128	1,060,215
Profit after tax	856,285	752,034
Dividend	90,065	126,432
Dividend per share	17.00FF	18.50FF
Earnings per share	84.31FF	86.46FF
Share capital	397,346	512,564
Shareholders funds	3,443,720	4,790,129

No of Employees: 23,031

PARIBAS, COMPAGNIE FINANCIÈRE DE,

5 rue d'Antin, 787-08 Paris Cedex 02
Tel: (1) 42.98.12.34
Cable: Paribas
Telex: 210041 PARB

Board of Directors: Michel François-Poncet (Chairman and Chief Executive Officer), Michel Albert, Claude Bebear, Michel Develle, Jean Gandois, Bernard Maitenaz, François Morin, Francois-Xavier Ortoli, Michel Pecqueur, Ambroise Roux, Hubert de Saint-Amand, Jean Dromer, Gérard Eskenazi, André Levy Lang
Officers of the Company: Michel François-Poncet (Chairman and Chief Executive Officer), François Morin (President), Hubert de Saint-Amand, Gilles Cosson (Senior Executive Vice President), Christian Manset (Senior Executive Vice President), Gilbert Granchet (Executive Vice President and

Company Secretary), André Battestini (Senior Executive Vice President)

Management Committee: Michel François-Poncet, François Morin, Hubert de Saint-Amand, Philippe Dulac, Patrick Deveaud, François Homolle, Gilles Cosson, Christian Manset, Alain de Wulf

PRINCIPAL ACTIVITIES: International banking services

Trade Names: PARIBAS

Subsidiary Companies: (100% owned unless stated): Banque Paribas (85.2%); Compagnie Bancaire (45.5%); du Nord; Banque Paribas Belgique SA (76%) (Belgium); Banque Paribas (Suisse) SA (65%) (Switzerland); Banque Paribas (Luxembourg) SA (90%) (Luxembourg); Banque Paribas (Nederland) NV (Netherlands); Banque Paribas Capital Markets Ltd (UK); OPB-Paribas (91.1%); CGIB (99.1%); Paribas International; OPFI-Paribas (71.5%); SOGEDIP; Cobepa (61.2%) (Belgium)

Financial Information:

	31.12.86 FF'000	31.12.87 FF'000
Operating income before tax	3,689,000	2,175,000
Profit after tax	3,232,000	3,291,000
Dividends	233,900	387,600
Dividends per common share	5.00FF	7.00FF
Dividends per preferred share	7.50FF	7.50FF
Earnings per share	36.00FF	32.60FF
Share capital	2,350,000	2,767,640
Customer deposits	162,393,000	175,870,000
Total assets	601,865,000	652,950,000

No of Employees: 27,500

PRÉSERVATRICE FONCIÈRE
PFA Tiard

Cedex 43, 92076 Paris-La-Defense
Tel: (1)42.91.10.10
Telex: PFA A 615030 F
Telefax: 42.91.12.10

Supervisory Board: Guy Verdeil (Président-Directeur Général), Gérard Boucher (Administrateur-Directeur Général), Jean Barnaud, Pierre Bazy, Nicholas Clive Worms, Jacques Legrand, Yves de Torsiac, Société Centrale Préservatrice Foncière Assurances representeé par Basile de Tiesenhausen

Management: Guy Verdeil (Président-Directeur Général), Gérard Boucher (Administrateur-Directeur Général) Pierre Guillon (Directeur Général Adjoint), François Pierre (Directeur Général Adjoint). Directeurs: Roger Benel, Louis-Pol Delestrée, Michel Frugier, Michel de Poix, Louis Prévost de la Boutetière, Christian Valin, Guy Camboulives, François d'Hennezel, Gérard Renucci

PRINCIPAL ACTIVITIES: Insurance and reinsurance and annuities

Trade Names: PFA (Préservatrice Foncière Assurances)

Parent Company: PFA (Préservatrice Foncière Assurances)

Subsidiary/Associated Companies: Précam (99.66%) (Belgium); Assurlux (47.71%) (Luxembourg); Rhea (99.99%); Saar Rhein (45%) (Germany); Atlantida de Seguros (99.47%) (Spain); Foncias (51%) (Burkina-Faso); Société Africaine d'Assurances et de Réassurances en République de Côte d'Ivoire (S.A.F.A.R.R.I.V.) (63.85%) (Ivory Coast); Omnium Gabonais d'Assurances et de Réassurances (O.G.A.R.) (63.94%) (Gabon); Société Nouvelle d'Assurances du Cameroun SNAC (65.77%) (Cameroun); Société Nouvelle d'Assurances du Sénégal (S.N.A.S.) (49%) (Senegal); Société Nouvelle d'Assurances de Guinée SONAG (51%) (Guinea)

Financial Information:

	31.12.86 FF'000	31.12.87 FF'000
Turnover	4,443,005	4,225,302
Profit before tax*	348,897	362,787
Profit after tax	176,522	214,573
Dividends	15,360	36,000
Dividends per share	3.84FF	9.00FF
Earnings per share net	14.13FF	53.64FF
Share capital	400,000	400,000
Shareholders funds	1,183,620	1,404,621

*before tax, salaries, amortisements and provisions

No of Employees: 2,333

PRETABAIL-SICOMI

82 rue de Courcelles, F-75008 Paris
Tel: (1) 47.66.02.53
Telex: PRETABA 648 365F

Board of Directors: J-C Genton (President and General Manager), H Martin (Vice President), A Brion, J-L Breuil-Jarrige, H de Carmoy, F Lepatre, C de Longevialle, Y Rapilly, V Sasson, OFP (represented by J-C Pineau), Caisse Centrale des Banques Populaires (represented by J-C Boyer), Caisse des Dépôts et Consignations (represented by B Boutrouille)

Management: J-C Genton (President and General Manager), P Vajda (General Manager), M Toussaint Du Wast (Deputy General Manager)

PRINCIPAL ACTIVITIES: Finance house with interests in property development for industrial and commercial use only

Subsidiary Companies: Omni-Energie

Financial Information:

	FF'000
Operating revenue	440,000

Principal Shareholders: OFP (18%)
No of Employees: 43

PROMODATA SA

83 ave de la Grande Armée, 75 782 Paris Cedex 16
Tel: (1) 45.01.54.13
Telex: 643039 F

Board of Directors: Robert Lafon (President-Directeur Général
Management: Robert Lafon (Président Directeur Général), René Aubron (Directeur Général), Bruno Berjal (Directeur Général Adjoint), Férat Dolabdjian (Directeur Technique), Christian Defarge (Directeur Marketing et Communications)

PRINCIPAL ACTIVITIES: Installation and leasing of computers; trade in new and second-hand computers

Trade Names: PROMODATA

Parent Company: Locafrance SA

Subsidiary Companies: TMIS; ETICA; ORDINABAIL; ORDIRENT

Principal Bankers: Banque Indosuez; BNP

Financial Information:

	FF'000
Sales turnover	1,952,000

No of Employees: 230

REVILLON LUXE SA

42 Rue la Boétie, F-75008 Paris
Tel: (1) 45.62.45.22
Cable: Nolliver Paris
Telex: 649106

Board of Directors: Jean-Claude Cathalan (Président)
Management: Robert Montagna (Directeur Général Fourrure), Patrick Maniotte (Directeur Général Parfums)
Senior Executives: Laure Chaffanjon (Presse et Relations Publiques), Ioana Ratill (Marketing), Maud Félice (Marketing Ingrid Millet), Yann Jaffeux (Juridique et Personnel)

FRANCE

PRINCIPAL ACTIVITIES: Manufacture of clothes including shirts, furs and leather goods; perfumes; interests in banking, printing and publishing, retailing through supermarkets; ready to wear

No of Employees: 8,660

SOCIÉTÉ COMMERCIALE DE RÉASSURANCE S.C.O.R.

1 av du Pdt Wilson Cédex 39, 92074 Paris la Défense
Tel: (1) 42.91.04.32
Cable: SCOREAS PARIS
Telex: 614151
Telecopier: (Grp II & III) (1) 47.67.04.09

Supervisory Board: Patrick Peugeot (President), Claude Bébéar (Groupe AXA-Président Directeur Général), Jacques Bonnot (Caisse Centrale de Réassurance-Président Directeur Général), André de Bretteville (Caisse Centrale des Mutuelles Agricoles-Président), Michel Freyche (Banque François du Commerce Extérieur-Président Directeur Général), Jacques-Henri Gougenheim (UAP International-Administrateur-Directeur Général), François Heilbronner (Groupe des Assurances Nationales-Président Directeur Général), Jean-Claude Jolain (Groupe des Sociétés Mutuelles Générales Françaises-Président Directeur Général), Michel Marchal (Abeille Paix Réassurance-Président Directeur Général), Roger Papaz (Assurances Générales de France-Directeur Général), Noël de Saint-Pulgent (Direction des Assurances-Sous-Directeur), Alain Tempelaere (Caisse Mutuelle d'Assurance et de Prévoyance-Directeur Général), Yves Thiré (Mutuelle Assurance Artisanale de France-Directeur Général), François Vilnet, Daniel Grao, Robert Joineau (Administrateurs élus représentant les salariés), Pierre de Vogüé (Président d'Honneur), Louis Franck (Président Honoraire)
General Management: Patrick Peugeot (Président Directeur Général), François Négrier (Directeur Général), Jacqueline Simon (Investissements et Finances, Directeur Général Adjoint), Serge Osouf (Centre de Résultats Paris, Directeur)
Senior Executives: Pierre Croizat (Directeur SCOR RE US), Yves Monmotton (Directeur Risques Techniques), Claude Zabiaux (Directeur de la Rétrocession), Klaus-Dieter Haussinger (Directeur Europe), François Reach (Directeur des Investissements), Claude Roy (Directeur des Relations Sociales)

PRINCIPAL ACTIVITIES: Reinsurance
Trade Names: S.C.O.R. (Société Commerciale de Réassurance)
Parent Company: Caisse Centrale de Réassurance (C.C.R.)
Subsidiary Companies: (100% owned unless stated): S.I.C.A.V. Placements Réassurance (98.68%); Réassurance et Finances (REAFIN) (99.99%); REAFIN-GESTION (65%); IMMOSCOR (99.99%); FORESCOR (99.99%); S.C.O.R. U.S. Corporation (88.14%) (USA); SCORFIN Incorporated (94.19%) (Canada); S.C.O.R. U.K. Group Ltd (UK); S.C.O.R. Reinsurance Australia Ltd (Australia); S.C.O.R. Far East Mmanagement Ltd (99.9%) (Hong Kong); S.C.O.R. Reinsurance Cy (Asia) Ltd (89.98%) (Hong Kong)
Principal Bankers: BNP; BFCE
Financial Information:

	31.12.87 FF'000
Gross premium income	4,441,000
Net premium income	3,414,000
Investment income	628,000
Profit after tax(1)	205,000
Retained profit	11,246
Dividend per share(2)	2.25FF
Profit per share net	15.36FF
Share capital	749,700
Equity capital(1)	1,622,000

(1) including minority interest
(2) including tax credit

Principal Shareholders: Caisse Centrale de Réassurance (22.28%); U.A.P. (14.95%); A.G.F. (14.07%); AXA (10.5%); Mutuelles (12.8%); Mutualité Agricole (4.12%); GAN (7.56%)
No of Employees: 464 (S.C.O.R. Paris); 686 (S.C.O.R. Group)

SOCIÉTÉ GÉNÉRALE

29 Blvd Haussmann, F-75009 Paris
Tel: (1) 40.98.20.00
Cable: SOGESAC
Telex: 212 127F
Telefax: (1) 40.98.70.39

Board of Directors: M Vienot (Président), Michel Albert, La Société Commercial Union Assurance Company PLC (représentée par Anthony B Wyand), La Socété Meiji Life, Georges Cailloue, Guy Dejouany, Jean-Paul Delacour, Jean Connedieu de Vabres, Jean-René Fourtou, Mme Michèle Goosaert, Ernest-Antoine Seillière de Laborde, Pierre Suard, Jean-Luc Wabant
Management Board: J-P Delacour (Directeur Général), Daniel Hua (Directeur Général Adjoint), Léopold Jeorger (Directeur Général Adjoint), Jean-Louis Truchement (Directeur Général Adjoint), Jean-Marie Weydert (Directeur Central), Louis-Moël Joly (Directeur Central)

PRINCIPAL ACTIVITIES: Banking
Trade Names: Société Générale
Subsidiary Companies: (100% controlling interest unless stated): Genebanque; Banque Gravereau; SG de Banque aux Antilles; SG Caledonienne de Banque; SCDB (86.11%); Calif (80%); Banque de Polynesie (80%); Sogenal (56.09%); BIP (49%); BAREP; Sogefim; Sogefinerg; Genecal (92.54%); Sogebail (28.52%); Sogefimur; Genefim (ex Sicotel) (52.55%); Diebold Computer Leasing; Auxibail; Cilomi; GML; Sofinabail; Solomateg (89.96%); Sofinauto; Sogemer; Sogenauto; Auxiloc; Auxilease; Solovam; Genefitec; Genefinance; Geneparts; Genefimmo; IEC; Credit Immobilier General; Sté Auxiliaire de Crédit; Sogefactoring; Sogexter; CREG (96.11%); Fimat (79%); Crédit General Industriel (64.03%); Finecorp; Geneval; Valminco; Financinvest; Fivamob Joubert; Sogapec; Sogevimmo; Ominco (51%); Sogeparts; Cargest SA; Salvepar (50.23%); Cofranteg; Air-Bail; SI Vergeze (60%); SIFP; Geninfo (ex Genefim); DC 2L; Diebold Location; Solobail; ECS (86.84%); Genecommerce (55%); Projis; Socogefi; Sogeplan; CFM; Ste Rue Edouard VII; Quintelec; SCA Paris Trocadero; Sogebanque; SG Canada; SG Australia Holdings; SG Bank Stockholm; SG Elsaessische Bank; Banco Supervielle SG (52.03%); SG Bank Nederland NV; SG Acceptance; Sogen Finanziaria; SG North America; Sogen Asia Ltd; SG Merchant Bank Ltd; Intersoge; SG Strauss Turnbull Holdings Ltd (ex SG Strauss Turnbull); SG Strauss Turnbull Securities Ltd (ex Strauss Turnbull and Co) (59.6%); Intergef SA; Sogecofi; Sogen Fiditalia; Sogefin Credito SA; Socgenlease Ltd; SG Financial Corp (ex Sogelease Corp); Sogelease Italia; Sogelease Hong Kong; Sogelease Japon; Sogelease Indonesia (65%); Sogenalease; Sogelease Pacifique; Surzur Overseas; Cia Mexicana de Leasing Intl (51%); Cia Panamena de Leasing Intl; Sogener; Sogen Finance Luxembourg; BAREP; Banque Byblos France (30%); Banque Franco-Yougoslave (25%); Francheque (45%); Polybail (80%); Gefibull (90%); Solovam (50%); Oxygaz (90%); Midi Parc Azur (90%); Delahaye-Ripault SA (29.99%); Les Accessoires Scientifiques (99.88%); Sogexport (90.6%); Sofrantem (32%); SG 2; Proland; Sogecap (97.89%); Quintelec; SN Noailles Immobiliere (99.97%); Societe Fonciere de Touraine (99.58%); STIH; SC Byron; Cofracib (50%); Odiprom; Sogecob (99%); Sogeprom (99%); Soginnove (22.67%); Valminco; Cargest Sa; Fivamob Joubert; Sogefirem; Ceramiques et Composites (49%); Air Liquide et Midical (26.67%); Industrie Chimique de Mulhouse Dornach (20.51%); SG International Inc; Sogen Securities (50%); Sogeko (50%); National SG Bank SAE (49%); Banco Sogeral (45%); SG de Banques au Senegal (37.93%); SG de Banques au Cameroun (37.8%); SG de Banques en Cote-D'Ivoire (37.19%); SG Marocaine de Banques (36.27%); SG de Banques en Guinee (34%); SG

Bank Nigeria (33.48%); SG Strauss Turnbull Securities Ltd (ex Strauss Turnbull and Co); Asia Credit Ltd (20%); Sogeliban (25%); European American Bancorp (23.15%); United Arab Bank (20%); Al Bank Al Ahli Al Omani (20%); Sogecofi (77.31%); Euroleasing (26.25%); Sogelease Maroc (40%); Sogefibail Cote-D'Ivoire (35.01%); Sogelease Cameroun (35.01%); Sogelease Industrial (50%); Arrendadora International (40%); Trilease (40%); Sogelease Malaysia (49%); Sogelease BV; Sogefin Leasing SA; Intertrail Rentals Ltd; Imfor; Generas; SG Asia Investments; Participaciones; Casadas; Finanverso; Frankrijk Holland; SG Thailand Holding; Valorban; Sagri Holding NV; Sofital; Thai Franco Holding (45%); Thai Franco Enterprise (45%)

Financial Information: Consolidated

	31.12.87 FF'000	31.12.88 FF'000
Profit before tax	3,743,000	5,227,000
Profit after tax	2,680,000	3,578,000
Retained profit	13,341,000	15,627,000
Dividends	531,000	727,000
Dividends per share	10.00FF	13.00FF
Earnings per share	42.50FF	54.40FF
Share capital	1,678,000	1,774,000
Shareholders funds*	21,412,000	24,676,000
Deposits	669,795,000	755,749,000
Total assets	817,139,000	942,071,000

*excluding minority interests

Principal Shareholders: The French State
No of Employees: 45,462

SOCIÉTÉ GÉNÉRALE ALSACIENNE DE BANQUE "SOGENAL"
8 rue Dôme, 67000 Strasbourg
Tel: 88.32.99.27
Cable: Sogenal
Telex: 870720 genalsa strbg
Telefax: 88.32.98.92

Administrative Board: René Géronimus (Président de la SOGENAL), Jean Brenas (Président des Houillères du Bassin de Lorraine), Claude Cotinaut (Cadre de la Succursale de Strasbourge Dorne de la SOGENAL), Daniel Hua (Directeur Général Adjoint de la Société Générale), Louis-Noël Joly (Directeur Central de la Société Générale), Antoine Veil (Administrateur Delégué de la Cie Internationale des Wagons Lits et du Tourisme), Henri Lachmann (Président de la Financière STRAFOR), Maurice Amiel (Président du Groupe TIMKEN Europe, Afrique et Asie de l'Ouest), Pierre Chambon (Professeur à l'Université Louis Pasteur, UER Sciences Medicales â Strasbourg), Gilbert de Dietrich (Président Directeur Général de la Société de Dietrich et Cie), Philippe Pfrimmer (Directeur de Groupe Honoraire), Marguerite Puhl-Demange (Président Directeur Général du Républicain Lorrain), Jacques Schaeffer (cadre de la Direction Générale de la SOGENAL), Jean-Marie Weydert (Directeur Central de la Société Générale)
General Management: René Géronimus (Président-Directeur Général), Albert Mursch (Directeur Général). Mandataires Generaux: Jean Busac (Directeur Général Adjoint, Direction des Relations Humaines), Patrice Bourson (Directeur Général Adjoint, Direction des Affaires Françaises), Henri Lassalle (Directeur Générale Adjoint, Etranger), Claude Wind (Secrétaire Général), François Wiederkehr (Directeur Central, Direction des Engagements et des Relations Commerciales des Agences Etrangères), Jean-Louis Zeisser (Directeur Central, Direction de la Gestion Administrative et Technique). Directeurs: Charles Kuhlmann (Direction des Affaires Juridiques et Contentieuses), Marcel Rehm (Direction de l'Etranger), Marcel Schmitt (Département des Traitements Centralisés). Directeurs Adjoints: Jean-Louis Bellone (Département de l'Action Commerciale), Max de Preval (Bourse et Gestion de Portefeuilles), Théo Bohrmann (Département de la Gestion Administrative), Robert Cudkowicz (Chargé de mission auprès du Président), Gérard

Schultheiss (Département Financier), Jean-Paul Wagner (Direction des Relations Humaines), Frederic Gérault (Département de l'Informatique), August Gruber (Département des Relations et Opérations Bancaire), Max de Préval (Bourse et Gestion de Portefeuille)

PRINCIPAL ACTIVITIES: All banking activities
Trade Names: SOGENAL
Parent Company: Société Générale (Paris)
Subsidiary Companies: Société Générale Financière d'Alsace et de Lorraine "Sogefinal" SA; Elsässiche Bank AG-Sogenal AG (Austria); Comptoir de Valeurs de Banque "COVALBA" SA (Luxembourg); Société Immobilière de Strasbourg SA (France); Sogenal Immobiliengesellschaft mbH (West Germany); Réalia SA (Belgium); Immobiliengesellschaft Bleicherweg AG (Switzerland); Sociéte Générale Elsässiche Bank & Co OHG (West Germany); Sogenal Vie-Strasbourg
Financial Information:

	31.12.86 FF'000	31.12.87 FF'000
Profit before tax	423,538	318,380
Profit after tax	90,153	91,146
Retained profit	63,195	53,773
Dividends	26,957	37,373
Dividends per share	20.50FF	2.90FF*
Earnings per share	234.77FF	18.05FF*
Share capital	263,000	322,184
Customer deposits	20,104,375	19,058,352
Total assets	48,685,161	60,460,992

*shares divided by eight

Principal Shareholders: Société Générale (52.58%)
No of Employees: 2,323

SOCIÉTÉ HOLDING BIS SA
See BIS, SOCIÉTÉ HOLDING, SA

SOCIÉTE LYONNAISE DE BANQUE SA
See LYONNAISE DE BANQUE, SOCIÉTE, SA

SOVAC CRÉDIT MOBILIER INDUSTRIEL
19-21 rue de la Bienfaisance, F-75008 Paris
Tel: (1) 40 08 28 28
Telex: 650920

Management: André Wormser (Président), Gérard Fabry, Yann L'Heveder, Antoine Bernheim
Senior Executives: Gérard Attard (Secrétaire Général)

PRINCIPAL ACTIVITIES: Banking; loan and credit services
Financial Information:

	31.12.87 FF'000
Total assets	22,814,000

SUEZ, COMPAGNIE FINANCIÈRE DE,
1 Rue D'Astorg, F-75008 Paris
Tel: (1) 42.66.47.87
Cable: Cofisu-Paris
Telex: 650323

Board of Directors: Renaud de la Genière (Président-Directeur Général), Roland Béjot, Pierre Bellon, Christophe Chun, Lucien Forget, Bertrand de Gallé, Antoine Jeancourt-Gallignani, Jean-Claude Milleron, Jérôme Monod, Mme Marie-José Monteils, Jacques Perget, Michel Richard, Pierre Suard, Jean-François Vincensini, Guy de Leusse (Secrétaire du Conseil)
Corporate Management: Renaud de la Genière (Président-Directeur Général), Bernard Egloff (Directeur Général Adjoint), Patrick Ponsolle (Directeur Général Adjoint), Gérard Worms (Directeur Général Adjoint, Délégué Général pour les Affaires Industrielles du Groupe), Eric Dufoix (Directeur), Thierry d'Hauteville (Directeur), Jean-Thomas Mandula (Directeur) François de Mortemart (Directeur), Henri Parent (Directeur), Claude Régnier (Directeur)

FRANCE

PRINCIPAL ACTIVITIES: Holding company with interests in banking, investment, insurance, industry, services, commerce, international transport and property development

Subsidiary and Associated Companies: (100% owned unless stated): Banks: Banque Indosuez; Banque Monod (99.6%); Banque Sofinco (60.8%); Banque Vernes et Commerciale de Paris; Banque La Hénin; Banque Parisienne de Crédit (51.1%); Compagnie Financière de Crédit Industriel et Commercial (4.3%). Financial Institutions: Société Auxiliaire pour l'Expansion de l'Industrie et du Commerce "Auxilex"; Factofrance Heller (33.2%). Insurance Companies: Compagnie Financière du Groupe Victoire (30.1%); Compagnie Industrielle (18%); Cecar (30%); La Hénin-Vie (62.8%). Financial Holding Companies: Suez International; Société d'Etudes de Participations et de Gestions Industrielles et Financières "Sepgif"; Société Parisienne de Documentation et de Gérance "Sopadog"; Société de Gestion Financière de Suez (99.7%); La Mure (98.3%); CGPFI (93.4%); SOFFO (47.3%); Compagnie Français d'Entreprises-CFE (99.4%); Comiphos (45.3%); CIDAF (45.2%); Centenaire Blanzy (41.3%). Industrial and Service Companies: Compagnie Industrielle Maritime (28.3%); Lyonnaise des Eaux (19%); Pathé Cinema (13.8%); Affichage Giraudy (12.8%); Compagnie Intelautomatisme (51%); Imétal (9.8%); Béghin-Say (7.4%); Crouzet (5%); Valéo (4.3%); Compagnie de Saint-Gobain (4%); Bouygues (4.9%); Roussel Uclaf (3.5%); International Matting Service Corporation (5%). Holdings: Parthena Investissement (37.6%); Cerus (10%); Compagnie d'Investissements Astorg (35.9%); Lille, Bonnières et Colombes (34.7%); Compagnie de Penhoët (28.1%); Cofide SpA (4.2%). Commercial and International Transport Companies: Compagnie Optorg (15.7%); Minemet SA (7.6%); Oxyde Chemicals BV (50%); International Métal Service (11.6%); Eurotunnel (5.7%); SAGA (45.5%); Compagnie Financière Delmas-Vieljeux (10%). Property Companies: Compagnie La Hénin (43.5%); Immobilière Surène Montalivet (15.2%). Capital Development Companies: Subarec; Suez Gestion; Suez Asia Development Capital

Financial Information:

	31.12.86	31.12.87
	FF'000	FF'000
Profit before tax	2,371,000	2,133,150
Dividends per share(1)(2)	18.40FF	6.00FF
Earnings per share net	162.43FF	33.83FF
Capital stock	3,048,379	4,766,270
Group assets	333,336,487	343,839,211

(1) Ordinary shares
(2) Parent Company

Principal Shareholders: French State
No of Employees: 191 Suez, 17,607 Group

TRAILOR SA

5 route nationale 10, Coignières, BP 49, 78311 Maurepas Cedex
Tel: 3050.61.26
Telex: 698 896 F
Telefax: 30.50.61/26

Board of Directors: J M Levêque, J Reinhold, Y Archer, SPT, SCSR

Management Board: J Reinhold (Président), A Roudil (Directeur Général), G Serre (Directeur Commercial et Marketing), P Regnault (Production), P Rocquemont (Location), B Harinck (Financier et Administratif)

PRINCIPAL ACTIVITIES: Trailers, semi-trailers, renting and leasing
Trade Names: Trailor
Parent Company: Société de Participations Trailor (SPT)
Principal Bankers: Crédit Lyonnais; BNP; Crédit Du Nord; Société Générale

Financial Information:

	1986	1987
	(11 mths)	(12 mths)
	FF'000	FF'000
Sales turnover	739,206	980,941
Profit before tax	109,239	85,656
Profit after tax	109,239	48,876
Earnings per share	390FF	174FF
Share capital	35,012	35,012
Shareholders funds	68,005	117,352

Principal Shareholders: Société de Participation TRAILOR
No of Employees: 1,258

UFB LOCABAIL

43 Quai de Grenelle, 75015 Paris
Tel: (33-1) 45.71.60.60
Telex: 200 015 LOCAB
Telefax: (33-1) 45.77.17.92; (33-1) 45.71.61.99

Board of Directors: Jean-Marie Bossuat (Président Directeur Général), Jean-Michel Bloch-Laine, André Lévy-Lang, François de Margerie, Pierre Martinaud, Boris Méra, François-Regis des Plas, Baudoin Prot, Gérard de Saint-Blanquat, Assurances Générales de France IART (représentées par M Roger Papaz), Crédit Lyonnais (représenté par Michel Renault), Société Générale (représentée par Jean-Paul Delacour)
Management: Jean-Marie Bossuat (Président Directeur Général), François-Régis des Plas (Directeur Général Adjoint), Jacques Brodu (Conseiller du Président), Jean de Ferluc (Directeur), Daniel Hoffsaes (Directeur), Roger Mangold (Directeur), Pierre Müller (Directeur), Jacques Mallet (Directeur), Philippe Bergeron (Contrôleur Général), François Dambrine (Directeur Sécretaire Général)

PRINCIPAL ACTIVITIES: Financing and leasing of equipment for industry, public works and transport; factoring
Parent Company: Compagnie Bancaire
Subsidiary Companies: Loca; Locabail Energie; Locabail International Finance; Locabail United Kingdom; Savelme; Socappa; Ufra-Services; UFB Asset Finance (UK); UFB Deutschland GmbH (Germany)
Financial Information:

	31.12.87
	FF'000
Turnover	8,099,000
Profit after tax	265,000
Dividends	42,820
Dividends per share	10FF
Earnings per share net	62FF
Share capital	4,281,995
Shareholders funds	2,015,000
Total assets	29,733,000

Principal Shareholders: Compagnie Bancaire (62.2%)
No of Employees: 985

UNION DE BANQUES ARABES ET FRANÇAISES
U.B.A.F.

190 ave Charles de Gaulle, 92523 Neuilly Cédex
Tel: (33 1) 47.38.01.01
Cable: UBAFRA
Telex: 610334
Telefax: (331) 47.38.13.88

Board of Directors and Censors: Aly Negm (Chairman), Dr Mohamed Abushadi (Honorary Chairman), Aboubaker A. Al-

Sherif (Vice Chairman), Bernard Thiolon (Vice Chairman), Jean Deflassieux (Honorary Vice Chairman). Shareholding Society: represented by: Hamza Abbas Hussein (Alahli Bank of Kuwait K.S.C.), Ali Al-Rashid Al Bader (Arab African International Bank), Abdul Majeed Shoman (Arab Finance and Holding Company), Rached Abdel Rahman Al-Zayani (Bank of Bahrain and Kuwait), Michel Marto (Bank of Jordan), Ahmed Bennani (Bank Al Maghrib), Mohamed Ould Nany (Banque Centrale de Mauritanie), Mourad Khellaf (Banque Extérieure d'Algérie), Patrick Haizet (Banque Française du Commerce Extérieur), Gérard Landry (Banque Générale du Phenix), Raymond Oidih Audi (Beryte Investment SA-Banque Audi), His Excellency Mahmoud Salah El Dine Hamed (Central Bank of Egypt), Mohamud Mohamed Nur (Central Bank of Somalia), Ali Ali Al-Nuseif (Central Bank of Yemen), Maen Raslan (Commercial Bank of Syria), Jacques Girault (Crédit Lyonnais), Bernard Jacquet (Crédit Lyonnais), Suleiman Bin Muhana El-Awady (Directorate General of Finance of the Sultanate of Oman), Michel Saliba (Jordan National Bank), Abdulhafid Zilitni (Libyan Arab Foreign Bank), Yousuf Hussain Kamal (Ministry of Finance and Petroleum State of Qatar), Clive Smith (National Bank of Abu Dhabi), Mohamed Ali Omaya (National Bank of Yemen), Tarek Al-Tukmachi (Rafidain Bank), El Rifai Kamal Salah Eisa (Riyad Bank), Mohamed Bouaouaja (Société Tunisienne de Banque), Hussain Abdel Gadir Hag Ahmed (Sudan Commercial Bank), Ahmed Adul Rahman Al-Samawy (Yemen Bank for Reconstruction and Development)

Management: Aly Negm (Chairman), Aboubakr A. Al Sherif (Vice Chairman), Bernard Thiolon (Vice Chairman), Jean-Marie Semonsu (General Manager)

Secretary General: Amin Serri Sadek Bassiouni. Senior Assistant General Managers: Roland Chalhoub (International Business), Vahé Kechichian (General Administration). Assistant General Managers: Jean Paul Thomas (International Business), Nouhad G Nassif (Strategic Planning and Budgeting), Jacques Martin-Tardivat (Capital Markets), Alfonse David (Banking Operations), Ghanim Faris (Advisor to the General Management for the Arab World). Second Assistant General Manager: Brahim Achour (U.B.A.F. Branches). Managers: Georges Zouein (Capital Markets-Foreign Exchange), Pierre Cressiot (Credit and Economic Studies), Georgette Pouhaer (Social Relations and Personnel)

PRINCIPAL ACTIVITIES: International commercial bank

Trade Names: U.B.A.F.

Parent Company: UBAC Nederland BV

Principal Associated Banks: ALUBAF Arab International Bank E.C. (Bahrain); UBAF (Hong Kong) Limited (Hong Kong); UBAF Bank Limited (London); UBAE Arab German Bank S.A. (Luxembourg/Frankfurt); UBAF Arab American Bank (New York); UBAE Arab Italian Bank S.p.A. (Rome/Milan); ALUBAF International Bank-Tunis (Tunis)

Principal Branches of the Bank: U.B.A.F. Bahrain Branch; U.B.A.F. Seoul Branch; U.B.A.F. Singapore Branch; U.B.A.F. Tokyo Branch; U.B.A.F. Osaka Branch

Financial Information: Consolidated figures

| | 31.12.86 | 31.12.87 |
	FF'000	FF'000
Profit before tax	86,880	81,152
Profit after tax	65,582	41,401
Retained profit	65,582	41,401
Earnings per share	70.90FF	44.76FF
Share capital	500,000	925,000
Shareholders funds	1,669,686	1,771,383
Deposits	44,304,627	42,637,846
Total assets	49,487,462	48,185,386

Principal Shareholders: UBAC Nederland BV (60%) (including 25 Arab institutions); Crédit Lyonnais (34.94%); Banque Française du Commerce Extérieur (4.32%); Banque Générale du Phénix (0.74%)

No of Employees: 497

UNION DES ASSURANCES DE PARIS

UAP

9 Place Vendôme, 7501 Paris 1er
Tel: (1) 42.60.33.40
Cable: Uapar Paris
Telex: 210798
Telefax: 47.74.28.22

Board of Directors: Chairman: Jean Dromer. Directors: René Barberye, Jean-Charles Boylaud, Roger Burnel, Noël Chamboduc de Saint-Pulgent, Guy Desnouhes, Michel Ferrant, Robert Fruton, Daniel Garnier, Jean-Louis Giral, Gilles Guitton, Philippe Jaffré, François Jourdan-Gassin, Jean-Pierre Landau, Roger Lebon, Pierre Le Clerc, Guy Mayaux, Jérôme Monod

Executive Committee: Chairman and Chief Executive Officer: Jean Dromer. Members: Didier Pfeiffer, Roland Lejart, Jacques-Henri Gougenheim, Roland Plazen. Secretary: Jean-Pierre Ruault

Principal Officers: Chairman and Chief Executive Officer: Jean Dromer. Chief Operating Officers: Didier Pfeiffer, Roland Lejart. Chief Group Supervisor: Roland Plazen. Secretary General: Jean-Pierre Ruault

PRINCIPAL ACTIVITIES: Insurance

Subsidiary Companies: UAP IARD; UAP Vie; UAP Capitalisation; UAP International; Banque Worms; UGIF; UGICOMI; ROMBAS

Financial Information:

	FF'000
Gross premium income	39,000,000

No of Employees: 10,230 Administrative; 6,506 Field staff; 1,709 General agents; 5,979 Other authorized agents

UNION INDUSTRIELLE DE CRÉDIT

UIC

8 Rue Lamennais, F-75008 Paris
Tel: (1) 45.61.98.88
Cable: INCREDUNI-PARIS 042
Telex: 280893F

Board of Directors: François Caries (Président), Christian de Longevialle (Vice Président Directeur Général)
Senior Executives: Pierre Lemas (Secrétaire Général)

PRINCIPAL ACTIVITIES: Financial credit services

No of Employees: 560

VALLOUREC

7 Place du Chancelier Adenauer, 75116 Paris Cedex 16
Tel: (1) 45.02.30.00
Telex: valre paris 611.906 f

Board of Directors: A Leenhardt (Président-Directeur Général), J C Cabre (General Manager), J Gandois, R Granier de Lilliac, J Guyot, R Hudry, J Laballery, M Lauré, R Loubert, F Mer, J Monod
Management: A Leenhardt (Président-Directeur Général)

PRINCIPAL ACTIVITIES: The Vallourec Group is present in three main areas of activity. It is the only seamless steel-tube manufacturer in France (producing carbon, alloy, stainless and non-ferrous pipes and tubes) and is a world leader in this field. In addition Vallourec is highly active in other diverse metallurgical operations and is a major force in construction, industrial installations and specialised engineering services

Subsidiary Companies: TUBES AND FINISHED ARTICLES: Valtubes: Anval Nyby Powder AB (Sweden); Cabval (50%) (USA); Escofier Inc (USA); Escofier Technologie (65%); Hughes Drill Pipe Assembly (50.8%); Interfit; Placements Vallourec Inc (Canada); Sanecap (63.3%) (Belgium); Sidro Rohrbogen (Consolidated) (Germany); Spécitubes; Starval; Valinox; Vallourec Holdings Inc (USA); Vallourec Inc (USA); Vallourec Industries; Vallourec Industries Inc (USA); Valmont; Valti (80%); Vam Premium Connections (51%) (Canada); Vam

FRANCE

PTS (51%) (USA). OTHER METALLURGICAL ACTIVITIES:
Sopretac; Affival; Affival Inc (USA); Biraghi Entrepose (75%);
Cerec; Tubauto; Industube (21.2%) (Morocco); No Sag (44%);
Souby (43.9%); Sotaci (25%) (Ivory Coast). CONSTRUCTION:
Valinco (50.5%); GTM-ENTREPOSE (21.7%)
Principal Bankers: Société Générale; Banque Nationale de Paris
Financial Information:

	31.12.86 FF'000	31.12.87 FF'000
Sales turnover	6,186,362	5,475,645
Share capital	508,964	508,964

Principal Shareholders: COFINAN (Groupe Nord Est); Usinor
No of Employees: 24,000 Group

Major Companies of
GERMANY

AACHENER UND MÜNCHENER LEBENSVERSICHERUNG AKTIENGESELLSCHAFT

Postfach 26, 5100 Aachen
Tel: (0241) 6001-0
Telex: 832346 amld d
Telefax: (0241) 6001-138

Supervisory Board: Dr Helmut Gies (Vors)
Management Board: Wilko H Börner (Vors), Dr Paul Bernhard Hagelschuer, Dr Michael Kalka, Walter Menz, Wilhelm Peter Stommel, Hans-Reinhardt Latten (Stv. V.)

PRINCIPAL ACTIVITIES: Life insurance
Parent Company: Aachener und Münchener Beteiligungs Aktiengesellschaft
Principal Bankers: Bank für Gemeinwirtschaft, Aachen; Commerzbank, Aachen; Commerzbank, Köln; Deutsche Bank, Aachen; Dresdner Bank, Köln; Postgiro, Köln
Financial Information:

	31.12.86 DM'000	31.12.87 DM'000
Premium income	1,153,800	1,360,000
Investment income	627,012	687,051
Profit before tax	320,803	341,861
Profit after tax	9,400	13,800
Retained profit	4,650	7,910
Dividend	4,400	5,200(1)
Dividend per share	10.00DM	10.00DM(2)
Earnings per share	20%	20%(3)
Share capital	22,000	30,000
Shareholders funds	64,000	70,000

(1) plus bonus DM2,600,000
(2) plus bonus DM5.00
(3) plus bonus 10%

No of Employees: 1,516

AACHENER UND MÜNCHENER VERSICHERUNG AG

Aureliusstr 2, Postfach 10, 5100 Aachen
Tel: (0241) 4560
Cable: Amvers Aachen
Telex: 0832872 amac d
Telefax: (0241) 68841

Supervisory Board: Dr Helmut Gies (Chairman), Joachim Maass (Deputy Chairman), Peter David, Monika Hendricks, Alan Arthur Horsford, Günter Markmann, Dr Bernhard Monheim, Dr Friedrich Nachtsheim, Dr Günther Radtke, Prof Dr jur Dr Ing Eh Reimer Schmidt, Hans-Werner Schüttler, Horst Schwieder
Management Board: Wolf-Dieter Bamgartl (Chairman), Arnulf Anstett, Wolfgang von Fuchs, Walter Menz, Dr Norbert Winter (Member), Dr Michael Rohe, Wilfried Simons, Dr Alfons Weiss (Deputy Member)
Senior Executives: Richard Valtin (General Representative), K-H Amendt (Director), R Fichtner (Director), Albert Schmitz (Director), Dr Hans Wille (Director), M Bittner (Deputy Director), W-G Loers (Deputy Director), A Poschen (Deputy Director), W Reinhard (Deputy Director), R Runte (Deputy Director), W Amelang (Head of Department), H J Garbe (Head of Department), Dr P Kriesendorf (Head of Department), F Jonas (Deputy Director), U Peters (Head of Department), E Prell (Head of Department), H-G Prien (Head of Department), J Wruck (Head of Department), A Ackermann (Executive), B Angenendt (Executive), N Barth (Executive), W Ehlen (Executive), G Guntermann (Executive), P-L Hubert (Executive), H Lüttgens (Executive), U Mader (Executive), A Mantel (Executive), G Nebelung (Executive), G Relitzki (Executive), P Savelsbergh (Executive), H-G Sievert (Executive), H Soiron (Executive), H Tüting (Executive), H D Wege (Executive)

PRINCIPAL ACTIVITIES: Insurance and re-insurance

Parent Company: Aachener und Münchener Beteiligungs-Aktiengesellschaft
Subsidiary Companies: A & M Verzekering Maatschappij NV, Rotterdam, London and Aachen; Munich Marine Insurance Co, London
Principal Bankers: Landeszentralbank; Dresdner Bank; BfG Bank
Financial Information:

	31.12.86 DM'000	31.12.87 DM'000
Premium income	1,159,660	1,201,193
Investment income	281,191	421,827
Profit before tax	114,800	123,821
Profit after tax	46,318	47,180
Retained profit	23,100	23,500
Dividends	15,840	17,600
Dividends per share	9.00DM	10.00DM
Earnings per share*	14.65DM	17.12DM
Share capital	88,000	88,000
Shareholders funds	251,179	280,759

*Without tax-credit

Principal Shareholders: Aachener und Münchener Beteiligungs-Aktiengesellschaft
No of Employees: 3,307

AGRIPPINA VERSICHERUNG AG

Postfach 140180, D-5000 Köln I
Tel: (0221) 77150
Telex: 8 885 391 agrv d
Telefax: (0221) 7715772

Supervisory Board: Fürsprecher Fritz Gerber, Arlesheim (Vors.), Elisabeth Becker-Töpfer (Düsseldorf)*, Rolf F Hüppi (Kilchberg), Herwig Scheffler (Kaarst, stv. Vors.)*, Wilfried Eule (Köln)*, Henri Francois-Poncet (München), Dr Manfred Gentz (Esslingen), Fürsprecher Hans-Peter Karlen (Erlenbach), Hans-Dieter Kaschig (Bergisch Gladbach)*, Prof. Dr. Drs. h.c. Peter Ludwig (Aachen), Hans-Rolf Schmitz (Köln)*, Torsten Tiefenbacher (Hamburg)*, *AN-Vertreter
Management Board: Dr Wilhelm Fritz (Vorsitzender), Dr Reinhard Fuchs, Dr Antonius Kemmermann, Dr Dr Gerold Loos, Ernst Meiser, Edilbert Müller

PRINCIPAL ACTIVITIES: Provision of general insurance services
Parent Company: Zürich Versicherungs-Gesellschaft, Zürich
Subsidiary Companies: Agrippina Lebensversicherung AG; Agrippina Rückversicherung AG; Agrippina Rechtsschutzversicherung AG; Patria Versicherung AG
Principal Bankers: Delbrück & Co; Deutsche Bank; Dresdner Bank; Commerzbank
Financial Information:

	31.12.86 DM'000	31.12.87 DM'000
Premium income	761,328	796,720
Investment income	117,155	116,937
Profit before tax	49,348	37,305
Profit after tax	17,898	16,348
Retained profit	20,676	19,599
Dividends	15,950	16,240
Dividends per share	29.00DM	29.00DM
Share capital	96,000	98,000
Shareholders funds	147,559	152,559

Principal Shareholders: "Zürich" Versicherungs-Gesellschaft, Zürich
No of Employees: 1,898

ALBINGIA VERSICHERUNGS-AG

Ballindamm 39, D-2000 Hamburg 1
Tel: (040) 30220
Telex: 0216774

Executive Board: J von der Meden (Chairman)

PRINCIPAL ACTIVITIES: Insurance

Subsidiary Companies: Albingia Lebensversicherungs-AG; Albinga Rechitsschutz-Versicherungs-AG; Albingia Verwaltungs-GmbH

Financial Information:

	DM'000
Premium income	1,400,000

No of Employees: Company 1,830

ALFRED C TOEPFER VERWALTUNGS GESELLSCHAFT GMBH

See TOEPFER, ALFRED C, VERWALTUNGS GESELLSCHAFT GMBH

ALLGEMEINE HYPOTHEKENBANK AG

Savignystrasse 43, 6000 Frankfurt am Main

Tel: (069) 7179-0

Telex: 04 13 996

Supervisory Board: Dr h c W Hesselbach (Vors), F Bode

Management Board: Klaus G Matthes, Dr Friedrich Pörings, Dr Bert Rauscher, Willi Stumpf

PRINCIPAL ACTIVITIES: Mortgage bonds and communal bonds, mortgage loans and communal loans

Principal Shareholders: Bank für Gemeinwirtschaft AG (25%); BHW-Gesellschaft für Wohnungswirtschaft mbH (25%); Deutsche Beamtenversicherung (25%); Volksfürsorge Lebensversicherung AG (25%)

No of Employees: 100

ALLIANZ AG HOLDING

Königinstrasse 28, Postfach 440124, 8000 München 44

Tel: (089)38000

Cable: Allianz München

Telex: 5230110 am d

Telefax: (089) 349941

Supervisory Board: Prof Dr Herbert Grünewald (Chairman), Gerhard Claasen, Dr Wilfried Guth, Dr Maximilian Hackl, Rudolf von Bennigsen-Foerder, Ewald Braden, Prof Dr. Ing. E.h. Werner Breitschwerdt, Dr-Ing e.h. Werner H Dieter, Dr Dr h.c. Klaus Liesen, Prof. Dr h.c. Hans L Merkle, Dieter Müller, Heinz Offermann, Walter Offtermatt, Dr Ing e.h. Bernhard Plettner, Dr Wolfgang Röller, Karl-Heinz Schubert, Lorenz Schwegler, Dr Hans Seyfried, Franz Stadler, Dr Rüdiger Stroebel

Board of Management: Dr Wolfgang Schieren (Chairman), Dr Jan Boetius, Dr Uwe Haasen, Dr Wolfgang Müller, Dr Friedrich Schiefer, Ernst Wunderlich

PRINCIPAL ACTIVITIES: Holding and reinsurance company of Allianz Group of insurance companies

Subsidiary Companies: Insurance Subsidiaries:- Germany: Allianz Versicherungs-AG (and its subsidiaries Frankfurter Versicherungs AG; Bayerische Versicherungsbank AG; Allianz Allgemeine Rechtsschutzversicherungs AG and others); Allianz Lebensversicherungs AG. Abroad: Cornhill Insurance PLC (UK); Riunione Adriatica di Sicurtà SpA (Italy) and other insurance subsidiaries in Austria, Chile, France, Italy, Luxembourg, Netherlands, Singapore, South Africa, Spain, Switzerland, UK, USA

Financial Information: Consolidated figures (1)

	31.12.86 DM'000	31.12.87 DM'000
Premium income(2)	10,010,402	10,599,937
Investment income	1,619,870	1,858,835
Profit before tax	923,760	775,283
Profit after tax	377,123	405,585
Retained profit	174,649	195,787
Dividends(3)	153,969	180,000
Dividends per share(3)	12DM	12DM(4)
Share capital	641,536	750,000
Shareholders funds	4,989,163	5,974,264

(1) excluding German life insurance company and foreign insurance companies
(2) on a gross basis; worldwide premium income 1986: DM19.2 billion; 1987: DM25.9 billion
(3) parent company
(4) recommended

Principal Shareholders: Münchener Rückversicherungs-Gesellschaft, Munich (25%)
No of Employees: 41,131 Group Worldwide

ALLIANZ LEBENSVERSICHERUNGS AG

Reinsburgstrasse 19, Postfach 534, D-7000 Stuttgart 1
Tel: (0711) 663-0
Cable: Allianzleben Stuttgart
Telex: 723571 ales d
Telefax: (0711) 663 2654

Supervisory Board: Dr jur Wolfgang Schieren (München, Vors) Ulrich Hahne (Stuttgart, stellv.Vors), Detlev Brandenburg, Hugo Klein, Jürgen Kunert, Hans Müller, Reinhold Pohl, Harold Törmer
Management Board: Dr jur Uwe Haasen (Vors), Dr jur Peter von Blomberg, Dr Dr jur Karl Reutter, Reinhold Hendricks (stellv.), Dr Henning Schulte-Noelle, Jürgen Eichelmann

PRINCIPAL ACTIVITIES: Capital life insurance; pension insurance; death-risk insurance
Trade Names: Allianz
Parent Company: Allianz AG Holding, München
Financial Information:

	31.12.86 DM'000	31.12.87 DM'000
Premium income	6,124,000	-
Profit after tax	67,050	68,700
Retained profit	31,650	34,350
Dividend	23,400	29,700
Share capital	130,000	165,000

Principal Shareholders: Allianz AG Holding München; Münchener Rückversicherungs-AG, München
No of Employees: 3,968 (full-time); 757 (part time)

BADEN-WUERTTEMBERGISCHE BANK AG
BW Bank

Kleiner Schlossplatz 11, 7000 Stuttgart 1
Tel: (0711)2094-0
Cable: BW-Bank
Telex: 721881 bws
Telefax: (0711) 2094-566

Supervisory Board: Dr Guntram Palm (Vorsitzender), Wolfgang Amend, Dr A Hetzel (Stellv Vorsitzender), Dr Erberhard Benz, Hans Hartmann, Ute-Susanne Herweg, Dr Fritz Hopmeier, Peter Keller, Rita Oster, Dr Dieter Paulus, Martin Pohl, Thomas Reinauer, Klaus Rücker, Rolf Schoeck, Professor Dr Robert Schwebler, Melitta Sieger, Hans Theuer, Johann Wieland, Dr Christoph Wocher, Carl Herzog von Württemberg

Management Board: Dr Otto K Deutelmoser, Dr Walter Küster, Dieter Maier, Dr Helmut Mattes, Dr Harro Petersen, Dr Manfred Prechtl, Manfred W Groos (stellv)

Senior Executives: Siegfried Huck (Generalbevollmaechtigter), Martin Pohl (Direktor) Heinz G Grimberger (Direktor), Kai Graf von der Recke (Generalbevollmächtigter), Eberhard Dietz (Direktor), Horst Marschall (Generalbevollmächtigter), Dr Rolf G Meyer (Direktor), Eckhard Neumann (Direktor), Wolfgang Sauter (Direktor), Rudolf Schmidt (Direktor), Jürgen Volk (Direktor), Jens Wünscher (Direktor), Rudolf Zipf (Direktor), Martin Zipperer (Direktor)

PRINCIPAL ACTIVITIES: All banking business

Subsidiary Companies: (100% owned unless stated): Anselm Verwaltungsunion GmbH (64%); Bankhaus Bensel GmbH; FAVOR Vermögensverwaltungs-GmbH; Grundstücksverwaltungsgesellschaft mbH; GVG Grundstücksvermittlungsgesellschaft mbH; Heilbronn Maschinenbau GmbH (99%); IVG Immobilien Vermittlungs-GmbH; Oberrheinische Vermögensverwaltungs-GmbH; Rodi & Wienenberger AG (99.4%); WKV Bank GmbH

Principal Bankers: Main bank connection- Landeszentralbank in Baden-Württemberg

Financial Information:

	31.12.86 DM'000	31.12.87 DM'000
Volume of business	12,556,000	13,421,000
Retained profit	35,920	33,800
Dividends	25,920	28,800
Dividends per share	12DM	12DM
Share capital	108,000	120,000
Share capital and reserves	425,000	509,000
Total deposits*	10,668,000	11,418,000
Total assets	11,299,085	12,129,747

*and bonds issued

No of Employees: 2,055

BAKOLA LEASING & BETEILIGUNGS GMBH
D-6800 Mannheim 1
Tel: 1 88-0
Telex: 462151 balea d

PRINCIPAL ACTIVITIES: Leasing services
Financial Information:

	31.12.87 DM'000
Total assets	25,845,000

BANK FÜR GEMEINWIRTSCHAFT AG
Theaterplatz 2, D-6000 Frankfurt (Main) 11
Tel: (69) 2580
Cable: Hauptbankwirt Frankfurtmain
Telex: 412210
Telefax: (69) 2587578

Management Board: Matthias Hoffmann-Werther, Gerhard Jakobi, Udo Knop, Rudiger Kreft, Dr Ralf Krüger, Melanie Radke, Erich Wilkee

PRINCIPAL ACTIVITIES: Banking activities
Subsidiary/Associated Companies: Allgemeine Hypothekenbank AG (25%); BSV Bank für Sparanlagen und Vermögensbildung AG (51%); Bau und Handelsbank AG (95%); WTB Westdeutsche Kreditbank GmbH (83.3%); BfG Luxembourg SA (Luxembourg); Hollandse Koopmansbank NV (33.3%) (Netherlands); BfG-Investment Fonds Gesellschaft mbH (95%); Deutsche Immobilien Fonds AG (50%); WKU Banque Universelle et Commerciale du Luxembourg SA (Luxembourg); BfG Finance Asia Ltd (Hong Kong); Israel Continental Bank (39.9%) (Israel)

Principal Shareholders: Beteiligungsgesellschaft für Gemeinwirtschaft AG (89.5%); Volksfürsorge Lebensversicherung AG (10%)
No of Employees: 9,605 (worldwide)

BAUSPARKASSE GDF WÜSTENROT GMBH
Hohenzollernstrasse 46, D-7140 Ludwigsburg
Tel: (07141) 16-1
Telex: 7264875 gdf d
Telefax: (0 71 41) 16-36 37

Supervisory Board: Staatsminister a.D.Dr.h.c. Rudolf Eberhard (München, Vorsitzender), Karl-Heinz Römer, Dr. h.c. Arno Paul Bäumer, Dr. Erich Bracher, Dr. Gerhard Götte, Erika Hochwimmer, Heinz Haas, Dr. Hellmuth Hahn, Dr. Alfred Hetzel, Ingeborg von Killisch-Horn, Heinrich Mohr, Wolfgang Pietsch, Klaus Rennebach, Ottmar Schnauber, Peter Schuhmacher, Artur Wolf
Management Board: Klaus W Rösch, Dr. Otto Schäfer, Dr. Walter Seuferle, Dr. Otfried Ulshöfer, Dr. Christoph Wocher (Sprecher), Hans Peter Kappes (stv)

PRINCIPAL ACTIVITIES: Building society
Parent Company: Wüstenrot Holding GmbH, Ludwigsburg
Financial Information:

	31.12.86 DM'000	31.12.87 DM'000
Profit before tax	98,448	89,812
Profit after tax	45,450	42,230
Retained profit	15,450	32,340
Share capital	404,250	404,250
Shareholders funds	1,074,000	1,084,000
Deposits	20,329,000	18,847,000
Total assets	22,392,000	20,894,000

Principal Shareholders: Wüstenrot Holding GmbH, Ludwigsburg
No of Employees: 4,181

BAUSPARKASSE SCHWÄBISCH-HALL AG
Crailsheimer Strasse 52, D-7170 Schwäbisch-Hall
Tel: (0791) 460
Telex: 74833

Management Board: Dr F Bieling

PRINCIPAL ACTIVITIES: Banking services

No of Employees: 3,240

BAYERISCHE BEAMTEN LEBENSVERSICHERUNG A.G.
Thomas Dehler-Str 25, 8000 München 83
Tel: (089) 67870
Cable: Beamtenleben Muenchen
Telex: 897220 BBVHV
Telefax: (089) 6787521

Supervisory Board: Dr jur Rainer Kessler (Vorsitzender), Peter Kroll (Stellv Vorsitzender)
Management: Dr jur Klaus D Schweickert (Vorsitzender), Rolf W Mintrop, Wolfgang Werner, Dr Heinz Scheller

PRINCIPAL ACTIVITIES: Life insurance
Subsidiary Companies: Bayerische Beamten Versicherung AG
Financial Information:

	31.12.86 DM'000	31.12.87 DM'000
Premium income	526,763	569,460
Investment income	325,494	324,066

Principal Shareholders: Mutual insurance corporation
No of Employees: 900

BAYERISCHE HANDELSBANK AG
Von-der-Tann Str 2, 8000 München 22
Tel: (089) 23041
Cable: HANDELSBANK
Telex: 522205 bhb d
Telefax: (089) 2304 304

Supervisory Board: Dr Helmut Scholz (Vors), Dr Albrecht Schmidt (stv Vors), Prof Dr Reinhold Kreile, Dr Heribald Närger, Johannes Fürst von Thurn und Taxis, Klemens

Wesselkock, *Hans Gammel, *Leonhard Wissmann, *Helga Weidinger, *Arbeitnehmervertreter

Management Board: Dr Walter Dieck, Dir Rudolf Hess, Dr Konrad Rüchardt

PRINCIPAL ACTIVITIES: Mortgage bank
Parent Company: Bayerische Vereinsbank AG
Financial Information:

	31.12.86	31.12.87
	DM'000	DM'000
Profit before tax	68,959	71,968
Profit after tax	30,218	30,839
Retained profit	15,516	15,955
Dividend	13,200	13,750
Dividend per share	12DM	12.50DM
Earnings per share	27.47DM	28.04DM
Share capital	55,000	55,000
Shareholders funds	382,316	399,955
Deposits	19,091,681	19,883,742
Total assets	19,569,254	20,378,545

Principal Shareholders: Bayerische Vereinsbank AG (76%)
No of Employees: 276

BAYERISCHE HYPOTHEKEN- UND WECHSEL-BANK AG

Theatinerstr 11, 8000 München 2
Tel: (089) 2366-1 (Theatinerstr address); (089) 9244-1 (Arabellastr address)
Cable: Hypobank München
Telex: 528650 hypod
Telefax: (089) 2366-2880 (Theatinerstr); (089) 9244-3188 (Arabellastr)

Supervisory Board: Dr Klaus Götte (Chairman), Edith Ruf (Deputy Chairman), Ernst Eigner, Ewald Matejka, Frieder Maxl, Dr. rer. pol. Hans Meinhardt, Dr. oec. publ. Günther Rid, Dr. jur. Walter Rieger, Dr. rer pol. Friedrich Schiefer, Dr. rer. pol. Ronaldo H Schmitz, Doris Schönbeck, Dr. rer. com. Hans Singer, Friedrich Stumpfegger, Adolf Wächter, Ferdinand Walter, Christel Welter
Board of Managing Directors: Dr. Hans Fey, Dr Hans-Hubert Friedl, Dr. Klaus Geiger, Dr. Joachim Hausser, Dr. Klaus Heiss, Dr. Peter Hoch, Dr. Eberhard Martini, Wilhelm Pfeiffer, Dr. Martin Schütte
Senior Executives: Hermann Bauer (Executive Vice President, International Department)

PRINCIPAL ACTIVITIES: Commercial and mortgage banking
Trade Names: Hypobank
Subsidiary Banks: (100% owned unless stated): Hypobank International SA (Luxembourg); Westfalenbank AG (over 99%) with its group companies: Westfalenbank International SA (Luxembourg); Gesellschaft für Grundbesitz mbH; Westfalen Kapitalverwaltungsgesellschaft mbH; Westfälische Hypothekenbank AG (75%); Württembergische Hypothekenbank AG (75%); Salzburger Kredit- und Wechsel-Bank AG (over 99%) (Austria); Holding-Salweba AG (Switzerland); Bayernhypo Finance NV (Netherlands); Hypo Trade Finance Ltd (UK); Hypo-Bank Verwaltungszentrum GmbH; Hypo-Bank Verwaltungszentrum GmbH & Co,KG; Hypo-Immobilien-Service Gesellschaft mbH; Portia Grundstücksverwaltungs Gesellschaft mbH (95%); Portia Grundstücksverwaltungsgesellschaft mbH & Co, Betriebs-KG (95%); Anwa Gesellschaft für Anlagenverwaltung mbH (95%); Diaterra Vermögensverwaltungs-Gesellschaft mbH; Fera Vermögensverwaltungs-Gesellschaft mbH; Imwa Gesellschaft für Immobilienverwaltung mbH (95%); Metaterra Gesellschaft für Immobilienverwaltung mbH; Synterra Gesellschaft für Immobilienverwaltung mbH; Meteor Immobilienverwaltungsgesellschaft mbH (80%); Mundus Immobilienverwaltungsgesellschaft mbH (80%)

Financial Information: Consolidated figures

	31.12.86	31.12.87
	DM'000	DM'000
Profit before tax	589,800	632,300
Profit after tax	238,600	250,200
Retained profit	177,600	190,000
Dividends per share	12.50DM	12.50DM
Earnings per share	75.70DM	71.80DM
Share capital	682,400	750,600
Shareholders funds	3,313,000	3,812,000
Deposits	114,901,000	120,216,000
Total assets	120,366,000	126,284,000

Principal Shareholders: shares widely held
No of Employees: 11,980 Parent Bank, 13,171 Group

BAYERISCHE LANDESANSTALT FÜR AUFBAUFINANZIERUNG (LFA)

Königinstrasse 15, D-8000 München 22
Tel: (089) 21 24-0
Telex: 524612
Telefax: 21 24 440

Supervisory Board: Anton Jaumann (Vorsitzender), Franz Josef Amm, Dr Karl Hillermeier, Albert Meyer, Franz Neubauer, Franz Olbert, Dr Wilhelm Wimmer, Joachim Schweinoch
Management Board: Udo Walter (Präsident), Hans Kellndorfer (Vizepräsident), Dr Richard Flessa, Jost Ritter, Dr Rupert Pfeffer, Dr Alois Langseder

PRINCIPAL ACTIVITIES: Banking and financial services
Financial Information:

	31.12.86	31.12.87
	DM'000	DM'000
Profit after tax	72,000	72,000
Retained profit	36,000	36,000
Share capital	300,000	300,000
Shareholders funds	666,000	711,000
Total assets	14,081,000	14,919,000

Principal Shareholders: State of Bavaria (100%)
No of Employees: 274

BAYERISCHE LANDESBANK, GIROZENTRALE

Brienner Strasse 20, Postfach 20 05 25, D-8000 München 2
Tel: 089-2171-01
Cable: bayernbank
Telex: 52862-70 gzm d
Faxa: (089) 2171-3579 (General); (089) 2171-5022 (Int Com Dept)

Board of Administration: Franz Neubauer (Chairman and Executive President of the Association of Bavarian Savings Banks), Dr h c Max Streibl (First Deputy Chairman, Bavaria State Minister of Finance), Erich Rambold (Second Deputy Chairman, Chief District Administrator and First President of the Association of Bavarian Savings Banks), August R Lang (Third Deputy Chairman, Bavaria State Minister of the Interior)
Board of Management: Dr Hans Peter Linss (Chairman), Alfred Lehner (Deputy Chairman), Prof Dr Gerhard Tremer, Dr Hubert Schmid, Dr Max Weiss, Dr Eberhard Zinn, Dr Peter Kahn, Werner Strohmayr
Senior International Officers: Dieter Grass (International Division), Alfred Müller (International Commercial Dept), Michael Stauder (International Finance Dept), Dr Jens Lindemann (International Network Dept), Günter Prechtel (International Control Dept), Gerhard Schott (Foreign Exchange Dept), Winfried Rohaus (New Issues and Credit Syndication Dept), Manfred J Baumann (International Trade Finance Dept), Bruno Lotz (Securities Division), Gerd Peiss (Equities-Domestic and International), Joachim Weinhold (International Bond and Equities Trading and Sales), Hartmund Hölzer (Economic Research)

PRINCIPAL ACTIVITIES: Commercial banking; investment banking; issuing activities; central banker to Bavaria and its municipalities; central banking institution of the Bavarian Savings Banks

Subsidiary Companies: Bayerische Landesbank International SA (100%) (Luxembourg)

Principal Bankers: London: Midland Bank PLC. New York: BancAmerica International; Citibank NA; Chemical Bank; Manufacturers Hanover Trust Company

Financial Information: Consolidated figures

	31.12.86 DM'000	31.12.87 DM'000
Profit before tax	260,290	254,342
Profit after tax	146,500	146,500
Retained profit*	80,000	80,000
Dividends	66,500	66,500
Capital stock	950,000	1,000,000
Shareholders funds	2,736,000	2,866,000
Deposits and bonds	96,312,180	111,167,283
Total assets	115,466,421	131,321,819

*allocation to capital reserves

Principal Shareholders: State of Bavaria (50%); Bavarian Savings Banks Association (50%)

No of Employees: 4,297

BAYERISCHE VEREINSBANK AG

1 und 14 Kardinal-Faulhaber-Strasse, 8000 Munich 2

Tel: (89) 2132-1

Cable: Vereinsbank

Telex: 52106-0 bvm d (foreign dept); 523768 bvm d (press dept)

Telefax: 21 326 182

Supervisory Board: Dr Werner Premauer (Chairman)

Management Board: Dr R Baur, Dr M Hackl, Prof Dr D Koellhofer, Dr A Puhlmann, P Reimpell, Dr A Schmidt, Dr H Scholz, Dr R Trautner, H Diehm, Dr E Prasch, H Bonn, Dr E Eisele, Hans-Detlef Bösel (Deputy)

Senior Executives: Bonds and Stock: Dr Christian Humbert, Gerold Lehmann (General Manager). International Division: Manfred Gottschaller, Caspar v Hauenschild (General Manager). Treasury and Foreign Exchange: Wolf Siekmann (General Manager). Corporate Finance: Dr Hans Freiherr von Schoen, John Freiherr von Twickel (General Manager)

PRINCIPAL ACTIVITIES: Commercial and mortgage banking (longterm/mortgage loan business)

Trade Names: Bayerische Vereinsbank

Subsidiary Banks: Bayerische Vereinsbank International SA, Luxembourg; Bayerische Vereinsbank SA (BV France), Paris; Bayerische Vereinsbank Finance Company BV, Amsterdam; BV Capital, New York; Wirtschafts-und Privatbank, Zürich

Financial Information:

	31.12.86 DM'000	31.12.87 DM'000
Profit before tax	442,000	459,000
Profit after tax	187,600	185,500
Retained profit	149,900	150,400
Dividends	26%	26%
Dividends per share	13.00DM	13.00DM
Share capital	572,000	574,000
Shareholders funds	2,610,800	2,662,300
Deposits	75,927,000	81,821,500
Total assets	81,466,300	87,553,200

No of Employees: 14,270 Group, 12,497 Parent Company

BAYERN-VERSICHERUNG ÖFFENTLICHE LEBENVERSICHUNGSANSTALT

Postfach 900204, D-8000 München 90

Tel: (089) 6236-0

Telex: 522431 bymue d

Management Board: Dr H Pestenhofer

PRINCIPAL ACTIVITIES: Life insurance

Financial Information:

	DM'000
Premium income	780,000

No of Employees: 1,570

BERLINER BANK AKTIENGESELLSCHAFT

Postfach 12 17 09, Hardenbergstrasse 32, 1000 Berlin 12

Tel: (030) 31 09-0

Cable: Bankkredit Berlin

Telex: 182010 bb d; 183445 beba d (Foreign Exchange); 1820144 bb d (International Division)

Telefax: 31092165; 31092168

Supervisory Board: Edzard Reuter (Chairman), Bernd Reinhard (Deputy Chairman), Norbert Anklam, Dipl Ing Hermann Franz, Klaus Hoffmann, Rüdiger Keller, Prof A Wilhelm Klein, Horst Kramp, Dr Wolfgang Peiner, Claudia Pienkny, Erich Rehm, Dr Günter Rexrodt, Lothar Schlinke, Johann Peter Sieveking, Hans-Jürgen Schulz, Renate Tröger, Günther Waschkuhn, Dr Wolfgang Watter, Hans Michael Wiebe, Karlheinz Woblick

Board of Management: Günther Bernt, Dr Knut Fischer, Hartmut Fuchs, Karl Lotter, Wolfgang Steinriede, Leopold Tröbinger

Executive Managers: Horst Buchholz, Robert K Gogarten, Andreas Graf von Hardenberg, Klaus A Heiliger, Jürgen Noack. Senior Managers: Dr Dietrich Beier, Helmut Beyer, Dr Jost Brehmer, Frank-Detlef Brenning, Heinz-Dieter Fuss, Lothar Heck, Claudio Jarczyk, Manfred Kutzki, Hans-Jürgen Meyer, Klaus Naumann, Rudolf Prast, Dr Joachim Preussner (Chief Legal Adviser), Helmut Ramthun, Bernd Scherenberger, Günther Schultze, Dr Helmuth Strothmann, Dr Helmut Wenzel, Dr Wilhelm Zülch. Branch Managers: Berlin: Rüdiger Keller, Gerhard Parpart, Manfred Staude. Düsseldorf: Dr Klaus Deeken, Günter Goldmann, Rainer Santowski. Frankfurt am Main: Dr Peter Baier, Horst Büttner, Dr Wolfgang Marquardt. Hamburg: Dr Rainer Flöge, Friedrich Höltke. Hanover: Norbert Kewitsch, Dr Wilfried Müller. London: Wolfdietrich F Brühl, Anthony J Price. München: Dieter Bleicher, Peter Rosenbauer, Manfred Weigand. Stuttgart: Andreas Fellmann, Wolfgang Krug. Corporate Banking: Joachim Rosenberger, Klaus Henner Schütt

PRINCIPAL ACTIVITIES: International banking

Principal Subsidiary Companies - Berliner Bank Group: (100% owned unless stated): Allgemeine Privatkundenbank AG (96.4%); Braunschweig-Hannoversche Hypothekenbank AG (60%); Berliner Bank International SA (Luxembourg); BB-Kapitalbeteiligungsgesellschaft mbH; BB-Unternehmensbeteiligungs-AG; BB-Leasing GmbH (95%); BB-DATA Gesellschaft für Informations- und Kommunikationssysteme mbH (80%); BB-Grundfonds Vermittlungs GmbH (50%); BB-Versicherungsservice GmbH (33.3%); Grundstücksgesellschaft "Berlin" mbH (BB-Hausverwaltung GmbH). Cooperation Partners: Gothaer Versicherungsbank VVaG; Gothaer Lebensversicherung AG

Financial Information:

	31.12.86 DM'000	31.12.87 DM'000
Profit before tax	42,338	80,973
Profit after tax	17,917	42,794
Consolidated profit	3,367	28,238
Dividends per share	-	8%
Share capital	250,000	250,000
Shareholders funds	751,813*	778,088*
Deposits	26,407,255	28,889,126
Total assets	28,141,586	30,681,003

*incl profit participation certificates

Principal Shareholders: City of Berlin (74%)

No of Employees: 4,726 Group

BERLINER HANDELS- UND FRANKFURTER BANK
BHF-Bank
Bockenheimer Landstr 10, Postfach 110311, 6000 Frankfurt 1

Tel: (069) 7180

Cable: bhf-bank

Telex: 411026 bhfbank d

Telefax: (069) 718 2296

Supervisory Board: Dr H.Ch.Schroeder-Hohenwarth (Chairman)

Managing Partners: Dr Wolfgang Graebner, Dr Peter Opitz, Dr Wolfgang Rupf, Wolfgang Strutz, Klaus Subjetzki, Rüdiger v Tresckow

Heads of Divisions: Dr Jörg Cramer, Sigurd Fischer-Güttich, Dr Hans-Peter Geitmann, Dr Klaus Gust, Jürgen Kallmeyer, Nikolaus Korsch, Dr Günther Mecklenburg, Tilo Paduch, Dieter Rampl, Dietmar Schmid, Karlheinz Schöffel, H Jürgen Schröter, Richard Spanner, Manfred Thamke, Dr Peter Wendt, Heinz Zollweg

PRINCIPAL ACTIVITIES: All banking activities

Subsidiary Companies: (100% owned unless stated): Credit and Finance Institutions: BHF-Bank Finance (Jersey) Ltd (Jersey); BHF-Bank International SA (Luxembourg); BHF-Bank (Jersey) Ltd (Jersey); BHF-Bank (Switzerland) Ltd (Switzerland); BHF Capital Markets Ltd (UK); BHF Securities (Asia) Ltd (50%) Hong Kong/Tokyo); BHF Finance (Delaware) Inc (USA); BHF SECURITIES Corporation (USA); BHF Securities (Asia) Ltd (50%) (Hong Kong/Tokyo); BHF TRUST Management Gesellschaft für Vermögensverwaltung mbH; Deutsche Hypothekenbank (Actien-Gesellschaft) (Hanover and Berlin) (over 50%); FRANKFURT-TRUST Investment-Gesellschaft mbH; Frankfurter Kreditbank GmbH (80%). Holding and Consultancy Companies: BHF FINANCE Ltd (Hong Kong); Frankfurter Beteiligungs-Treuhand GmbH; Frankfurter Vermögens-Treuhand GmbH; Industrie-Beteiligungs-Gesellschaft mbH. Real Estate Companies: Berliner Gesellschaft für Vermögensverwaltung mbH; Mittelrheinische Grundstücks-GmbH; NIDDA Grundstücks-und Beteiligungs-GmbH; NIDDA Grundstücks-und Beteiligungs-GmbH & Co im Westend. Other Affiliated Companies: AQUA Butzke-Werke AG (over (50%)

Principal Bankers: All principal banks

Financial Information:

	1986	1987
	DM'000	DM'000
Profit after tax	87,816	87,277
Dividends(1)	55,283	53,372
Dividends per share(1)	12.00DM(2)	12.00DM
Share capital(1)	197,438	222,384
Shareholders funds(1)	829,660	878,109
Deposits	15,815,575	17,646,584
Total assets	27,520,674	30,579,964

(1) Parent Company only

(2) Plus 2.00DM Bonus

No of Employees: 3,051

BERLINISCHE LEBENSVERSICHERUNG AG
Gustav Stresemann Ring 7-9, D-6200 Berlin

Tel: (06121) 7731

Telex: 4186 850

Management Board: Dr Gerhart Gomm (Vorsitzender), Hans Joachim Sappke, Dr Franz Weber, Günter Wüst, Klaus-Dieter Blasche

PRINCIPAL ACTIVITIES: Life insurance

Financial Information:

	DM'000
Premium income	560,000

No of Employees: 1,230

BHW BAUSPARKASSE GMBH
Beamtenheimstättenwerk Gemeinnützige Bausparkasse für den Offentlichen Dienst GmbH
Lubahnstrasse 2, D-3250 Hameln 1

Tel: (05151) 181

Telex: 092899

Telefax: (05151) 18 3001

Supervisory Board: Alfred Krause, Gustav Fehrenbach (im jährlichen Wechsel)

Management Board: Dr jur L Storck, Dr jur B Künstler, F Jordan, D J Gumpel, G Hörter

PRINCIPAL ACTIVITIES: Banking services

Trade Names: BHW Bausparkasse GmbH

Subsidiary Companies: BHW-Gesellschaft für Wohnungswirtschaft mbH; BHW Bank AG

Principal Shareholders: Beteiligungsgesellschaft für Gemeinwirtschaft AG, Frankfurt/Main; Betriebs-und Anlagegesellschaft mbH, Bonn

No of Employees: 4,800

BREMER LANDESBANK
Domshof 26, 2800 Bremen 1

Tel: (0421) 332 0

Cable: Landesbank Bremen

Telex: 2402234; 2402240 blbd

Telefax: (0421) 332 23 22

Supervisory Board: Dr Dietrich Hoppenstedt (Chairman)

Management Board: Dr Peter Hasskamp (Chairman), Axel Weber (Vice Chairman), Horst-Günter Lucke, Dr Manfred Schröder, Fritz Lütke-Uhlenbrock

PRINCIPAL ACTIVITIES: Banking activities

Subsidiary/Associated Companies: (100% owned unless stated): Bremer Landesbank Finance (Curaçao) NV (Netherlands Antilles); Deutsche Schiffahrtsbank AG (25.06%); NordFinanz Bank AG (50%); Beteiligungs- und Verwaltungs-Ges Bremen-München mbH i L (40%); Bremer Fondsverwaltung GmbH; Bremer Spielcasino GmbH & Co Kommanditgesellschaft (24%); Bremer Toto und Lotto GmbH (33.33%); Bremerhavener Eiswerke GmbH (30%); Bremerhavener Kühlhäuser GmbH (50%); Bremische Grundstückgesellschaft mbH; Datenverarbeitungsgesellschaft der nieders Sparkassenorganisation mbH & Co OHG (20%); Gemeinnützige Braker Bau- und Wohnungsgesellschaft mbH (25.1%); Gemeinnütziger Bremer Bauverein (26.73%); Gemeinnützige Wohnungsbaugesellschaft Ammerland mbH (32.26%); Haus- und Boden-Anlagefonds Dr Adolf Darjes & Co KG Bonn-Center (27.75%); Haus- und Boden-Anlagefonds Dr Adolf Darjes & Co KG 6-Städte-Fonds (30.6%); Haus- und Boden-Anlagefonds Dr Adolf Darjes & Co KG Geschäfts- und Gewerbebau (21.16%); Nordwestdeutsche Kapitalbeteiligungsgesellschaft mbH (40%); Nordwestdeutsche Vermögensverwaltungs-Gesellschaft mbH; PIKAN Verwaltungs GmbH & Co Vermietungs-KG; Unterstützungseinrichtung der Bremer Landesbank Kreditanstalt Oldenburg GmbH; Verwaltungsgesellschaft Geestemünder Kühlhaus GmbH & Co (95%); VVG Versorgungsschiff Verwaltungs-Gesellschaft mbH (21.2%); Wesermünder Grundstücksgesellschaft mbH; Wirtschaftsförderungsgesellschaft Weser-Jade mbH; Wirtschaftsförderungsgesellschaft für den Landkreis Oldenburg (Oldb) mbH (23.75%); Wohnungsbaugesellschaft Jade mbH (27.27%)

Principal Bankers: Landeszentralbank

Financial Information:

	31.12.86 DM'000	31.12.87 DM'000
Business volume	27,848,000	29,646,000
Profit after tax	-	-
Retained profit	-	-
Shareholders funds	674,500	674,500
Deposits	23,899,219	25,872,390
Total assets	26,242,000	28,228,000

Principal Shareholders: Norddeutsche Landesbank, Freie Hansestadt Bremen
No of Employees: 1,280

COLONIA VERSICHERUNG AG

Postfach 80 50 50, Colonia-Allee 10-20, 5000 Köln 80
Tel: (0221) 690-01
Cable: Coloniavers
Telex: 8815-0

Supervisory Board: Dr N Graf Strasoldo (Chairman), K-M Bornfleth (Vice-Chairman)
Executive Board: D Wendelstadt (Chairman), Dr A Biagosch, G Brandt, G Kutz, H W Pieper, F-K Rauch, A Suhr, E Fhr von Schorlemer, A von Krosigk

PRINCIPAL ACTIVITIES: Insurance
Subsidiary/Associated Companies: Colonia Lebensversicherung AG; Colonia Krankenversicherung AG; Kölnische Hagel-Versicherungs-Gesellschaft; Colonia Bausparkasse AG; Wikinger Lloyd Versicherungs-AG; Kölnische Beratungsges f. Risk Management mbH; Kölnische Rückversicherungs-Gesellschaft; Nordstern Allgemeine Versicherungs-AG; Rheinisch-Westfälische Boden-Credit-Bank; Colonia Anlage-Dienst Vermittlungsges mbH; Colonia Gundbesitz-Vermietungs-und Verwaltungs-Gesellschaft GbR; Colonia Versicherung Verwaltungs-und Vertriebsgesellschaft mbH; Concordia Versicherungsbeteiligungs-Aktiengesellschaft (Austria); Colonia Life Holdings Ltd (Canada); Colonia Insurance Co (UK) Ltd; Colonia Finance Holdings BV
Financial Information:

	31.12.86 DM'000	31.12.87 DM'000
Premium income	2,251,765	2,291,229
Investment income	153,569	128,523
Profit before tax	72,331	48,833
Profit after tax	31,610	22,850
Retained profit	16,625	15,920
Dividends	15,100	15,800
Dividends per share	12DM*	12DM*
Earnings per share	41.28DM	38.78DM
Share capital	130,000	130,000
Shareholders funds	472,000	479,000

*12DM on 120 million DM,
14DM on 10 million DM

Principal Shareholders: Bankhaus Sal Oppenheim jr & Cie (under 25%); Kölnische Verwaltungs AG (over 25%)
No of Employees: 4,663

COMMERZBANK AG

32-36 Neue Mainzer Strasse, PO Box 10 05 05, D-6000 Frankfurt a.M. 1
Tel: 69-1362 (Ext 0)
Cable: Commerzbank Peterson International 4th Edition
Telex: 4152530; Swift: COBADEFF

Supervisory Board: Dr Raban Freiherr von Spiegel (Chairman), Herbert Bayer, Reinhold Borchert, Erhard Bouillon, Hugo Eberhard, Prof Dr Herbert Grünewald, Dr Uwe Haasen, Ehrensenator Dr Carl H Hahn, Gerald Herrmann, Hans-Georg Jurkat, Dr Hans-Jürgen Knauer, Peter Kretschmer, Gabi Locher, Dr Torsten Locher, Horst Sauer, Hans-Georg Stritter, Dr-Ing Hanns Arnt Vogels, Dipl-Ing Heinrich Weiss, Dr Gerd Wollburg, Wolfgang Ziemann

Management Board: Dr Walter Seipp (Chairman), Dr Erich Coenen, Dietrich-Kurt Frowein, Dr Kurt Hochheuser, Goetz Knappertsbusch, Martin Kohlhaussen, Klaus Mueller-Gebel, Juergen Reimnitz, Dr Kurt Richolt, Dr Axel Freiherr von Ruedorffer, Dr Juergen Terrahe
Executive Vice Presidents: Alfred Knör, Klaus-Peter Müller, Wolfgang Otto, Herbert Peters, Manfred Wilsdorf. Chief Legal Adviser: Dr Helmut Becker

PRINCIPAL ACTIVITIES: Commercial and investment banking
Subsidiary Companies: Rheinische Hypothekenbank (Frankfurt); Berliner Commerzbank (Berlin); Commerz-Credit-Bank Aktiengesellschaft Europartner (Saarbrücken); Commerzbank International SA (Luxemburg); Commerzbank (Nederland) NV (Amsterdam); Commerzbank (Schweiz) AG (Zürich); Commerzbank (South East Asia) Ltd (Singapore); Commerzbank Overseas Finance NV (Curacao); Commerzbank Investment Management GmbH (Frankfurt); Commerz International Capital Management GmbH (Frankfurt); Commerz Securities (Japan) Company Ltd (Hong Kong/Tokyo); EuroPartners Securities Corporation (New York). The bank also has 882 branches
Financial Information: Consolidated figures

	31.12.86 DM'000	31.12.87 DM'000
Profit before tax	726,000	747,000
Profit after tax	408,000	423,000
Retained profit	299,000	335,000
Dividends	299,000	335,000
Share capital	1,038,000	1,040,000
Shareholders funds	4,353,000	5,078,000
Deposits	140,328,000	153,511,000
Total assets	148,150,000	161,731,000

Principal Shareholders: approximately 160,000 shareholders
No of Employees: 26,640

CONTINENTALE KRANKENVERSICHERUNG AG

Ruhrallee 92, Postfach 1343, D-4600 Dortmund 1
Tel: (0231) 12011
Telex: 822515

Management Board: H Bach (Vorsitzender), W Witte (Stellv Vorsitzender), H Hoffmann, G Rehermann

PRINCIPAL ACTIVITIES: Insurance
Subsidiary Companies: Conti Versicherungsvermittlungs GmbH; Continentale Sachversicherung AG (74%); Continentale Lebensversicherung AG
Principal Bankers: Deutsche Bank
Financial Information:

	DM'000
Premium income	1,200,000

No of Employees: 1,600

DEBEKA KRANKENVERSICHERUNG VAG

Ferdinand Sauerbruch-Strasse 18, 5400 Koblenz
Tel: (0261) 498-0
Telefax: (0261) 41402

Supervisory Board: Hermann Lange (Vorsitzender), Paul Hoffmann (stv Vorsitzender), Georg Bachmaier, Dieter Berg, Prof Joseph Breuer, Dr Otto Fuhrmann, Werner Gebhard, Manfred Heil, Paul Hoffmann, Johannes Hollering, Willi Klee, Peter Kurepkat, Ellen Öztürk, Prof Norbert Plassmann, Prof Dr Reinhold Rörig, Gerhard Schröder, Charlotte Sefeloge, Ingrid Simka, Hans Weber
Management Board: Peter Greisler (Vorsitzender), Joachim Rudolph, Jürgen R.Rudolph
Senior Executives: Heinrich Goronzy (Abt Dir), Herbert Grohe (Abt Dir), Gerhard Hof (Abt Dir), Georg Huber (Org Dir)

PRINCIPAL ACTIVITIES: Medical insurance such as insurance for medical expenses, hospital expenses, hospital daily allowances and sickness daily allowances, insurance for nursing care expenses

Subsidiary Companies: Debeka Allgemeine Versicherung AG
Principal Bankers: Sparkasse Koblenz; LZB in Rheinland-Pfalz (Koblenz); Postgiro Köln
Financial Information:

	31.12.86 DM'000	31.12.87 DM'000
Premium income	1,468,557	1,535,516
Investment income	234,677	245,872
Profit after tax	156,152	110,702

No of Employees: 4,562

DEUTSCHE ANLAGEN-LEASING GMBH

Wilhelm-Theodor-Römheld Str 30, 6500 Mainz 1
Tel: (061) 31 80 11
Telex: 4187637

PRINCIPAL ACTIVITIES: Property management; leasing of industrial equipment and machinery
Financial Information:

	DM'000
Sales turnover	1,000,000

DEUTSCHE AUSGLEICHSBANK

Wielandstrasse 4, 5300 Bonn 2
Tel: (0228) 831-0
Telex: 885422
Telefax: (0228) 831-255

Supervisory Board: Dr Bert Even (Vors), Dr Claus Fricke (stellv. Vors), Ernst-Otto Sandvoss (stellv. Vors), Roland Bergsträsser, Dr Herbert Czaja, Kurt Fiedler, Gerd Fischwasser, Dr Werner Funke, Michael Glos, Helmut Haun, Egon Herfeldt, Werner Heubaum, Wolfgang Hinrichs, Wilhelm Hoffmann, Dr Knut Kage, Dr Klaus-Joachim Kübler, Heinz Rapp, Wolfgang Roth, Otto-Ernst Sievers, Dr Hermann-Otto Solms, Dr Horst Waffenschmidt, Dr Georg Frhr V Waldenfels
Management Board: Dr Kurt Andreas (Vors), Dr Walter Althammer (stellv Vors), Klaus Nötzel

PRINCIPAL ACTIVITIES: Banking services
Financial Information:

	31.12.86 DM'000	31.12.87 10,000
Profit after tax	10,000	10,000
Share capital	120,000	120,000
Deposits	11,886,842	11,948,414
Total assets	13,399,694	13,310,320

Principal Shareholders: Federal Republic of Germany
No of Employees: 472

DEUTSCHE BANK AG

Taunusanlage 12, PO Box 100601, D-6000 Frankfurt am Main
Tel: (069) 71500
Cable: Deutbank
Telex: 417300 Deutschbank Frankfurt

Honorary President: Hermann J Abs (Franfurt am Main); Chairman of Supervisory Board: Wilfried Guth
Board of Managing Directors: Rolf-E Breuer, Horst Burgard, Ulrich Cartellieri, Alfred Herrhausen, Eckart van Hooven, Hilmar Kopper, Georg Krupp, Ulrich Weiss, Herbert Zapp, Michael Endres (Deputy), Jürgen Krumnow (Deputy), Ellen R Schneider-Lenné (Deputy)
Executive Vice Presidents: Dr Klaus Gaertner, Rudolf Habicht, Dr Frank Heintzaler, Dr Klaus Juncker, Dr Hanns Kippenberger, Christoph Könneker, Heinrich Kunz, Günter Olf, Axel Osenberg, Barthold von Ribbentrop, Hans Rosentalski, Dr Hans Walter Schlöter, Gerd Schmitz-Morkramer, Dr Werner Schwilling, Günter Sonnenburg, Helmut Trötscher, Christian L Vontz, Johann Wieland, Dr Karl Friedrich Woeste, Wilhelm Schlaus (General Counsel), Dr Franz-Josef Trouvain (Chief Economist)

PRINCIPAL ACTIVITIES: All banking activities
Consolidated Companies: (100% owned unless stated): COMMERCIAL BANKS: Deutsche Bank Berlin AG; Deutsche Bank Saar AG (69.2%); Handelsbank in Lübeck AG (90.2%); Banca d'America e d'Italia SpA (98.3%) (Italy); Deutsche Bank (Asia) AG; Deutsche Bank (Asia Credit) Ltd (Singapore); Deutsche Bank Australia Ltd (Australia); Deutsche Bank (Canada); Deutsche Bank Luxembourg SA (Luxembourg); DB (Belgium) Finance SANV (Belgium); DB UK Finance Ltd (UK). MORTGAGE BANKS: Deutsche Centralbodenkredit-AG (89.9%); Frankfurter Hypothekenbank AG (92.2%); Lübecker Hypothekenbank AG. INVESTMENT BANKING AND CAPITAL INVESTMENT COMPANIES: Deutsche Gesellschaft für Fondsverwaltung mbH; DWS Deutsche Gesellschaft für Wertpapiersparen mbH (57%); Deutsche Bank Capital Corporation (USA); Deutsche Bank Capital Markets Ltd (UK); Deutsche Bank (Suisse) SA (Switzerland); Deutsche Capital Management Australia Ltd (Australia); Deutsche Capital Markets Australia Ltd (Australia). INSTALMENT FINANCING COMPANIES AND SPECIALIZED INSTITUTIONS: Deutsche Bank Bauspar-AG; Deutsche Kreditbank für Baufinanzierung AG; EFGEE Gesellschaft für Einkaufs-Finanzierung mbH; GEFA Gesellschaft für Absatzfinanzierung mbH; Deutsche Credit Corporation (USA). LEASING COMPANIES: Deutsche Gesellschaft für Immobilien-Leasing mbH (95%); GEFA-Leasing GmbH; DB Export-Leasing GmbH; BAI Leasing SpA (Italy). INTERNATIONAL FINANCING COMPANIES: Deutsche Bank Finance NV (Netherlands Antilles); Deutsche Bank Financial Inc (USA). OTHER DOMESTIC AND FOREIGN COMPANIES: Alma Beteiligungsgesellschaft mbH; Elektro-Export-Gesellschaft mbH; Hessische Immobilien-Verwaltungs-Gesellschaft mbH; Matura Vermögensverwaltung mbH; Süddeutsche Vermögensverwaltung GmbH; Trinitas Vermögensverwaltung GmbH; Deutsche Credit Services Inc (USA); Deutsche Portfolio Corporation (USA)
Financial Information:

	31.12.86 DM'000	31.12.87 DM'000
Business volume	164,094,000	168,723,000
Profit after tax	824,000	425,431
Retained profit	272,000	-
Dividends	552,000	425,431
Dividends per share	12DM*	12DM
Share capital	1,623,365	1,772,631
Reserves	6,658,218	8,003,822
Provisions and reserves	13,526,054	14,344,204
Deposits	140,341,324	144,576,955
Total assets	159,927,755	165,197,588

*plus 5DM bonus

Number of Shareholders: 320,000
No of Employees: 43,951 (Deutsche Bank AG), 54,579 (Group)

DEUTSCHE BEAMTEN-VERSICHERUNG ÖFFENTLICHRECHTLICHE LEBENS- UND RENTEN-VERSICHERUNGSANSTALT

DBVL

Postfach 2109, Frankfurter Strasse 50, 6200 Wiesbaden
Tel: (06121) 3631
Cable: debevau Wiesbaden
Teletex: 6121946
Telefax: (06121) 363-359

Supervisory Board: Dr Walter Hesselbach (Vors), Heinz Frieser (stellv. Vors), Otto Regenspurger, Herbert Bergs, Emil Bock, Friederich Bode, Gisbert Kaub, Dr Bodo Künstler, Karl Kunrath, Richard Müller, Rolf Preiss, Heinz Putzhammer, Dr Louis Storck, Willi Tilger, Wolfgang Warburg, Martin Wurm
Management Board: Manfred Broska (Vors), Dr Folk Fuss, Dr Hans-Peter Nickisch, Helmut Holländer, Karl-Heinz Röhling, Manfred Ruhe, Joachim Wieland

PRINCIPAL ACTIVITIES: Life insurance
Subsidiary Companies: (100% owned unless stated): Deutsche Beamten-Versicherung AG; APK Krankenversicherungs-AG

für den öffentlichen Dienst (74%); Partner-Gruppe Lebensversicherung AG; Partner-Gruppe Allgemeine Versicherung AG (79.8%); Partner-Gruppe Krankenversicherung AG (60%); DBV+PARTNER Rechtsschutz-Versicherung AG (90%); DBV+PARTNER Rückversicherungs-AG; DBV+PARTNER Vermittlungsgesellschaft mbH für Versicherungen und Vermögensbildung; Quelle+Partner Lebensversicherung AG (50%); Quelle+Partner Versicherung AG (50%); GERMANA Belgische Lebensversicherung (99.9%) (Belgium); HEROLD Belgische Allgemeine Versicherungsgesellschaft (99.1%) (Belgium)

Financial Information:

| | 31.12.86 | 31.12.87 |
	DM'000	DM'000
Premium income	827,595	882,719
Investment income	485,059	382,744
Profit before tax	43,629	25,025
Profit after tax	25,866	17,482
Share capital	70,000	100,000
Shareholders funds	137,703	153,544

No of Employees: 2,700

DEUTSCHE BUNDESBANK

14 Wilhelm-Epstein-Str 14, Frankfurt/Main 50

Tel: (069) 158-1

Cable: Notenbank Frankfurtmain

Telex: 414431 (foreign); 41227 (national)

Telefax: (069) 560 10 71

Directorate: K O Pöhl (President), Prof Dr Drs h.c. H Schlesinger (Deputy President), J W Gaddum, Prof Dr Dr h.c. L Gleske, Prof Dr C Köhler, O Werthmöller, Dr G Storch

Managing Boards of the Land Central Banks: Baden-Württemberg: Prof.Dr.Dr.h.c.Norbert Kloten (President), Helmut Schieber (Vice President), Helmut Holzmaier. Bavaria: Lothar Müller (President), Rudolf Ströhlein (Vice President), Dr.Walter Gulden. Berlin: Dr.Dieter Hiss (President), Dietger Oberndorfer (Vice President). Bremen: Dr. Kurt Nemitz (President), Johannes Kremers (Vice President). Hamburg: Dr.Wilhelm Nölling (President), Gerhard Jennemann (Vice President). Hessen: Dr.Alfred Härtl (President), Walter Kulla (Vice President), Heinz-Georg Völlgraf. Lower Saxony: Prof Dr Helmut Hesse (President), Roman Flesch (Vice-President). North Rhine-Westphalia: Hans Wertz (President), Helmut Röthemeier (Vice-President), Edgar Krug. Rhineland-Palatinate: Dr Heinrich Schreiner (President), Eberhard Weiler (Vice President). Saarland: Hans Gliem (President), Erhard Dauzenroth (Vice President). Schleswig-Holstein: Dr.Johann Baptist Schöllhorn (President), Peter Titzhoff (Vice President). Representative Offices: Horst Rinke (New York Manager), Klaus Paeben (Tokyo Manager)

Heads of Departments: Manfred J Körber (Press and Public Relations, Library and Archives, Foreign Language Service), Ekkehard Bauer (Legal), Wolfgang Leue (Audit), Dr Klaus Hanau (Research), Peter Walter (Cash), Rudi Klepzig (Payments and Accounting, Balance Sheet), Dr Karl Thomas (Credit), Franz Scholl (Foreign), Dr Wolfgang Rieke (International Monetary Questions, Organisations and Agreements), Wendelin Hartmann (Organisation), Werner Lockemann (Administration and Construction), Richard Lang (Banking and Minimum Reserves), Thomas Buch (Personnel)

PRINCIPAL ACTIVITIES: Central bank of the Federal Republic of Germany

Subsidiary Companies: Representative Offices: New York, Tokyo

Financial Information:

| | 31.12.86 | 31.12.87 |
	DM'000	DM'000
Profit after tax	7,795,842	337,545
Retained profit	7,795,842	337,545
Capital	290,000	290,000
Total assets	220,851,242	230,474,636

No of Employees: 15,707

DEUTSCHE BUNDESPOST

Heinrich-von-Stephan-Strasse 1, Postfach 80 01, 5300 Bonn 2

Tel: (0228) 14-0

Teletex: 22859=BPM

Telex: 8861 101 bpm d

Telefax: (0228) 14-8872

Administrative Council of the Deutsche Bundespost: H Windelen (Vorsitzender), K van Haaren, G O Pfeffermann

Senior Executives: Federal Minister of Posts and Telecommunications: Dr Ch Schwarz-Schilling, Parliamentary State Secretary: W Rawe, State Secretary: Dr W Florian

PRINCIPAL ACTIVITIES: Operation of national postal and telecommunications services; operation of postal savings bank

Subsidiary Companies: (100% owned unless stated): Postal and Telecommunications Companies: Deutsche Postreklame GmbH; Deutsche Fernkabel-Gesellschaft mbH; TKS, Telepost Kabel-Servicegesellschaft mbH (54%); Deutsche Telepost Consulting GmbH (DETECON) (30%); DANET-Gesellschaft für Beratung und Softwareentwicklung mbH (30%); KABELCOM Braunschweig, Gesellschaft für Breitbandkabel-Kommunikation mbH (24%); KABELCOM Wolfsburg, Gesellschaft für Breitbandkabel-Kommunikation mbH (24%); KABELCOM Osnabrück, Gesellschaft für Breitbandkabel-Kommunikation mbH (24%); GSI-DATEL-Gesellschaft für Datenfernverarbeitung mbH (5%). Housing Companies: Gemeinnützige Postbaugesellschaft mbH (99.9%); Gemeinnützige Post-Bau- und Wohnungsgesellschaft Köln mbH (99.9%); DAHEIM, Gemeinnützige Wohnungsbaugesellschaft mbH (99%); Gemeinnützige Wohnungsbau Aktiengesellschaft Rhein-Main (54.4%); Gemeinnützige Deutsche Wohnungsbaugesellschaft mbH Organ der staatlichen Wohnungspolitik (41.7%); Gemeinnützige Wohnungsgesellschaft Baden-Württemberg AG (33.5%); Aktiengesellschaft Hellerhof (20.6%); Postbaugenossenschaft Baden-Württemberg eG (20.6%); GEWOBAU, Gemeinnützige Wohnungsbaugesellschaft mbH (7%); Landes-Bau-Genossenschaft Württemberg eG (4%); Gemeinnützige Baugenossenschaft Arlinger eG (0.01%). Transport Enterprises and Utilities: Vereinigte Bundesverkehrsbetriebe GmbH (47.2%); Deutsche Lufthansa AG (1.8%); Gleisgemeinschaft Süd, Dieselstrasse (44%); Bahnbetriebsgesellschaft Borsigwalde GmbH (3.4%); Fernkälte, Geschäftsstadt Nord, Gesellschaft bürgerlichen Rechts (12%); 4 electricity cooperatives; 4 water supply cooperatives; Wasserverband Grosser Feldberg, Schmitten (Hochtaunuskreis) (31.5%); Wasserbeschaffungsverband Wilhelmsdorf, Weilrod (Hochtaunuskreis) (5.2%). Banking Enterprises: Deutsche Pfandbriefanstalt, Wiesbaden/Berlin (8.2%); 17 Post-Spar- und Darlehnsvereine with DM20 each; Post-Spar- und Darlehnsgenossenschaft Saarbrücken eG

Financial Information:

| | 31.12.86 | 31.12.87 |
	DM'000	DM'000
Sales turnover	49,640,900	52,050,100
Profit after tax	3,269,500	2,702,900

No of Employees: 559,101

DEUTSCHE CENTRALBODENKREDIT AG

Kaiser-Wilhelm Ring 27-29, D-5000 Köln 1

Tel: (0221) 57211

Telex: 8882925

Telefax: (0221) 5721 505

Supervisory Board: Georg Krupp (Vorsitzender), Gerd Schmitz-Morkramer (stv. Vorsitzender), Helmut Hoffrichter, Dipl Volksw Paul Kottendorf, Dr Klaus Liesen, Paul Heinz Matthias, Dr Nikolaus Graf Strasoldo, Ingrid Thiemann, Dr Paul-Robert Wagner

Management Board: Dr Michael Gellen, Dr Jürgen Huvendick, R A Detlev Rode

PRINCIPAL ACTIVITIES: Banking services

Financial Information:

	31.12.86	31.12.87
	DM'000	DM'000
Profit before tax	106,000	117,000
Profit after tax	43,300	45,200
Retained profit	21,300	23,200
Dividend	17,300	20,160
Dividend per share	12DM	12DM
Share capital	72,000	72,000
Shareholders funds	674,000	700,000
Total assets	27,471,000	29,206,000

No of Employees: 451

DEUTSCHE GENOSSENSCHAFTS-HYPOTHEKENBANK AG

Ost-West Strasse 91, D-2000 Hamburg 11
Tel: (040) 361481
Telex: 02 12 743

Management Board: B Schramm (Vors), H Gisch, H Schmuck, H Wriedt

PRINCIPAL ACTIVITIES: Mortgage bank
Parent Company: Deutsche Genossenschaftsbank

No of Employees: 460

DEUTSCHE GIROZENTRALE/KOMMUNALBANK

PO Box 110542, D-6000 Frankfurt-am-Main 11
Tel: 69 26 93
Telex: 414168; 4189141
Telefax: 69 26 93 (Ext 490)

Management Board: E-O Sandvoss (Chairman), M Zass, Dr W Hennicke, Dr E Zinn, H J Reichert
Senior Executive: C D Homann (International), A Lindauer (Foreign Exchange), H-H Kotz (Public Relations)

PRINCIPAL ACTIVITIES: All banking activities
Subsidiary Companies: Deutsche Girozentrale International SA; DEKA Deutsche Kapitalanlage GmbH; DESPA Deutsche Sparkassen-Immobilien-Anlage-GmbH; Deutsche Landesbankenzentrle AG; LBG Leasing Beteiligungs GmbH
Financial Information:

	31.12.86	31.12.87
	DM'000	DM'000
Share capital	560,000	560,000
Deposits	35,276,309	37,170,743
Total assets	36,479,657	38,409,080

No of Employees: 300

DEUTSCHE HYPOTHEKENBANK FRANKFURT-BREMEN AG

Postfach 106347, D-2800 Bremen
Tel: (0421) 36391
Telex: 02 44 590

Management Board: Dr Claus Bingold, Dr Rainer Jaeckle, Paul-Ernst Penndorf, Hermann Schnüll

PRINCIPAL ACTIVITIES: Banking
Parent Company: Dresdner Bank

Financial Information:

	31.12.87
	DM'000
Profit after tax	59,500
Dividend	13,500
Dividend per share	10DM*
Share capital	54,000
Capital and disclosed reserves	494,000
Total assets	24,700,000

*plus bonus of DM2.50

No of Employees: 280

DEUTSCHE KRANKENVERSICHERUNG AG

Aachener Strasse 300, D 5000 Koeln 41
Tel: (0221) 578-0
Cable: deukravers koeln
Telex: 08-881 634
Telefax: (0221) 578 3694

Supervisory Board: Klemens Wesselkock (Vorsitzender), *Albert Klein (stellv.Vorsitzender), Hans-Dieter Sellschopp (stellv.Vorsitzender), Dr Henning Schulte-Noelle (stellv.Vorsitzender), Peter Kakies, Heinrich Birr, *Erika Humm, *Helmut Kliemann, *Richard Sommer, *Franziska Wiethold, *Max Sassenfeld, Hans Hugo Schlitzberger, *von den Betriebsangehörigen gewählte Mitglieder
Management Board: Hans Georg Timmer (Vorstandsvorsitzender), Horst Lange (ordentl. Vorstandsmitglied) Georg Ludwig (ordentl. Vorstandsmitglied), Bernhard Schattenberg (ordentl. Vorstandsmitglied), Dr Roland Delbos (stellv. Vorstandsmilglied), Dr Bernhard Schareck (stellv. Vorstandsmitglied)
Senior Executives: E Klein, Dr R Matzel, W Seyfert (Titular-Direktoren), R Dahl, G Dibbern, Dr Emde, Dr J Hegen, H Kliemann, H Langohr, R Maas, H-J Pankow, A Sonnemann, W Unrath, G Zimmer, E Zoller (Abteilungsdirektoren)

PRINCIPAL ACTIVITIES: Private health insurance
Trade Names: DKV-, dkv-
Subsidiary Companies: (100% owned unless stated): GEMEDA Gesellschaft für medizinische Datenerfassung und Auswertung sowie Serviceleistungen für freie Berufe mbH; dkv-International S.A. (99.99%) (Belgium); N.V. Verzekeringsmaatschappij Rijnmond (Netherlands); SANA Kliniken-GmbH (22.64%), Quarta-Vermögensgesellschaft mbH (10.269%)
Principal Bankers: Dresdner Bank AG; Westdeutsche Landesbank
Financial Information:

	31.12.86	31.12.87
	DM'000	DM'000
Premium income	2,386,712	2,619,195
Investment income	439,532	477,696
Profit before tax	75,448	96,018
Profit after tax	22,102	34,314
Retained profit	16,000	28,000
Total dividends	6,102	6,314
Dividends per share	5.50DM	5.69DM
Dividends per 50DM share	1.28DM	1.32DM
Earnings per share	24.56DM	38.13DM
Share capital	45,000	45,000
Shareholders funds	33,000	34,152

Principal Shareholders: Allianz Aktiengesellschaft Holding; Hamburg-Mannheimer Vers.-AG; Muenchener Rueckvers.-Ges
No of Employees: 6,163

DEUTSCHE LEASING AG

Hungerner Strasse 6-12, D-6000 Frankfurt 60
Tel: (0611) 15291
Telex: 4 14 098
Telefax: (069) 5970 220

Management Board: Dr Albrecht Dietz, Dipl.-Kfm. Horst Figge, Dipl.-Kfm. Reinhard Balzereit, Franz Pogodda

PRINCIPAL ACTIVITIES: Leasing of machinery and equipment; financial services
Principal Bankers: Landesbanken
Financial Information:

	DM'000
Sales turnover	1,000,000

DEUTSCHE PFANBRIEFANSTALT HAUPVERWALTUNG

Postfach 2169, 6200 Wiesbaden
Tel: (06121) 3480
Telex: 4186572; 4186857
Telefax: (06121) 348549

Management Board: Dr jur L Schork, H Wilhelm, Dr jur Thilo Köpfler, E Meister, Dr jur C Steiner
Senior Executives: U Burow, Dr G Unckel, F-J Vollmer

PRINCIPAL ACTIVITIES: Banking services; building finance
Subsidiary Companies: (100% owned unless stated): Deutsche Bau- und Bodenbank AG (92.53%); Bau- und Bodenverwaltungsgesellschaft GbR (99%); Deutsche Wohnstätten-Hypothekenbank AG; GEV Grundstückswerwerbs- und Verwertungs GmbH; Deutsche Bau- und Grundstücks AG (92.53%); BHG Gesellschaft zur Betreuung von Haus- und Grund-besitz mbG (92.53%); BAUBO Bau- und Bodenverwertungs- und -verwaltungsgesellschaft mbH (92.53%); Terrain-Aktiengesellschaft Herzogpark (90.91%); Terrain-Verwaltungs-GmbH Herzogpark (90.91%)
Financial Information:

	31.12.87 DM'000
Profit after tax	47,0340
Retained profit	29,010
Share capital	243,910
Total assets	68,621,368

No of Employees: 763

DEUTSCHE SCHIFFSBELEIHUNGS-BANK AG

Postfach 11 19 13, Katharinenstrasse 13, 2000 Hamburg 11
Tel: (40) 37 69 9-0
Cable: Schiffsbank
Telex: 2 14 029; 2 15 315 sbank d
Telefax: (40) 37 699178

Supervisory Board: Klaus Müller Gebel (Vorsitzender), Dr Klaus Plett (stellv. Vorsitzender), Dr Gebhard Dirksen, Eberhard Rainer Luckey, Wolfgang Trillmich, Inge-Christa Oltenau, Hylke Boerstra, Sebastian Scharf
Management Board: Dr Conrad v. Sydow, Jens Blöcker. Direktors: Rolf Hilmer, Anton Schepers, Hans Christian Sienknecht (Senior General Manager)
Senior Executives: Lothar Ahrens, Michael von Appen, Uwe Bauer, Peter Krüger, Hans-Jürgen Schulte, Eckhard Retzlaff

PRINCIPAL ACTIVITIES: Financial services for the shipping industry
Principal Bankers: Commerzbank AG.; Hamburgische Landesbank -Girozentrale-; Vereins- und Westbank AG
Financial Information:

	31.12.86 DM'000	31.12.87 DM'000
Share capital	35,000	35,000
Shareholders funds	41,000	41,000
Total assets	2,382,836	2,595,632

Principal Shareholders: Vereins- und Westbank AG; Commerzbank AG; Hamburgische Landesbank-Girozentrale
No of Employees: 66

DEUTSCHE VERKEHRS-KREDIT-BANK AG

23-25 Untermainkai, PO Box 110532, D-6000 Frankfurt Main 11
Tel: (069) 2648-0
Teletex: 69 9777
Telefax: (069) 2648444

Supervisory Board: Wilhelm Paellmann (Chairman)
Management Board: Arno Grunhold, Dr Klaus J Menche, Dr Franz Schlossnikl
Senior Executives: Manfred Löw (Foreign Division), Manfred Begert (Credit Division), Werner Jung (Domestic Money), Hans Bunse (Public Relations), Andreas Kaup (Foreign Exchange)

PRINCIPAL ACTIVITIES: Bank
Principal Bankers: Correspondents worldwide
Financial Information:

	31.12.86 DM'000	31.12.87 DM'000
Dividends	6DM	6DM
Share capital	75,000	75,000
Deposits	4,893,955	5,113,609
Total assets	5,152,661	5,361,889

Principal Shareholders: Deutsche Bundesbahn (German Federal Railway), Frankfurt (75.1%)
No of Employees: 914

DEUTSCHER HEROLD LEBENSVERSICHERUNG AG

Postfach 1448, Poppelsdorfer Allee 25-33, D-5300 Bonn 1
Tel: (0228) 26801
Telex: 8 86 419

Supervisory Board: Fritz Schleifer (Vors)
Management Board: Wilhelm Ewert (Sprecher), Hans Dieter Ritterbex (Sprecher), Gerhard Birkental (stellv Vorst), Ludwig van Hamme, Herbert E Raab, Helmut Ungerathen

PRINCIPAL ACTIVITIES: Life insurance
Principal Bankers: Dresdner Bank; Sparkasse

No of Employees: 550

DEUTSCHER KRAFTVERKEHR ERNST GRIMMKE GMBH + CO. KG

Vogelsanger Weg 49, 4000 Düsseldorf
Tel: (0211) 6392-0
Cable: dkv d
Telex: 8586696
Telefax: (0211) 6392-192

Management Board: J O Berkau, Dr W E Becker, K H Henke

PRINCIPAL ACTIVITIES: Filling stations; credit-card system for international transport companies
Trade Names: DKV europa service
Principal Bankers: Commerzbank; Dresdner Bank; Bankhaus von der Heydt Kersten & Söhnne
Financial Information:

	1986 DM'000	1987 DM'000
Sales turnover	1,410,038	1,418,641

No of Employees: 163

DEUTSCHER LLOYD LEBENSVERSICHERUNG AG

Karlstrasse 10, D-8000 München 2
Tel: (089) 5908-1
Telex: 523986

Management Board: Dr H Kluge (Vorsitzender), Dr M Bachmann, M Clausen, E Stein

PRINCIPAL ACTIVITIES: Life insurance
Financial Information:

	DM'000
Premium income	510,000

No of Employees: 1,000

DEUTSCHER RING LEBENSVERSICHERUNG AG

Postfach 112040, D-2000 Hamburg 11
Tel: (040) 35 991
Telex: 0215 545

Management Board: Hans Grutschus, Dr Karl-Dieter Koch, Viktor Müller, Heinz Scheloske, Edmund Schmidl, Dr Ulrich Rumm

PRINCIPAL ACTIVITIES: Life insurance
Financial Information:

	DM'000
Premium income	630,000

No of Employees: 2,000

DG BANK
Deutsche Genossenschaftsbank

Am Platz der Republik, PO Box 100651, D-6000 Frankfurt am Main 1
Tel: (069) 744701
Cable: Deegeebank
Telex: 412291 dgd
Telefax: (069) 7447-1685

Supervisory Board: Bernhard Schramm (Chairman), Dr Gerhard Barner (Vice Chairman), Dr h c Arnold Kremer (Vice Chairman)
Board of Managing Directors: Helmut Guthardt (Chairman), Karl Fehrenbach (Vice Chairman), Dr Karl-Herbert Schneider-Gaedicke (Vice Chairman), Roland Bergstraesser, Dr Armin Herrmann, Dr Alfons Humpert, Dr Johann Remmers, Dr Wulf v Schimmelmann, Gunther Schmidt-Weyland
Senior Executives International Division: Johann-Gottlieb von Suesskind-Schwendi, Klaus-Dieter Reiter, Alfred E Mallmann

PRINCIPAL ACTIVITIES: Wholesale commercial and investment bank, central bank of the German cooperative banking system
Consolidated Banking Subsidiaries: Deutsche Genossenschafts-Hypothekenbank AG, Hamburg/Berlin; DG BANK Luxembourg SA, Luxembourg; DG BANK (Schweiz) AG, Zurich; London & Continental Bankers Ltd (LCB), London; DG Capital Company Ltd, Hong Kong; DG BANK GZB (Asia) Ltd, Singapore; DG Diskontbank AG, Frankfurt am Main
Financial Information:

	31.12.86 DM'000	31.12.87 DM'000
Profit before tax	320,324	359,725
Profit after tax	192,921	175,635
Retained profit	107,063	70,000
Dividends	55,071	56,832
Dividends per share	5%	5%
Share capital	1,116,420	1,200,000
Shareholders funds	2,370,420	2,724,000
Deposits	65,225,000	76,284,000
Total assets	111,951,038	127,416,013

Principal Shareholders: Five regional cooperative banks; Landwirtschaftliche Rentenbank
No of Employees: 4,717

DR AUGUST OETKER KG
See OETKER, DR AUGUST, KG

DRESDNER BANK AG

Jürgen-Ponto-Platz 1, PO Box 110661, D-6000 Frankfurt am Main 11
Tel: (069) 263-0
Cable: DRESDBANK
Telex: 415240 dr d
Telefax: (069) 263-4831

Supervisory Board: Rolf Diel (Chairman), Helmut Kahler (Deputy Chairman), Prof. Dr. rer. pol. Horst Albach, Klaus Carlin, Karl Fenn, Hannelore Hornburger, Dr. h.c. Horst K. Jannott, Cornelius J van der Klugt, Dr. Wolfgang Leeb, Dipl.-Kfm. Uwe Plucinski, Gunter Rose, Sultan Salam, Prof. Dr. rer. nat. Dr.-Ing. E. h. Rolf Sammet, Bernhard Schneider, Dr. Dieter Spethmann, Dipl.-Ing. Emmanuel Tesch, Dr. Dr,-Ing. E. h. Dr. phil. h.c. Kurt Werner, Ludwig Winter, Christian Zahn
Board of Managing Directors: Dr. Wolfgang Röller (Chairman), Dr. Hans-Gunther Adenauer, Meinhard Carstensen, Dr. Christoph von der Decken, Piet-Jochen Etzel, Dr. Werner Funke, Kurt Morgen, Jürgen Sarrazin, Dr. Christian Seidel, Dr. Alfons Titzrath, Dr. Bernd Voss, Gerhard Eberstadt (Deputy Managing Director), Bernhard Walter (Deputy Managing Director)
Heads of Divisions: Hans-Dieter Geller (Domestic), Wolfgang Höschele (Domestic), Dr. Erich Karsten (Domestic), Udo H Gubitz (International), Werner A Hundt (International), Hans Olivier (New Issues and Underwriting), Dr. Günther Radtke (New Issues and Underwriting), Heinz Kramer (Securities and Investment), Gerhard Baier (Securities and Investment), Günter Spatz (Securities and Investment), Dr. Gerd Trabant (Securities and Investment), Rolf Willi (Treasury), Helmut Morsbach (Accounting and Controlling), Adam Eitenmüller (Auditing), Dr. Adolf-Friedrich Jacob (Corporate Planning), Dr. Manfred Schaudwet (General Secretariat, Economic Research and Public Relations), Prof. Dr. Theodor Heinsius (Legal Services), Walter Gruhn (Operations and Administration), Tronje Schrader (Personnel), Hugo Chill (Risk Management), Dr. Horst Müller (Risk Management)

PRINCIPAL ACTIVITIES: Banking and financial services
Subsidiary Companies: (100% owned unless stated): Germany:- Bank für Handel und Industrie AG; Bankhaus Reuschel & Co (over 50%) with WKV-Bank GmbH; Deutsch-Südamerikanische Bank AG; Oldenburgische Landesbank AG (60.5%); Deutsche Hypothekenbank Frankfurt-Bremen AG (86%); Hypothekenbank in Hamburg AG (82.8%); Norddeutsche Hypotheken-und Wechselbank AG (58.2%); Pfälzische Hypothekenbank AG (87.2%); Deutscher Investment-Trust Gesellschaft für Wertpapieranlagen mbH (90%); dresdnerbank investment management Kapitalanlagegesellschaft mbH; DEGI Deutsche Gesellschaft für Immobilienfonds mbH (55%). Abroad:- ABD Securities Corporation (75%) (USA); Banque Veuve Morin-Pons (France); Compagnie Luxembourgeoise de la Dresdner Bank AG-Dresdner Bank International (Luxembourg); Dresdner-ABD Securities Ltd (50% of the voting capital) (Hong Kong, Japan); Dresdner Bank Canada-Banque Dresdner du Canada (Canada); Dresdner Bank (Schweiz) AG (Switzerland); Dresdner Forfaitierungs Aktiengesellschaft (DFA) (Switzerland); Dresdner (South East Asia) Ltd (Singapore); Dresdner Finanziaria SpA (Italy); Dresdner Finance BV (Netherlands); Dresdner US Finance Inc (USA)
Financial Information:

	31.12.86 DM'000	31.12.87 DM'000
Net income for the year	504,447	466,771
Consolidated profit	301,485	308,126
Dividends	20%	20%
Dividends per share*	15.63DM	15.63DM
Share capital	1,242,505	1,311,573
Deposits and bonds	185,854,000	195,675,000
Total assets	196,819,571	206,938,296

*including tax-credit

No of Employees: 38,116 (Dresdner Bank plus consolidated companies)

DSL BANK
Deutsche Siedlungs- und Landesrentenbank

Kennedyallee 62-70, Postfach 200 310, 5300 Bonn 2
Tel: (228) 889-0
BTX-Nr: 67889
Telex: 8869973 dsl d
Telefax: (228) 889624; Teletex: 228324 dsl Bank

Supervisory Board: Dr. Karl Gross (Vorsitzender), Dr. Joachim Henke (stellv. Vorsitzender), Dr. Hans Detzer, Dr. h.c. Josef Ertl, Hans-Günter Hoppe, Dr. Dionys Jobst, Karl-Heinz Koch, Dr. Andor Koritz (stellv Mitglied), Dr. Erwin Möller, Harro Moller-Racke, Dr. Bernd Otto, Randolf Rodenstock, Friedhelm Rentrop, Wolfgang Roth, Dr. Dr. Wolfram Ruhenstroth-Bauer (stellv. Mitglied), Dr. R Martin Schmidt, Hans Peter Schmitz, Dr. Rudolf Schnieders, Dr. Franz Schoser, Werner Schulz, Dr. Dieter Soltmann, Dr. Sönke Traulsen, Anton Weiler

Management Board: Dr jur. Hermann Sehrbrock (Vorsitzender), Dr jur. Dieter Goose, Karl-Ernst Löbbe, Prof Dr. jur. Dirk Schmidt, Gerhard Tobeschat

PRINCIPAL ACTIVITIES: Credit business; securities and money market business; bank funding; Berlin-Loans; commercial real estate financing; agricultural loans; consolidation funding; cooperation with banks; discount credits; export credits; international loans; investment financing; leasing financing; syndicated loans; tax-incentive project financing

Subsidiary Companies: SILA Grundstückgesellschaft

Principal Bankers: Landeszentralbank Bonn

Financial Information:

	31.12.86	31.12.87
	DM'000	DM'000
Profit before tax	78,691	92,086
Profit after tax	35,000	37,000
Retained profit	8,750	9,250
Share capital	256,000	256,000
Shareholders funds	589,251	621,570
Total assets	40,531,399	43,577,642

Principal Shareholders: Federal Republic of Germany (99%)

No of Employees: 799

FRANKFURTER HYPOTHEKENBANK AG

Junghofstrasse 5-7, Postfach 10 08 48, 6000 Frankfurt/Main 1

Tel: (0611) 298980

Cable: Hypotheken Bank Frankfurt Main

Telex: 411608 fhbf d

Supervisory Board: Dr H Burgard (Chairman), Dr U Weiss (Deputy Chairman), Dr H U Brauner, E Fischer, Dr R Freudenberg, R Gessert, W Haag, M Hauck, Dr H Messer, Dr M Otto, H H Reschke, M Rüb

Management Board: Dr D Boschert, Dr B W Rohrer, Dr H Schuck

Executive Vice President: W H Jordan. Senior Executives: J Gasparini, H-J Horstkotte, H Kleine-Büning, H C H Kreusler, R Oberbarnscheidt, Dr Pannwitz, W Rasch, N Neuhaus

PRINCIPAL ACTIVITIES: Bank specialising in mortgage and local authority loans

Parent Company: Deutsche Bank AG

Subsidiary/Associated Companies: Frankfurter Gesellschaft für Vermögensanlagen GmbH; Frankfurt Kassenverein AG; Interessengemeinschaft Frankfurter Kreditinstitute GmbH; Nass Heimstätte GmbH

Principal Bankers: Deutsche Bank AG

Financial Information:

	31.12.86	31.12.87
	DM'000	DM'000
Profit after tax	44,968	51,504
Retained profit	24,968	51,504
Dividends	-	21,504
Dividends per 50DM share	12DM*	12DM
Share capital	76,800	89,600
Total assets	29,241,848	29,392,382

*plus 1DM Bonus

No of Employees: 439

FRANKFURTER SPARKASSE VON 1822

Neue Mainzer Strasse 49-53, D-6000 Frankfurt (Main) 1

Tel: (069) 2641-0

Cable: FRASPA FRANKFURT MAIN

Telex: 411506 fspa d

Telefax: 2641-2900

Management Board: K Wächter (Chairman), Dr H Kolbeck, R Roller

PRINCIPAL ACTIVITIES: Banking

Financial Information:

	31.12.87
	DM'000
Profit before tax	74,313
Profit after tax	35,691
Capital	360,535
Total assets	8,536,000

No of Employees: 1,843

FRANKFURTER VERSICHERUNGS AG

Taunusanlage 18, 6000 Frankfurt am Main 1

Tel: (069) 71261

Cable: Frankallianz Frankfurtmain

Telex: 0411 376

Telefax: Kalle-Infotec

Supervisory Board: Ernst Wunderlich (Vorsitzender), Dieter H Alberg (stv Vorsitzender), Dr Klaus Gerathewohl (stv Vorsitzender), Gert Becker, Heinz Diehl, Heinz Dürr, Michael Falkenstein, Doris Fürbeth, Dr Klaus Götte, Dr Hans-Michael Höcherl, Hanspeter Jeuther, Dr Werner Lamby, Professor Dr Hans Joachim Langmann, Gisbert Malitz, Manfred Mörseburg, Otto Rettberg, Bernhard Stöver, Klaus Subjetzki, Dieter Ullsperger, Alfons Werner

Management Board: Dr H Schmeer (Vors), Dr O Griem, J T Kuhn, H Röder, Dr H Winter, V Hirsch (stv. Dir)

PRINCIPAL ACTIVITIES: Insurance services including fire, property and accident cover

Parent Company: Allianz AG

Principal Bankers: Deutsche Bank AG; Dresdner Bank AG; Commerzbank AG; Berliner Handels-Frankfurter Bank; Bankhaus Hauck & Sohn

Financial Information:

	31.12.86	31.12.87
	DM'000	DM'000
Premium income	1,323,216	1,400,973
Investment income	127,341	133,065
Profit before tax	105,028	69,530
Profit after tax	37,690	24,500
Retained profit	16,990	12,300
Dividend	8,790	9,600
Dividend per 50DM share	10DM	10DM
Share capital	48,000	48,000
Shareholders funds	313,590	329,300

Principal Shareholders: Allianz Versicherungs-AG; Münchener Rückversicherungsgesellschaft

No of Employees: 3,342

FRANKONA RÜCKVERSICHERUNGS-AG

Maria-Theresia-Strasse 35, D-8000 München 80

Tel: (089) 92 28-0

Telex: 5 22 531

Telefax: (089) 9228-395

Supervisory Board: Dr. Hans Gerling (Cologne, Chairman), Dr. Joachim Theye (Bremen, Deputy Chairman), Dr. jur. Hermann Flath (Hanover), *Nicole Graf (Munich), Dr. jur. Wilhelm Henning (Hanover), *Franz Herbst (Munich), Georg Krupp (Düsseldorf), Rudolf-August Oetker (Bielefeld), *Gabriele Sagarakes (Munich), Rudolf Schlenker (Cologne), *Simon Schüller (Munich), *Uwe Splett (Munich), Dr Alfons Titzrath (Düsseldorf), Günter Völker (Wuppertal), Günter Vogelsang (Düsseldorf), *Employees' representatives

Management Board: Dr. Achim Kann (General Manager), Klaus Bultmann (Deputy General Manager), Kristian Gross (Member), Dr. Reinhard Hinne (Member), Dr. Günter Jacobs (Member), Dieter W Lüer (Member)

PRINCIPAL ACTIVITIES: All classes of reinsurance
Subsidiary Companies: Maria-Theresia-Grundstücksverwaltungs-GmbH; Frankona-Unterstützungs GmbH; Frankona America Life Reassurance Co (USA); Frankona America Service Co (USA); Frankona Contact Office Ltd (UK); Frankona Reinsurance Co (UK) Ltd (UK)
Financial Information:

	30.6.86	30.6.87
	DM'000	DM'000
Premium income	1,473,700	1,486,500
Investment income	124,100	106,300
Profit before tax	14,800	22,600
Profit after tax	7,000	8,000
Retained profit	7,000	8,000
Dividend	12%(1)	14%
Dividend per share	6DM(2)	7DM
Share capital	85,500	85,500
(1) plus bonus 2%		
(2) plus 1DM		

Principal Shareholders: Gerling-Konzern Versicherungs-Beteiligungs AG
No of Employees: 380

FRIED. KRUPP GMBH
See KRUPP, FRIED., GMBH

GEBR RÖCHLING
See RÖCHLING, GEBR,

GEFA LEASING GMBH
Laurentiusstrasse 19-21, 5600 Wuppertal 1
Tel: (0202) 38 20
Telex: 08 591 602
Telefax: (0202) 38 23 22

Supervisory Board: Dr Eckart van Hooven, Dr Herbert Zapp, Hans Rosentalski, Dr Klaus Juncker, Dr Hans Walter Schlöter
Management Board: Richard Bernhardt, Dr Uwe Kayser, Dr Peter Lambeck

PRINCIPAL ACTIVITIES: Leasing of industrial equipment
Parent Company: GEFA Gesellschaft für Absatzfinanzierung mbH
Subsidiary Companies: ALD AutoLeasing D GmbH
Principal Bankers: Deutsche Bank
Financial Information:

	1987	1988
	DM'000	DM'000
Sales turnover	705,000	725,000
Profit before tax	19,800	21,100
Share capital	25,000	25,000
Shareholders funds	55,000	55,000

Principal Shareholders: Deutsche Bank (100%)

GENOSSENSCHAFTLICHE ZENTRALBANK AG
PO Box 59, Heilbronner Strasse 41, 7000 Stuttgart 1
Tel: (0711) 2040-1
Cable: Genozentralbank; SWIFT: DGZB DESG
Telex: 72244350 gz d
Telefax: (0711) 223207

Supervisory Board: Ernst Geprägs (Chairman), Dr. Rainer Märklin (Deputy Chairman), *Ewald Barth, *Erich Czotscher, Helmut Bucher, Alfred Gresser, *Emil Hengerer, Dr. Heinz Hörner, Dr. Hermann Hohner, *Wolfgang Jung, Dr. Reinhold Kissling, Martin Maier, Heinz Metzger, Werner Remmele, *Guido Schaller, Dr. Erhard Sommer, Oskar Stollsteiner, *Rolf Storz, *employee representative

Management Board: Hansgeorg Degen, Helmut Koschka, Rudi Schühle, Dieter Wössner
Senior Executives: Helmut Aydt, Horst Fröhlich, Werner Klenk, Carl Alfons Maser, Franz Miller, Manfred W Schmid, Armin Jäger

PRINCIPAL ACTIVITIES: All banking activities
Trade Names: GZB Stuttgart
Subsidiary/Associated Companies: DG Bank Deutsche Genossenschaftsbank; Bausparkasse Schwäbisch Hall AG; R + V Versicherungsgruppe; DG HYP Deutsche Genossenshafts-Hypothekenbank AG; Münchener Hypothekenbank eG; Union-Investment-Ges mbH; DIFA Deutsche Immobilien Fonds AG; DEVIP Deutsche Gesellschaft für Investment-Fonds GmbH; BWG Baugesellschaft Württ. Genossenschaften mbH; GVG Geno Vertriebsgesellschaft für EDV-Leistungen mbH; GENO Leasing GmbH; SDK Süddeutsche Krankenversicherung aG
Financial Information:

	31.12.87	31.12.88
	DM'000	DM'000
Dividends	15,000	15,000
Dividends per share	8%*	8%*
Share capital	150,000	150,000
Reserves	386,600	434,000
Total assets	17,286,600	19,720,700
*plus 4%		

No of Employees: 456

GERLING-KONZERN ALLGEMEINE VERSICHERUNGS AG
von Werthstrasse 4-14, Postfach 10 08 08, D-5000 Köln 1
Tel: (0221) 1441
Cable: Gerlingkonzern
Telex: 88110 gk d

Management Board: A Weiler (Chairman), E Gottschalk, G Heyer, W Kalenberg, Dr G Luttmer, G Ohligschlaeger, H Schilling, Dr P-R Wagner, F Kethers

PRINCIPAL ACTIVITIES: Insurance
Parent Company: Gerling-Konzern Versicherungs-Beteiligung AG
Subsidiary Companies: Gerling Konzern Rechtsschutz Versicherungs AG; Gerling Institut fur Schadenforschung und Schadenverhuetung GmbH
Principal Bankers: ZB; Postscheck Koeln
Financial Information:

	DM'000
Premium income	2,000,000

No of Employees: 6,020

GOTHAER VERSICHERUNGSBANK VVAG
Kaiser Wilhelm Ring 23-25, 5000 Köln 1
Tel: (0221) 5746-00
Teletex: 17 221 341+; 17 221 305+

Supervisory Board: Dr Hans Vossloh (Vorsitzender), Heinz Moeller (stellv Vorsitzender), Dr Günter Heidecke, Dr Jörg Mittelsten Scheid, Franz-Josef Röllgen, Dr Manfred Scholz. Arbeitnehmervertreter: Franz Bolder, Dieter Hoffmann, Jürgen Pinnow
Management Board: Prof A Wilhelm Klein (Vorsitzender), Franz-Josef Geimer (stellv Vorstizender), Herbert Rilke, Dr Heinrich Brands, Rainer P Buchheld, Manfred Schobert

PRINCIPAL ACTIVITIES: All classes of insurance
Subsidiary Companies: Gothaer Rückversicherung AG; Gothaer Krankenversicherung AG; Gothaer Hellas Insurance Co SA (Greece)
Principal Bankers: Berliner Bank AG; Deutsche Bank AG; Dresdner Bank AG; Kölner Bank von 1867; LZB; Westdeutsche Landesbank; Landeszentralbank

Financial Information:

	31.12.86 DM'000	31.12.87 DM'000
Premium income	1,442,902	1,526,072
Investment income	140,460	120,402
Profit before tax	48,346	39,722
Profit after tax	20,057	15,248
Retained profit	20,035	15,220
Shareholders funds	230,480	275,700
Total assets	2,186,485	2,263,025

No of Employees: 2,751

GROSSVERSANDHAUS QUELLE GUSTAV UND GRETE SCHICKEDANZ KG

Schickedanz-Gruppe

Grossversandhaus Quelle Nürnbergerstr 91-95, D-8510
 Fürth/Bay
Tel: (911) 7421
Cable: Schickedanz Fürth/Bay
Telex: 626851-0
Telefax: 708580

Management Board: Dr K Zumwinkel (Vorsitzender), H Bittlinger
 (Universalversand), H Schroff (Spezialversand), H Kiel
 (Logistik und DV), Dr Lorentz (Personal und Finanzen), H
 Janssen (Fachfilialen), H Ammon (Technik und Elektronik)

PRINCIPAL ACTIVITIES: Mail order and retail sales of all types
Trade Names: Privileg, Universum, Practica, Revue, Simone,
 Helix, Euroval, Mars, Weekend, Uniropa etc
Subsidiary/Associated Companies: Schöpflin; Foto-Quelle; Noris
 Verbraucherbank; Möbel Hess; Quelle France; Quelle Austria;
 Peter Hahn; Zenker Häuser; Apollo Optik; Quelle Belgium;
 Sinn AG
Principal Bankers: Dresdner Bank AG
Financial Information:

	31.1.87 DM'000	31.1.88 DM'000
Sales turnover	8,930,000	9,210,000
Shareholders funds	475,000	475,000

No of Employees: Group 32,000

GZS GESELLSCHAFT FÜR ZAHLUNGSSYSTEME MBH

Theodor-Heuss Allee 80, 6000 Frankfurt am Main 90
Tel: (069) 7933-0
Telex: 4189 067
Telefax: (069) 79 33 123

Supervisory Board: Dr Jürgen Terrahe (Chairman), Gustav Adolf
 Schröder (First Vice Chairman), Wolfgang Grüger (Second
 Vice Chairman), Dr Ulrich Weiss, Gregor Böhmer, Helmut
 Bonn, Dr Lothar Faisst, Dr Rolf Lürig, Josef Meyer, Kurt
 Morgen, Karl Rühle, Dr Wulf von Schimmelmann, Jürgen
 Sengera, Dr Peter Hoch, Helmut Steiner
Management Board: Dr Friedrich Bösel (Chairman), Herbert
 Dorner, Michael Neese, Detlev Buchal
Senior Executives: Hans-Ulrich Maass (Director Marketing and
 Advertising), Rudolf Wirmer (Director, Payment Systems
 Development), Heinrich Endter (Senior Manager-
 Administration and Personnel), Hans-Erich Rebscher (Senior
 Manager-Operations), Werner Seitz (Senior Manager-
 eurocheque Division), Peter Gerz (Senior Manager-
 Controlling)

PRINCIPAL ACTIVITIES: GZS Gesellschaft für Zahlungssysteme
 mbH issues the EUROCARD on behalf of all German banks
 and savings banks. It also clears the eurocheques which
 German travellers abroad write in the specific local currencies
 and those which foreign visitors make out in D-marks in
 Germany. Furthermore, GZS is developing new card-based
 cashless payment systems e.g. the use of the eurocheque
 card without cheque forms at electronic cash registers in the
 retail and service industries

Trade Names: eurocheque, EUROCARD
Subsidiary Companies: Eurocard Service GmbH
Principal Bankers: Deutsche Genossenschaftsbank; Hessische
 Landesbank; Commerzbank AG; Deutsche Bank AG
Financial Information:

	31.12.87 DM'000
Sales turnover	3,200,000
Profit before tax	10,286
Profit after tax	3,579
Retained profit	100%

Principal Shareholders: All German banks and savings banks
No of Employees: 586

HAFTPFLICHTVERBAND DER DEUTSCHEN INDUSTRIE VERSICHERUNGSVEREIN AUF GEGENSEITIGKEIT

Riethorst 2, Postfach 510369, D-3000 Hannover 51
Tel: (0511)645-1
Telex: 922678 hdi d
Telefax: (0511) 645-4545

Supervisory Board: Prof. Dr F Thomée (Chairman)
Management Board: A Morsbach (Chairman), Dr H Reichmann
 (Deputy Chairman), R Claus Bingemer, Dr H Dickmann, Dr J
 Schmidt-Salzer

PRINCIPAL ACTIVITIES: All classes of accident and loss
 insurance
Subsidiary Companies: HANNOVER Rückversicherungs-AG;
 EISEN UND STAHL Rückversicherungs-AG;
 Schadenschutzverband GmbH; HRV HANNOVER
 RECHTSSCHUTZ Versicherungs-AG; HANNOVER Allgemeine
 Versicherungs-AG; HANNOVER Finanz GmbH; HANNOVER
 Umwelttechnik GmbH; Marcard, Stein & Co Bankhaus;
 HANNOVER International Insurance (Nederland) N.V.
 (Netherlands); HANNOVER International (Belgique) S.A.
 (Belgium); HANNOVER International (France) (France);
 HANNOVER International Aktiengesellschaft für
 Industrieversicherungen (Austria); HANNOVER Seguros SA
 (Brazil); HANNOVER Internacional Argentina SA (Argentina)
Principal Bankers: Norddeutsche Landesbank
Financial Information:

	31.12.86 DM'000	31.12.87 DM'000
Premium income	1,423,110	1,526,719
Investment income	165,205	161,419
Profit before tax	49,708	47,577
Profit after tax	25,000	20,000
Retained profit	25,000	20,000
Shareholders funds	265,000	285,000

No of Employees: 2,130

HAMBURGER SPARKASSE

Ecke Adolphspl/Gr Burstah, Postfach 11 15 49, 2000 Hamburg
 11
Tel: (040) 379-0
Cable: Haspa
Telex: 0211827
Telefax: (040) 379-3418

Management Board: P Mählmann, Dr K-J Dreyer, Dr E Geginat,
 B Schölermann, W Matthews
Senior Executives: G Klenke (Manager), K Nielsen (Deputy
 Manager)

PRINCIPAL ACTIVITIES: Bank activities
Financial Information:

	1986 DM'000	1987 DM'000
Deposits	19,897,200	20,606,200
Total assets	24,214,500	25,437,500

No of Employees: 6,000

GERMANY

HAMBURGISCHE LANDESBANK GIROZENTRALE

PO Box 102820, Gerhart-Hauptmann-Platz 50, 2000 Hamburg 1
Tel: (040) 33 33 2985
Cable: Landesbank Hamburg
Telex: 2-161632; 2-161792 (Foreign Exchange)
Telefax: (040) 33 33-27 07 3/a

Board of Managing Directors: Dr H Fahning, Dr K Plett, Ilse
  Klinger, P Schmidt, H P Becker, Heide-Christiane Baruschke

PRINCIPAL ACTIVITIES: All banking activities including general
  credit, base credit, marine credit, international finance, stocks,
  investment, advisory services and foreign transactions
Subsidiary/Associated Companies: (100% owned unless stated):
  Deutsche Schiffsbeleihungs-Bank AG (28.6%); Hamburg LB
  International Ltd (Hong Kong); Hamburgische
  Betriebsverwaltungs-Gesellschaft am Gerhart-Hauptmann-
  Platz mbH; Kommanditgesellschaft Archimedes Gesellschaft
  für Computerhandel mbH & Co; Kommanditgesellschaft
  Altstadt Verwaltungsgesellschaft & Co
  Grundstücksgesellschaft; Offentliche Bausparkasse Hamburg
  (50%); Hamburgische Wohnungsbaukreditanstalt (24%)
Financial Information:

	31.12.86 DM'000	31.12.87 DM'000
Share capital	447,000	447,000
Total assets	36,457,900	40,454,662

No of Employees: 1,150

HAMBURG-MANNHEIMER VERSICHERUNGS AG

Überseering 45, Postfach 60 10 60, D-2000 Hamburg 60
Tel: (040) 63761
Cable: Hamavag Hamburg
Telex: 02174600 hmd

Executive Board: G Kalbaum (Chairman), Dr J Gassmann, P
  Kakies, H Rothfuss, W Uszko, Dr J Mees

PRINCIPAL ACTIVITIES: Life and accident insurance
Subsidiary/Associated Companies: Bausparkasse Heimbau AG;
  Hamburg-Mannheimer-Sachversicherungs AG (33.23%);
  HANSA Ges für Betriebsberatung mbH; Alters-
  Versorgungdurch Eigenwohnung 1955 Vermittlungs GmbH;
  Hansa-Betriebs-Treuhand GmbH; Bremer Bestattungsanstalt
  GmbH
Principal Bankers: Dresdner Bank; Deutsche Bank
Financial Information:

	DM'000
Premium income	2,600,000

No of Employees: 10,120

HANNOVER RÜCKVERSICHERUNGS-AG

Postfach 61 03 69, D-3000 Hannover 61
Tel: (0511) 5604-0
Telex: 922599 hres
Telefax: (0511) 5604-188

Supervisory Board: Adolf Morsbach, Dr Wilhelm Henning,
  François Drouault, Dr Heinz Reichmann, Horst Brode, Werner
  Burghardt
Management Board: R Claus Bingemer (Chairman), Dieter
  Pfaffenzeller, Theodor Dielmann, Dr jur Andreas-Peter
  Hecker, Udo Schubach, Dr jur Jürgen Brenzel, Winfried
  Krüger
Senior Executives: Underwriting: André Arrago (Romance and
  Arabian Countries, Middle East), Wolfgang Bartkowiak
  (German assumed business), Georges Constanty (South
  America), Andreas Grabi (Retrocession), Jürgen Lang (US
  business), Gonda Lauritzen-Grupe (Ireland, South Africa), Dirk
  Lohmann (US business, Canada, Bermuda), Paul Louven
  (Scandinavia, Benelux, Spanien), Clemens Möller (Facultative
  Underwriting), Dr Adolf Helmuth Reich (Life and Health
  Underwriting), Harald Schenk (Life Underwriting), Dietmar

Stenzel (Great Britain, Asia, Australia). Administration:
  Eckhard Eschke (Organisation, EDP), Heinrich Knappe
  (Personnel), Klaus-Heinz Kunze (Claims), Herbert Scheel
  (Investment), Siegfried Schultchen (Controlling), Bernd Spier
  (Administration), Wolfgang Strietzel (Technical Accounting),
  Berthold Trappe (Auditing, Training), Carsten Weichert
  (Accounting)

PRINCIPAL ACTIVITIES: Insurance and reinsurance
Trade Names: HR, HANNOVER Re, hanr, HANNOVER Rück
Parent Company: Haftpflichtverband der Deutschen Industrie
  VaG
Subsidiary Companies: Insurance Corporation of Hannover
  (ICH) (Chicago); The Seven Continents Insurance Company
  (Hamilton, Bermuda); International Insurance Company of
  Hannover Ltd (IH) (London); HANNOVER Services Ltd (HS)
  (London); Hollandia Holdings (Pty) Ltd (HH) (Johannesburg);
  HANNOVER National Beteiligungs GmbH (Hannover);
  HANNOVER International Beteiligungs GmbH (Hannover);
  HANNOVER Dienstleistungs-Beteiligungs GmbH (Hannover);
  GbR HANNOVER Rückversicherungs-AG/EISEN UND STAHL
  Rückversicherungs-AG-Grundstücksgesellschaft (Hannover);
  HANNOVER Ri Servizi Tecnici srl (Italy), HDI Immobilien-Fond
  Nr 4, Nürnberg Bucher KG (München)
Principal Bankers: Deutsche Bank AG, Hannover; Morgan
  Guaranty Trust Co of New York; Morgan Guaranty Trust
  Company Branch London, London; Banque Paribas, Paris;
  Banca Commerciale Italiana SpA, Milan; Amro Bank
  Amsterdam-Rotterdam Bank NV, Amsterdam; Société
  Générale de Banque, Bruxelles; Nedbank International
  Branch, Johannesburg; Westpac Banking Corporation,
  Sydney
Financial Information:

	31.12.86 DM'000	31.12.87 DM'000
Premium income gross	1,473,313	1,543,246
Premium income net	1,205,799	1,258,182
Investment income	222,225	300,498
Profit before tax	115,174	92,605
Profit after tax	109,925	73,338
Retained profit	107,225	70,571
Dividend	2,700	2,767
Dividend per share	8%	8%
Share capital	50,000	75,000
Shareholders funds*	353,381	473,952

*including transfer to
  contingency fund

Principal Shareholders: Haftpflichtverband der Deutschen
  Industrie VaG
No of Employees: 236

HANNOVERSCHE LEBENSVERSICHERUNG A.G.

Postfach 2529, 3000 Hannover 1
Tel: (0511) 8007-1
Telefax: (0511) 8007-666

Supervisory Board: Erhard Bouillon (Vorsitzender), Adolf
  Morsbach (stellv Vorsitzender), Kurt Brückner, Prof Dr jur
  Carsten P Claussen, Prof Dr rer pol Gerhard Fels, Dr jur
  Manfred Gentz, Dieter Krüger, Dr Erwin Möller, Burkhard
  Scherrer
Management Board: Dr jur Wilhelm Henning (Vorsitzender),
  Gerhard Grass-Loheide, Dipl Math Karl-Heinz Minkwitz, Dr jur
  Eckhart Frhr von Uckermann (stellv)

PRINCIPAL ACTIVITIES: Life insurance
Financial Information:

	31.12.86 DM'000	31.12.87 DM'000
Premium income	367,899	435,068
Profit after tax	6,000	7,000
Shareholders funds	27,000	34,000

No of Employees: 292

HESSISCHE LANDESBANK-GIROZENTRALE-
Helaba Frankfurt

Junghofstrasse 18-26, Postfach 110833, D-6000 Frankfurt am
Main 11
Tel: (069) 132-01
Cable: HELABA FRANKFURT. SWIFT: HELA DE FF
Telex: 415291-0 gfd
Telefax: (069) 1322000

Board of Managing Directors: Herbert J Kamierzak, Hans-Karl
Nelle, Erich Dreher, Horst Herberholz, Hermann-Adolf
Kunisch, Theo Meinz, Josef Meyer, Walter Schäfer

PRINCIPAL ACTIVITIES: Central bank
Subsidiary Companies: Helaba Luxembourg, Hessische
Landesbank International S.A.
Principal Bankers: New York: Bankers Trust Co., Chase
Manhattan Bank NA; Chemical Bank; Citibank NA; Irving
Trust Co; Manufacturers Hanover Trust Co.; Morgan
Guaranty Trust Co.; Swiss Bank Corp. London: Lloyds Bank
Plc; Barclays Bank Plc; Midland Bank Plc; National
Westminster Bank Plc

No of Employees: 2,796

HUK-COBURG
Haftpflicht-Unterstützungs-Kasse
Kraftfahrender Beamter Deutschlands aG in
Coburg

Bahnhofsplatz Postfach 402, 8630 Coburg
Tel: (09561) 96-0
Cable: HUK-Coburg
Telex: 0663414 hukco d
Telefax: (09561) 96 36 36

Supervisory Board: Dr B Schröder (Chairman)
Management Board: Dr G Schramm, Dr H-J Drews, F Beck, W
Michel, R-P Hoener, Dr W Wesler

PRINCIPAL ACTIVITIES: Insurance
Subsidiary Companies: HUK-Coburg-Allgemeine; HUK-Coburg-
Leben; CORAG
Principal Bankers: Bayerische Landesbank Girozentrale
Financial Information:

	1986 DM'000	1987 DM'000
Premium income	1,587,460	1,720,682
Profit after tax	37,560	16,902
Share capital	474,745	491,647

No of Employees: 3,567

HYPOTHEKENBANK IN HAMBURG AG

Hohe Bleichen 17, 2000 Hamburg 36
Tel: (040) 359 10 0
Telex: 213 856 hahyp d

Supervisory Board: Dr C von der Decken, Dr W Leeb, U
Bandow, Dr M Hohage, G Ohmsen, R Schmidt, D Schrödter,
S Schumann, W Valk
Management Board: Dr C Leibrock, Dr M Meissner, Dr W
Weitze

PRINCIPAL ACTIVITIES: Mortgage bank
Parent Company: Dresdner Bank AG, Frankfurt

No of Employees: 202

IDUNA ALLGEMEINE VERSICHERUNG AG,
HAMBURG

Neue Rabenstrasse 15-19, 2000 Hamburg 36
Tel: (040) 41 24 0
Cable: Iduna Hamburg
Telex: 17 40 20 52
Telefax: (040) 41 24-2958

Supervisory Board: G Wäger (Vorsitzender), P Schnitker (stellv
Vorsitzender), H Landré, H Möllmann, G E Pommeranz, J
Ring, Friedel Schapals, H Westerich, H Neuhaus
Management Board: H Becker (Vorsitzender), K O Appel, J Bär,
U Bodenburg, W Ehrig, C Krüger, G Kühn, Dr H Schlee, M
Schönrock, M Schröder, M Timm

PRINCIPAL ACTIVITIES: Insurance in the following sectors:
general accident, general liability, motor car, aviation, legal
aid, fire, robbery, water damage, plate glass, storm damage,
householder's comprehensive, technical marine, non life
Trade Names: Iduna Versicherungen
Parent Company: Iduna Vereinigte Lebensversicherung AG für
Handwerk, Handel, Gewerbe
Participations: ADLER Feuerversicherung AG; ADLER
Lebensversicherungs-Aktiengesellschaft; Iduna Bausparkasse
AG; VdK Versicherung der Kraftfahrt AG; HANSAINVEST
Hanseatische Investmentgesellschaft mbH; VÖDAG
Versicherung für den Öffentlichen Dienst AG im Adler-Iduna-
Verbund
Principal Bankers: Vereins- und Westbank
Financial Information:

	31.12.86 DM'000	31.12.87 DM'000
Premium income	469,935	470,905
Investment income	47,876	48,961
Profit before tax	11,041	10,995
Profit after tax	6,113	6,107
Retained profit	113	107
Dividends	6,000	6,000
Dividends per share	15DM	15DM
Share capital	40,000	40,000
Shareholders funds	36,580	39,178

Principal Shareholders: Iduna Vereinigte Lebensversicherung eG
für Handwerk, Handel und Gewerbe
No of Employees: 1,081

IDUNA VEREINIGTE LEBENSVERSICHERUNG
AG FÜR HANDWERK, HANDEL UND
GEWERBE, HAMBURG

Neue Rabenstrasse 15-19, 2000 Hamburg 36
Tel: (040) 41 24 0
Cable: Iduna Hamburg
Telex: 17 40 20 52
Telefax: 40 20 52

Supervisory Board: Gerd Wäger (Vorsitzender), Dr F Conzen
(stellv. Vorsitzender), W Arndt, H Bunde, F Eble, H
Frommknecht, M Herrscher, A Borgstadt, R Ruf, W Hinrichs,
H Wiedemann, F Brakemeier, H Stubbe
Management Board: H Becker (Vorsitzender), K O Appel, J Bär,
U Bodenburg, W Ehrig, C Krüger, G Kühn, Dr H Schlee, M
Schönrock, M Schröder, M Timm

PRINCIPAL ACTIVITIES: Life assurance, pension schemes,
insurance, savings and investment
Trade Names: Iduna Versicherungen
Subsidiary Companies: (100% owned unless stated): Iduna
Allgemeine Versicherung AG (96.02%); Iduna Bausparkasse
AG (51%); VdK Versicherung der Kraftfahrt AG (94.7%);
Hansa-Beteiligungs-GmbH (50%); ADLER Feuerversicherung
AG (99.97%); ADLER Lebensversicherung AG (74%); VÖDAG
Versicherung für den Öffentlichen Dienst AG im Adler-Iduna-
Verbund (90%)
Principal Bankers: Vereins- und Westbank Hamburg
Financial Information:

	31.12.86 DM'000	31.12.87 DM'000
Life premium income	1,677,000	1,835,000
Investment income	800,824	872,686
Profit before tax	10,267	14,207
Profit after tax	3,000	4,000
Shareholders funds	48,600	51,600

No of Employees: 4,600

INDUSTRIEKREDITBANK AG-DEUTSCHE INDUSTRIEBANK, DÜSSELDORF/BERLIN

Karl-Theodor-Strasse 6, 4000 Düsseldorf 1
Tel: D'dorf (0211) 8221-0, Berlin (030) 31009-0
Telex: 8582791 ikbd; 8582617 ikbw; 8581847 ikbg; 184376 indbk
Telefax: 82 21-559

Supervisory Board: Dr Dieter Spethmann (Vorsitzender)
Management Board: Dr Hans Jürgen Amelung, Dr Eckhard Gottschalk, Dr Walter Krüger, Dr Alexander von Tippelskirch, Reinhart Hundrieser
Senior Executives: Joachim Neupel (Düsseldorf, Generalbevollmächtigter), Edwin Brecht (Düsseldorf, Generalbevollmächtigter), Rolf Dern (Düsseldorf, Generalbevollmächtigter), Dr Siegfried C Cassier (Düsseldorf, Generalbevollmächtigter), Dr Kurt Rieger (Düsseldorf, Generalbevollmächtigter), Dr Franz Peter Groh (Stuttgart, Generalbevollmächtigter, Leiter d Niederl. Baden-Württemberg), Dr Rolf Toepfer (Frankfurt/M. L, Generalbevollmächtigter, Leiter d Niederl. Hessen, Rheinland-Pfalz und Saarland), Franz-Josef Dick (Düsseldorf, Generalbevollmächtigter), Karlheinz Nellessen, Dr Gerhard Rutkowski (Leiter d Niederl. Nordrhein-Westfalen), Wolf-D Schermer (Berlin, Direktor, Leiter d Niederl. Berlin), Bernd Skopalik (München, Direktor, Leiter d Niederl. Bayern), Georg-Jesko von Puttkamer (Hamburg, Direktor, Leiter d Niederl. Norddeutschland)

PRINCIPAL ACTIVITIES: Industrial medium and long term credit
Subsidiary Companies: (100% owned unless stated): IKB International (Luxembourg); IKB Leasing GmbH; IKB Immobilien-Leasing GmbH (50%); IKB Consult GmbH (50%); Auto Kredit Bank AG (60%); IKB Finance BV (Amsterdam); IKB Beteiligungsgesellschaft mbH
Financial Information: Consolidated figures (1)

	31.3.86 DM'000	31.3.87 DM'000
Profit before tax	103,021	114,931
Profit after tax	40,922	45,706
Retained profit	12,000	12,000
Dividends	28,800	33,600
Dividends per share	8.00DM	8.00DM
Share capital	180,000	210,000
Shareholders funds	608,595	752,623
Deposits (2)	3,375,712	4,325,577
Total assets	17,643,832	19,187,735

(1) Including IKB & IKB International & IKB Leasing GmbH
(2) Mostly long-term liabilities to non-banks

Principal Shareholders: Stiftung zur Förderung der Forschung für die gewerbliche Wirtschaft (21%); HOSTRA Beteiligungsgesellschaft mbH (26%)
No of Employees: 787

INDUSTRIEVERWALTUNGS GMBH

Zanderstr 5, Postfach 200886, D-5300 Bonn 2 (Bad Godesberg)
Tel: (02228) 844-0
Cable: Ivauge Bonn 2
Telex: 0885435 ivg d
Telefax: (02228) 844-107

Management Board: Dr Günter Nastelski, Dipl Kfm Fried Scharpenack

PRINCIPAL ACTIVITIES: Acquisition and administration of industrial firms, shares, partnerships and all related businesses, real estate, buildings, machinery, transport and other fitments; acquisition and storage of petrol and other substances
Parent Company: BRD-Bund
Subsidiary/Associated Companies: Fernleitungs-Betriebsgesellschaft mbH; Industrieanlagen-Betriebsgesellschaft mbH; Motorenwerk Bremerhaven GmbH; Industriebeteiligungsgesellschaft mbH; Prakla-Seismos GmbH; Steigerwald Strahltechnik GmbH; Tankterminal Wilhelmshaven GbH; Deutsche Zeppelin-Reederei Vermögensverwaltungs GmbH; Steigerwald Beteiligungs-und Verwaltungs GmbH; Uberseeische Industrie-und Handels GmbH; Deutsche Gesellschaft ZumBau und Betrieb von Endlagern für Abfallstoffe mbH
Principal Bankers: Bayerische Vereinsbank; Commerzbank; Dresdner Bank AG; Westdeutsche Landesbank Girozentrale
Financial Information:

	DM'000
Sales turnover	160,000

No of Employees: 650

ITT GESELLSCHAFT FÜR BETEILIGUNGEN MBH

Guerickestr 7, 6000 Frankfurt/Main 90
Tel: (069) 761095
Telex: 4152550

Management Board: H Rössle (Chairman), Dr H v Grünberg, Dr R Berger, Dr W Berthold, J Meins, L Micic, Dr K Tröster

PRINCIPAL ACTIVITIES: Holding company with interests in automotive equipment, electrical and electronic equipment, air conditioning, metal products, trading and insurance companies
Parent Company: ITT Industries Inc, USA (100%)
Direct Subsidiary Companies: (100% owned unless stated): Alfred Teves GmbH; Deutsche ITT Handels GmbH; Galfer SpA; Transatlantische Rückversicherungs-AG (59.9%); Alfred Teves SA; Müller & Weigert GmbH (53.85%); Alfred Teves GmbH & Co OHG (85.5%); ITT Electronic Components Distribution GmbH & Co OHG (99.9%); Cannon Electric GmbH; Elkose GmbH Elektronic Vertrieb (49.75%); ITT Medienverlag GmbH & Co KG (80%)
Financial Information:

	31.12.86 DM'000	31.12.87 DM'000
Sales turnover	3,728,000	3,982,000
Profit after tax	82,000	9,000
Retained profit	86,000	92,000
Dividends	80,000	98,000
Share capital	472,000	472,000

No of Employees: 23,400

KÖLNISCHE RÜCKVERSICHERUNGS-GESELLSCHAFT AG

Theodor-Heuss-Ring 11, D-5000 Köln 1
Tel: (0221) 7759 0
Cable: colognere koeln
Telex: 08 885 231-0 krv d

Supervisory Board: Dr N G Strasoldo (Chairman)
Management Board: Dr J Zech (Chairman), Prof Dr H Hauptmann, Dr G Hofmann, Dr L Meyer, Dr A Saacke, W Zeller, G Lorenz
Senior Executives: E Mai, R Isringhaus, H Scharfeld, K W Steurer, Dr K Sticker, Dr W Weiler

PRINCIPAL ACTIVITIES: All classes of re-insurance
Trade Names: Cologne Re
Subsidiary Companies: The Cologne Re has subsidiaries in the USA, Great Britain, Bermuda and Australia. It is also represented in France, Colombia, Mexico, New Zealand, Sweden, Singapore, Spain, Hong Kong and Japan
Principal Bankers: Sal Oppenheim Jr & Cie
Financial Information:

	DM'000
Group premium income	1,900,000

Principal Shareholders: Colonia Versicherung AG; Nordstern Allgemeine Versicherungs AG
No of Employees: 370

KREDITANSTALT FÜR WIEDERAUFBAU (KFW)

Palmengartenstrasse 5-9, Postfach 11 11 41, D-6000 Frankfurt am Main 11
Tel: (069) 74310
Cable: Kreditanstalt Frankfurtmain
Telex: 4152560 kwd; Teletex: 699804
Telefax: 7431-2944

Supervisory Board: Dr.Gerhard Stoltenberg (Vorsitzender), Dr.Martin Bangemann (Stellv.Vorsitzender), Dr.h.c.Hermann J Abs (Ehrenvorsitzender), Dr.Hans Jürgen Amelung, Ernst Breit, Dr.h.c.Helmut Geiger, Hans-Dietrich Genscher, Michael Geuenich, Helmut Guthardt, Heinz Heckmann, Constantin Freiherr Heereman von Zuydtwyck, Prof.Dr.Reimut Jochimsen, Heinrich Jürgens, Ignaz Kiechle, Hans Klein, Ursula Konitzer, Dr Klaus-Joachim Kübler, Hans Matthöfer, Dr.h.c.Tyll Necker, Dr.Theodor Paul, Karl Otto Pöhl, Klaus Richter, Dr. Wolfgang Röller, Dr.-Ing.E.h Günther Sassmannshausen, Herbert Schmalstieg, Dr.Hans Walter Schmidt, Dr.Ulrich Steger, Dr.h.c.Franz Josef Strauss, Dr.Jürgen Warnke
Board of Managing Directors: Dr.Richard Brantner, Dr.Gerhard Götte, Dr. Heinrich Harries, Dr.Manfred Schüler, Dr.Gert Vogt
Senior Executives: Hansjürgen Bäkow, Friedrich Forell, Werner Frank, Dr.Werner Hammel, Klaus Hennig, Rudolf Klein, Jürgen Krombach, Hans-Dieter Krukenberg, Peter von Lindeiner-Wildau, Hans Reich, Rüdiger Sass, Ralf Schauer, Gunnar Seibert, Wolfgang Siegel, Dr.Klaus Sturm, Roland Uecker, Detlef Vogt, Dr.Günter Wolf, Louis Graf v. Zech-Burkersroda

PRINCIPAL ACTIVITIES: Bank with economic-policy objectives. Promotion of German economy by granting investment loans for small to medium-sized companies, for the protection of the environment, for structural adjustment of certain sectors of the economy, for regional development purposes, for innovation and by giving guarantees. In addition, long term finance, mainly for exports to developing countries, ship supplies and sale of civil aircraft and untied loans for securing raw materials. Assistance to developing countries through loans and grants within the framework of the Federal Republic of Germany's policy of Financial Cooperation ("FZ") with developing countries
Principal Bankers: Deutsche Bundesbank, Frankfurt
Financial Information:

	31.12.86 DM'000	31.12.87 DM'000
Business Volume	97,140,000	100,187,000
Annual surplus*	163,253	132,542
Share capital	1,000,000	1,000,000
Liable equity capital	2,646,000	2,848,000
Total assets	93,572,000	96,835,655

*transferred to special reserves

Principal Shareholders: Federal Republic of Germany (80%); German Federal States (20%)
No of Employees: 905

KREISSPARKASSE KÖLN

Postfach 101108, Neumarkt 18-24, 5000 Köln 1
Tel: (0221) 2085561
Cable: Kreispakoe
Telex: 8883226

Supervisory Board: Dr H-J Möhle (Chairman)
Management Board: W Wasser, T Hellmann, H-P Krämer, Dr I Ellgering, H Höfer, W Kürten (Manager Foreign Department)

PRINCIPAL ACTIVITIES: Banking activities
Financial Information:

	31.12.86 DM'000	31.12.87 DM'000
Deposits	9,656,390	10,212,267
Total assets	10,305,453	10,859,234

KRUPP, FRIED., GMBH

Altendorfer Str 103, Postfach 10, 4300 Essen 1
Tel: (0201) 188-1
Cable: Krupp Essen
Telex: 0857385 fkes d
Telefax: (0201) 188-41 00

Supervisory Board: Dr h.c. B Beitz (Chairman), E Breit (Deputy Chairman), Ebrahim Arabzadeh Djamali (Tehran), K-O Göbert, H Grönhoff, G Ippers, W Iserlohn, Dr-Ing. Manfred Lennings, Prof. Dr-Ing. Dr h.c. H Leussink, Prof. Dr G Lorenz, L Mernizka, Dr-Ing Mohamad-Mehdi Navab-Motlagh (Tehran), Prof. Dr-Ing. E.h. W Niefer, Dr rer. pol. Wolfgang Röller, F Rutetzki, Dipl.-Ing. Dr-Ing. E.h. Franz J Spalthoff, H Stader, T Steegmann, H Wagendorf, H Zeretzke
Executive Board: Dr rer. pol. W Scheider (Chairman), Dr jur. Gerhard Cromme, Dr rer. pol. A Lukac, Dipl.-Ing. E.h. G Neipp, J Rossberg, Dr rer. pol. H J Schröder, Prof. Dr-Ing. K F Triebold
Senior Executives: Heads of Staff Departments at Group Headquarters: C Krutoff (Central Bureau), R Lommatzsch (Public Affairs), Dr. jur. H-M Ringleb (Legal Affairs), Dr. rer. pol. T Geer (Corporate Development), Dr. rer. pol. R S. Glück (Controlling), Dipl.-Kfm. M Dickel (Auditing), K Diekmann (Accounting), K-H Günther (Financing), B Jonas (Taxes), Dr. sc. pol. H W Busch (Personnel)

PRINCIPAL ACTIVITIES: Mechanical engineering, plant construction, electronics, steel, trading and services
Corporate Companies (100% owned unless otherwise stated): Mechanical Engineering:- Krupp Maschinentechnik GmbH: Krupp Bellaplast Maschinenbau GmbH (75%); Krupp Corpoplast Maschinenbau GmbH; Krupp Formaplast Maschinenbau GmbH; Krupp Kautex Maschinenbau GmbH; Krupp Ibérica SA (Spain); Krupp Industries Ltd (UK); Krupp Rubber Machinery Inc (USA). Krupp MaK Maschinenbau GmbH: FWI Fahrzeuginstandsetzungswerk Ichendorf GmbH (51%); Krupp MaK Diesel Inc (Canada); Krupp MaK Scandinavia A/S (Denmark); MaK Méditerranée SARL (France); Krupp MaK (London) Ltd (UK); Maschinefabriek Bolier BV (44%) (Netherlands). Werner & Pfleiderer GmbH (50.1%): WP-Industriebeteiligungsgesellschaft mbH; Werner & Pfleiderer GmbH & Co; Werner & Pfleiderer Haton BV (Netherlands); Werner & Pfleiderer (UK) Ltd (UK); Werner & Pfleiderer AG (Austria); Werner & Pfleiderer Corporation (USA); Werner & Pfleiderer SPRL (Belgium); Werner & Pfleiderer France SARL; Oy Werner & Pfleiderer AB (Finland); Sherkate Sahami Khass Werner & Pfleiderer Iran (Tehran); Werner & Pfleiderer South Africa (Pty) Ltd (South Africa); Werner & Pfleiderer AB (Sweden); Werner & Pfleiderer Italia Srl (Italy); Werner & Pfleiderer do Brasil (Fornos) Ltda (Brazil); Werner & Pfleiderer Bakery Equipment Inc (USA). Krupp Widia GmbH: Widia Heinlein GmbH; Krupp Medizintechnik GmbH; Krupp Widia Corporation (USA); Dentex Research Development Inc (USA); ULTRA-MET Manufacturing Comp (USA); Krupp Widia Belgium SA (Belgium); Krupp Widia Ibérica SA (Spain); Herko Vitoria SA (Spain); Krupp Widia Japan Ltd (Japan); Krupp Widia Nederland BV (Netherlands); Krupp Widia S E Asia (Pte) Ltd (Singapore); Krupp Widia SPA (Italy); Krupp Widia UK Ltd (UK); Krupp Widia Vertriebsges mbH (Austria); Meturit AG (Switzerland); Widia India Ltd (51%) (India); Krupp Widia France SA (99.9%) (France); Krupp Widia (Schweiz) AG (99.3%) (Switzerland); Laboratoires des Produits Dentaires Odoncia SA (France). Plantmaking:- Krupp Industrietechnik GmbH: Krupp Stahlbau Altbach GmbH; Krupp Stahlbau Hannover GmbH; Supratron F J Zucker GmbH; Krupp Stahlbau Berlin GmbH & Co KG; Stahlwerk Augustfehn GmbH; Krupp Technica Horizontal-Stranggiessanlagen GmbH; Siebenhaar Antriebstechnik GmbH; Krupp Stahltechnik GmbH (50%); Krupp Industries Ltd (Hong Kong); Krupp Hellas Industrial-Technical and Commercial SA (Greece); Krupp South Africa (Pty) Ltd (South Africa); Krupp Industries Inc (USA); Krupp (Australia) Pty Ltd (Australia); Krupp Canada Inc (Canada); Krupp Techniques Industrielles SA (France); Bobach GmbH (30%); ELA Elektroanlagen GmbH (25%);

GERMANY

Rhein Biotech Gesellschaft für neue biotechnologische Prozesse und Produkte mbH (24.5%); Buckau-Wolf India Ltd (49.9%) (India); H B Estructuras Metálicas SA (26%) (Colombia). Krupp Koppers GmbH: Hartung, Kuhn & Co Maschinenfabrik GmbH; Krupp Wilputte Corp (USA); Koppers France SA (France); Nippon Koppers YK Tokyo (99.6%) (Japan). Krupp Polysius AG (82.5%): Polysius SA (Spain); Polysius Corp (USA); Polysius SA (France); Polysius Ltd (UK); Polysius Projetos Industriais Ltda (Brazil); Polysius (Pty) Ltd (South Africa); Meura Invest SA (Belgium); Nomon SA de CV (70%) (Mexico). Electronics:- Krupp Atlas Elektronik GmbH: Krupp Atlas Datensysteme GmbH; Melchert Elektronik GmbH (85%); Melchert Electronics Ltd (Eire); Suprenum-Ges. für numerische Superrechner mbH (54%); C Plath GmbH-Nautisch-Elektronische Technik (51%); Krupp Atlas Elektronik Norge A/S (Norway); Fried. Krupp Japan Y K (Japan); Krupp Atlas Elektronik UK Ltd (UK). Steel:- Krupp Stahl AG (70.4%): Krupp Stahl Vertriebsgesellschaft mbH; Krupp Stahl Kaltform GmbH; Krupp Hüttenwerke Wohnungsbau gGmbH; Hohenlimburger Kleinbahn GmbH; FORTUNA-Stahlexportgesellschaft mbH; Krupp Brüninghaus GmbH (98%); Gerlach-Werke GmbH (58.2%); Krupp Pulvermetall GmbH (51%); Krupp Stahltechnik GmbH (50%); Hansa Rohstoffe GmbH, Düsseldorf (49%); Krupp Stal Danmark ApS (Denmark); Aciers Krupp SARL (France); SRL Aciers Krupp Staal BVBA (Belgium); Krupp Steel Co Ltd (UK); Krupp Edelstahl Vertriebsges. mbH (Austria); Acciai Krupp SRL (Italy); Krupp Stal AB (Sweden); Mure SA (Spain); Mavilor SA (99.9%) (France); Schmiedewerke Krupp-Klöckner GmbH (50%); Hellweg Liegenschaften GmbH & Co KG (50%); "Constantin" Liegenschaften GmbH & Co KG (50%); VACMETAL Gesellschaft für Vakuum-Metallurgie mbH (50%); P W Lenzen GmbH & Co KG (33.3%); Metall-Spezialrohr GmbH (50%); Rohstoffhandel GmbH (20%); Aceros Fortuna SA de CV (49%) (Mexico); Smitfort Staal BV (40%) (Netherlands); Grupo Solar SA de CV (25.1%) (Mexico); Ertsoverslagbedrijf Europoort CV (20%) (Netherlands). Trading:- Krupp Lonrho GmbH: Krupp Grafische Betriebe GmbH; Krupp Handel Metall und Rohr GmbH; Panopa Verkehrs-Gesellschaft mbH; Hansa Rohstoffe GmbH (51%); Krupp Energiehandel GmbH (51%); Stefan Keith GmbH; Helene-Amalie Kohlenhandelsgesellschaft mbH; Huck Brennstoffe GmbH; Oberhausener Kohlen- und Eisenhandelsgesellschaft mbH; Blumenfeld Brennstoffe GmbH (51%); Isar Mineralölhandel GmbH (50%); Transatlantic Bulk Carriers Inc (Liberia); Krupp-Steag Strahlmittel GmbH (50%); Neska Schiffahrts- und Speditionskontor GmbH (50%); WK Wärmetechnik Kundendienst GmbH (50%); Aceros Fortuna SA (43.7%) (Argentina); Krupp Steel & Eng Ltd (40%) (Nigeria); Leong Jin Corp (26%) (Singapore). Other Member Companies:-Buckau-Walther AG (93.9%): Total Walther Feuerschutz GmbH; Metz Feuerwehrgeräte GmbH; Cosmos Feuerlöschgerätebau GmbH; Interbrandschutz GmbH; Helmut Geissler Glasinstrumente GmbH (95%); Aug. Hoenig GmbH; Total Walther SA (Spain); Total Walther Feuerschutz AG (Switzerland); Walther Incendie SARL (France); Total Walther Feuerschutz Ges.mbH (95.3%) (Austria); Total Mexicana SA (93%) (Mexico); Walther Brandbeveiliging BV (Netherlands); Robert Reichling & Comp GmbH. Krupp Metalúrgica Campo Limpo Ltda (59.8%) (Brazil); Westdeutsches Assekuranz-Kontor GmbH; Krupp Wohnungsbau gGmbH (99.9%); Krupp Pulvermetall GmbH; GST Gesellschaft für Systemtechnik mbH; Krupp Iran GmbH (Tehran); Krupp Steel Products Inc (USA); Fried. Krupp Finance BV (Netherlands); Iran Krupp Investment AG (50%) (Switzerland)

Financial Information:

	31.12.86	31.12.87
	DM'000	DM'000
Group sales turnover	15,846,805	14,105,496
Profit after tax	126,228	41,773
Retained profit*	40,000	10,000
Share capital	700,000	700,000
Shareholders funds	2,041,496	1,967,170

*Fried Krupp GmbH

Principal Shareholders: Public benefit foundation "Alfried Krupp von Bohlen und Halbach-Stiftung" (74.99%); Islamische Republik Iran (25.01%)
No of Employees: 65,205

KÜHNE & NAGEL (AG & CO)

Wilhelm Kaiser Brücke 1, D-2800 Bremen 1
Tel: (0421) 36050
Telex: 2 461 610
Telefax: (0421) 3 60 55 04

Unlimited Partner: Kühne & Nagel AG, Luxembourg. Limited Partners: Kühne & Nagel Speditions-AG, Bremen; Klaus-Michael Kühne, Hamburg; Lonrho Plc, London
Chief Executives: Delegates of the Board of Directors: Roland W Rowland, Klaus-Michael Kühne
Management: Hans Friedrich Barmbold, Jürgen Koop, Dr K D Pietsch, P Wagner

PRINCIPAL ACTIVITIES: International freight forwarding (air, sea, overland); warehousing/distribution; port operations; seaworthy packing; insurance broking; travel services
Trade Names: Kühne & Nagel
Parent Company: Kühne & Nagel AG, Switzerland
Subsidiary Companies: Cargopack Verpackungsgesellschaft für Industriegüter mbH; Jordaberg Hafenbetrieb GmbH; Kühne & Nagel Luftfracht GmbH; Transpetrol GmbH
Principal Bankers: Deutsche Bank
Financial Information: Group worldwide

	DM'000
Sales turnover	5,700,000

No of Employees: 8,600 worldwide

LANDESBANK RHEINLAND-PFALZ-GIROZENTRALE

Grosse Bleiche 54/56, Postfach 2969, 6500 Mainz
Tel: (06131) 130
Cable: LANDESBANK
Telex: 4187858 gmz d
Telefax: (06131) 132724; 132729

Supervisory Board: Dr Carl-Ludwig Wagner (Chairman), Dr Dieter Braun-Friderici (Deputy Chairman), Helmut Badenbach, Dr Peter Barth, Heinz Blum, Kurt Bockmann, Jürgen Creutzmann, *Hans-Christian Doell, Dr Sigurd Dülz, *Hans-Adolf Fischer, Jockel Fuchs, Rudi Geil, Konrad Hartel, *Dr Manfred Hartmann, Karl Hoppe, Willi Hörter, Sigfried Klaassen, Dr Heinrich J Klein, Dr Georg Klinkhammer, *Bernd Kunze, Michael Kutscheid, Wolfgang Langguth, *Sigrid Müller-Gessinger, Karl-Adolf Orth, *Günther Pönitz, Ernst Radermacher, Michael Reitzel, Dr Christian Rosskopf, *Josef Stingl, *Manfred Töpel, *appointed by the employees
Board of Managing Directors: Dr jur Paul Wieandt (Chairman), Klaus G Adam, Hermann-Josef Bungarten, Franz-Karl Eichborn
Executive Managers: Werner Arnolds, Helmut Heinz, Jürgen Kösters, Dr Manfred Mühl. Division Executives: Friedhelm Ehlers, Caspar Giani, Norbert Havenith, Helmut Hoock, Anton Issel, Klaus Jaeger, Dr. Franz Orth, Roland Poczka, Dr. Jürgen Schiel, Klaus-Dieter Scholz, Michael Tänzer, Hans Urban, Rüdiger Vehof, Winfried Wagner, Dr. Harald O Witte

PRINCIPAL ACTIVITIES: All banking activities
Subsidiary Companies: Landesbank Rheinland-Pfalz International SA (Luxembourg); Rheinland-Pfalz Finanz AG (Zürich, Switzerland)
Financial Information:

	31.12.86	31.12.87
	DM'000	DM'000
Profit after tax	15,000	77,016
Capital	520,000	520,000
Shareholders funds	959,000	1,040,375
Deposits	35,593,000	36,477,000
Total assets	40,452,000	41,833,569

Principal Shareholders: Land Rheinland-Pfalz (50%); Sparkassen-und Giroverband Rheinland-Pfalz (50%)
No of Employees: 1,967

LANDESBANK SAAR GIROZENTRALE

Ursulinenstrasse 2, Postfach 2 18, 6600 Saarbrücken
Tel: (0681)3006-00
Cable: SaarLB
Telex: 17 681 986
Telefax: (0681) 3006-202

Supervisory Board: Werner Klumpp (Präsident)
Management Board: Herbert Weber (Vorstandsvorsitzender), Ernst Lenz (stv. Vorstandsvorsitzender), Dr Georg Gräsel, Arno Neu, Gustav Volz
Senior Executives: Fritz Anschütz, Ernst Clodo, Dr Albert Hammes, Hans Jörg Huber, Karl Heinz Hubig, Oskar Kolles, Jürgen Müsch, Hans Rass, Armand Robert, Dr Joachim Rudert, Manfred Welker, Peter Zielke

PRINCIPAL ACTIVITIES: All banking business

Subsidiary Companies: SKG Bank GmbH; Saarländische Investitionskreditbank AG
Principal Bankers: Landeszentralbank
Financial Information:

	1986	1987
	DM'000	DM'000
Profit before tax	19,596	18,353
Profit after tax	7,600	6,900
Capital	85,000	85,000
Capital and reserves	189,910	226,675
Deposits	3,503,814	3,682,605
Total assets	9,707,514	9,752,431

Principal Shareholders: Sparkassen und Giroverband Saar (100%)
No of Employees: 622

LANDESBANK SCHLESWIG-HOLSTEIN GIROZENTRALE

Martensdamm 6, D-2300 Kiel
Tel: (0431) 900 01
Cable: Landesbank Kiel
Telex: 292601; 292657; 1743154105
Telefax: (0431) 900 2446; Teletex: 43154105

Supervisory Board: Björn Engholm (Vorsitzender), Dr Stefan Pelny, Dr Jürgen Miethke, Wolfgang Stut, Heide Simonis, Uwe Thomas, Dr Klaus Murmann, Dr Dr Jürgen Pratje, Klaus Gärtner, Heinrich Schröder, Dr Dietrich Schulz, Jan Sierks, Dr Fritz Süverkrüp, Karl Heinz Luckhardt, Geerd Bellmann, Günter Blöcher, Heinrich Soldwedel, Dr Gernot Korthals, Johann Graf zu Rantzau, Klaus Spiegel, Werner-Heinz Stummer, Georg Wedegärtner-Johannes Bebensee, Wolfgang Delfs, Waltraut Fuhrmann, Günter Gerson, Hans-Heinrich Neutmann, Horst Paustian, Karl-Heinz Ravn, Hans-Otto Rohweder, Wolfgang Sander, Dieter Struck
Management Board: Gerd Lausen (Vorsitzender), Dr Werner Heller (stellv. Vorsitzender), Peter Pahlke, Dr Rembert v Rehren, Ernst Schröder

PRINCIPAL ACTIVITIES: All banking activities

Subsidiary Companies: DVG Kiel-Datenverarbeitungsgesellschaft mbH der Landesbank; LB Schleswig-Holstein Finance BV (Netherlands); Landesbank Schleswig-Holstein International SA (Luxembourg); W Jacobsen AG; Schiffshypothekenbank zu Lübeck AG; mbH; Schleswig-Holsteinische Kapital-Beteiligungsgesellschaft mbH;
Principal Bankers: Landeszentralbank; Postscheckamt Hamburg

Financial Information:

	31.12.86	31.12.87
	DM'000	DM'000
Profit after tax	26,000	14,500
Retained profit	11,000	14,500
Dividends	11,000	-
Share capital	245,000	245,000
Shareholders funds	340,000	355,000
Deposits	8,174,000	8,763,000
Total assets	30,170,000	32,373,000

Principal Shareholders: Land Schleswig-Holstein; Sparkassen und Giroverband für Schleswig Holstein
No of Employees: 1,404

LANDESGIROKASSE ÖFFENTLICHE BANK UND LANDESPARKASSE

Postfach 386, Königstrasse 3-5, D-7000 Stuttgart 1
Tel: (0711) 20611
Cable: LANDESGIROKASSE
Telex: 0723868 LG D

Board of Managing Directors: Dr Walther Zügel (Chairman and Chief Executive Officer), Albrecht Egerer (Vice Chairman), Helmut Roth (Managing Director), Otto Seiter (Managing Director), Dieter Stahl (Managing Director), E Reinhold Schreiner (Deputy Managing Director)

PRINCIPAL ACTIVITIES: Banking services
Subsidiary Companies: Schwäbische, Finanzierungsgesellschaft mbH Teilzahlungsbank (100%); Städt Pfandleihanstalt AG (99.7%)

No of Employees: 3,812

LANDESKREDITBANK BADEN-WÜRTTEMBERG

Schlossplatz 10/12, PF 4049, 7500 Karlsruhe
Tel: (0721) 150-0 (Karlsruhe); (0711) 122-0 (Stuttgart)
Teletex: 721 192 LKBBWKa
Telefax: 150-1001

Supervisory Board: Dr G Palm (Vorsitzender), D Schlee (stellv Vorsitzender), M Herzog (stellv Vorsitzender)
Management Board: R Schoeck (Vorsitzender), H-D Saue (stellv Vorsitzender), G Goll

PRINCIPAL ACTIVITIES: Banking services
Subsidiary/Associated Companies: (100% owned unless stated): Baden-Württembergische Gesellschaft für Beteiligungen mbH (33.33%); ERBA AG für Textilindustrie (15.83%); Gesellschaft für Kultur- und Wissenschaftsförderung Baden-Württemberg mbH; LKB-Bankbeteiligungsgesellschaft mbH (BBG); LKB-Energiebeteiligungsgesellschaft mbH (EBG); Mittelständische Beteiligungsgesellschaft Baden-Württemberg GmbH (MBG) (25.68%); Landesentwicklungsgesellschaft Baden-Württemberg für Städtebau und Wohnungswesen mbH (LEG) (41.71%); Landeswohnungsbau-und Städtebaugesellschaft Baden-Württemberg mbH (LSG) (35.49%); Gemeinnützige Landeswohnungsbau GmbH (24.72%); Gemeinnützige Wohnungs-und Siedlungsbau GmbH Baden (14.71%); Siedlungsgesellschaft für das Doggererzgebiet Oberbaden mbH (66.67%); Württembergische Immobilien GmbH (11.11%)

Principal Shareholders: State of Baden-Württemberg
No of Employees: 1,041

LANDWIRTSCHAFTLICHE RENTENBANK

Hochstrasse 2, PO Box 101445, D-6000 Frankfurt 1
Tel: (69) 21070
Telex: 0411 414

Management Board: Prof Dr E Pabsch, H Schildhauer, G W Seidel

PRINCIPAL ACTIVITIES: Bank loans to industry, commerce and agriculture
No of Employees: 180

GERMANY

LANDWIRTSCHAFTLICHER VERSICHERUNGSVEREIN MÜNSTER AG

Kolde-Ring 21, 4400 Münster
Tel: (0251) 7020
Cable: lvmversicherung 44 münsterwestf
Telex: 8 92 560 lvm d
Telefax: (0251) 702 1099

Supervisory Board: H Ostrop (Vorsitzender), W Werdeling (stellv Vorsitzender), P-J Etzel, Dr H-D Westerhoff, E Meyer, U Kersten
Management Board: K-A Loskant (Vorsitzender), G Kettler, J Schulze Thering (stellv Vorsitzender), Dr F R Leffler, R Bayha, K Bewerunge, J Borchert, J Herwig (Generalbevollmächtigter)
Senior Executives: Th Awater, Dr H Born, B Gauselmann, P Linkert, G Rensing, O Egginger, H Fölling, G May, G Meyr, H Schelten, J Fortströer, H Arz

PRINCIPAL ACTIVITIES: Insurance
Subsidiary Companies: LVM Lebensversicherungs-AG; LVM Rechtsschutzversicherungs-AG; LVM Krankenversicherungs-AG
Principal Bankers: Stadtsparkasse Münster
Financial Information:

| | 31.12.86 | 31.12.87 |
	DM'000	DM'000
Premium income	1,184,034	1,270,145
Investment income	155,055	143,844
Profit after tax	32,000	33,000
Open reserves	340,000	373,000
Total assets	1,767,567	1,866,384

No of Employees: 1,814

LEHNKERING MONTAN TRANSPORT AG

Schifferstr 26, Postfach 10 09 51, D-4100 Duisburg
Tel: (0203) 3188-0
Cable: Schiffahrt Duisburg
Telex: 855879 lmd d
Telefax: (0203) 3188-213

Management Board: Heribert Becker, Wolfgang Lehr, Mathias Germelmann (Deputy)

PRINCIPAL ACTIVITIES: Inland waterway and coastal transport with own and chartered vessels; special transports of liquid chemicals and shipbrokerage for coasters and barges, clearance and agency; transshipment, stevedoring, storage, grain transports; truck transports; gravel dredging, production and sale of building materials, waste management; forwarding; handling of export, import and transit cargo, barge shipments, storage; heavy lift transport; railway, truck, and container transport, air freight; packing, insurance, customs clearance; physical distribution studies, transport engineering
Parent Company: Metallgesellschaft AG
Principal Bankers: LZB; Deutsche Bank; Dresdner Bank; Commerzbank
Financial Information:

	DM'000
Sales turnover	307,000

Principal Shareholders: Metallgesellschaft AG, Frankfurt/Main (95.7%)

MANNHEIMER VERSICHERUNG AG

Postfach 1021 61, Augusta-Anlage 65, D-6800 Mannheim 1
Tel: (0621) 4571
Telex: 0463269
Telefax: (0621) 457-243

Management Board: Dr jur Rudolf Schuttze, Dr jur Klaus-Peter Jörg, Dipl-Kfm Dr oec Siegfried Schmidt, Dipl-Kfm Heinz Bloch, Dipl-Volksw Dr rer pol Lothar Stöckbauer, Dipl Psych Hans Schreiber

Senior Executives: Heinz-Walter Berens, Günter Ernst, Leo Franken, Klaus Hantsch, Ralf Hempel, Heinz Friedrich Marten, Dieter W Müller, Günther Orth, Herbert Schmidt, Manfred W O Schultz

PRINCIPAL ACTIVITIES: Provision of a wide range of insurance services
Principal Bankers: Deutsche Bank
Financial Information:

	DM'000
Premium income	2,340,000

No of Employees: 10,000

METALLGESELLSCHAFT AG

Reuterweg 14, Postfach 10 1501, D-6000 Frankfurt am Main 1
Tel: (69) 159-0
Cable: Metalag
Telex: 412250 mgf d
Telefax: (69) 159-2125

Supervisory Board: Dr Wolfgang Roeller (Vorsitzender), *Waldemar Rudewig (Stv. Vorsitzender), Prof Dr-Ing E h Werner Breitschwerdt, Dr Horst Burgard, Rolf Dollmann, *Dr Rolf-J Freyberg, Dr Uwe Haazen, Walter Janal, Karl Lichtenstein, *Horst Meusch, Dr Heribald Närger, Karl Gustaf Ratjen, *Wolfgang Reiber, *Maria-Barbara Schauss, Dr Roland Schelling, Lothar Schlünkes, Peter Schuhmacher, *Dr Manfred Schumann, Fahed Majed Al-Sultan, Sir Bruce Watson, *Arbeitnehmervertreter
Board of Management: Dr Dietrich Natus (Vorsitzender), Dr Otto W Asbeck (Kfz-Teile), Werner Busch (Finanzen), Dr Heinrich Goetz (Arbeitsdirektor), Dr Horst G Karus (Chemie), Dr Heinz Schimmelbusch (Rohstoffe), Dr Meinhard Forster (stv. Finanzen), Dr Jens-Peter Schaefer (stv. Anlagenbau)
Senior Executives: Dr Clemens von Arnim, Hubert Buss, Helmut Maczek, Guenther Mannes, Ernesto Melber, Dr Walter Sies, Dr Harald Rieger, Wolfgang Wink

PRINCIPAL ACTIVITIES: Extraction, production, treatment, processing, and trading of ores, other minerals, metals, metallic and chemical products, raw materials to the ferrous industry, energy raw materials; development and exploitation of technical processes, industrial plant contracting, banking and financial transactions, insurance arrangements, shipping, transportation, and forwarding; participation in enterprises concerned with similar activities
Trade Names: Zamak, Silumin, Silumin Beta, Synkrolith, Pyrox
Subsidiary/Associated Companies: Metallgesellschaft GmbH has subsidiary and associated companies in 11 countries including: Sachtleben Bergbau GmbH; Metall Mining Corp; Metallgesellschaft Ltd; "Berzelius" Metallhutten-GmbH; Ruhr-Zink GmbH; VDM Nickel-Technologie AG; Rheinzink GmbH; Kolbenschmidt AG; Lurgi GmbH; Lurgi Corporation; Sachtleben Chemie GmbH; CHEMETALL GmbH; Synthomer Chemie GmbH; Eisenerz-GmbH; Sté Continentale Parker SA; Lehnkering Montan Transport AG; DEKA Transport BV; Unterweser Reederei GmbH; Norddeutsche Affinerie AG; Metallgesellschaft Corp.; The Ore & Chemical Corp; Metallgesellschaft Trading Corp; Metallbank GmbH; Cetema BV; MG Services Company; First Boston Metallgesellschaft; Metallgesellschaft Far East Ltd; Haendler & Natermann GmbH
Principal Bankers: Deutsche Bank AG; Dresdner Bank AG
Financial Information: World figures

| | 30.9.86 | 30.9.87 |
	DM'000	DM'000
Sales turnover	14,905,846	13,329,348
Profit after tax	76,477	78,565
Dividends per share	6DM	6DM
Earnings per share	12DM	15DM
Share capital	280,000	280,000

Principal Shareholders: Algemeine Verwaltungsgesellschaft für Industriebeteiligungen mbH, Munich (over 25%); Kuwait Investment Authority (20%)
No of Employees: Worldwide 24,394

MÜNCHENER RÜCKVERSICHERUNGS-GESELLSCHAFT

Königinstrasse 107, Postfach 40 13 20, D-8000 München 40
Tel: (089) 3891-0
Telegram: Münchenerrück. Cable: Munichre Munich
Telex: 5 215233-0 mr d; Teletex: 897660=mrd
Telefax: (089) 399056

Supervisory Board: Dr jur Dieter Spethmann (Vorsitzender),
Helmut Haeusgen (stellv Vorsitzender), Dr jur Wolfgang
Schieren (stellv Vorsitzender), Prof Dr Dieter Farny, Dr jur
Maximilian Hackl, Dr rer pol Alfred Herrhausen, Eberhard von
Kuenheim, Prof Dr rer pol Rolf Rodenstock, Prof Dr rer nat
Matthias Seefelder, Otto Wolff von Amerongen, Dr rer nat
Gerhard Berz, Günther Hoowaarts, Annemarie Kroiss, Ludwig
Lommer, Vera Zielke
Board of Management: Dr iur hc Horst K Jannott (Vorsitzender),
Dr jur Klaus Gerathewohl, Dr jur Hans-Jürgen Schinzler,
Klaus Conrad, Rudolf Ficker, Dieter Göbel, Dr oec publ Edith
Lukas, Dieter Nonhoff, Hans Raffler, Ralph Roth, Hans-Dieter
Sellschopp
Senior Executive Managers: Dr jur Wolf Otto Bauer, Dr jur
Eckart Beulke, Dr jur Hans Freund, Werner Graf, Sven
Hallstroem, Luise Himmelseher, Dr oec publ Ernst Hosp,
Heinz Hüppmann, Wolfram W Karnowski, Werner Katti, Dr jur
Udo Knoke, Max K Lang, Anton Madlinger, Herbert Waldemer

PRINCIPAL ACTIVITIES: All branches of reinsurance
Financial Information:

	30.6.86	30.6.87
	DM'000	DM'000
Premium income	11,803,916	11,946,548
Investment income	1,553,016	1,526,821
Profit before tax	181,835	207,240
Profit after tax	56,463	60,766
Retained profit	57,183	63,500
Dividends	18%	20%
Dividends per share	9DM	10DM
Earnings per share	14.40DM	17.11DM
Share capital	590,000	590,000
Shareholders funds	1,367,348	1,367,348
Retained profit and reserves	834,532	840,848

No of Employees: 1,719

NASSAUISCHE SPARKASSE

Rheinstrasse 42-44, D-6200 Wiesbaden
Tel: (06121) 3 64-0
Telex: 04186726

Management Board: W Herms (Vors), Dr H J Beatus, Dr G P
Glomb, H C Lucht, M von Uslar

PRINCIPAL ACTIVITIES: Banking services

Principal Shareholders: State of Hesse
No of Employees: 2,310

NORDDEUTSCHE GENOSSENSCHAFTSBANK AG

Schiffgraben 53-57, Postfach 249, 3000 Hannover 1
Tel: (0511) 1227-0
Telex: 922751
Telefax: (0511) 1227 155

Supervisory Board: Karl-Heinz Scharmann (Vorsitzender),
Sönke Paulsen (stellv Vorsitzender), Ludof von Veltheim
(stellv Vorsitzender), Holger Bajorat, Hans-Joachim Ewald, Dr
Klaus F Geiseler, Dr Dietrich Hahlbrock, Carsten Hansen,
Rudolf Heckerodt, Horst Ihssen, Hans Kruse, Gerd Lange,
Heinrich Pralle, Reinhold Schack, Helmut Schade, Rolf
Schmiester, Friedrich Schreiner, Manfred Schütze, Rudolf
Sölter, Paul Sommer
Management Board: Dr G Barner (Vorsitzender), Dr K P
Caspritz, Dr A Erdland, Dr Ch Pleister, Dr K Preiss, E-A Rust,
Dr S Henning (stellv)

Senior Executives: Jens Stein-Angel (Head of International
Division and Head of Correspondent Banking)

PRINCIPAL ACTIVITIES: Banking

Subsidiary Companies: Weka-Kredit-Bank AG (98%);
Landkreditbank Schleswig-Holstein AG (90.66%); (HML-Bank)
Hallbaum, Maier & Co AG (66.48%)

Financial Information:

	31.12.86	31.12.87
	DM'000	DM'000
Profit after tax	40,800	50,200
Dividend	5.12%*	5.12%*
Share capital	230,000	230,000
Shareholders funds	580,600	612,200
Deposits	-	15,024,000
Total assets	17,270,100	16,790,000

*8% including tax credit

No of Employees: 957

NORDDEUTSCHE LANDESBANK GIROZENTRALE
NORD/LB

Georgsplatz 1, Postfach 290, 3000 Hannover 1
Tel: (0511) 103-0
Cable: NORD/LB
Telex: 921620
Telefax: 103-2502

Supervisory Board: Birgit Breuel (Chairman), Dr D Hoppenstedt
(Vice Chairman), M Borg, P Brenner, H Droste, K Fischer, G
Glogowski, Dr H Haller, W Hirche, E Jürges, W Karmann, K
Lauenstein, H Müller, P Pfad, M Riegert, H Rienäcker, H-J
Röhrs, Dr Ing E h Sassmannshausen, Dr W P Schmidt, W
Schulze, H Werner

Board of Management: Dr B Thiemann (Chairman), Dr G
Dirksen (Vice Chairman), Dr H A Curdt, Dr J Runge, C V
Hammerstein-Loxten, Dr H Vieregge, K Schiersmann

Executive Managers: D Fürst, H Korte, P Lückenotto, D
Pannenberg, H F Saure, W Schildt, H-P Schott

PRINCIPAL ACTIVITIES: All types of banking business,
including agricultural credits, bank boxes, bonds, building
society savings, business loans of all types, coins and
medals, deposit accounts, discounting of bills, estate
administration, factoring, foreign exchange dealings, Giro
transactions, gold bars, hire purchase loans, import and
export financing, incentive savings, interim credits, investment
advice, junior savings, leasing, letters of credit, loans to
public authorities, mortgages, overdraft facilities, overseas
remittances, personal current accounts, personal loans,
premium bonds, saving accounts, saving bank certificates,
saving club facilities, securities, shares, ships financing, target
saving, travellers' cheques

Trade Names: NORD/LB

Consolidated Companies: (100% owned unless stated):
NORD/LB Norddeutsche Landesbank Luxembourg SA
(Luxembourg); Bremer Landesbank (75%); Bremer
Landesbank Finance (Curaçao); Niedersächsische Bank
GmbH; NORD/LB Finance (Curaçao); Braunschweig GmbH;
Bürohaus Petuelring GmbH & Co Vermietungs-KG (99.9%);
Bürohaus Ricklinger Kreisel GmbH & Co Vermietungs-
Kommanditgesellschaft (99.7%); DURUM AG (Switzerland);
Fürstenberg Ehemalige Herzoglich Braunschweigische
Porzellanmanufaktur (97.9%); LBS-Immobilien GmbH;
Niedersächsische Immobilien-Anlagegesellschaft-Fonds 2-Dr
Müller KG; Niedersächsische Immobilien-Anlagegesellschaft-
Fonds 7-Dr Focken KG (55.4%); Norddeutsche Gesellschaft
für kommunale Anlagen mbH; Norddeutsche Immobilien-
Anlagegesellschaft mbH

Financial Information: Consolidated figures

	31.12.86	31.12.87
	DM'000	DM'000
Profit before tax	105,000	181,000
Profit after tax	37,000	61,000
Retained profit	25,000	42,000
Dividends	19,000	19,000
Capital	2,120,000	2,162,000
Deposits and bonds	88,291,000	93,972,000
Total assets	96,513,000	102,068,000

Principal Shareholders: State of Lower Saxony; Lower Saxon Savings Bank Association

No of Employees: 6,072

NORDSTERN ALLGEMEINE VERSICHERUNGS-AG

Gereonstr. 43-65, Postfach 10 13 68, D-5000 Köln 1

Tel: (0221) 148-1

Cable: Nordsternfeuer

Telex: 08882714 nst d

Telefax: (0221) 12 1461; Teletex: 2214179 nstd

Supervisory Board: Dr N Graf Strasoldo (Vorsitzender)

Management Board: C Kleyboldt (Vorsitzender), Dr W Asmus, G Aymanns, D Bierbaum, Dr R Kürten, V Meis, W Gemünd (stv), H Kühl (stv)

Head Office Senior Executives: Direktoren:- Dr U Bertermann, J B Cords, B Langöhrig, F Lürbke, K J Nerlich, Ch Petzold, F Pfeiffer, Dr R Rex, G Schweikert. Stellvertretende Direktoren: V Diffring, N Werkmeister

PRINCIPAL ACTIVITIES: National and international, direct and indirect insurance and reinsurance; life- and health insurance as reinsurance only; supplying insurance of any kind

Subsidiary Companies: Nordstern Lebensversicherungs-AG; Union Actien-Gesellschaft für Versicherungen; Nordstern Grundbesitz-Vermietungs-u Verwaltungs GesbR; Nordstern Versicherungs-AG (Austria); Nordstern Verwaltungs AG (Austria); Nordstern Service-Gesellschaft mbH (Austria); Star Assurance Society Ltd (UK); Nordstern Service International (US) Inc. (USA); Nordstern Insurance Company of America (USA); Nordstern-Colonia-Hellas AG (Greece); Corona AG (Liechtenstein); Familia Sicherheits-, Versicherungs-und Sparberatungsgesellschaft mbH (Austria); Parana CIA de Seguros Germano-Brasileira (Brazil); Colonia-Nordstern Schadeverzekering NV (Netherlands); Victoria-Meridional, S.A., De Seguros Generales (Spain); Interstar-Service Vermittlungs-u Verwaltungs GmbH (Switzerland) and Branch offices in: Belgium, France, Greece, Italy, Switzerland and Turkey

Principal Bankers: Deutsche Bank AG; Dresdner Bank AG; Sal Oppenheim jr & Cie; Westdeutsche Landesbank

Financial Information:

	1986	1987
	DM'000	DM'000
Premium income	1,011,490	1,028,202
Investment income	123,356	124,160
Profit before tax	61,569	68,492
Profit after tax	27,164	27,444
Retained profit	12,659	12,003
Dividends	7,700	8,800
Dividends per share	14%	16%
Earnings per share	54.76DM	54.10DM
Share capital	55,000	55,000
Shareholders funds	240,959	259,603

Principal Shareholders: Colonia Versicherung AG (57.68%); Winterthur Schweizerische Versicherungs-Ges (37.2%) (Switzerland)

No of Employees: 2,527

NORD-WEST-RING SCHUH-EINKAUFSGENOSSENSCHAFT EG

Schaumainkai 69, Postfach 700 868, D-6000 Frankfurt (Main) 70

Tel: (069) 60 67-1

Cable: Nordw estring

Telex: 0412906

Telefax: (069) 62 42 00

Supervisory Board: Fr Schüttfort (Chairman), Dipl Kfm Klaus Söldner (Deputy Chairman), Werner Dielmann, Jochen Galipp, Günter Graf, Theodor Grosse-Kreuel, Georg Mayer, J J Rohof, Dipl Kfm Peter Suyer, Manfred Voigt

Executive Board: Dr Georg C Neumann (Chairman, Marketing), Lothar Schlentrich (Deputy Chairman), Thorsten Schmidt, Manfred T Wellenbeck

PRINCIPAL ACTIVITIES: Wholesale footwear, purchasing cooperative with marketing services

Subsidiary Companies: NORD-WEST RING BANK eG; FACH-SPORT Handels GmbH; QUICK-SCHUH GmbH; CONVENTA Treuhand GmbH; NORD-WEST-RING Austria GesmbH (Austria); QUICK-SCHUH GesmbH (Austria); QUICK-AG (Switzerland)

Financial Information:

	DM'000
Sales turnover	1,000,000

No of Employees: 1,515

NÜRNBERGER ALLGEMEINE VERSICHERUNGS-AG

Rathenauplatz 16/18, D-8500 Nürnberg 21

Tel: (0911) 531-0

Cable: Lebensbank Nuernberg

Telex: 622163 nver d

Telefax: (0911) 531-3206

Supervisory Board: Konsul Senator W Braun (Vorsitzender), F Haberl (stellv. Vors.), L Edler v Braun, A W Graf v Faber-Castell, F Fertig, H Gropp, E Lorenz, H Ravenborg, T Schoeller

Management Board: Konsul Dr G Bayer (Vorsitzender), K-H Diedrich, K Falk, W Leiber, H-P Schmidt

Senior Executives: Prokuristen: H Potzler, G Riedel, H Hauschild, H Berger, H Haas, E Schön, H Stenzel, H Selch, G Lipp, H Wagner, D Günther, G Herrmann

PRINCIPAL ACTIVITIES: All classes of insurance

Parent Company: Nürnberger Lebensversicherung AG

Subsidiary Companies: (100% owned unless stated): Nürnberger Beamten Allgemeine Versicherung AG; GARANTA Versicherungs-AG (74%); INSURANCE Internationale Versicherungsmakler GmbH; Nürnberger Versicherungen Verwaltungsgebäude Nunnenbeckstrasse GbR (50%); Nürnberger Versicherung AG (50%)

Principal Bankers: Dresdner Bank; Deutsche Bank; Commerzbank; Bay. Hypo-Bank; Postgiroamt; Landeszentralbank; Bay Vereinsbank; Stadtsparkasse

Financial Information:

	31.12.86	31.12.87
	DM'000	DM'000
Premium income	682,786	756,768
Investment income	50,014	40,165
Profit before tax	36,519	20,385
Profit after tax	13,308	7,790
Retained profit	7,378	4,848
Dividends	1,120	1,120
Dividends per share	14%	14%
Earnings per share*	140DM	140DM
Share capital	8,000	8,000
Shareholders funds	112,300	119,000

*1,000 DM nominal value

No of Employees: 1,262 salaried staff, 8,336 agents

NÜRNBERGER LEBENSVERSICHERUNG AG

Postbox 210180, Rathenauplatz 16/18, D-8500 Nuernberg 21
Tel: (0911)5310
Cable: Lebensbank Nbg.
Telex: 622163 nver d
Telefax: (0911) 531-3206

Supervisory Board: W Braun (Vorsitzender), H Stoll (stellv.
 Vorsitzender), F Haberl (stellv. Vorsitzender), W Bauditz, L E
 v Braun, A W Graf v Faber-Castell, H Forstmeier, H Krapf, H
 Ravenborg, Th Schoeller, M Schweiger, G Wanka
Management Board: Dr G Bayer (Vorsitzender), P Burghard, W
 Leiber, H-P Schmidt
Senior Executives: Prokuristen: A Brunner, M Christlieb, K-H
 Eberle, D Facius, H Forstmeier, A Holzberger, Dr W
 Kaufmann, A Kraus, W Kreil, D Kuegel, D Leuzinger, H
 Maedl, H Maralt, P Merz, H Rascher, E Schertlin, Dr H
 Schnittger, H-J Steier, R Volkert, G Wagner , K Weigel, H
 Zahorik

PRINCIPAL ACTIVITIES: Life insurance
Subsidiary Companies: Nürnberger Allgemeine Versicherungs-
 AG; Nürnberger Grundstuecks-u Verwaltung GmbH;
 Nürnberger Beamten Lebensversicherung AG für den
 öffentlichen Dienst; Nürnberger Versicherung AG Salzburg
Principal Bankers: Dresdner Bank; Deutsche Bank;
 Commerzbank; Bay Hypobank; Bay Vereinsbank
Financial Information:

	31.12.86	31.12.87
	DM'000	DM'000
Premium income	1,034,959	1,179,193
Investment income	534,585	543,352
Profit before tax	16,582	19,009
Profit after tax	6,680	6,680
Retained profit	6,680	6,680
Dividends	1,680	1,680
Dividends per share	14%	14%
Earnings per share	14.0DM	14.0DM
Share capital	12,000	12,000
Shareholders funds	51,680	56,680

No of Employees: 1,225 salaried staff, 8,336 agents

OETKER, DR AUGUST, KG

Lutterstr 14, Postfach 21, 4800 Bielefeld
Tel: (0521) 1550
Telex: 0932945 oebf

Supervisory Board: Rudolf August Oetker
Management Board: A Oetker, Dr G Sandler, R Stelbrink, Dr J
 H de La Trobe, Dr D Holzinger, K P Erbrich, Dr E F Schröder

PRINCIPAL ACTIVITIES: Food; beer and soft drinks; shipping;
 hotels; banking and insurance interests
Trade Names: Dr Oetker, Langnese Honig, Ültje, Clausthaler,
 Römer Pils, Selters, Dortmunder Actien-Pils, Berliner Kindl
Subsidiary/Associated Companies: FOOD: Dr August Oetker; Dr
 Oetker Eiskrem GmbH; Dr Oetker Tiefkühlkost GmbH; Dr
 Oetker Tiefkühl-Backwaren GmbH; Oskar Beck GmbH & Cie;
 DIBONA Markenvertrieb KG; Einkaufszentrum Bau- und
 Betriebs-Ges mbH & Co Zabel-Krüger-Damm KG; ETO
 Nahrungsmittelfabriken Richard Graebener; Fleischer GmbH;
 Heinz Grünewald; Hanseatische Hochseefischerei AG;
 Hanseatische Nahrungsmittel-Fabrik Bey & Co; Import/Export
 GmbH; Lindenberg & Co (Fisch- und Hummerhandel);
 Hermann Meyer & Co KG; Meyer & Beck Handels-
 Kommanditgesellschaft; Mönkeländer Fleischwarenfabrik
 GmbH & Co; Reese Gesellschaft Nährmittelfabrik;
 Schwarzwald-Konditorei Dr August Oetker GmbH; St Magnus
 Gesundkost-Vertriebs-GmbH; Tiedemann & Meier (GmbH &
 Co); ültje GmbH; Edelfisch Fischspezialitäten-
 Handelsgesellschaft mbH. SHIPPING: Atlas Levante Linie
 GmbH; Bahia Bulkcarrier GmbH; Bock, Godeffroy & Co;
 Caravelle Assekuranz-Vermittlungs- und Schadenskontor
 GmbH; Columbus Line Reederei GmbH; Deutsche Levante

Linie GmbH; Deutsche Nah-Ost Linien GmbH & Co KG;
Hamburg-Südamerikanische Dampfschifffahrts-Gesellschaft
Eggert & Amsinck; Hamburg-Süd Reiseagentur GmbH;
Hanseatische Hafenbetriebsgesellschaft Eggert & Amsinck;
H.S.K. Hanseatische Seefrachtenkontor GmbH; Hanseatische
Seeverkehrs-Gesellschaft mbH; Nah-Ost Schiffahrtsagentur
GmbH; NAUTITANK Schiffahrtsgesellschaft mbH;
CONTAINERSCHIFF-REEDEREI Gesellschaft mbH MS; "Cap
Polonio"; CONTAINERSCHIFF-REEDEREI Gesellschaft mbH
MS "Cap Trafalgar"; Containerschiffsreederei MS "Columbus
Olinda" Dr August Oetker KG; Containerschiffsreederei MS
"Columbus Ohio" Dr August Oetker KG; Olinda Bulkcarrier
GmbH; Rudolf A Oetker; Schiffahrtsgesellschaft Columbus
Line GmbH; Schiffahrtsgesellschaft NAVITANK GmbH; TECO
Trans Europe Container Operators GmbH. OTHER
INTERESTS: Allgäuer Brauhaus AG; Allgäuer
Erfrischungsgetränke GmbH; Andreas Brauerei; AGRICHEMA
Materialflusstechnik GmbH; Baugesellschaft Sparrenberg
Rudolf A Oetker; Baugesellschaft Vorwärts Rudolf A Oetker;
Brenner Hotel; Ceres-Verlag Rudolf August Oetker KG;
Chemische Fabrik Budenheim Rudolf A Oetker; Deutsche
Hyperphosphat GmbH; Freiherr von Maltzan
Immobilienverwaltungsgesellschaft KG; Gasthausgesellschaft
Gambrinus GmbH; Gesellschaft für Absatzberatung und
Marketing der Oetker-Gruppe mbH; Gilbert J McCaul & Co
GmbH; GWP Gesellschaft mbH für Weiterbildung und
Projektmanagement; Handelsgesellschaft Sparrenberg mbH;
Hanseatische Nahrungsmittelfabrik Bey & Co KG; Hussmann
KG; Max Jordan Bauunternehmung; Michelmann & Schlüter
GmbH; Dr Oetker Haushaltsgeräte OHG; Dr August Oetker
Finanzierungs- und Beteiligungs-Gesellschaft mbH; Oetker
Aussenhandelsgesellschaft mbH; Omnia Werbegesellschaft
mbH & Co KG; Omnia Werbegesellschaft mit beschränkter
Haftung; Orimex Handelsgesellschaft mbH; Schwarzwald-
Klinik in Villa Stephanie GmbH; SIMTRA Trading Handel mit
Chemieprodukten GmbH. Companies no longer consolidated:
Braufinanz, Beratungs- und Management GmbH
Financial Information:

	31.12.86	31.12.87
	DM'000	DM'000
Sales turnover	2,992,358	2,914,013

No of Employees: 8,800

OPPENHEIM, SAL, JR & CIE

Unter Sachsenhausen 4, Postfach 101310, D-5000 Köln 1
Tel: (0221) 145-01
Cable: Neptuno Köln; SWIFT Code: SOPP DE 3K
Telex: 8882547
Telefax: (0221) 1651-512

Partners: Will Marx, Alfred Freiherr von Oppenheim, Manfred
 Freiherr von Oppenheim, Dr Nikolaus Graf Strasoldo, Hans-
 Ulrich Trippen, Dr Karl-Heinz Wessel, Dr Thomas Bscher,
 Matthias Graf von Krockow
Executive Managers: J F Graf von Oppersdorff, Dr W Damm, W
 Matthias, W Meissner, P Porzelt
Managers: H Büttgenbach, R Dreeser, K Fränken, K Heisch, A
 Hubbe, Dr J Kauffmann, E Krichler, Dr H Leutner, Dr D
 Rehbein*, H Rössinger, B J Rotter, B Röttgers, R Schell, Dr J
 W Schmitz, L Schubert, M Schütz, W Freiherr von Uslar-
 Gleichen. Deputy Managers:- J Anders, L Berger, H
 Bossbach, J Bratengeier, G Freiherr von Eyb, H Hilgert, W
 Hoffmann, W Holthausen, R Kessler, R Klostermann, W
 Krakau, N Mayer, Dr H-T Panek, H-J Plum, K Priesmann, K
 Satzger, F-M Stallmann, F J Theisen, K Vilbrandt, A von
 Warburg, I Wester, W Wolber *Legal Adviser

PRINCIPAL ACTIVITIES: Banking
Subsidiary Companies: Kölnische Verwaltungs-
 Aktiengesellschaft für Versicherungswerte; Neptuno
 Verwaltungs- und Treuhand-GmbH; OPPENHEIM
 Kapitalanlagegesellschaft mbH; RHEINISCHE
 KAPITALANLAGEGESELLSCHAFT mbH

GERMANY

Financial Information:

	1986 DM'000	1987 DM'000
Deposits	2,870,580	2,861,433
Total assets	3,520,513	3,513,607

No of Employees: 503

OST-WEST HANDELSBANK AG

Stephanstrasse 1, 6000 Frankfurt am Main
Tel: (069) 2168-0
Cable: Ostwestbank
Telex: 416636 ostw d; 4189224 ofx d; 416651 owdep d
Telefax: (069) 2 16 83 19; SWIFT: OWHB DE FF

Supervisory Board: Victor V Geraschenko (Chairman, Moscow),
Wladimir G Bolchakov (Deputy Chairman, Cologne), Vladimir
G Malinin (Deputy Chairman, Moscow), Vladimir N Goriounov
(Zurich), Reinhard Meissner (Frankfurt/Main), Hannelore
Schulist (Mörfelden-Walldorf)
Management Board: Valeri W Ljuljtschew (Chairman, Bad
Homburg), Ildar K Nigmetzanov (Deputy Chairman,
Oberursel), Dipl-Voksw Günther Hammerschmitt (Bad
Homburg), Diether K Pohlmeyer (Neu-Isenburg)
Senior Executives: Division Directors: Gerhard Becker, Hermann
Merle, Hans-Georg Neumann, Peter Seiffert, Günter Schulze,
Valeri I Zaitsev. General Directors: Gerd Wellershoff, Vladislav
J Khrebtichtchev

PRINCIPAL ACTIVITIES: All banking functions and services,
except investment; trade between Germany and communist
countries in particular
Principal Bankers: London Correspondents: Moscow Narodny
Bank; National Westminster Bank; Midland Bank; New York
Correspondents: Chase Manhattan Bank; Citibank; Irving
Trust; Girard International Bank; Mellon Bank International
Financial Information:

	31.12.86 DM'000	31.12.87 DM'000
Business volume	2,050,000	1,950,000
Profit for the year	5,450	5,553
Retained profit	2,670	1,980
Share capital	65,000	65,000
Capital and reserves	94,220	96,200
Borrowed funds and deposits	1,850,000	1,750,000
Total assets	1,961,051	1,870,000

Principal Shareholders: State Bank of the USSR (9.54%); Bank
for Foreign Economic Affairs of the USSR (9.23%); Bank of
Savings and Crediting for the Population (8.92%); V/O
Almazjuvelirexport (8.92%); V/O Exportles (8.92%); V/O
Promsyrioimport (8.92%); V/O Sojuzgazexport (8.92%); V/O
Sojuzkoopvneshtorg (8.92%); V/O Sudoimport (8.92%); V/O
Techmashimport (8.92%) all in Moscow
No of Employees: 10091

PFÄLZISCHE HYPOTHEKENBANK AG

An der Rheinschanze 1, 6700 Ludwigshafen
Tel: (0621) 59971

Management Board: Dr B v Brevern, Dr H R Heigl, Dr J Jacobi,
R Neff

PRINCIPAL ACTIVITIES: Banking

PROVINZIAL LEBENSVERSICHERUNG HANNOVER

Schiffgraben 4, 3000 Hannover 1
Tel: (051 1) 102-0
Telex: 923 111

Supervisory Board: Adolf Freiherr von Wangenheim
(Vorsitzender)
Management Board: Dr Hermann Flath (Vorsitzender), Hans-
Peter Bender, Dr Hubert Diekstall, Klaus Dill, Dr Robert
Polhausen, Hans Weichsler

PRINCIPAL ACTIVITIES: Life insurance
Parent Company: Landschaftliche Brandkasse Hannover,
Hannover
Financial Information:

	DM'000
Premium income	370,000

No of Employees: 170

R & V VERSICHERUNGSGRUPPE

Taunusstrassee 1, 6200 Wiesbaden
Tel: 533-0
Cable: Raivo
Telex: 04 186 819; 04 186 803; BTX: *34567#
Telefax: (06121) 533-4500; (06121) 533-369

Management Board: Dr Peter C von Harder (Vorsitzender), Kurt
E Boehme, Goetz Engelhardt, Dr Werner Fauth, Robert
Ludwig, Dr Juergen Loehr, Manfred Schlottke, Dr Gerhard
Schubert, RA Hans-Dieter Wehlmann, Dr Eckhardt Wilkens,
Klaus R Zimmermann, Dr Bernhard Zloch (stellv. Vorstand)
Senior Executives: RA Dr Manfred Antoni, Willi Reeh,
(Generalbevollmächtigte des Vorstandes)

PRINCIPAL ACTIVITIES: All forms of insurance and reinsurance
Group Companies: R & V Allgemeine Versicherung AG; R & V
Lebensversicherung aG; R & V Pensionsversicherung aG;
Vereinigte Tierversicherung Gesellschaft aG; R & V
Rechtsschutzversicherung AG; Rhein-Main
Rueckversicherungs-Gesellschaft AG; R & V
Krankenversicherung AG; R + V Versicherung Vermittlungs-
GmbH
Principal Bankers: Landeszentralbank Wiesbaden; Deutsche
Genossenschaftsbank, Frankfurt/Main; Wiesbadener
Volksbank
Financial Information:

	1986 DM'000	1987 DM'000
Premium income	3,541,000	3,818,000
Capital investment	14,419,000	15,865,000

No of Employees: 6,502

RHEINHYP AG
RHEINHYP Rheinische Hypothekenbank AG

Taunustor 3, Postfach 160655, D-6000 Frankfurt/Main 1
Tel: (069) 23 82-1
Telex: 413203 rhyph d
Telefax: 2382202/infotec 6000

Supervisory Board: Dr Walter Seipp (Vorsitzender)
Management Board: Klaus Adlung, Volker Bärschneider,
Burkhard Kamps, Dr Karsten von Köller
Senior Executives: Wolfgang Ahr, Dr Gerd Koide

PRINCIPAL ACTIVITIES: Mortgage bank

RÖCHLING, GEBR,

R Wagner Str 9, Postfach 10 12 63, 6800 Mannheim 1
Tel: (0621) 4402-0
Cable: Röchling
Telex: 463165
Telefax: (0621) 4402 284

Advisory Board: Kurt Wigand Frhr. von Salmuth (Vorsitzender),
Dr. Kai Werner (stellv. Vorsitzender), Dr. Friedrich Wilhelm
Clauser, Arthur Graf Saurma-Jeltsch, Dr. Peter Bally, Renate
Schuster, Dipl.-Phys. Werner Thürmel, Georg Sigismund Frhr.
von Salmuth, Dr. Jochen Mackenrodt, Dirk Zumkeller
Managing Partners: Arthur Graf Saurma-Jeltsch, Dr. Volker
Rusbüldt, Fred W Freese
Senior Executives: Heinz Georg Möller (Rechnungswesen,
Finanzen und Personalwesen), Kurt Lauer (Planung,
Investition), Dr. Bernd Michael Hönle (Revision,
Gesellschaftsrecht und Steuerrecht), Rainer Gebhardt
(Datenverarbeitung, Organisation), Dr. Gerd Coring
(Vertragsrecht, Immobiliarrecht)

PRINCIPAL ACTIVITIES: Wholesale coal and fuel oils; building materials; plant protection products; fittings; raw materials; production of plastics for various applications; malt; sheet metal; banking interests

Subsidiary/Associated Companies: (100% owned unless stated): Röchling Eisenhandel KG; Karl Hornung KG; Röchling KG; Röchling Seehafenverkehrsgesellschaft; Röchling Haren KG; Röchling Sustaplast KG; Röchling Kaltwalzwerk KG; Röchling Metallwaren KG; Röchling'sche Treuhandgesellschaft mbH; WKP Württembergische Kunststoff-Plattenwerke GmbH & Co KG (25%); Röchling Mannheim GmbH; Depalor SA (75%) (France); Röchling Austria Röchling GmbH & Co KG (Austria); Röchling AG (Switzerland); Seeber Srl (over 50%) (Italy)

Financial Information:

	31.12.86 DM'000	31.12.87 DM'000
Sales turnover	883,000	702,000
Shareholders funds	122,565	127,438

No of Employees: 2,240

SAL OPPENHEIM JR & CIE
See OPPENHEIM, SAL, JR & CIE

SGZ BANK
Suedwestdeutsche Genossenschafts-Zentralbank AG
PO Box 101906, 46 Bockenheimer Anlage, 6000 Frankfurt/Main 1
Tel: (069) 71390
Cable: Zentralkasse
Telex: 415251-0 sz d; SWIFT: GENO DE FS
Telefax: 7139309

Supervisory Board: Siegfried Gitzinger (Chairman), Johannes Heintz (Vice Chairman), Willi Himmelmann (Vice Chairman)
Board of Managing Directors: Dr Arnold Kremer (Chairman), Dr Ulrich Brixner (Vice Chairman), Gerhard Barth, Robert Markloff, Dr Karl-Heinz Vollmer, Dr Manfred Waechtershaeuser (Deputy)
Senior Executives: Klaus Stoerzbach (Senior Vice President International Division), Horst Pieper (Manager Correspondent Banking)

PRINCIPAL ACTIVITIES: Central banking functions for cooperative banks under the name of Volksbank or Raiffeisenbank in the Southwest of Germany
Principal Bankers: All the principal banks in London and New York
Financial Information:

	31.12.86 DM'000	31.12.87 DM'000
Profit before tax	159,000	162,969
Profit after tax	77,000	61,600
Retained profit	40,000	31,600
Dividends	7.5%	7.5%
Share capital	240,000	240,000
Shareholders funds	1,064,000	1,120,000
Deposits	17,947,000	20,547,588
Total assets	20,183,000	23,482,682

Principal Shareholders: 568 Volksbanks and Raiffeisenbanks
No of Employees: 753

SIGNAL KRANKENVERSICHERUNG AG
Postfach 237, 4600 Dortmund 1
Tel: (0231) 10860
Telex: 8 22 231

Management Board: Heinrich Frommknecht (Vorsitzender)

PRINCIPAL ACTIVITIES: Insurance including health insurance
Financial Information:

	DM'000
Premium income	904,000

No of Employees: 4,100

SPARKASSE BOCHUM
Postfach 102469, 4630 Bochum 1
Tel: (0234) 6111410
Telex: 825777 bocs d

PRINCIPAL ACTIVITIES: Banking

SPARKASSE DER STADT BERLIN WEST
Bundesallee 171, 1000 Berlin 31
Tel: (030) 86901
Cable: Westsparkasse Berlin
Telex: 183844 bls d
Telefax: 8692270

Executive Board: Hubertus Moser (Chairman), Horst Gust, Josef Mohren, Jochem Zeelen

PRINCIPAL ACTIVITIES: Banking activities
Financial Information:

	31.12.86 DM'000	31.12.87 DM'000
Profit before tax	157,161	132,063
Profit after tax	63,000	63,000
Retained profit	63,000	63,000
Shareholders funds	702,000	755,000
Deposits	15,992,656	16,378,805
Total assets	17,539,169	18,488,810

No of Employees: 3,980

SPARKASSE IN BREMEN (DIE)
Am Brill 1, D-2800 Bremen
Tel: (0421) 1790
Cable: Sparkasse Bremen
Telex: 17 4212010 brms d
Telefax: (421) 179 2196

Management Board: Dr. jur. Heinrich Frick (Managing Director), Ulrich Nölle (Managing Director), Friedrich Rebers (Managing Director), Horst Schöfisch (Managing Director)
Senior Executives: Gerhard Puschmann (Director International Division)

PRINCIPAL ACTIVITIES: Banking
Financial Information:

	31.12.87 DM'000
Total assets	9,236,000

SPARKASSEN-VERSICHERUNG AG
Löwentorstrasse 65, D-7000 Stuttgart 50
Tel: (0711) 81061
Telex: 722904

Management Board: Dieter Ehmann

PRINCIPAL ACTIVITIES: Insurance
Parent Company: Wurttemburgische Kumm Landesbank Girozentrale
Financial Information:

	DM'000
Premium income	360,000

No of Employees: 250

SR BETEILIGUNGEN AG
Leopoldstrasse 4, D-8000 München 40
Tel: (089) 33606166
Telex: 5216953 srb d

PRINCIPAL ACTIVITIES: Insurance
Financial Information:

	DM'000
Premium income	4,000,000

No of Employees: 7,360

STADTSPARKASSE HANNOVER

Raschplatz 4, Postfach 145, D-3000 Hannover 1
Tel: (0511) 346-1
Telex: 9 23209

Senior Executives: Kurt Fischer, Dr Paul Helmut Schmitz, Gunther Rohlf, Hans Zipper

PRINCIPAL ACTIVITIES: Banking
Financial Information:

	31.12.87
	DM'000
Total assets	7,432,000

STADTSPARKASSE DÜSSELDORF

Berliner Allee 33, Postfach 7707, D-4000 Düsseldorf
Tel: (0211) 8781
Cable: 8582509
Telex: 85 82509

Management Board: Fritz Kulins (Managing Director), Wolfgang Dornseifer (Managing Director), Anton Mainz (Managing Director)
Senior Executives: Mrs Margot Kratz (Manager Foreign Department)

PRINCIPAL ACTIVITIES: Banking
Financial Information:

	31.12.87
	DM'000
Total assets	76,300,000

STADTSPARKASSE KÖLN

2-12 Habsburgerring, 5000 Koln 1
Tel: (0221) 2261
Cable: Stadtsparkasse
Telex: 8882020 klnf d
Telefax: (0221) 240 1324

Supervisory Board: Norbert Burger (Chairman), Richard Blömer (First Deputy Chairman), Walter Rempe (Second Deputy Chairman), 14 other members, 14 deputy members
Management Board: Fritz Hermanns (Chairman), Wilhelm Lohner, Dr Gerhard Herbst, Jörg Dieter, Klaus Kons; Deputy Board Members: Heribert Bastians, Winfried Drewes
Senior Executives: General Managers: Hans-Peter Klein, Klaus Wolf, Hans Hollmann; Assistant General Managers: Willi Fassbender, Karl Heimig, Rolf Jansen, Oskar Staab, Hans Tillmanns, Peter Müllejans (Responsible for Foreign Business), Dr Eberhard Wolff, Rudi Adolph, Bernd Kortmann, Heinz Dörfler

PRINCIPAL ACTIVITIES: A full range of banking services
Trade Names: Commercial & Savings Bank Founded 1826
Financial Information:

	31.12.86	31.12.87
	DM'000	DM'000
Profit after tax	36,204	33,653
Retained profit	22,204	21,653
Shareholders funds	504,773	535,357
Deposits	11,097,775	12,887,226
Total assets	14,198,217	15,091,662

Principal Shareholders: The City of Cologne
No of Employees: 3,336

STADTSPARKASSE MÜNCHEN

PO Box 247, Sparkassenstrasse 2, D-8000 München 1
Tel: (089) 21670
Telex: 5216200

Supervisory Board: Josef Turiaux
Management Board: Georg Himmel, Anton Mauerer
Senior Executives: Werner J Hiering (Head of International Division)

PRINCIPAL ACTIVITIES: Banking services

Financial Information:

	31.12.87
	DM'000
Total assets	10,847,000

SÜDDEUTSCHE BODENCREDITBANK AG

PO Box 249, Ottostrasse 21, D-8000 München 2
Tel: (089) 5112 0
Telex: 5 23 554
Telefax: (089) 51 12-3 65

Management Board: Dr F May, Dr G Fischer, J Weber

PRINCIPAL ACTIVITIES: Mortgage bank
Principal Bankers: Landeszentralbank in Bayern, München

No of Employees: 210

SÜDWESTDEUTSCHE LANDESBANK GIROZENTRALE

PO Box 10 60 49, Lautenschlagerstr. 2, D-7000 Stuttgart-10
Tel: (0711) 127-0 (Stuttgart); (0621) 428-0 (Mannheim)
Cable: Landesbank (Stuttgart and Mannheim)
Telex: 72519-0 gzs d (Stuttgart); 463 143 gzma d (Mannheim)
Telefax: (0711) 127-3278 (Stuttgart); (0621) 415754 (Mannheim)

Board of Management: Werner Schmidt (Chairman of the Board), Dr Karl Heidenreich (Vice Chairman of the Board), Alfred Baumann, Hans Eberhard Berndt, Dr Claus Helbig, Rolf Limbach, Dieter Münch. Deputy Members of the Board: Karl Blattner, Heinz Fritz, Dr Richard Moritz, Dietrich Weishäupl, Hermann Zondler
Heads of International Division: Stuttgart: Dietrich Weishäupl. Mannheim: Michael Schmidt. International Business and Departments: Stuttgart: Correspondent Banking: Werner Seeliger (Vice President). Commercial Banking: Reinhold Welz (Vice President), Emil Frank (Vice President). International Finance: Martin Halder (Senior Vice President). Treasury Division: Heiko Laib (Senior Vice President). Foreign Exchange: Walter Maier (Vice President). Bond Dealing: Roland Banzhaf (Vice President). Mannheim: Commercial Banking: Peter Leitz (Vice President). International Finance: Martin Baugard (Vice President), Helmut Knobloch (Vice President)

PRINCIPAL ACTIVITIES: All banking services: The bank was established in 1988 by a merger between Landesbank Stuttgart and Badische Kommunale Landesbank, Mannheim
Subsidiary Companies: Bakola Leasing-u Beteiligungsgesellschaft mbH; Süd-Leasing GmbH; Süd Mobilien-Leasing GmbH; Württembergische Kapitalbeteiligungs-GmbH; Süd-Factoring GmbH; Südwestdeutsche Landesbank International SA (Luxembourg); Südwestdeutsche Landesbank (Schweiz) AG (Switzerland)
Correspondent Bankers: Worldwide
Financial Information:

	31.12.87
	DM'000
Business Volume	70,200,000
Bond issued	35,600,000
Capital and reserves	1,350,000
Total assets	68,700,000

Principal Shareholders: The 92 Savings Banks in Baden-Württemberg
No of Employees: 2,084

TELENORMA

Telefonbau und Normalzeit Lehner & Co

Mainzer Landstr 128-146, Postfach 10 21 60, D-6000 Frankfurt am Main 1
Tel: (069) 2661
Cable: Telenorma Frankfurtmain
Telex: 411141 tnfm d
Telefax: (069) 266 2233

Executive Officers: Dr. Michael Schwarzer (Vorsitzender), Dr H-L Habicht, Jürgen Krieg, H O Kullmann, Dr. Dieter Lennertz, P Sperl, Otto-Felix Weigel

PRINCIPAL ACTIVITIES: Active in the development, production, sale, installation, maintenance and insurance of systems and equipment in the following areas: Private Communications Systems: Switching systems for voice, text, data and images; communications terminals; private networks. Public Communications Systems: Switching Systems: mobile telephone systems; line networks, network termination equipment. Information Systems: Office and data systems; integrated application solutions; telefax and videotex systems; data acquisition systems; time management systems; time service systems. Security Systems: Alarm systems; TN Security Service; display systems

Trade Names: Telenorma

Parent Company: Robert Bosch GmbH, Stuttgart

Group Companies: Telenorma Telefonbau und Normalzeit GmbH; Friedrich Merk-Telefonbau GmbH; TN Saarbruecken GmbH; Elektra Versicherungs-AG; GeTeTe Versicherungs-Vermittlungs GmbH; Telenorma SA (Belgium); Telenorma Sarl (France); Telenorma Hellas AE (Greece); Telenorma SpA (Italy); Telenorma BV (Netherlands); Gesellschaft für automatische Telefonie AG (Austria); Telenorma SA (Spain); Telenorma CA (Venezuela)

Principal Bankers: Dresdner Bank; Landeszentralbank

Financial Information: Figures for Group Worldwide

	31.12.86	31.12.87
	DM'000	DM'000
Group turnover	2,082,014	2,147,940
Profit before tax	114,041	53,758
Profit after tax	78,732	34,112
Retained profit	56,868	25,216
Share capital*	73,755	73,755
Shareholders funds	336,717	341,810

*Parent Company TN Lehner & Co only

Principal Shareholders: Robert Bosch GmbH (100%)

No of Employees: Group 17,475

THYSSEN INDUSTRIE AG

Am Thyssenhaus 1, Postfach 10 37 45, 4300 Essen 1

Tel: (0201) 1061

Cable: Thyssenind Essen

Telex: 857 757-0 ti d

Telefax: (0201) 23 84 75

Supervisory Board: Dr D Spethmann (Chairman), G Hamacher (Vice-Chairman)

Management Board: Dr W Bartels (Chairman), K Bax, U Berntzen, Dr H-E Forster, W Haastert, Dr C Hendricks, Dr E Höffken, F Hoppe

PRINCIPAL ACTIVITIES: Machine tools; industrial engineering and construction, environmental protection equipment, power plant equipment; structural steel work; locomotives, railway passenger cars, gears and couplings; shipbuilding and repairs; plastics; metal working; forgings; foundry products

Parent Company: Thyssen AG

Subsidiary Companies: Thyssen-MAN Aufzüge GmbH; Thyssen Guss AG; Lintorfer Eisengiesserei GmbH; Thyssen Engineering GmbH; "Ofu" Ofenbau-Union GmbH; Thyssen Wassertechnik GmbH; Thyssen Nordseewerke GmbH; Thyssen Maschinenbau GmbH; Hüller Hille GmbH; Nothelfer GmbH; Maschinenfabrik Diedesheim GmbH; Thyssen Polymer GmbH; AST Automation und Steuerungstechnik GmbH; Blocher-Motor GmbH & Co KG; Herbert Maschinen und Anlagen GmbH; Waggon Union GmbH; TGW Thyssen Getriebe-und Kupplungswerke GmbH; several companies outside West Germany

Financial Information:

	30.9.86	30.9.87
	DM'000	DM'000
Sales turnover worldwide	5,424,000	5,213,000
Sales turnover net(1)	4,787,399	4,478,019
Profit before tax	100,975	109,371
Profit after tax(2)	44,524	50,815
Share capital	470,000	470,000
Shareholders funds	715,298	717,100

(1) domestic consolidation
(2) according to profit and loss transfer agreement with Thyssen AG

Principal Shareholders: Thyssen AG

No of Employees: 34,969

TOEPFER, ALFRED C, VERWALTUNGS GESELLSCHAFT GMBH

Ballindamm 2-3, 2000 Hamburg 1

Tel: (040) 334020

Telefax: (040) 33 83 20

Supervisory Board: A Toepfer (Chairman)

Management Board: Dr R Oertel (General Manager), H Toepfer (General Manager)

Senior Executives: Dr H Fiebiger (Manager)

PRINCIPAL ACTIVITIES: Financial holding company with whose interests include wholesale milling products and animal feedstuffs

Subsidiary/Associated Companies: Club-Kraftfutterwerke GmbH; Muehle Rueningen AG; Alfred C Toepfer Schiffahrts GmbH; Union Kraftfutterwerke GmbH

Principal Bankers: Deutsche Bank

Financial Information:

	1986	1987
	DM'000	DM'000
Sales turnover	720,000	600,000
Share capital	80,000	80,000

Principal Shareholders: Stiftung F.V.S., Hamburg

No of Employees: 650

TRINKAUS & BURKHARDT KGAA

Königsallee 21/23, Postfach 11 08, 4000 Düsseldorf 1

Tel: (0211)8311

Telex: 8581490 tub d

Telefax: (0211) 8312691a

Managing Partners: Herbert H Jacobi (Sprecher), Ernst W Brutsche, Dr Christoph Niemann, Dr Sieghardt Rometsch, Wolfgang von Waldthausen, Gerhard Winkel

Supervisory Board: Dr Bernhard Freiherr von Falkenhausen (Vorsitzender des Aufsichtsrates), Hervé de Carmoy (stellvertretender Vorsitzender des Aufsichtsrates und Vorsitzender des Aktionärsausschusses), Rodolfo Bogni, Léon Bressler, *Heinz Dolle, Wolfgang Haupt, *Reinhard Matten, *Brita Wenke, Dr Hanns-Hermann Winkhaus, *staff representatives

Shareholders Committee: Hervé de Carmoy (Vorsitzender), Léon Bressler (stellvertretender Vorsitzender), Rodolfo Bogni, Dr Bernhard Freiherr von Falkenhausen

PRINCIPAL ACTIVITIES: Bank services

Subsidiary Companies: Trinkaus & Burkhardt Gesellschaft für Bankbeteiligungen mbH; Trinkaus & Burkhardt (International) SA (Luxembourg); Trinkaus & Burkhardt (Schweiz) AG (Switzerland); Trinkaus Interinvestments (Switzerland); Trinkaus & Burkhardt Finance NV (Netherlands); Internationale Kapitalanlagegesellschaft mbH; TGS Vermögensverwaltungsgesellschaft mbH; J H Vogeler GmbH & Co KG; Trinkaus & Burkhardt Immobilien GmbH; Gesellschaft für industrielle Beteiligungen und Finanzierungen mbH; Grundstücksgesellschaft Trinkausstrasse KG; Burkhardt & Co Grundstücks-Kommanditgesellschaft

Financial Information:

	31.12.86 DM'000	31.12.87 DM'000
Profit before tax	54,615	49,992
Profit after tax	24,928	25,200
Dividends	16,200	16,200
Dividends per share	9DM	9DM
Share capital	90,000	90,000
Shareholders funds	252,500	258,500
Deposits	3,840,029	4,133,733
Total assets	4,686,526	4,889,133

Principal Shareholders: Midland Bank
No of Employees: 658

VEREINS- UND WESTBANK AG

Alter Wall 22, D 2000 Hamburg 11
Tel: (40) 36 92-01
Cable: vereinwest
Telex: 215 164-0 vh d
Telefax: (40) 36 92-2870

Supervisory Board: Dr. Maximilian Hackl (München, Chairman), Günter Nawrath (Aumühle, Deputy Chairman), *Christoph Schmidt (Schleswig, Deputy Chairman), Horst Becker (Rosengarten), Konsul Dr. Folkert Bellstedt (Ingelheim/Rhein), *Jürgen Betz (Hamburg), Hans Dethleffsen (Flensburg), *Bernd Fritze (Hamburg), *Hans-Werner Hakenholt (Hamburg), *Karl Kaula (Hamburg), *Gisela von Mahren (Hamburg), *Herbert Singer (Wohltorf-Aumühle), *Employees' representative

Board of Management: Udo Bandow, Rolf Kirchfeld, Eberhard-Rainer Luckey, Dr. Ulrich Meincke, Dr. Hans-Curt v. Pannwitz, Dr. Ulrich Ivo v. Trotha, Heinz Arno Wascheck

Senior Executives: Hans-Jürgen Schüder (Senior Executive Vice President), Dr. Karl Hansch (Executive Vice President), Hartmut Heinicke (Executive Vice President), Karl-Heinz Holst (Executive Vice President), Klaus-Ludolf Huschke (Executive Vice President), Christoph R Kremp (Executive Vice President), Dr Kay Ranniger (Executive Vice President), Klaus Reszczynski (Executive Vice President), Manfred Schulz (Executive Vice President), Hartwig Stapelfeldt (Executive Vice President), Dieter Thormählen (Executive Vice President), Peter Ullmann (Executive Vice President), Egon Vorwälder (Executive Vice President), Wolfram Bertz (Senior Vice President), Richard Brehm (Senior Vice President), Klaus Büttner (Senior Vice President), Joachim Czarnetzki (Senior Vice President), Egon Dieck (Senior Vice President), Hans-Joachim Fischer (Senior Vice President), Kurt Foitschik (Senior Vice President), Dr Christof Fritzen (Senior Vice President), Uwe Heck (Senior Vice President), Bernd Kiso (Senior Vice President), Jürgen Köpcke (Senior Vice President), Jürgen Liebscher (Senior Vice President), Dieter Meyer (Senior Vice President), Claus Mutschke (Senior Vice President), Günter Nickel (Senior Vice President), Peter Petersen (Senior Vice President), Uwe Rummert (Senior Vice President), Peter Schlichting (Senior Vice President), Günter Schnell (Senior Vice President), Dr Stephan Schüller (Senior Vice President), Burkhard Skibbe (Senior Vice President), Wolfgang Steger (Senior Vice President), Dr Ulrich Stipke (Senior Vice President), Uwe Susemihl (Senior Vice President), Klaus Wehrt (Senior Vice President), Axel Kaupisch (General Legal Counsel), Josef Geiger (General Legal Counsel)

PRINCIPAL ACTIVITIES: Banking and finance
Subsidiary Banks and Companies: included in the Consolidated Financial Accounts: Bankhaus Neelmeyer Aktiengesellschaft (78.94%); Beteiligungs- und Handelsgesellschaft in Hamburg mbH and its subsidiaries MERKURHOF Grundstücksgesellschaft mbH. and UNION Verwaltungs- und Treuhandgesellschaft mbH; Boden GmbH in Hamburg; Conrad Hinrich Donner with its Subsidiary Bankhaus Schliep & Co; Geestemünder Bank Aktiengesellschaft (82%) with its subsidiary Grundstücksgesellschaft Geestbank-Haus;

Hanseatische Investitions-Bank GmbH (66.67%) with its subsidiary Hanseatische Leasing GmbH; HDV Hamburger Datenverarbeitungs-Gesellschaft mbH; HKV Hanseatiche Kreditbank GmbH; NORDINVEST Norddeutsche Investment-Gesellschaft mbH; SKV KREDIT-BANK GmbH (66.67%); VEREINS- UND WESTBANK INTERNATIONALE SA (Luxembourg); VereinWest Overseas Finance (Jersey) Ltd (Jersey, CI); Verwaltungsgesellschaft Katharinenhof mbH; Verwaltungsgesellschaft Katharinenhof mbH & Co KG Westerland

Principal Bankers in Hamburg: Landeszentralbank Hamburg
Financial Information: Consolidated figures

	31.12.86 DM'000	31.12.87 DM'000
Profit before tax	137,954	137,754
Profit after tax	66,019	62,474
Retained profit	56,323	66,356
Dividends	37,680	37,680
Dividends per share	12DM	12DM
Share capital	157,000	157,000
Shareholders funds	627,501	635,574
Deposits*	15,017,292	15,152,343
Total assets	17,897,638	18,451,055

*incl deposits from banks

Principal Shareholders: Bayerische Vereinsbank AG, Munich (over 25%)
No of Employees: 4,121

VEREINSBANK IN NÜRNBERG AG

PO Box 4249 Marienstrasse 3, 8500 Nürnberg
Tel: (0911) 202710
Telex: 622523

Supervisory Board: Dr Helmut Scholz (Vorsitzender)
Management Board: Dr E Hermann, Dr J Francke, Dr J Parchmann
Senior Executives: Dr G Seebach, S Blos, P Dahlke, R Heuberger, R Stark, W Wieninger

PRINCIPAL ACTIVITIES: Banking services
Parent Company: Bayerische Vereinsbank
Principal Bankers: Landeszentralbank, Nürnberg

Principal Shareholders: Bayerische Vereinsbank, München-Nürnberg; M M Warburg; Brinckmann, Wirtz & Co, Hamburg
No of Employees: 160

VEREINTE KRANKENVERSICHERUNG AG

Fritz-Schäfferstrasse 9, D-8000 München 83
Tel: (089) 6785-0
Telex: (089) 521 57 21
Telefax: (089) 67 85-6523

Supervisory Board: Dr Rolf Gamper (Vorsitzender), Reinhold Schick (stv Vorsitzender), Dr Christoph Blanc, Dr Peter Frey, Dr Udo Knoke, Hans Peter Lenssen, Dr Edith Lukas, Horst-Günther Rohde, Dr Werner G Seifert, Erich Steierer, Norbert Trautwein, Franz Weber
Management Board: Dr Hans Karl Jäkel (Vorsitzender), Dr Christoph Dorschel (stv Vorsitzender), Werner Fertl, Dr Hans Feyock, Wolfgang Hall, Peter Pfeil, Götz Reichart, Günter Schemmann, Dr Helmut-Joachim Werner

PRINCIPAL ACTIVITIES: Insurance
Parent Company: Schweizer Rück Beteiligungen AG
Financial Information:

	31.12.86 DM'000	31.12.87 DM'000
Premium income	1,995,266	2,069,810
Profit after tax	16,880	14,300
Retained profit	8,880	7,150
Share capital	45,000	45,000
Shareholders funds	160,592	168,592

No of Employees: 2,663

VEREINTE VERSICHERUNGEN
Postfach 20 25 22, D-8000 München 2
Tel: (089) 6785-0
Telex: 5-215 721
Telefax: (089) 6785 6523

Supervisory Board: Dr H K Jäkel (Vors.), Dr Chr. Dorschel (stv. Vors.), Horst Dietz, Dr H Feyock, W Hall, E Puschmann, W Fertl, G Reichart, P W Pfeil, Dr H J Werner, G Schemmann

PRINCIPAL ACTIVITIES: All types of insurance
Parent Company: Schweizer Rück Beteiligungen AG
Subsidiary Companies: Companies of the Vereinte Versicherungen: Vereinte Versicherung AG; Vereinte Krankenversicherung AG; Vereinte Lebensversicherung AG
Financial Information:

	DM'000
Total premium income	3,185,000

No of Employees: 8,000

VICTORIA LEBENS-VERSICHERUNGS-AG
Victoriaplatz 1, Postfach 11 16, D-4000 Düsseldorf
Tel: Berlin (030)884 16-0; D'dorf (0211)8280
Telex: Berlin 183477 vvbd; D'dorf 8582984 vvdd
Telefax: Berlin (030) 88416207; D'dorf (0211) 828 2222

Supervisory Board: Dr E Overbeck (Vorsitzender)
Management Board: Dr Edgar Jannott (Vorsitzender), Dr Werner Fricke, Dr Martin Horstmann, Dr Günther Köhler, Gert Schlösser, Manfred Stoll

PRINCIPAL ACTIVITIES: Life insurance and pensions
Subsidiary Companies: Victoria Feuer-Versicherungs-Aktien-Gesellschaft; Victoria Rückversicherungs-Aktien-Gesellschaft; Vorsorge Grundstücks Aktiengesellschaft formerly Versorge Lebensversicherungs-Aktiengesellschaft; D.A.S. Deutscher Automobil Schutz Allgemeine Rechtsschutz-Versicherungs-AG; D.A.S. Deutscher Automobil Schutz Versicherungs-Aktiengesellschaft
Principal Bankers: Bank für Handel und Industrie; Berliner Bank; Berliner Commerzbank; Deutsche Bank Berlin; Dresdner Bank; Deutsche Bank; Commerzbank; Trinkaus & Burkhardt; Westdeutsche Landesbank, LZB in Nordrhein-Westfalen
Financial Information:

	DM'000
Premium income	1,820,000

No of Employees: Group 11,920

VOLKSFÜRSORGE DEUTSCHE LEBENSVERSICHERUNG AG
An der Alster 57-63, 2000 Hamburg 1
Tel: (040) 28650
Fax: (040) 28 65-3369
Telex: 2112440
Teletex: 40 20 02-0

Supervisory Board: H Matthöfer (Vors), G Volkmar (stellv. Vors.), G Fehrenbach, Prof Dr H Aufenanger, H Bednarek, H Berger, H Bleicher, S Döding, U Girmond, K Göttsch, W Hamann, K-H Hoffmann, A Horné, S Mahrt, N Rath, W Schneider, G Stückl, H Teitzel, M Wiesmeier, L Zimmermann
Management Board: W Schwickart (stellv Vors), E Berger, S Bussjäger, G H Maack, E Milau, L Pieschl, H Stenzaly, F Völker

PRINCIPAL ACTIVITIES: All classes of insurance
Subsidiary Companies: (100% owned unless stated): Volksfürsorge Deutsche Sachversicherung AG (99.9%); Volksfürsorge Rechtsschutzversicherung AG (99.9%); Volksfürsorge Krankenversicherung AG (99.9%); Hamburger Internationale Rückversicherung AG; BBG Beteiligungs- und Betreuungsgesellschaft mbH; KATHARINA

Grundstücksverwaltungs- und Betreuungsgesellschaft mbH; IRIS Grundstücksverwaltungs- und Betreuungsgesellschaft mbH; Allgemeine Hypothekenbank AG (25%); VBV Vermittlungs-Büro für Versicherungen GmbH (40%); MIB Mietservice und Immobilien-Betreuungsgesellschaft mbH; Hamburg-Bremer Vermögensverwaltungsgesellschaft mbH; HIR Grundstücksverwaltungs AG (94.8275%)
Financial Information:

	1986 DM'000	1987 DM'000
Premium income	2,704,000	3,033,000
Investment income	1,783,000	1,700,000
Profit after tax	52,900	27,000
Retained profit	52,900	27,000
Dividends per share	18%	18%
Share capital	150,000	150,000
Shareholders funds	254,000	254,000

Principal Shareholders: VHU-Vermögensholding u Unternehmensbeteiligungsgesellschaft mbH; Vermo Vermögensverwaltungsgesellschaft mbH
No of Employees: 2,051

WERHAHN, WILH, KG
Königstr 84, D-4040 Neuss 1
Tel: (02101) 23021
Telex: 8517045

Supervisory Board: Paul Werhahn, Heribert Werhahn, Dr Joachim Gerhard, Wilhelm Werhahn
Management Board: Dr Walter Drummen (Vorsitzender), Hermann Friedrich Offermann (stellv Vorsitzender)

PRINCIPAL ACTIVITIES: Holding company with a wide range of interests including cereal processing, manufacture of timber products and banking
Principal Bankers: Bankabteilung der Firma Werhahn Neuss
Financial Information:

	DM'000
Sales turnover	3,440,000

No of Employees: 1,300

WESTDEUTSCHE GENOSSENSCHAFTS-ZENTRALBANK EG
Ludwig-Erhard-Allee 20, Postfach 9027, D-4000 Düsseldorf 1
Tel: (211) 778-00
Cable: WGZ Bank
Telex: 8581900 WGZ D
Telefax: (211) 778-12 77

Supervisory Board: H Kloten (Chairman), P Schnitker (Deputy Chairman), B Bottermann, E Daute, H Eckardt, M Jorris, N Köchling, W Schlenkenbrock, A Thiele
Management Board: E Heinke, D Pahlen, W Thiemann, G Unckell, H-G Monssen (Deputy)
Senior Executives: F D Jaechel (Director International Division, Duesseldorf), W Wolff (Director Securities/Bonds), Dr Jacobs (Director Loan Department, Duesseldorf), Hr Korte (Director Loan Department, Muenster)

PRINCIPAL ACTIVITIES: All banking activities; central institution of 616 "Volksbanks and Raiffeisenbanks" in Rhineland and Westphalia
Subsidiary Companies: DG Bank Deutsche Genossenschaftsbank; DG HYP Deutsche Genossenschafts-Hypothekenbank; Münchener Hypothekenbank eG; Bausparkasse Schwäbisch-Hall AG; R + V Versicherungsgruppe; Union Investment Gmbh; DEVIF Deutsche Gesellschaft für Investment-Fonds GmbH; DIFA Deutsche Immobilien Fonds AG; VR-Leasing Gesellschaft der Volksbanken und Raiffeisenbanken mbH; DG Immobilien Leasing GmbH

Financial Information:

	31.12.86 DM'000	31.12.87 DM'000
Profit before tax	195,500	154,700
Profit after tax	61,600	58,100
Retained profit	45,000	40,000
Dividends	8.5%	8.5%
Capital and reserves	1,023,900	1,098,000
Deposits	23,047,300	24,063,000
Total assets	26,044,600	27,573,400

No of Employees: 1,238

WESTDEUTSCHE LANDESBANK GIROZENTRALE

WestLB

Herzogstrasse 15, 4000 Düsseldorf
Tel: (D'dorf) (0211) 826-01; (Münster) (251) 412-01
Cable: Landesbank Düsseldorf; Landesbank Münster Westfalen
Telex: (D'dorf) 8582605 gzd d; (Münster) 892823 gzms d
Teletex: (0211) 17211304 Firmengeschäft

Supervisory Board: Dr Helmut Kessler
Management Board: Friedel Neuber (Chairman), Dr Ludwig Trippen (Vice Chairman), Wolfgang Burda, Dr Axel Kollar, Dr Johannes Ringel, Dr Wolfgang Prautzsch, Hans-Peter Saettele, Jürgen Sengera, Dr Eberhard Weiershauser

PRINCIPAL ACTIVITIES: Universal banking
Trade Names: WestLB
Subsidiary Companies: WestLB International SA, Luxembourg; Banque Franco-Allemande SA, Paris; BKA Bank für Kredit und Aussenhandel AG, Zürich; West LB Securities Pacific Ltd, Tokyo; WestLB UK Ltd, London
Financial Information:

	31.12.86 DM'000	31.12.87 DM'000
Profit before tax	317,463	282,166
Profit after tax	123,917	99,381
Retained profit	84,000	84,000
Declared reserve*	2,245,746	2,281,944
Dividends	91,062	84,250
Share capital	1,815,316	1,815,315
Shareholders funds	4,044,000	4,061,000
Deposits	137,498,000	141,088,000
Total assets	148,218,591	152,503,849

*including group profit

Principal Shareholders: The State of North Rhine-Westphalia; The Regional Associations of the Rhineland and of Westphalia-Lippe; The Sparkassen Associations of the Rhineland and of Westphalia-Lippe
No of Employees: 7,531

WESTFALENBANK AG

Huestrasse 21-25, Postfach 10 2710, 4630 Bochum 1
Tel: Bochum (0234) 616-0, Düsseldorf (0211) 8227-0
Cable: Westfalenbank
Telex: 825825 wbal (Bochum); 8582833 wbdf (Düsseldorf)
Telefax: (0234) 616400 (Bochum); (0211) 8227 258 (Düsseldorf)

Supervisory Board: Dr Klaus Marquardt (Chairman), Dr Wilhelm Arendts (Deputy Chairman), Heinz Becker, Dr Ernst Denzel, Dr Klaus Geiger, Dr h.c. Klaus Kuhn, Dr Alfred Lukac, Dr Josef Rust, Rainer Hammerschmidt, Wolfgang Redenz, Otto Sollböhmer, Rolf Vallender
Management Board: Jürgen Kleppa, Claus Schmidt, Dieter Pöhlmann, Dr Birger Lassen
Senior Executives: H G Falk (Manager International Division), Ch Seiler (Deputy Manager International Credit Division), E Steinke (Deputy Manager Foreign Exchange Division), D Nickel (Deputy Manager Commercial Transactions)

PRINCIPAL ACTIVITIES: All banking services
Parent Company: Bayerische Hypotheken- und Wechsel-Bank AG, München

Subsidiary Companies: Westfälische Hypothekenbank Aktiengesellschaft, Dortmund; Westfalenbank International S.A., Luxembourg; Gesellschaft für Grundbesitz mbH, Bochum
Principal Bankers: LZB
Financial Information:

	31.12.86 DM'000	31.12.87 DM'000
Deposits	4,618,787	4,745,393
Total assets	5,191,084	5,284,167

No of Employees: 705

WESTFÄLISCHE HYPOTHEKENBANK AG

Postfach 717, 4600 Dortmund 1
Tel: (0231) 10820
Telex: 08 22 827
Telefax: (0231) 1082257

Supervisory Board: Dr Hans Fey (Vors), Dieter Pöhlmann (1 stellv Vors), Dr Klaus Heiss (2 stellv Vors), Wilhelm Ewert, Jürgen Kleppa, Hans Dieter Ritterbex, Dr Bernd Thiemann, Erich Wilke. Arbeitnehmervertreter: Karl-Günther Ehrich, Paul-Josef Gödde, Dieter Kuratis, Birgit Maevus
Management Board: Dr Rudolf Nolting-Hauff (Sprecher), Dipl.-Kfm Walter Dietrich Poetsch, Dipl.-Kfm Franz Huber
Senior Executives: D Schachel, Dr H Heerma, D Stratmann

PRINCIPAL ACTIVITIES: Wide range of banking services including mortgages and loans to industry, commerce and farming; issue of bonds
Trade Names: WESTHYP
Parent Company: Bayerische Hypotheken- und Wechselbank, AG, München
Financial Information:

	31.12.86 DM'000	31.12.87 DM'000
Profit before tax	61,576	60,765
Profit after tax	32,700	31,420
Retained profit	20,750	19,420
Dividend	14,750*	13,420
Dividend per share	25%*	22%
Share capital	59,000	63,000
Shareholders funds	308,000	344,000
Total assets	16,322,409	17,639,226

*plus 5% of Centenary Bonus

Principal Shareholders: Westfalenbank AG, Bochum (50%); Bayerische Hypotheken und Wechsel Bank AG, München (25%); Deutscher Herold Allgemeine Versicherungs AG, Bonn (12.5%); Deutscher Herold Lebensversicherungs AG, Bonn (12.5%)
No of Employees: 270

WESTFÄLISCHE PROVINZIAL FEUERSOZIETÄT

Postfach 59 60, Bröderichweg 58, 4400 Münster
Tel: (02 51) 2 19-0
Telex: 08 92 822 prova d
Telefax: (02 51) 2 19-23 00

Supervisory Board: Verwaltungstrat: Herbert Neseker (Münster) (Vors)
Management Board: R Brauner (Vors), H-J Heinrichs, D Heumann, H Hillebrands, J Kropp, Dr W Lange

PRINCIPAL ACTIVITIES: Insurance
Trade Names: Westfälische Provinzial-Feuersozietät Versicherung der Sparkassen
Subsidiary Companies: Westfälische Provinzial-Lebensversicherungsanstalt
Principal Bankers: Westdeutsche Landesbank Münster

Financial Information:

	1986 DM'000	1987 DM'000
Premium income	1,007,000	1,060,000

No of Employees: 1,794 (including subsidiary)

WILH WERHAHN KG

See WERHAHN, WILH, KG

WIRTSCHAFTLICHE LANDHANDELSVEREINIGUNG GMBH

Vereinigte Landwarenkaufleutein Norddeutschland

Langer Garten 17-19, 3200 Hildesheim
Tel: (05121) 744-0
Cable: WLV Hi
Telex: 0927202
Telefax: (05121) 744169

Senior Executives: A Wanza, K P Duve, G Kleymann, C Jaser, H Neitzel

PRINCIPAL ACTIVITIES: Wholesale agricultural produce, cereals, fodder, fertilisers, plant protecting products, seeds etc
Principal Bankers: Nordd Landesbank; Commerzbank; Deutsche Bank
Financial Information:

	1987 DM'000
Sales turnover	690,000
Dividends	9.38%
Share capital	8,500

No of Employees: 340

WOHNUNGSBAUFÖRDERUNGSANSTALT DES LANDES NRW

Karl Arnold Platz 1, D-4000 Düsseldorf 30
Tel: (0211) 4304-1

Supervisory Board: Dr C Zöpel
Management Board: Dr K H Engelbergs (Vors), F Kinnigkeit, E Ullrich

PRINCIPAL ACTIVITIES: Banking

No of Employees: 180

WUPPERTALER STADTWERKE AG

Bromerger Str 39-41, Postfach 201616, 5600 Wuppertal 2
Tel: (0202) 569-1
Cable: Wuppertal-Barmen
Telex: 8591788 wsw d

Management Board: Dr Kurt Sunkel (Chairman), Friedrich Senger (Deputy)

PRINCIPAL ACTIVITIES: Production and supply of electricity, heating and gas; the acquisition and supply of water; construction and operation of transport facilities, in particular trams and suspension railways as well as the transport of people by motor vehicle
Principal Bankers: LZB; Deutsche Bank; Commerzbank; Dresdner Bank; Credit- u Volksbank; Bank für Gemeinwirtschaft; Stadtsparkasse; Westd Landesbank; Sparkasse Velbert
Financial Information:

	DM'000
Sales turnover	810,000

No of Employees: 3,500

WÜRTTEMBERGISCHE FEUERVERSICHERUNG AG

Johannesstrasse 1-7, Postfach 60, D-7000 Stuttgart 1
Tel: (0711)662-1
Cable: wuerttfeuer stuttgart
Telex: 723553 wvers d
Telefax: (0711) 662520

Supervisory Board: Otto Julius Maier (Chairman), Manfred Stahl (Deputy Chairman), Victor Frey, Gerhard Hauber, Heinz Idelberger, Roland Klett, Dr phil Herbert Nierhaus, Dr jur Paul-Francis Niquille, Dr oec Herbert Schoenenberger, Werner Schulze, Dr jur Walter Sigle, Dr jur Paul A Stein
Management Board: Dr jur Georg Buechner (Chairman), Lutz Doblaski, Dr phil Ludwig Fusshoeller, Georg Mehl, Dr jur Dieter Paulus, Rolf-Dieter Schulte (Deputy Member)
Senior Executives: Dr O Bechert (Manager), K Nachtrieb (Manager), F R Vaessen (Manager), G Hoefer (Deputy Manager), H Schwenk (Deputy Manager), Dr W Schweikert (Deputy Manager), H Krämer (Deputy Manager), K-J Walter (Deputy Manager), Dr K-H Weinmann (Deputy Manager)

PRINCIPAL ACTIVITIES: Fire and marine insurance; property insurance (household, homeowners etc); general third party liability and personal accident insurance; motor insurance; reinsurance
Subsidiary Companies: Württembergische Rechtsschutzversicherung AG; Württembergische und Badische Versicherungs-AG; Nord-Deutsche Versicherungs AG; Folgate Insurance Co Ltd; Levensverzekering Maatschappij Erasmus NV (Netherlands)
Principal Bankers: Postscheck Stuttgart; Landeszentralbank
Financial Information:

	31.12.86 DM'000	31.12.87 DM'000
Premium income	1,237,510	1,286,286
Investment income	172,929	163,497
Profit before tax	74,809	58,621
Profit after tax	27,678	23,078
Retained profit	14,078	14,078
Dividends	14,078	14,078
Dividends per share	10DM	10DM
Earnings per share*	21.08DM	19.11DM
Share capital	70,400	70,400
Shareholders funds	346,700	355,700
*excluding tax credit		

Principal Shareholders: Bayerische Rückversicherungs AG (25%); Allgemeine Rentensanstalt, Lebens-und Rentenversicherungs AG (20%); Schweizer Rück Holding AG (7%); Robert Bosch GmbH (10%); Baden-Württembergische Bank AG (10%)
No of Employees: 3,317

WWK-LEBENSVERSICHERUNG AG

Marsstrasse 42, D-8000 München 2
Tel: (089) 5114-1
Telex: 0524 921

Management Board: Franz Josef Schwarz (Vorsitzender), Walter Binder, Felix Fessl, Manfred Weckert

PRINCIPAL ACTIVITIES: Life insurance
Financial Information:

	DM'000
Premium income	680,000

No of Employees: 1,850

Major Companies of GREECE

AGRICULTURAL BANK OF GREECE
23 Panepistimiou Street, 105 64 Athens
Tel: 3230521 27
Telex: 215810

Governor: Vassilis A Kafiris (also Chairman)
Directors: J Daskalopoulos (Head of International Division)

PRINCIPAL ACTIVITIES: Banking
Financial Information:

	31.12.87
	Dr'000
Total assets	2,151,568

BANK OF CRETE SA
22 Voukourestiou St, Athens 106 71
Tel: 01-3606511/15
Cable: Kretabank Athens
Telex: 21-8633 BOFK-GR
Telefax: 01-3644832

Management: Panayotis Vakalis, Achilles Kyriakidis

PRINCIPAL ACTIVITIES: Bank
Trade Names: Bank of Crete SA
Principal Bankers: Irving Trust, New York; Barclays, London; Deutsche Bank, Hamburg

No of Employees: 850

BANK OF GREECE
Banque de Grèce
21 Venizelos Ave, PO Box 105, Athens 133
Tel: 320-1111; 3230-010 (Foreign Exchange)
Cable: Bankgrèce
Telex: 215102; 218348 (Foreign Exchange)

Governor: Dimitrius Alikias; Deputy Governor: Sbiros Papankolaou

PRINCIPAL ACTIVITIES: Bank

COMMERCIAL BANK OF GREECE SA
11 Sophokleous, Athens 102 35
Tel: 3210-911
Cable: Commerzbank Athens
Telex: 216545-9

Chairman: D Pepelassis; General Manager: P Poulis
Directors: Mr Lambrou (Managing Director)

PRINCIPAL ACTIVITIES: Bank
Associated Banks: Ionian & Popular Bank of Greece; Bank of Attica; Bank of Pireaus; Investment Bank
Financial Information:

	31.12.86	31.12.87
	Dr'000	Dr'000
Profit before tax	5,766,934	3,147,721
Profit after tax	5,572,786	2,452,808
Retained profit	3,434,197	163,426
Dividends	2,139,789	2,292,632
Dividends per share	280Dr	300Dr
Earnings per share	729Dr	321Dr
Share capital	4,203,158	4,203,158
Deposits	545,617,391	656,083,013
Total assets*	881,626,332	985,050,796

*including contra accounts

No of Employees: 6,885

CREDIT BANK A.E.
40 Stadiou Street, 10252 Athens
Tel: 326 0000
Cable: Creditbank
Telex: 218691 CCB GR; 218959 CCB GR (Foreign Exchange)
Telefax: 3224522

Chairman and General Manager: Y S Costopoulos
Deputy General Manager: Constantine A Kyriacopoulos
Senior Executives: Pantelis D Petroutsopoulos (Manager, Foreign Relations Department)

PRINCIPAL ACTIVITIES: Bank
Subsidiary Companies: Ependyseis Pisteos A.E.; Alfa Computers A.E.; Emporiki Insurance Company; Generali Life A.E.; Delta Informatics A.E.
Financial Information:

	31.12.86	31.12.87
	Dr'000	Dr'000
Profit before tax	1,841,021	3,203,028
Profit after tax	929,372,977	499,718,085
Retained profit	390,000	460,000
Dividends	792,000,000	1,650,000
Dividends per share	450Dr	750Dr
Share capital	2,640,000	3,300,000
Deposits	232,335,755	306,271,857
Total assets	439,485,778	554,217,206

No of Employees: 3,001

IONIAN BANK
Ionian & Popular Bank of Greece SA
45 El Venizelou Avenue Athens 132
Tel: 3225-501
Cable: Ionpopbank
Telex: 215269

Board of Directors: Prof Panayotis G Korliras (Chairman), Emmanuel Niadas (Vice Chairman), Eugene Calafatis (General Manager of the Ionian Bank), Panayotis Poulis (General Manager of the Commercial Bank), Evangelos Mitilinaios (Chairman of the Piraeus Chamber of Small Industry), Georgis Koropoulis, Ioannis Alafouzos, Ioannis Markakis (Representative of Ionian Bank Employees), Ioannis Chrisikos (Representative of Ionian Bank Employees)

Executive Management: Prof Panayotis G Korliras (Chairman), Eugene Calafatis (General Manager)

Senior Executives: Stephanos Rivans (Audit Department) Panayotis Zacharopoulos (Branch Administration Department), Andreas Vourvopoulos (Corporate Banking Department), Nikolaos Kontothanassis (Delayed and Troubled Loans Department), Eleftherios Theodorou (Financial and Economic Analysis Department), Nikolaos Kavouras (Financial Services Department), Nikolaos Kantiotis (General Accounting Department), Georgios Koundouriotis (Management Information Systems Department), Antonios Fillipides (International Department), Constantinos Gontikas (Legal Services Department), Theodoros Kanellakis (Operations Department), Ioannis Papandreou (Personnel Department), Dimitrios Andriolas (Technical Services Department), Georgios Michopoulos (Treasury Department), Dionysios Maroussis (Marketing and Public Relations Department), Panayotis Lainas (Main Branch Manager), Vassilios Tsekouras (Legal Advisor), Prof Alexandros Christofelis (Technical Advisor). Regional Management: Georgios Baloyannis (Regional Office of Piraeus and Aegaean Islands), Andreas Xanthopoulos (Regional Office of Northern Greece), Nikolaos Zacharopoulos (Regional Office of Peloponnese and Western Greece), Menelaos Panayotopoulos (Regional Office of Central Greece), Vassilios Anegnostopoulos (Regional Office of Crete)

PRINCIPAL ACTIVITIES: Commercial bank
Trade Names: Ionian Bank
Parent Company: Commercial Bank of Greece SA
Subsidiary Companies: Ionian Hotel Enterprizes; Phosphoric Fertilizers Industry; Investment Bank; Commercial Bank of Greece (Germany); Neorion Syros Shipyards SA; Eleusis Shipyards

No of Employees: 2,988

GREECE

NATIONAL BANK OF GREECE SA

86 Eolou St, Athens 102 32
Tel: 3210-411; 3210-501; 3210-601
Cable: Ethnotrapeza Athens
Telex: 0214931/8; 0215385/6; 0218771/2; 0219417; 0215209
Telefax: 3251-072

Governor: S Panagopoulos
Deputy Governors: C Georgoutsakos, G Kasmas, C
  Georgacopoulos
Directors: H E the Archbishop of Athens and All Greece, J
  Pliakos, S Calogeropoulos-Stratis, E Venizelos, P
  Lambropoulos, S Metaxas, A Panagopoulou, E
  Papadopoulos, N Papantoniou, N Piskopos, A Potamianos, G
  Raftopoulos, G Tsouyopoulos, H Zarifopoulos, G Anomeritis

PRINCIPAL ACTIVITIES: Commercial bank
Associated Banks: Atlantic Bank of New York (USA); National
  Investment Bank for Industrial Development; Traders Credit
  Bank, Athens; South African Bank of Athens; Banque
  Nationale de Grèce (France); National Bank of Greece
  (Canada); National Housing Bank Athens
Correspondents: All principal Banks
Financial Information: Non Consolidated figures-NBG SA only

	31.12.86	31.12.87
	Dr'000,000	Dr'000,000
Profit before tax	3,681,000	4,053,000
Profit after tax	3,521,000	3,923,000
Retained profit	369,000	210,000
Dividends	3,142,000	3,709,000
Dividends per share	720Dr	850Dr
Share capital	19,636,000	19,636,000
Reserves	22,168,000	22,387,000
Deposits	2,159,576,000	2,592,464,000
Total assets*	2,557,735,000	2,934,371,000

*excluding contra accounts

Principal Shareholders: Public Sector (56.9%)
No of Employees: 16,059

Major Companies of
ITALY

ABEILLE COMPAGNIA ITALIANA DI ASSICURAZIONI (L')

Via Leopardi n. 15, 20123 Milano
Tel: (02) 8589.1
Cable: ASSICABEILLE
Telex: 316029 ABEASS
Telefax: (02) 8589.331

Administrative Board: Antonio Sozzani (Presidente), Dr Pierre Mercier (Amministratore Delegato), Ing Jean Albert Arvis, Avv Michele Castelnuovo Tedesco, Dr Philippe Chareyre, Ing Jack Frances, Ing Robert Gachet, Dr Jean Paul Galbrun, Ing Giampiero Pesenti, Dr Jean Marc Vernes, Dr Michel Marchal (Presidente Onorario), Rag Giancarlo Canevari (Segretario del Consiglio)
Management Board: Dr Pierre Mercier (Direttore Generale), Rag Giancarlo Canevari (Direttore Centrale)

PRINCIPAL ACTIVITIES: Insurance and re-insurance
Parent Company: Compagnie Financière Du Groupe Victoire Paris
Affiliated Companies: Società Finanziaria di Partecipazioni; Indosuez Italia Holding SpA; Société Franco-Belge de Participations Mobilières; Abeille Vie; Abeille Assurances; Risim Immobiliare Srl; CARD Spa
Principal Bankers: Banca Commerciale Italiana; Cassa di Risparmio delle Provincie Lombarde; Banque Indosuez-Italia
Financial Information:

	31.12.86	31.12.87
	L'000,000	L'000,000
Premium income	123,371	137,484
Investment income	278,868	312,087
Profit before tax	20,414	20,884
Profit after tax	15,727	15,451
Retained profit	12,169	11,181
Dividends	3,558	4,270
Dividends per share	1,300L	1,560L
Earnings per share	5,745.8L	5,644.8L
Share capital	8,211	8,211
Shareholders funds	63,085	74,978

Principal Shareholders: Compagnie Financière du Groupe Victoire, Paris; La Fondaria, Florence
No of Employees: 359

ALLEANZA ASSICURAZIONI SPA

Viale Luigi Sturzo 37, I-20154 Milano
Tel: (02) 62961
Cable: Asvita
Telex: 331-303 asvita
Telefax: (02) 653 718

President: Prof Dr Libero Lenti; Vice-President: Cav. del Lav. Avv. Enrico Randone
Board of Directors: Dr Adolfo Baldissera, Avv. Luigi Chiaraviglio, Dr Ing. Arch. Giuseppe Chiodi, Dr Aldo Cicoletti, Dr Giulio Colella, Dr Alfonso Desiata, Avv. Emilio Dusi, Dr Paolo Iona, Dr Fabio Padoa, Dr Angelo Saraceno, Dr Maurizio Bonsi
Senior Executives: Dr Alberto Giorgetti (General Manager), Dr Dante Lamperti (General Manager), Dr Valerio Florio (Central Manager), Dr Giuseppe Francese (Central Manager), Dr Ing. Antonio Lotto (Central Manager)

PRINCIPAL ACTIVITIES: Industrial life insurance, annuities and accident insurance together with related financial activities
Trade Names: Alleanza Assicurazioni SpA
Parent Company: Assicurazioni Generali
Subsidiary/Associated Companies: (100% owned unless stated): Edicom SpA; Agricola S. Giorgio SpA; Covadonga SA (90%) (Spain); Euralliance SA (15%) (France); Banca Mercantile Italiana SpA (7.7%)
Principal Bankers: All major Italian Banks

Financial Information:

	31.12.86	31.12.87
	L'000,000	L'000,000
Premium income	491,733	609,700
Profit after tax	49,371	52,231
Share capital	120,000	120,000

Principal Shareholders: Generali
No of Employees: 2,622

ASSICURAZIONI GENERALI SPA

See GENERALI, ASSICURAZIONI, SPA

ASSITALIA
Assicurazioni d'Italia SpA

Corso Italia 33, 00198 Roma
Tel: (06) 84831
Cable: Assitalia Roma
Telex: 611051 Asital I

Chairman: G Pieraccini
Deputy Chairman: P Cassietti. Managing Directors: M Fornari, A Fornarini. General Manager: I Morganti. Deputy General Managers: D Di Iullo, V Mungari, G Pettinelli

PRINCIPAL ACTIVITIES: Insurance and reinsurance
Financial Information:

	L'000,000
Premium income	1,190,000

Principal Shareholders: Istituto Nazionale Delle Assicurazioni; Istituto Mobiliare Italiano
No of Employees: 1,950

BANCA CATTOLICA DEL VENETO S.P.A.

Via S Corona 25, PO Box 781, I-36100 Vicenza
Tel: (0444) 519111
Cable: Venetdige
Telex: 480082 bcv ho; 431262 bcv id
SWIFT: CAVEIT 2V

Board of Directors: Chairman: *Feliciano Benvenuti; Vice Chairmen: *Paolo Biasi, *Candido Fois; Directors: *Gianfranco Barbato, Bruno Bolla, Rina Brion, Piergiorgio Coin, Alessandro Faedo, *Angelo Ferro, Giancarlo Ligabue, Virgilio Marzot, Roberto Riccoboni, Sebastiano Rumor, Tiziano Treu, *Gino Trombi. Secretary to the Board: Paolo Stocchiero, *Members of the Executive Committee
General Management: Domenico Spedale (Chief Executive and Managing Director), Mario Calamati (Deputy Managing Director), Paolo Stocchiero (Senior Executive). International Division: Agostino Poncato (Executive Area Manager), Philip Pigozzo (Area Manager), Barbara Wörsdörfer (Area Manager), Fabrizia Zambonin (Area Manager). Foreign Exchange and Euro-Currency Deposits: Gianni Gavagnin (Manager Chief Dealer), Piercarlo Pellicchiero (Principal-Euro-Currency Deposits), Pasquale Cilento (Procurist-Foreign Exchange), Paolo Baccarin (Procurist-Foreign Exchange). Representative Offices: Domenico Ciampani (Rome Representative), Peter Giannotti (New York Representative), Marco Ferrario (Hong Kong Regional Representative). Treasury and Securities Division: Sandro De Bernardis (Senior Vice President). Foreign Securities: Carlo Galeandro (Executive Manager)

PRINCIPAL ACTIVITIES: All banking services
Parent Company: Nuovo Banco Ambrosiano
Associated Companies: Mediocredito delle Venezie (Venice); Fiscambi Leasing SpA (Milan); La Centrale Servizi SpA (Milan); AV Commerciale Srl (Venice) (all over 10% shareholding)
Principal Correspondents: Banque Nationale de Paris (France); Deutsche Bank (Germany); The Sanwa Bank (Japan); Lloyds Bank Plc (UK); Citibank (USA); Irving Trust Co (USA); Union Bank of Switzerland (Switzerland)

Financial Information: Not consolidated

	31.12.86	31.12.87
	L'000,000	L'000'000
Profit before tax	145,898	122,034
Profit after tax	80,030	72,134
Dividends	21,429	31,900
Capital stock	75,000	75,000
Bank deposits(1)	942,526	1,680,408
Customer deposits(2)	5,545,962	5,896,412
Total assets	8,880,861	10,023,670

(1) and current accounts
(2) and savings accounts

No of Employees: 3,458

BANCA COMMERCIALE ITALIANA

Piazza della Scala 6, 20121 Milano
Tel: 88501 (45 lines direct dialling in)
Cable: DIRCOMIT
Telex: 310080 BCI HO I
Telefax: 88503026

Board of Directors: Enrico Braggiotti (Chairman)
Management Board: Mario Arcari (Managing Director), Sergio
  Siglienti (Managing Director), Lino Benassi (General
  Manager), Enrico Beneduce (General Manager), Ferdinando
  Bonicalza (General Manager), Luigi Fausti (General Manager),
  Giorgio Ferretti (General Manager), Pietro Grandjacquet
  (General Manager), Alberto Abelli (Deputy General Manager),
  Aldo Civaschi (Deputy General Manager), Riccardo Ferrari
  (Deputy General Manager), Antonio Masala (Deputy General
  Manager), Gian Paolo Minardi (Deputy General Manager),
  Franco Riolo (Deputy General Manager), Italo Rossi (Deputy
  General Manager), Carlo Tamborini (Deputy General Manager)

PRINCIPAL ACTIVITIES: Commercial banking, with branches
  and representative offices worldwide. Branches abroad in
  Abu Dhabi, Chicago, Cairo, Frankfurt/Main, Hong Kong,
  London, Los Angeles, Madrid, New York, Singapore, Tokyo.
  Representative offices abroad: Amsterdam, Ankara, Athens,
  Beirut, Belgrade, Berlin, Bombay, Brussels, Buenos Aires,
  Cairo, Caracas, Lisbon, Mexico City, Moscow, Munich,
  Osaka, Paris, Peking, Shanghai, Sydney, Tehran, Tokyo,
  Warsaw, Washington
Main Subsidiary Companies: Banca di Legnano SpA; Banco di
  Chiavari e della Riviera Ligure SpA; Comit Factoring SpA;
  Comit Leasing SpA; COMITSIEL SpA FIN COMIT SpA;
  Società Italiana di Revisione e Fiduciaria-S.I.RE.F. SpA
Financial Information:

	31.12.87	31.12.88
	L'000'000	L'000'000
Profit before tax	375,354	633,350
Profit after tax	314,354	501,350
Retained profit	119,054	306,050
Dividends	195,300	195,300
Dividends per ordinary share	180L(1)	180L
Dividends per savings share	210L(2)	210L
Earnings per share	319L	477L
Ordinary share capital	840,000	840,000
Savings share capital	210,000	210,000
Shareholders funds(3)	2,122,891	2,428,148
Deposits	56,726,990	63,104,585
Total assets(4)	66,268,389	74,130,376

(1) 18% on a par value of
  L5,000
(2) 21% on a par value of
  L1,000
(3) including retained profit
  but excluding share capital
(4) excluding contingent and
  contra accounts

Principal Shareholders: Istituto per la Ricostruzione Industriale-
  IRI (a state holding) (59.33%)
No of Employees: 19,030

BANCA D'ITALIA

Via Nazionale, I-00184 Roma
Tel: 47921
Telex: 622697
Telefax: (06) 474 7820 (Group 3)

Governor: dott. Carlo Azeglio Ciampi
Directors: dott Lamberto Dini (Director General), dott Antonio
  Fazio, dott Tommaso Padoa-Schioppa
Senior Executives: Fabrizio Saccomanni (International Division)

PRINCIPAL ACTIVITIES: Banking
Financial Information:

	31.12.86	31.12.87
	L'000,000	L'000'000
Deposits	85,518,857	166,183,767
Total assets	203,959,320	223,990,410

BANCA NAZIONALE DEL LAVORO

Via Vittorio Veneto 119, 00187 Roma
Tel: (06)47021
Cable: Dirbancoper; SWIFT: BNLI IT RR
Telex: 616116; 680342; 610622

Chairman: Dr Nerio Nesi; *Vice Chairman:* prof Giuseppe Ricci
Directors: G Pedde (Managing Director and Chief General
  Manager)
Senior Executives: dott Fabio Laratta (Senior Deputy General
  Manager), E Tartaglia (Senior Deputy Chief General Manager),
  C Morselli, G Pedde, V Postiglione (Deputy Chief General
  Managers)

PRINCIPAL ACTIVITIES: Banking, including finance for public
  works, mortgage credit and credit for cooperatives
Subsidiary Companies: Banc Nazionale Del Lavoro International
  (Luxembourg); Lavoro Bank AG (Switzerland); Banque Lavoro
  SA (Switzerland); LOCAFIT (Deutschland) Anlagen Leasing
  GmbH (Germany); LOCAFIT FRANCE SA (France); Banca
  Nazionale Del Lavoro of Canada (Toronto, Canada); Banca
  Nazionale Del Lavoro of Canada (Montreal, Canada); Banca
  Nazionale del Lavoro of Canada (Vancouver, Canada);
  T.I.E.C. The Italian Economic Corporation (USA); BNL US
  Corporation (USA); Banca Nazionale Del Lavoro SA
  (Argentina); Lavoro Bank Overseas NV (Curaçao). Branches
  in Atlanta, Barcelona, Chicago, Hong Kong, London, Los
  Angeles, Madrid, Miami, New York, Paris and Singapore.
  Representative Offices: in Bahrain, Beijing, Brussels, Cairo,
  Caracas, Düsseldorf, Frankfurt/Main, Houston, Kuala Lumpur,
  Mexico City, Munich, Sao Paulo, Sydney, Tehran and Tokyo
  and participations Worldwide

Principal Shareholders: Italian Treasury
No of Employees: 25,379

BANCA NAZIONALE DELL'AGRICOLTURA

Via Salaria 231, 00199 Roma
Tel: 85881
Cable: DIRAGRICOLTURA
Telex: 612121; 612098
Telefax: 85883663

Chairman: G A Armenise
Administration: F Pepe (Managing Director), Dott Guiseppe
  Pellegrini (Head of International Division), Romeo Mulacarne
  (Deputy Chief International Division), Bruno Avallone
  (Manager International). International Banking: Alessandro
  Franchi (Manager). Regional Departments: Central Europe,
  UK, Eire: Giancarlo Bonini, Franco Patacchiola, Aldo Maggi,
  Romolo Speranza. Middle East, Eastern Europe, India, Africa,
  Scandinavia, Switzerland: Antonio Di Thiene, Uberto Secco,
  Pietro Foti, Roberto D'Ascia. Americas, Far East, Indian Sub-
  Continent, Australasia: Sergio Nardini, Piero De Angelis,
  Francesco Macioce, Tiziana Sgreccia. Italy: Benito Neri.
  Capital Markets: Sergio Felici (Manager), Umberto Colucci,
  Gianfranco Lami, Ugo Bernabucci, Renato Palmieri, Luciano
  Petracchini. Foreign Exchange and Money Market: Giacomo
  De Angelis. Dealers: Donato Di Nunzio, Paolo M Macorig,

Adriano Ventura, Massimo Carducci, Francesco Chiricozzi, Carlo Baroni

PRINCIPAL ACTIVITIES: Bank

Subsidiary Companies: Gestifondi SpA (70%); Nagrafin Bank Ltd (Cayman Islands); London & Continental Bankers Ltd (UK); Banca Gatto & Porpora SpA; Interaudit SpA; Società per Azioni Montecavo; Banca Nazionale dell'Agricoltura Inter-Holding SA (Luxembourg); Bank Europaeischer Genossenschaftsbanken AG (Switzerland); Associated Companies: Interbanca-Banca per Finanziamenti a Medio e Lungo Termine SpA; Agrigest-Società di Gestione del Fondo di Previdenza Agricoltori SpA; Bolefin Factor SpA; SG2 Italia SpA; Siam Leasin SpA

Principal Shareholders: Bonifiche Siele Finanziaria SpA (39.6%); Federazione Italiana dei Consorzi Agrari (12.7%)

No of Employees: 8,045 (worldwide)

BANCA POPOLARE DI BERGAMO
Piazza Vittorio Veneto 8, 24100 Bergamo
Tel: (035) 392111
Telex: 300410
Telefax: (035) 392480

Chairman: Emilio Zanetti
Directors: G A Banfi, G Riva, A Gibellini

PRINCIPAL ACTIVITIES: Banking
Parent Company: Gruppo Nord Est
Subsidiary Companies: Credito Varesino Spa
Financial Information:

	31.12.87	31.12.88
	L'000,000	L'000,000
Profit before tax	135,622	152,657
Profit after tax	81,434	81,452
Retained profit	19,515	19,531
Dividend	61,919	61,921
Dividend per share	1,100L	1,100L
Earnings per share	1,447L	1,445L
Share capital	28,145	28,146
Shareholders funds	725,455	742,841
Deposits	6,680,161	7,257,343
Total assets	10,197,101	10,953,193

No of Employees: 2,748

BANCA POPOLARE DI MILANO SCARL
Piazza F Meda 4, 20121 Milano
Tel: 7700.1 (sel pass) 77.00
Cable: Popolbanca Milano
Telex: 310202 popban I
Telefax: (02) 770 01 (Group 3)

President: Piero Schlesinger
Directors: Aldo Cova (General Manager), Carlo Cattaneo (Central Manager, Head of International Division), Giovanni Giannini (Head of Correspondent Banking Division), Guiseppe Rivoltella (Head of Foreign Exchange Division), Lanfranco Ferrari (Head of Foreign Department)

PRINCIPAL ACTIVITIES: Banking activities
Financial Information:

	31.12.86	31.12.87
	L'000,000	L'000'000
Deposits	14,120,504	17,845,081
Total assets	51,596,948	54,246,740

No of Employees: 4,221

BANCA POPOLARE DI NOVARA
Via Carlo Negroni 12, I-28100 Novara
Tel: 4451
Cable: Banca Novara-Novara
Telex: 200 371 NOVDIG I
Telefax: 391020

Board of Directors: Roberto Di Tieri (Chairman), Cajo Enrico Balossini, Alberto Ricevuti, Lino Venini (Vice Chairmen), Piero Bongianino, Carlo Piantanida (Managing Directors), Giulio Cesare Allegra, Sergio Baratti, Ettore Bermani, Achille Boroli, Giovanni Brignone, Marco Broggi, Luigi Buzzi, Alessandro Cicogna Mozzoni, Claudio Cocito, Edo Deagostini, Edoardo Gregotti, Federico Guasti, Mario Pavesi, Pietro Stella, Sergio Pininfarina

General Management: Carlo Piantanida (Managing Director), Piero Bongianino (Managing Director)

PRINCIPAL ACTIVITIES: All banking activities
Subsidiary Companies: Istituto Nazionale di Credito Edilizio (I.N.C.E.) SpA (99.33%); Banca Interpopolare SA (Switzerland) and its subsidiary Teliber SA (Switzerland); Società di Gestione Fondi Comuni Banche Popolari (SOGEPO) SpA (50%); Compagnia Finanziaria Ligure Piemontese (COFILP) SpA (61.1%)
Financial Information:

	31.12.86	31.12.87
	L'000,000	L'000,000
Profit before tax	373,000	359,000
Net profit after tax	144,058	145,429
Dividends	94,250	94,256
Dividends per share	1,000L	1,000L
Earnings per share	1,528L	1,543L
Share capital	47,125	47,128
Reserves	933,756	953,135
Shareholders funds	611,441	645,903
Deposits*	17,004,825	18,053,463
Total assets	48,832,453	55,176,220

*and current accounts

No of Employees: 7,313

BANCA SAN PAOLO DI BRESCIA
Corso Martiri della Libertà 13, 25100 Brescia
Tel: (030) 29921
Telex: 300010
Telefax: (030) 299 23 38 (Group 1)

President: Dr. Ing. Adolfo Lombardi
Directors: Dr Alberto Valdembri (General Manager), Vittorio Vincenzi (Head International Division), Cesletino Roscio (Manager International Division), Fabrizio Valzelli (Chief Dealer Foreign Exchange Department), Giuseppe Bravi (Foreign Correspondent Banking)

PRINCIPAL ACTIVITIES: Banking
Financial Information:

	31.12.86	31.12.87
	L'000,000	L'000'000
Deposits	44,673,295	5,420,038
Total assets	50,022,873	15,259,942

BANCA TOSCANA SPA
Via Leone Pancaldo 4, 50127 Florence
Tel: (055) 43 911
Cable: BANTOSCANA
Telex: 570507 BT DIRE I
Telefax: (055) 4739030; SWIFT; TOSC IT 3F

President: Giuseppe Bartolomei
Directors: Marcello Fazzini (Managing Director), Lorenzo Cappelli (General Manager), Sergio di Peppo (Chief Manager, International)

PRINCIPAL ACTIVITIES: Banking
Parent Company: Monte dei Paschi di Siena Banking Group
Subsidiary Companies: Banca di Valle Camonica

Principal Shareholders: Monte Paschi di Siena (91.641%)

ITALY

BANCO DI NAPOLI

Via Toledo 177/178, 80132 Napoli
Tel: 081/7911111
Cable: Dirbanco
Telex: 720526 BN EXA I
Telefax: 081/5520118

Board of Directors: Prof Luigi Coccioli (Chairman), Vincenzo
Scarlato (Deputy Chairman), Sergio Abis, Andrea Amatucci,
Antonio Argento, Vittorio De Nigris, Angelo Mancusi,
Francesco Manfredi, Raffaele Perrone Capano, Paolo
Ranuzzi De Bianchi, Giovanni Somogyi. Secretary: Giovanni
Spina

Executive Committee: Luigi Coccioli (Chairman), Vincenzo
Scarlato (Deputy Chairman), Ferdinando Ventriglia (Chief
Executive), Petro Giovannini, Girolamo Marsocci, Raffaele
Minicucci, Carlo Pace

Management: Ferdinando Ventriglia (Chief Executive), Franco
Serpieri (Deputy Chief Executive), Giovanni Acampora (Deputy
Chief Executive), Arturo Giorgio (Deputy Chief Executive),
Giovanni Giorgio (Deputy Chief Executive), Gustavo Di Cesare
(Central Management), Vittorio Morelli (Central Management),
Raffaele Perrone (Central Management), Benito Plotino
(Central Management), Martino Sferruzza (Central
Management), Paolo Spadaro (Central Management),
Francesco Tedesco (Central Management)

PRINCIPAL ACTIVITIES: Banco Di Napoli conducts business as
a short-term credit institution, but also acts as a medium and
long-term lending institution through a number of specialized
Divisions. It offers, by a group of majority-owned subsidiaries,
a wide range of qualified financial services and a large
number of innovation-intensive activities

Subsidiary Companies: Finban SpA; Innovare SpA; Datitalia
Processing SpA; Leasimmobili SpA; Mediterranea SpA; SEM
"Il Mattino"; Banco di Napoli Internationa (Luxembourg);
Sofiban SpA; BN Factoring; BN Leasing; Finrete; Gestiban
SpA; Effeppi SpA

Financial Information:

	31.12.86	31.12.87
	L'000,000	L'000,000
Gross profit	523,000	432,000
Net profit	55,000	62,000
Share capital and reserves	1,129,000	1,170,000
Deposits*	51,785,000	59,153,000
Total assets	60,430,000	69,301,000
*and borrowed funds		

No of Employees: 13,107

BANCO DI ROMA

Viale u Tupini 180, 00144 Roma
Tel: 54451
Cable: Cenbanroma
Telex: 616184

Chairman: Romeo dalla Chiesa; Vice Chairman: Mario Arcelli;
Managing Directors: Ercole Ceccatelli, Marcello Tacci

Directors: Renato Cassaro, Pietro Ciucci, Gabriele Di Palma,
Giacomo Figliola Baldieri, Antonio Marzano, Pietro Marzotto,
Franco Nobili, Ambrogio Puri, Renato Riverso. Secretary:
Antonio Izzi

Central Management: Managing Directors: Ercole Ceccatelli,
Marcello Tacci. Central Managers: Vincenzo Astolfi, Mario de
Luca, Pietro Gori, Giuseppe Greco, Antonio Nottola, Oddone
Pinto, Roberto Paolo Rossi, Eliseo Timo', Giovanni Ungaro.
Deputy Central Managers: Sergio De Nicolais, Mario Gabriele,
Filippo Garibbo, Antonio Izzi, Marco Martini, Nicola Stock,
Leopoldo Tripepi

PRINCIPAL ACTIVITIES: Domestic and international banking
activities

Subsidiary Companies: (100% owned unless stated): Domestic
Network: 350 branches and agencies, 1 banking subsidiary
with 76 branches and agencies: -Banco di Perugia;
International Network: 15 branches and agencies, 16

representative offices, 5 banking subsidiaries with 30
branches and agencies:- Banco di Roma International SA
(Luxembourg); Banco di Roma (France); SERIT-Società
Esattoriale Riscossione Imposte e Tesorerie SpA; SEF-
Società Esattoriale Fasanese SpA; Banco di Roma per la
Svizzera SA (25.1%) (Switzerland); Banco di Roma (Belgio)
SA (30%) (Belgium); Consorzio 3 BIN (50%); Credito Fondiario
SpA (26.3%); Europartners (Australia) Pty Ltd (33.33%)
(Australia); Figeroma-Fiduciaria e di Gestione Roma SpA
(94.06%); Finroma SpA (99.32%); Romagest SpA (78.53%);
Romaleasing SpA (96.61%); Société de Gestion du Rominvest
International Fund SA (65.09%) (Luxembourg); SAIM-Società
Amministrazione Immobili SpA (20%); SPAGET-Gestione
Esattorie e Tesorerie SpA (50%); SPI-Servizi e Prodotti per
l'Informatica SpA (65%); Uniroma SpA (50%); Banco di Roma
(Chicago) (USA)

Principal Bankers: All main banks in the world
Financial Information:

	31.12.87
	L'000,000
Deposits	50,982,977
Total assets	151,254,320

Principal Shareholders: I.R.I. (Istituto per la Ricostruzione
Industriale)
No of Employees: 14,600

BANCO DI SANTO SPIRITO S.P.A.

Largo Angelo Fochetti 16, 00154 Roma
Tel: (06) 51721
Cable: Spirdir
Telex: 610162 BSS DIR I; 612178 BSS DIR I
Telefax: 5172/2850

Board of Directors: Rodolfo Rinaldi (Chairman), Gianfranco
Imperatori, (Vice Chairman) Elio Tartaglia (Managing Director),
Giacomo Antonelli, Fabrizio Antonini, Mario Are, Renato
Cassaro, Pietro Ciucci, Umberto Granati, Enrico Micheli,
Francesco Romano. Secretary of the Board: Edoardo
Carmagnola

Management: Vice-General Managers: Paolo Accorinti, Angelo
Tommasini. Central Managers: Marcello Asteriti, Luigi
Carignani, Mario Petroni, Cesare Torregiani, Giovan Battista
Zucchetti. Central-Joint Managers: Massimo Amari, Mario
Cannizzaro, Giacomo Maria Cavallucci, Walter Corradini,
Mario Giannini, Michele Imperiali, Mario Martello, Arnaldo
Melica, Riccardo Tristano

PRINCIPAL ACTIVITIES: Commercial banking
Subsidiary Companies: Assoleasing S.p.a.; Microleasing S.p.a.;
Banco di Santo Spirito (Luxembourg) S.A.
Financial Information:

	31.12.86	31.12.87
	L'000,000	L'000,000
Profit before tax	99,143	55,824
Profit after tax	47,063	46,624
Retained profit	945	1,238
Dividends	22,200	21,000
Dividends per share	50L	35L
Earnings per share	10%	7%
Share capital	300,000	300,000
Shareholders funds	982,246	1,009,104
Deposits	18,589,949	19,099,439
Total assets	48,035,269	53,199,595

Principal Shareholders: IRI (Istituto per la Ricostruzione
Industriale) (85.65%)
No of Employees: 5,680 (worldwide)

BANCO DI SARDEGNA

36 Viale Umberto, Sassari 07100
Tel: (079) 279111
Telex: 790049
Telefax: (079) 27 91 11 (Group 3)

President: Dr Angelo Solinas (also Chairman)

Directors: Dr Angelo Giagu de Martini (General Manager)
Senior Executives: Alberto Pinna (Head of International Division)

PRINCIPAL ACTIVITIES: Banking
Financial Information:

	31.12.87
	L'000,000
Total assets	7,186,100

BANCO DI SICILIA
Via Generale Magliocco 1, 90141 Palermo
Tel: (091) 587933-274111
Telex: 910012

President: G Parravicini; Vice President: dott Guido Savagnone
Directors: Dr Ottavio Salamone (Managing Director)
Senior Executives: dott Giacomo Perticone, prof Salvatore La Francesca

PRINCIPAL ACTIVITIES: Banking activities with financial sections for real estate, industrial and mining credit, public works, agriculture and fishing
Subsidiary Companies: Euramerica Finanziaria Internazionale SpA (82%); Euramerica Fiduciaria e di Revisione SpA (50%)

No of Employees: 9,200 (worldwide)

BREDA, FINANZIARIA ERNESTO, S.P.A.
Piazza della Repubblica n. 32, 20124 Milano
Tel: (02) 6556841/654861
Cable: BREDA MILANO
Telex: 310050 BREDA I
Telefax: (02) 6700 498

Board of Directors: Dott Giuseppe Cosentino (Presidente), Dott Sergio Maggi (Vice Presidente), Dott Luigi Roth (Amministratore Delegato), Avv. Paride Accetti (Amministratore), Ing Arnaldo Antichi (Amministratore), Ing Claudio De Albertis (Amministratore), Prof Ernesto Lapenna (Amministratore), Ing Umberto Marino (Amministratore), Dott Claudio Torneo (Amministratore), Prof Carlo Tribuno (Amministratore), Prof Victor Uckmar (Amministratore). Collegio Sindacale: Dott Cesare Frassineti (Presidente), Dott Adolfo Book (Sindaco Effettivo), Dott Piero Gnudi (Sindaco Effettivo), Rag Massimo Protasi (Sindaco Supplente), Dott Gian Galeazzo Tesei (Sindaco Supplente)
Management: Dott Ernesto Calaprice (Direttore Generale)

PRINCIPAL ACTIVITIES: Assumption of shareholding; technical and financial coordination of controlled companies whose interests include the iron and steel industry, railway rolling stock, diesel engines, turbines and industrial machinery, heating and ventilation, defence and armaments, tyres, paper, electronics and real estate
Subsidiary Companies: Breda Fucine Spa; Breda Fucine Meridionali Spa; Breda Meccanica Bresciana Spa; OTO Melara Spa; OTO Trasm Spa; OTO Breda Sud Spa; Termomeccanica Italiana Spa; Officine Galileo Spa; S.I.G.M.A.-Società Immobiliare Generale Milanese Azionaria Spa; Microcontrol Spa
Financial Information:

	L'000,000
Turnover	1,000,000

Principal Shareholders: EFIM
No of Employees: Group 7,547

CASSA DI RISPARMIO DE GENOVA E IMPERIA
Via Cassa di Risparmio 15, 16123 Genova
Tel: (10) 20911
Telex: 270089 Carigi i
Telefax: (10) 280013

President: Avv G B Gianni Dagnino
Directors: Dr Giovan Battista Villa (General Manager)
Senior Executives: Rag Luigi Aliberti (Head of International Division)

PRINCIPAL ACTIVITIES: Banking
Financial Information:

	31.12.87	31.12.88
	L'000,000	L'000'000
Total assets	7,438,626	8,220,301

CASSA DI RISPARMIO DELLE PROVINCIE LOMBARDE
CARIPLO
Via Monte di Pieta 8, 20121 Milano
Tel: (02) 88661
Cable: Cariplo Milano
Telex: 310280 CARIPLO
Telefax: (02) 88665225

Board of Directors: Roberto Mazzotta (Chairman), Camillo Ferrari (Vice Chairman), Carlo Polli (Vice Chairman); Angelo Albini; Giovanni Battistini; Angelo Caloia; Gabriele Sergio Cioccarelli; Dario Damiani; Gianfranco Maj; Riccardo Marchioro; Giorgio Milani; Luigi Penna; Sergio Radaelli; Francesco Ciro Rampulla; Roberto Sarfatti; Giorgio Tavecchio; Ivanoe Vaini; Giuseppe Vimercati; Albino Zucca
Senior Management Board: Sandro Molinari (Chief Executive), Pierluigi Novello (Deputy Chief Executive), Romano Pesci (Deputy Chief Executive), Angelo Roncareggi (Deputy Chief Executive), Giorgio Rossi (Deputy Chief Executive). General Managers: Luciano Brizzi; Agostino Colace; Gianfranco De Notte; Carlo Ferraris; Luigi Mombelli; Bruno Spadoni; Luigi Tarenzi; Gianfranco Testa; Francesco Trazzi
Department Managers: Luciano Brizzi (Accounting); Luigi Mombelli (Agricultural Credit); Carlo Bianchi (Cashier's); Renato Villa (Charity); Angelo Pellegrini (Collection and Payments); Rodolfo Turati (Data Processing); Pier Giulio Cottini (Development); Arturo Garbagnati (Foreign); Marcello Valfredi (Inspection); Adalberto Urzi' (Legal); Francesco Trazzi (Loans); Bruno Spadoni (Management Services); Marcello Affri (Mortgage Credit); Carlo Ferraris (Mortgage Loans); Gianfranco De Notte (Personnel); Gianguido Milanesi (Property Maintenance); Graziano Grignoli (Public Agencies); Bernardino Perugini (Public Works); Michele Oddone (Real Estate Management); Alberto Mauri (Research and Planning); Agostino Colace (Secretary's); Benito Covolan (Securities); Mario Zorzoli (Shareholdings); Bruno Parini (Steward's)

PRINCIPAL ACTIVITIES: Banking activities
Trade Names: CARIPLO
Subsidiary Companies: (100% owned unless stated): Istituto Bancario Italiano; Mediocredito Lombardo (65.49%); Leasindustria (73.24%); Mediofactoring (54.12%); Magazzini Generali Fiduciari Cariplo (99.98%); Fondigest (54.50%); Bossi 2 (80%); Vapra; Ente Lombardo Potenziamento Zootecnico; Immobiliare Porretta; Centro Lombardo Incremento Floro-orto-frutticoltura
Financial Information: Group figures

	31.12.86	31.12.87
	L'000,000	L'000'000
Profit before tax	1,050,183	1,036,315
Profit after tax	584,112	543,927
Retained profit	550,114	500,165
Dividends	33,998	43,762
Reserves	3,685,807	4,181,156
Deposits	53,230,523	59,885,117
Total assets	67,952,300	75,422,204

No of Employees: 14,293

CASSA DI RISPARMIO DI BOLOGNA
Via L C Farini 22, I-40124 Bologna
Tel: (051) 339111
Cable: CASSARISPARMIO
Telex: 510282 CRBO; Swift: CRBOIT2B

Chairman: Gianguido Sacchi Morsiani
Management: Leone Sibani (General Manager), Franco F Ferri (Vice General Manager), Aere Ferrari (Vice General Manager),

Giovanni Bassu (Central Manager), Alberto Gozzi (Central Manager)

PRINCIPAL ACTIVITIES: Banking
Financial Information:

	31.12.86	31.12.87
	L'000,000	L'000,000
Profit before tax	135,777	125,476
Profit after tax	68,371	70,953
Dividends	-	2,483
Dividends per share	-	8,300L
Share capital	309,700	340,700
Shareholders equity	132,335	242,513
Deposits	4,168,387	4,229,325
Total assets	6,344,721	6,916,668

No of Employees: 1,531

CASSA DI RISPARMIO DI FIRENZE

4/6 Via Bufalini, 50122 Florence
Tel: 27801
Telex: 572391 CRFIES (foreign department)

President: Lapo Mazzei (also Chairman)
Directors: Luigi Tinti (General Manager)
Senior Executives: Paolo Campaioli (Manager Foreign Department), Guiseppe Manuelli (Deputy Manager Foreign Department), Romano Martini (Chief Dealer Foreign Exchange), Guiseppe Pacini (Correspondent Banking), Marcello De Ciutis (Swift. Liaison)

PRINCIPAL ACTIVITIES: Banking
Financial Information:

	31.12.87
	L'000,000
Total assets	8,407,800

CASSA DI RISPARMIO DI PADOVA E ROVIGO

22 c.Garibaldi, 35122 Padova
Tel: (049) 844611
Telex: 430074 Crest 1
Telefax: (39-49) 663568

President: Prof. Ettore Bentsik
Directors: Antonio Finotti (General Manager)
Senior Executives: Rag Piero Zanettin (Head of International Division), Dott L Albrigo (Chief Dealer)

PRINCIPAL ACTIVITIES: Banking
Financial Information:

	31.12.87
	L'000,000
Total assets	7,385,850

CASSA DI RISPARMIO DI ROMA

Via del Corso 320, 00186 Rome
Tel: (06) 67071
Cable: Carisparmio Roma
Telex: 613541-630134 CRRDIR
Telefax: (06) 67073783; (06) 67073731

Board of Directors: Prof. Pellegrino Capaldo (Chairman), Emmanuele Emanuele (Vice Chairman), Paolo Emilio Ascoli, Camillo Barluzzi, Novello Cavazza, Nicola Ciarrocca, Giuseppe Di Nardi, Candeloro Mignano, Arturo Rossignoli, Claudio Bianchi, Gatti Serafino. Matteoni Osvaldo, Mario Picchi
Executive Committee: Pellegrino Capaldo (Chairman), Emmanuele Emanuele (Vice Chairman), Carlo D'Amelio, Mario Castiello D'Antonio, Mario Ercolani, Nazzareno Ferri, Cesare Geronzi (General Manager)
Senior Executives: Cesare Geronzi (General Manager), Lucio Veneziani (Deputy General Manager), Rodolfo Corcione (Deputy General Manager)

PRINCIPAL ACTIVITIES: Banking

Subsidiary Companies: (100% owned unless stated): CORNICE IMMOBILIARE SpA; Banque Generale du Commerce (99.5%); CR Roma Factoring SpA (90%); Sistemi Informativi SpA (60%); Nicro SpA (51%)
Financial Information:

	31.12.87	31.12.88
	L'000,000	L'000,000
Profit before tax	200,743	211,749
Profit after tax	84,743	97,249
Retained profit	84,743	97,249
Capital	415,919	488,909
Deposits	14,894,386	17,152,628
Total assets	23,768,233	26,962,570

No of Employees: 4,409

CASSA DI RISPARMIO DI TORINO

Via XX Settembre 31, 10121 Torino
Tel: 57.661
Telex: 220350/221225 CRTOI
Telefax: 5766577

Chairman: Prof. Enrico Filippi
Chief General Manager: Giorgio Giovando. Deputy Chief General Managers: Alessandro Conforti, Natale Monzeglio, Gianfranco Penone

PRINCIPAL ACTIVITIES: Banking
Financial Information:

	31.12.86	31.12.87
	L'000,000	L'000,000
Profit before tax	200,783	215,992
Profit after tax	171,945	178,038
Retained profit	153,945	156,038
Deposits	13,905,667	15,936,685
Total assets	17,861,267	20,242,145

CASSA DI RISPARMIO DI VERONA VICENZA E BELLUNO

Via Garibaldi 1, 37121 Verona
Tel: (045) 936111
Telex: 480056 VEREST
Telefax: (45) 591516

Board of Directors: Avv Alberto Pavesi (Presidente), Prof Alfredo Baldani Guerra (Vice Presidente), PI Aldo Pivetti (Vice Presidente), Avv Armando Cremonese, Dott Eugenio De Mas, Avv Luigi Righetti
Board of Management: Rag Antonio Finotti (Direttore Generale), Rag Giancarlo Garino (Vice Direttore Generale)

PRINCIPAL ACTIVITIES: Banking
Financial Information:

	31.12.87
	L'000,000
Net interest income	445,991
Profit before tax	260,182
Shareholders funds	851,497
Deposits(1)	8,361,200
Total assets(2)	12,160,740

(1) including Interbank
(2) less contra accounts

No of Employees: 2,926

CONDOTTE D'ACQUA SPA, SOCIETÀ ITALIANA PER,

Viale Liegi 26, 00198 Roma
Tel: (06) 85181
Cable: Condotte Roma
Telex: 61515

Board of Directors: M de Sena (Chairman and Member of Executive Committee), I G Caiati (Vice Chairman and Member of Executive Committee), Dott F Pacces (Vice Chairman), G Paolella (Managing Director), Dott M Amari, F Mezzalama, L

Migliorini, Ing P Peddis, C Berardi (and Member of Executive Committee), F Berlinguer, S Stefanini, L Maranca (Secretary)

PRINCIPAL ACTIVITIES: Building industry; civil engineering; finance and property

Subsidiary Companies: (100% owned unless stated): Condag-Impresa Lavori Marittimi SpA (52.5%); Condil-Condotte Lavor; Idraulici SpA; Con-Piem-Condotte Piemonte Impresa Generale di Costruzioni SpA (51%); Impresa Pietro Cidonio SpA; Metro roma SpA; Spamo-Sta Partecipazioni Mobiliari SpA; Condotte International Holding SA (Luxembourg); Italcontractors Consortium of Condotte and Partners-Partnership Co (Iran) (90%); Panedile Argentina SA (99.3%) (Argentina); Sté Francaise Condotte d'Acqua SA (99.9%) (France); Condotte International Holding SA (Luxembourg); Italcontractors Consortium of Condotte and Partners (90%) (Iran)

Financial Information:

	L'000,000
Sales turnover	316,000

No of Employees: 1,790 group

CREDITO ITALIANO SPA

Piazza Cordusio 2, 20123 Milano
Tel: 02/8862 1
Cable: Credirezio
Telex: 312401; 310103 CRITMM I
Telefax: 02/8862 3524

Directors: Natalino Irti (Chairman), Carmelo Petyx (Deputy Chairman), Leo Solari (Deputy Chairman), Lucio Rondelli (Managing Director), Pier Carlo Marengo (Managing Director)

General Managers: Mario Arbuffo, Egidio G Bruno, Vittorio Laviano, Attilio Molendi, Gregorio Pascazio, Gianni Testoni. Deputy General Managers: Gianfranco Brusco, Pietro Calella, Roberto De Simone, Egidio Lorenzi, Aldo Malferrari, Antonio Palazzo, Angelo Torchio

Divisional Managers: Carlo Barassi (Chief Manager EDP), Carlo Bianchini (Chief Manager Marketing), Maurizio Cartocci (Chief Manager Organisation Systems), Tommaso Cartone (Chief Manager Investment Holdings and Company Secretarial), Enrico Catelani (Chief Manager Management Control System), Paolo Crescini (Chief Manager Premises), Achille Donini (Chief Manager Public Relations and Secretariat), Marco Fantazzini (Chief Manager Legal), Gerardo Guida (Chief Manager H.O. & Foreign Branches Personnel), Luigi Mendolia (Chief Manager Domestic Personnel), Emilio Merciai (Chief Manager Organisational Development), Pietro Modiano (Chief Manager Economic Research and Planning), Francesco Monti (Chief Manager Domestic Network), Sabino Murgese (Chief Manager Internal Auditing Technical and Administration), Uberto Scaramuzza (Chief Manager International Relations), Stefano Sini (Chief Manager Credit Control and Monitoring), Dino Tiezzi (Chief Manager Credits), Raffaello Vignoletti (Chief Manager Accounting). Divisional Managers Finance: Claudio Brovedani (Chief Manager Treasury and Securities Back-Office), Fausto Petteni (Chief Manager Primary Market Italian and Eurosecurities), Francesco Schepes (Chief Manager Forex, Treasury-Foreign Currencies), Guido Sporeni (Chief Manager Treasury-Italian Lira), Franco Vignotti (Chief Manager Secondary Market-Italian and Eurosecurities)

PRINCIPAL ACTIVITIES: Credit company; banking services
Trade Names: Credito Italiano S.p.A.
Subsidiary/Associated Companies: Banca dei Comuni Vesuviani S.p.A.; Cordusio Società Fiduciaria per Azioni; Fincor Merchant Credit S.p.A.; Credit Factoring International S.p.A.; Creditwest S.p.A.; Data Management S.p.A.; Credit Leasing S.p.A.; Mediobanca S.p.A.; Credito Fondiario S.p.A.; Gesticredit S.p.A.; Servizi Fiduciari SEFI S.p.A.; Locat Locazione Attrezzature S.p.A.; Selma Società Esercizio Locazione Macchine Attrezzature S.p.A.; Credito Italiano Finance Corporation Limited (Bahamas); Banque Transatlantique S.A (France); Beal Banque Européenne pour l'Amerique Latine S.A. (Belgium); Libra Bank PLC (UK);

Credito Italiano International Ltd (UK); Credito Italiano Delaware Inc (USA); Capital Advisors SA (Luxembourg)
Principal Bankers: All leading banks in all countries
Financial Information:

	31.12.86	31.12.87
	L'000,000	L'000'000
Total income	4,836,659	4,744,482
Profit before tax	380,083	158,541
Profit after tax	207,785	138,399
Retained profit	393	178
Dividends	93,000	124,350
Dividends per share(1)	90L	75L
Dividends per share(2)	105L	90L
Share capital	500,000	800,000
Shareholders funds	2,457,115	3,083,219
Deposits	54,739,585	51,442,140
Total assets	61,298,986	57,876,864
(1) ordinary		
(2) savings		

Principal Shareholders: IRI - Istituto per la Ricostruzione Industriale (67.13%)

CREDITO ROMAGNOLO SPA

Via Zamboni 20, 40126 Bologna
Tel: (51) 338111
Telex: 510148
Telefax: 338491-379563

Board of Directors: Prof. Gerardo Santini (Presidente), Mario Fantini (Direttore Generale), Dr. Sebastiano Ciacci, Prof. Francesco Corrado, Cav. Lav. Dr. Luigi Derserti, Dr. Giuliano Graziosi, Avv. Antonio Griziotti, Cav. William Lazzarini, Cav. Lav. Dr. Achille Maramotti, Giorgio Seragnoli, Dr. Augusto Serra Zanetti, Fabrizio Zanon di Valgiurata

Management: Mario Fantini (Direttore Generale), Dr. Antonio Mazzanti (Vice Direttore Generale), Dr. Guido Bondi (Direttore Centrale), Rag. Antonio Donati (Direttore Centrale), Dr. Marco Nonni (Direttore Centrale)

PRINCIPAL ACTIVITIES: All banking operations; exchange services; foreign services; financial support services
Subsidiary Companies: Banca Agricola Commerciale Della Repubblica Di San Marino; Banca Venturi; Rololeasing; Rolotel; Rolofond
Principal Bankers: Credito Romagnolo

No of Employees: 4,145

EUROPA ASSICURANZIONI SPA

Via Balbi 2, I-16100 Genova
Tel: (010) 26931
Telex: 270285 LEVANT I

PRINCIPAL ACTIVITIES: Provision of a wide range of insurance services
Principal Bankers: Dirpopolare; Comitbanca; Banca Novara
Financial Information:

	L'000,000
Premium income	1,600,000

FINANZIARIA ERNESTO BREDA S.P.A.

See BREDA, FINANZIARIA ERNESTO, S.P.A.

FONDIARIA COMPAGNIA DI ASSICURAZIONI E RIASSICURAZIONI SPA (LA)

Piazza della Libertà 6, Firenze
Tel: 47941
Telex: 57430 assifond

Board of Directors: Michele Castelnuovo-Tedesco (Presidente Onorario), Raul Gardini (Presidente), Mario Schimberni (Vice Presidente), Alfonso Scarpa (Amministratore Delegato), Hanspeter Bruderer, Luigi Cappugi, Lino Cardarelli, Camillo De Benedetti, Michele de Michelis, Guiseppe Garofano, Ginolo Ginori Conti, Paolo Jovenitti, Vincenzo Maranghi,

Mario Mauri, Alberto Montanari, Enrico Pecci, Vittorio E Rimbotti, Jean Marc Vernes, Horst Detlev Von Der Burg
Management: Alfonso Scarpa (Direttore Generale), Sergio Chiostri (Condirettore Generale)
Senior Executives: Mario Marinesi (Segretario del Consiglio)

PRINCIPAL ACTIVITIES: Insurance of all types and reinsurance
Subsidiary Companies: (100% owned unless stated): La Previdente SpA (58.1%); Banca Mercantile Italiana SpA (60.6%); Card SpA (64.9%); GEAS SpA (69.8%); Finanziaria Media Italia-FIMEDIT SpA; Dominion Insurance Ltd (UK); La Fenice RI SpA (60%); Milano Assic SpA (57.9%)
Financial Information:

	L'000,000
Premium income	640,000

Principal Shareholders: Iniziativa Meta (38.5%); Medio Banca (15%)
No of Employees: 1,107

GENERALI, ASSICURAZIONI, SPA
2 Piazza Duca degli Abruzzi, 34100 Trieste
Tel: (040) 6711
Cable: Generali
Telex: 460190 GRALI I
Telefax: (040) 671600

Honorary Chairman: C Merzagora; Chairman: E Randone (also Managing Director)
Directors: C De Benedetti (Vice Chairman), M Luzzatto (Vice Chairman), A Rosa (Vice-Chairman), A Desiata (Managing Director), E Coppola di Canzano (Managing Director), A Baldissera, R de Banfield-Tripcovich, A Bernheim, F Cingano, L Coccioli, E Dusi, G Fanelli, M Monti, G Rossi, H Treichl, R Ossola, F Padoa, E Romanin Jacur
Senior Executives: U Della Casa, C Giussani, A Molinari (General Managers), A Bernardi, G Brugnoli, F Fegitz, G Gutty, P Iona (Central Managers)

PRINCIPAL ACTIVITIES: Insurance and reinsurance of all types
Trade Names: Assicurazioni Generali
Major Subsidiary Companies: Italy: Aurora Assicurazioni; Alleanza Assicurazioni; Europ-Assistance Italia; Friuli-Venezia Giulia Assicurazioni "La Carnica"; Gefina; Genedil; Genagricola; Genimmobil; San Luca Settantanove; Edicom; La Venezia Assicurazioni; Lotto Tre; Agricoltura Assicurazione; Navale Assicurazioni; Trieste e Venezia Assicurazioni; Unione Mediterranea di Sicurtà; Sementi Dotto; Agricola S Giorgio; Società Italiana Assicurazioni Danni; Albana; Prunus; Immobiliare Centro Direzionale; Cascinotto. France: Generali France; La Concorde; Compagnie Continentale d'Assurances; La Fédération Continentale; L'Equité I.A.R.D.; La Lutèce; Euralliance; Européenne de Protection Juridique; Europ Assistance S.A.; Compagnie Dakar Saint Louis; Generali France at Cie; S.A.I. 154/156 Boulevard de la Gare; S.I. du Parc de la Noue; Société Immobilière Centrale; Sotrimlo. Germany: Generali Lebensversicherung; Deutscher Lloyd Lebensversicherung; Deutscher Lloyd Versicherung; Dialog Versicherung; Dialog Lebensversicherung; EA Rechtsschutz; Europ Assistance Versicherung; Lloyd Immobilien. Belgium: Generali Belgium; Generali Belgium Invest. Netherlands: Levensverzekering de Nederlanden van 1870; Schadeverzekering de Nederlanden van 1870; Algemene Holding en Financiering; Levensverzekering de Eerste Hollandsche; Schadeverzekering de Eerste Hollandsche; Verzekerings Groep de Nederlanden van 1870; Beleggingsmaatschappij de Grachten; Graafschap Holland; Lepaka; Nedasinvest; Transhol; Nieuw Welgelegen. UK: Dog Breeders' Insurance Co; Europ Assistance Ltd; Gefina International; Northern Star Insurance Co; Europa Insurance Co. Austria: Erste Allgemeine Versicherung; Generali Allgemeine Lebensversicherung; Generali Rückversicherung; Wiener Allianz Versicherung; Eigenvorsorge Lebensversicherung; Estra Liegenschaftsvermietung; EBA Vermögensverwaltung; Allgemeine Immobilien; Opernring-Hof Bau- und Betriebsges; Braun & Co; Europäische Güter-

u.Reisegepäck-Vers.; Brewo Grundstücksverwaltung. Spain: Caja de Previsión y Socorro; Covadonga; Europ Assistance España. Greece: Generali Life. Switzerland: Union Suisse; Albula; Holdux; Pardux; USA: Transocean Holding Corp; Europ Assistance Worldwide Services. Netherlands Antilles: Gefina International NV; Holdux NV. Brasil: Generali do Brasil; Transocean do Brasil; São Marcos. Colombia: Cia. Granadina de Seguros; Cia. Granandina de Seguros de Vida. Guatemala: Aseguradora General. South Africa: The Standard General Insurance Co; S.G.I. Properties. Australia: Vanguard Insurance Co
Principal Bankers: Banca Commerciale Italiana; Banca Mercantile Italiana; Banca Nazionale del Lavoro; Credito Italiano; Samuel Montagu London; Morgan Guaranty Trust-N.Y.; Fuji Bank Tokyo; BNF Frankfurt
Financial Information:

	1986	1987
	L'000,000	L'000,000
Premium income	3,569,100	4,428,419
Investment income	809,323	904,687
Profit after tax	218,042	254,242
Dividends	103,817	125,084
Share capital	350,000	420,000*

*L1,060,000 million in 1988
Shareholders funds

Principal Shareholders: Mediobanca; Euralux; Banca D'Italia; IMIGEST; Fonditalia; La Fondiaria
No of Employees: 18,801 Group in Italy and abroad

INA
Istituto Nazionale delle Assicurazioni
Via Sallustiana 51, 00100 Roma
Tel: (06) 47221
Cable: Assistato Roma
Telex: 610336 INA RM I
Telefax: (06) 47224559

Supervisory Board: Prof Antonio Longo (Chairman), Prof Giuseppe Ammassari, Avv Pier Luigi Cassietti, Dott Mario Fornari, Dott Sergio Maggi, Avv Antonio Marotti, Dott Filipp Marzano, Dott Giacinto Militello, Dott Oreste Piemontese, Dott Ernesto Sciommeri, Sen Renzo Sclavi, Dott Roberto Tronchetti Provera, Avv Mario Tuccillo, Dott Aldo Vannini, Dott Domenico Fortini, Avv Luigi Pallottino
Management: M Fornari (General Manager), M Bartolozzi (A. General Manager), C Caccialanza (A. General Manager), F Pietrobono (A. General Manager), E Tenderini (A. General Manager)

PRINCIPAL ACTIVITIES: Life insurance
Subsidiary Companies: Assitalia; Praevidentia
Financial Information:

	31.12.86	31.12.87
	L'000'000	L'000,000
Premium income	1,381,773	1,802,452
Investment income	775,495	765,832
Profit after tax	54,222	62,980

Principal Shareholders: State owned company
No of Employees: 1,407

INTERBANCA BANCA PER FINANZIAMENTI A MEDIO E LUNGO TERMINE SPA
C.so Venezia 56, I-20121 Milano
Tel: (02) 7731.1
Cable: INTERBANCA
Telex: 312649 INTBAN I
Telefax: (02) 784321

Chairman: Dr Alberto Riccardi
Directors: Dr Giampero Auletta Armenise, Atos Bagnoli, Giancarlo Buscarini, Giorgio Cappon, Amato Ciocchetti, Giorgio Ferrari, Marco Gavazzi, Pietro Gnisci, Gustavo Greco, Gianemilio Osculati, Federico Pepe, Carlo Rivano, Giulio Rovelli, Giorgio Sella

Senior Executives: Dr Emanuele De Bernardi (General Manager), Dr. Dardano Maglio (Deputy General Manager), Dr Giovanni Zammarano (Senior Manager Foreign Department)

PRINCIPAL ACTIVITIES: Medium and long term financing in favour of industrial, commercial and service sectors in Italy; medium and long term export credits; issuing of bonds and certificates of deposit on a medium term basis

Financial Information:

	31.12.87
	L'000,000
Profit before tax	45,460
Profit after tax	29,260
Dividends	10,955
Share capital	68,467
Share capital and reserves	284,793
Total assets	6,288,568

Principal Shareholders: Banca d'America e d'Italia; Banca Nazionale dell Agricoltura, Istituto Centrale Banche e Banchieri; Credito Romagnolo; Banca Sella; Istituto di Credito delle Casse Rurali ed Artigiane; Banca del Friuli; Banco di Desio e della Brianza; Banco S Geminiano e S Prospero; Credito Emiliano

No of Employees: 326

IRI
Istituto Ricostruzione Industriale
Via Veneto 89, 00187 Roma
Tel: 47271
Cable: Istiri Roma
Telex: 610468 istiri i

PRINCIPAL ACTIVITIES: Operations in the fields of credit, iron and steel making, mechanical engineering, electronics and data processing, telecommunications, shipbuilding, food industry, sea and air transport, roads and highways, infrastructures, radio and television, professional training etc, through the control of five hundred firms

Principal Bankers: Banca Commerciale Italiana; Credito Italiano; Banco di Roma; Banco di San Spirito; Mediobanca. Main Subholdings: Finmeccanica; STET; Finsider; Italstat; SME; Fincantieri; Finmare; Finsiel; Alitalia

Financial Information:

| | L'000,000 |
| Group sales turnover | 47,000,000 |

No of Employees: 440,000

ISTITUTO BANCARIO ITALIANO SPA
Via Manzoni 3, 20121 Milano
Tel: 88901
Cable: Ibidir
Telex: 310676 IBIMI
Telefax: 8890 4520

Chairman: Giampiero Cantoni
Directors: Carlo Giltri (General Manager)
Senior Executives: Aldo Parmigiani (Head of International Division)

PRINCIPAL ACTIVITIES: Banking
Parent Company: Cassa di Risparmio delle Provincie Lombarde
Correspondent Banks in Europe and North America:
Oesterreichische Laenderbank, Austria; Générale Bank, Belgium; Canadian Imperial Bank of Commerce, Canada; Den Danske Bank af 1871 A/S, Denmark; Kansallis-Osake-Pankki, Finland; Banque Nationale de Paris, France; Deutsche Bank AG, Germany; Ionian and Popular Bank of Greece, Greece; National Irish Bank, Eire; Kredietbank SA Luxembourgeoise, Luxembourg; Amsterdam-Rotterdam Bank N.V., Netherlands; Bergen Bank, Norway; Banco Portugues do Atlantico, Portugal; Banco Bilbao Vizcaya, Spain; Skandinaviska Enskilda Banken, Sweden; Crédit Suisse, Switzerland; Midland Bank plc, UK; Manufacturers Hanover Trust

Company, USA and others in Australia, India, Japan and Singapore

Financial Information:

	31.12.87
	L'000,000
Profit after tax	21,373
Surplus profit and reserves	530,260
Capital stock	150,000
Deposits	7,481,491
Total assets*	36,630,475
Total assets net	9,251,044
*including contra/contingent accounts	

Principal Shareholders: Cassa di Risparmio delle Provincie Lombarde
No of Employees: 2,995

ISTITUTO BANCARIO SAN PAOLO DI TORINO
Piazza San Carlo 156, 10121 Torino
Tel: (011) 5551
Cable: ISTSANPAOL
Telex: 212040 DRPAOL I
Telefax: (011) 548469

Chairman: Prof. Gianni Zandano
Directors: Geom. Enrico Salza (Deputy Chairman), Dott. Claudio Bellavita, Prof. Mario Boidi, Dott. Tommaso Capitanio, Prof. Terenzio Cozzi, Rag Marco Desiderato, Dott. Renato Tullio Ferrari, Avv. Carlofelice Rossotto, Gianni Enrico Scerni, Prof. Pietro Verzeletti
Senior Executives: Prof Zefferino Franco (General Manager)

PRINCIPAL ACTIVITIES: Banking activities
Subsidiary Companies: (100% owned unless stated): Italy: Banca Provinciale Lombarda SpA; Banco Lariano SpA; Eurosystem-Compagnia Europea di Organizzazione SpA; Eurofactoring SpA; Sanpaolo Invest SpA; Eurofond SpA; Leasint SpA; Pitagora SpA; Sicilsud Leasing SpA; Sanpaolo Finance SpA; Fispao SpA; Locat SpA; Sanpaolo-Sviluppo SpA; Sanpaolo-Hambros SpA; Sanpaolo Formazione SpA; Flucogesta SpA; Sanpaologest Fiduciaria SpA; Sanpaolo Esattorie SpA. Abroad: Bankhaus Brüll & Kallmus AG; San Paolo Bank Bahamas Ltd; San Paolo-Lariano Bank; San Paolo US Financial Co; San Paolo US Holding Co; First Los Angeles Bank (by means of San Paolo US Holding Co); Melita Bank International Ltd (Malta); Hambros PLC (8.34%)
Financial Information: Consolidated data

	31.12.86	31.12.87
	L'000,000	L'000,000
Profit before tax	1,055,065	958,906
Profit after tax	588,328	565,209
Retained profit	553,328	528,209
Shareholders funds	3,171,044	3,691,980
Deposits	61,004,163	67,263,196
Total assets	79,562,572	85,916,724

No of Employees: 12,400 (Group: 18,000)

ISTITUTO MOBILARE ITALIANO
IMI
25 Viale dell'Arte, 00144 Roma
Tel: 54501
Cable: Mobilital
Telex: 610256 IMIROM
Telefax: 54 50 38 88

Board of Directors: Luigi Arcuti (Chairman), Mario Ercolani (Deputy Chairman), Piero Barucci, Francesco Bignardi, Piero Bongianino, Adelmo Brustia, Luigi Coccioli, Giuseppe Falcone, Gian Stefano Frigerio, Vito Gamberale, Antonio Longo, Roberto Mazzotta, Rino Onofri, Giuseppe Pasqua, Antonio Pedone, Giuliano Segre, Alberto Tomassini, Claudio Varrone, Gianni Zandano, Umberto Zanni
General Management: Giuseppe Saracini (General Manager), Cesare Barbieri (Deputy General Manager), Luciano Martino

(Deputy General Manager), Vittorio Serafino (Deputy General Manager), Enrico Fioravanti (Secretary-General)

PRINCIPAL ACTIVITIES: International bank

Subsidiary Companies: (100% owned unless stated): Banca Manusardi & C SpA; Credito Navale IMI (51%); Fideuram SpA; Fideuram Assicurazioni SpA (62.66%); Fideuram Fiduciaria SpA; Fideuram Vita SpA; Studi Finanziari SpA; Fidimi Consulting SpA (97.33%); Finimis SpA (99%) IMI Bank (Cayman); IMI Commercial Paper (USA); IMI International SA (Luxembourg); Euram Securities Corporation (USA); Euramex SA (Mexico); Finimi BV (Netherlands); Fonditalia SA (Luxembourg); IMI Capital Markets USA Corp (USA); IMI Capital Markets SA (Luxembourg); IMI Capital Markets UK Ltd (UK); IMI Finance Belgium SA (Belgium); IMI Securities Ltd (UK); Imil Ltd (UK); Interfund Advisory Co SA (Luxembourg); Int Fin Product Sales Ltd (Cayman Island); Int Securities Advisory SA (Luxembourg); SIGE International Ltd (UK); Turis AG (Switzerland); IMIGEST SpA (89.1%); SIGE SpA; Duomo Srl; Eurovenco SpA; Majno Dodici Srl; Sige Capital Markets SpA; Sige Fiduciaria SpA; Sigeco SpA; Tradital SpA; SPEI Leasing SpA; Edilspei Srl; Spei Factoring SpA. Affiliated Companies: Arcotecnica & Rolf Siegenthaler SpA; ASH SpA; Benati USA Inc (25%) (USA); Buffetti SpA (20%); Centro Vittora SpA; Comau Finanziaria SpA (6.95%); Consorzio Bancario SIR SpA (27.72%); Crystal Palace SpA; Easy SpA; Eidos SpA; Elettrocarbonium SpA (13.13%); Engineering SpA (20%); Fime Consulting SpA (50%); Finanziaria Ferrone SpA (20%); Finbancaria SpA (20%); Finban SpA (25%); Finec SpA (30%); Finnat (20%); Finsessanta Srl (50%); Finsystem SpA (20%); Fornara Fin e di Partec SpA (16.93%); Fispao SpA (20%); France Alfa SA (25%) (France); Green Has Italia SpA; Immobiliare Crisanzio Srl (50%); Immobiliare S Marco Srl (50%); IM Intermetro SpA (11.1%); IMIBJ SpA; (50%); Ince SpA; Intelesis SpA; Inglen SpA (20%); ITAB Group Ltd (28%) (UK); Ital Design SpA; ITALOX Ltd; IVG SpA (20%); Leasing Pubblico SpA (40%); Nuova Valvotecnic SpA (16%) (Switzerland); Leasingroup Sicilia SpA (25%); Place Vittoria Inc (37.2%) (Canada); Pontello Finanziaria srl (20%); Profilacer SA (25%) (France); Pronim SpA (22.5%); Salvagnini SpA (20%); Produits d'Acier SA; S.E.F.I. Export SpA (25%); SIM SpA (20%); Società SPS S Cons pA (15%); Zwahlem & Mayr SA (14%) (Switzerland)

Financial Information: Consolidated highlights

	31.3.87	31.3.88
	L'000,000	L'000,000
Profit before tax	995,900	646,400
Net income	524,400	419,000
Share capital	1,650,000	1,650,000
Shareholders equity	3,782,900	4,099,800
Total assets	30,674,100	33,559,000

Principal Shareholders: Cassa Depositi e Prestiti, Rome
No of Employees: 941

ITALIA ASSICURAZIONI S.P.A.

Via Fieschi 9, I-16121 Genova
Tel: (010) 53801
Cable: Italiassicur
Telex: 270136 italass i
Telefax: (010) 592856

President: Alfonso Scarpa; Vice Presidente: Carlo Galeazzi
Directors: Claudio Reichlin (Amministratore Delegato), Sergio Chiostri, Piero Castelli, Gian Vittorio Cauvin, Giacomo Clerici, Mauro De Andrè, Mario Marinesi, Rinaldo Piaggio, Victor Uckmar (Consiglieri)
Senior Executives: Bruno Mondini (Direttore Generale), Giorgio Oppezzi (Condirettore Generale)

PRINCIPAL ACTIVITIES: Insurance and reinsurance of all classes

Trade Names: Lloyd Italico
Parent Company: Compania di Assicurazioni di Milano SpA, Milan

Subsidiary/Associated Companies: (100% owned unless stated): BASFIN Srl; Immobiliare Montebignone Srl; Finitalia SpA (50%); Casavita Srl; Immobiliare Nuova Italia Srl; Immobiliare Plumbago Srl

Financial Information:

	31.12.86	31.12.87
	L'000'000	L'000,000
Premium income	449,991	502,424
Investment income	38,339	42,461
Profit before tax	7,048	7,014
Profit after tax	4,797	5,214
Retained profit	247	2,051
Dividends	3,000	3,000
Dividends per share	100L	100L
Earnings per share	159.9L	173.8L
Share capital	30,000	40,000
Shareholders funds	67,767	69,415

No of Employees: 834

MEDIOBANCA SPA

Via Filodrammatici 10, 20121 Milano
Tel: 8829-1
Cable: Mediobanca Milan
Telex: 311093 Meban I

Chairman: Antonio Maccanico
Directors: S Salteri (Managing Director), G Agnelli, E Braggiotti, R Cassaro, E Ceccatelli, F Cingano, E Cuccia, W Graebner, P C Marengo, V Maranghi, L Pirelli, L Rondelli, S Siglienti, M Tacci

PRINCIPAL ACTIVITIES: Joint stock merchant bank providing medium and long term credit

Subsidiary/Associated Companies: (100% owned unless stated): Compage SpA; Compass SpA; F.B.M.-Hudson Italiana SpA (67.35%); Data Leasing SpA (70%); Intersomer, Società Mercantile Internazionale SpA; Ricerche e Studi SpA; Selma, Società Esercizio Locazione Macchine Attrezzature SpA (80%); Spafid, Società per Amministrazioni Fiduciarie SpA; TRADEVCO, The Liberian Trading and Development Bank Ltd (60%) (Liberia); Graucom AG (Switzerland); Incar (Côte d'Ivoire) SA (Ivory Coast); Incar Tanzania Ltd (Tanzania) (66.67%); Incar (Zambia) Ltd (Zambia); Incar Zimbabwe (Private) Ltd (Zimbabwe); Intersomer (Zambia) Ltd (Zambia); Nuclital SpA; Promotex-Promozione Esportazione Impianti; Seteci-Esportazione Tecnologie e Componenti Industrali; Tecnesa SpA (10%)

Financial Information:

	30.6.86	30.6.87
	L'000,000	L'000,000
Profit after tax	172,842	113,000
Dividends	34,000	34,000
Dividends per share	2,000L	2,000L
Share capital	170,000	170,000
Shareholders funds	1,234,116	1,281,900
Total assets	14,130,530	14,659,197

Principal Shareholders: Banca Commerciale Italiana (20.06%); Credito Italiano (20.06%); Banco di Roma (16.78%)
No of Employees: 2,478 (Group); 278 (Mediobanca)

MONTE DEI PASCHI DI SIENA BANKING GROUP

Piazza Salimbeni 3, 53100 Siena
Tel: 294111
Cable: Monte Paschi
Telex: 570080 PASDIR I; 572346 PASEST I (International Division)
Telefax: (Group 3) (0577) 29 41 11

Chairman: Piero Barucci
Senior Executives: C Zini (Chief Executive and General Manager), A Bordogna (General Manager, International Division)

PRINCIPAL ACTIVITIES: Banking

Subsidiary Banks: Credito Commerciale SpA; Credito Lombardo SpA; Banca Toscana SpA; Italian International Bank Plc; Banca di Messina SpA

No of Employees: 16,029

MONTEDISON SPA

Foro Buonaparte 31, 20121 Milano
Tel: (02) 63331
Cable: Montedison
Telex: 310679 montedis

Board of Directors: *Raul Gardini (President and Chairman), *Alexander F Giacco (Chief Executive Officer, Vice Chairman and Managing Director for Strategy and Industrial Management), *Giorgio Porta (Vice Chairman for General Affairs and Industrial Activities), *Sergio Cragnotti (Vice Chairman for Special Affairs), *Lino Cardarelli (Managing Director for Financial Policy), Enrico Pizzi (Secretary), *Mario Conde, *Mauro De Andrè, *Arturo Ferruzzi, Giuseppe Garofano, Fabio Inghirami, Salvatore Ligresti, Adone Maltauro, *Renato Picco, Enrico Randone, Lino Rondelli, Paul Tjepkema, Victor Uckmar, Carlo Vannini, Jean Marc Vernes, Horst Detlev von der Burg, *Members of the Executive Committee

Statutory Audit Committee: Prof Luigi Guatri (President). Standing Statutory Auditors: Enrico Gianzini, Giuseppe Pietrasanta, Sergio Pivato, Roberto Magnani. Substitute Statutory Auditors: Luigi Capè, Ferdinando Superti Furga

Senior Executives: Managers by business sector: Giuseppe Bencini (Fibres), Gian Carlo Cimoli (Energy), Ezio Colombo (Functional Chemicals), Ettore Dell'Isola (Agroindustry), Gian Luigi Diaz (Consumer and Fabricated Products), Giovanni Di Drusco (Advanced and Composite Materials), Paolo Morrione (Pharmaceuticals), Giuseppe Garofano (Services), Andrea Mattiussi (Petrochemicals and Plastics), Leonard Rosenblatt (Specialty Chemicals and High Performance Materials). Department Heads: Gian Carlo Ciaccia (Legal and Corporate Affairs), Alvise Conciato (Administration and Control), Giorgio Rigamonti (Personnel)

PRINCIPAL ACTIVITIES: The Montedison Group is a leading international diversified chemical group and is the second largest private sector industrial concern in Italy in terms of turnover. It is engaged in a wide range of businesses which are divided into four segments: Chemicals which includes petrochemicals and plastics, specialty chemicals and high performance materials, fibres, agroindustry products, consumer and fabricated products, functional chemicals and other businesses. Energy which includes petroleum and aromatics products, electric power and oil and gas exploration and production. Pharmaceuticals which includes the manufacturing and marketing of ethical pharmaceuticals, specialty bulk products, diagnostics and reagents and over-the-counter products. Services which includes retailing; insurance; property; engineering, construction and telecommunications; financial services; and other businesses

Trade Names: Moplen, Mopen R, Adriblastina, Daunoblastina, Adriamycin, Epirubicin, Fomblin, Algoflon, Tecnoflon, Terital, Leacril, etc

Holding Company: Montedison SpA

Subsidiary Companies: Montedison is the parent company of the Montedison Group and holds, directly or indirectly, shares in a large number of subsidiaries and associated companies in many countries. The most important are: Acna, Agrimont, Antibioticos, Ausimont, Erbamont, Farmitalia-Carlo Erba; Iniziativa ME.T.A.; Montedipe; Montefibre, Rol, SELM, Standa

Financial Information:

	31.12.86	31.12.87
	L'000,000	L'000,000
Sales turnover	12,834,000	13,791,000
Profit before tax	641,000	874,000
Profit for the year	320,000	409,000
Dividends	115,000	151,000
Dividends per ord share	40L	40L
Dividends per savings share	60L	60L
Share capital	2,568,000	2,705,000

Principal Shareholders: The Ferruzzi Group
No of Employees: 66,785 Group

NUOVO BANCO AMBROSIANO SPA

Piazza Paolo Ferrari 10, Casella postale 1235, 20100 Milano
Tel: (2) 859 41
Telex: 3335687 NBADIG I
Telefax: (2) 807276

Chairman: Giovanni Bazoli; Vice Chairman: Paolo Baratta (Chairman, Crediop), Francesco Paolo Mattioli (Central Manager, Fiat), Giorgio Zanotto (Chairman, Banca Popolare di Verona)

Directors: Sergio Ceccuzzi (Managing Director, SMI Soc Metallurgica Italiana), *Antonio Ceola (General Manager, Banca Popolare Veneta), Romano Ceroni (General Manager, Credito Romagnolo), Giorgio Cigliana (General Manager, Crediop), Aldo Cova (General Manager, Banca Popolare di Milano), Dino Marchiorello (Chairman, Banca Antoniana di Padova e Trieste), Giuseppe Nardini (Chairman, Credito Romagnolo), Giampiero Pesenti (Managing Director and General Manager, Italcementi), Giorgio Piantini (Chairman, La Centrale-Compagnia Finanziaria), Maria Teresa Salvemini (Director, Crediop), *Piero Schlesinger (Chairman, Banca Popolare di Milano), Alberto Valdembri (General Manager, Banca San Paolo di Brescia), Felice Vitali (General Manager, Gemina), Tommaso Zerbi (Director, SIDIEF SpA), *Members of the Executive Committee

Senior Executives: Chief Executive and General Manager: Gino Trombi. Deputy General Managers: Nereo Dacci, Ferruccio Codeluppi. International Division: Gianluigi Pagliani (Director and Head). Correspondent Banking: Camillo Celenza (Senior Executive Area Manager), Flavio Ramella (Executive Area Manager), Vincenzo Parlante (Area Manager). Foreign Exchange and Euro-Currency Deposits: Mario Zucca (Senior Executive Manager), Mario Mascheroni (Chief Dealer), Giuseppe Di Vietri (Deputy Chief Dealer). Treasury and Securities Division: Poliuto Boaretto (Director and Head), Angelo Ciocchini (Manager, Treasurer), Franco Panareo (Manager, Foreign Securities), Alessandro Villa (Manager, Italian Securities). Advertising/Public Relations: Mrs Gigliola Zecchi (Director and Head)

PRINCIPAL ACTIVITIES: Banking

Subsidiary Companies: (100% owned unless stated): Banca Cattolica del Veneto SpA (42.63%); Fiscambi Holding SpA (66.54%); Fiscambi Leasing SpA (indirect 64%); P. Leasing SpA (indirect 51%); Fiscambi Factoring SpA (indirect 100%); Ambrofid SpA (64%); Italfid-Italiana Fiduciaria SpA; La Centrale Fondi SpA (direct 60%, indirect 10%); La Centrale-Compagnia Finanziaria SpA (direct 50%, indirect 50%); La Centrale Formazione Srl (direct 40%, indirect 60%); Assiprogetti SpA (50.28%); C.S.O. SpA (ex G.S.I. SpA) (direct 49%, indirect 51%); Ambro Italia SpA

Financial Information: consolidated figures

	31.12.86	31.12.87
	L'000,000	L'000,000
Profit before tax	227,304	188,376
Profit after tax	70,692	61,765
Dividend	23,209	23,209
Dividend per ordinary share	70L	70L
Dividend per preference		
share	90L	90L
Earnings per share	-	162.8L
Share capital	425,879	425,879
Shareholders funds	1,051,126	1,079,836
Deposits	8,439,410	7,276,173
Total assets	34,992,828	20,839,505

Principal Shareholders: Banca Popolare di Milano (13.45%); Crediop-Consorzio di Credito per le Opere Pubbliche (12.97%); Gemina-Generale Mobiliare Interessenze Azionarie (14.09%); Banca Popolare di Verona (7.6%); Banca San Paolo di Brescia (6.41%); Banca Popolare Veneta (3.14%); Banca Antoniana di Padova e Trieste (3.64%); Credito Romagnolo (2.95%); Banca Popolare Vicentina (2.69%); Individual shareholders (approx 50,000) (33.06%)
No of Employees: 7,000

RAS

Riunione Adriatica di Sicurtà SpA

Corso Italia 23, 20122 Milano
Tel: (02) 88441; 88421
Telex: 320065 RAS DG I
Telefax: 0874229/8900740

Honorary Chairman: E Lolli; Chairman: U Zanni
Directors: Detlev Von Der Burg (Deputy Chairman), Dario Gaverbi (Managing Director), Angelo Marchiò (Managing Director)
Senior Executives: G Baseggio (Deputy General Manager), A Lentati (Deputy General Manager), E Orlando (Deputy General Manager), F Prusso (Deputy General Manager)

PRINCIPAL ACTIVITIES: Insurance
Trade Names: RAS
Subsidiary Companies: (100% owned unless stated): Lavoro & Sicurtà; Unione Subalpina di Assicurazioni (69.2%); L'Italica (94.8%); L'Italica Vita; Compagnia di Genova (99.8%); Mutua Assicuratrice Cotoni; Compagnia Europea d'Assicurazione; Lloyd 1885 (99.6%); Compagnie de Réassurances de Paris (99.92%) (France); Cresa (Spain); Portugal Previdente (68.42%) (Purtugal); Sark Sigorta (33.59%) (Turkey); Jefferson (USA); Canadian Home (99.4%) (Canada); Adriatic Insurance Co of Canada (99.92%) (Canada); Aseguradora Cuauhtemoc (12.39%) (Mexico); Compania Aseguradora Argentina (19.9%) (Argentina); Companhia Adriatica de Seguros Gerais (80%) (Brazil); El Sol (19.23%) (Peru); Adriatica de Seguros (17.5%) (Bolivia); Adriatic Life (99.99%) (Australia)
Financial Information: Consolidated figures

	L'000'000
Premium income	1,747,000

SAI

Società Assicuratrice Industriale SpA

Corso Galileo Galilei 12, 10126 Torino
Tel: (011) 65621
Cable: Securus Turin
Telex: 212080 Saito I
Telefax: (011) 6562685

Chairman: Salvatore Ligresti
Directors: C Caruso (Deputy Chairman), A La Russa (Deputy Chairman), F Rapisarda (Managing Director), G Beltrami, L Betti, G Cantanzaro, V La Russa, P R N Payne, A Talarico, E Trupiano, A Ligresti, P Agostoni, S Brunotti, A Riffeser
Senior Executives: G Brinatti (General Manager)

PRINCIPAL ACTIVITIES: Insurance

Trade Names: SAI-Società Assicuratrice Industriale SpA
Subsidiary Companies: Società Finanziaria Pozzi-Richard Ginori; SAI-Agricola; ATAHOTELS; Italcementi; Italmobiliare; Siat; Eurofond; Saifond; Saifinaziaria; Pirelli; Progestim; Finsai International; Sim Etoile; Capotaormina; Il SOLE - Ass.Ni e Riass.Ni; Ausonia Vita FIDA Finanziaria D'Affari; S.LUCA Settantanove Srl; CO.FI.DI Leasing; SARP - Soc. Ass.ne Rischi Persone; Previndustria - Fid. Prev. Imprenditori
Principal Bankers: Credito Italiano; Banca Commerciale Italiana; Banco Di Roma; Istituto Bancario S.Paolo; Banca Nazionale Del Lavoro; Banco Di Napoli; Banco Di Santo Spirito; Istituto Bancario Italiano; Banca Sella
Financial Information:

	1986	1987
	L'000'000	L'000'000
Premium income	1,323,238	1,512,158
Investment income	275,712	297,936
Profit before tax	115,167	92,292
Profit after tax	100,955	87,596
Retained profit	75,861	58,292
Dividend	23,692	28,092
Dividend per ordinary share	160L	160L
Share capital	137,500	165,000
Shareholders funds	479,680	729,399

Principal Shareholders: Interbaros International Holdings, Amsterdam (22.22%); Premafin Finanziaria SpA (43.64%); Imigest SpA (2.05%); Primigest SpA (1.25%); Instituto Bancario Italiano (1.1%)
No of Employees: 2,759

SICILCASSA

Cassa Centrale di Risparmio V.E. per le Province Siciliane

Via F Cordova, 76-90143 Palermo, Sicily
Tel: (091) 6291111
Cable: Sicilcassa Dirgen, Palermo
Telex: 910029 CARIPA I
Telefax: (091) 6292550

Chairman: Giovanni Ferraro
Directors: Agostino Mulé (General Manager)
Senior Executives: Pasquale Salamone (Deputy General Manager), Francesco D'Amico (Deputy General Manager), Girolamo Maltese (Head of Correspondent Banking, Treasury, Investment Banking, Leasing and Trade Finance Divisions), Armando Spinella (Head of Foreign Exchange, Syndicated Loans Division), Erasmo Grutta (Head of Private Banking)

PRINCIPAL ACTIVITIES: Banking
Financial Information:

	31.12.87	31.12.88
	L'000,000	L'000,000
Profit before tax	59,881	25,124
Profit after tax	35,285	13,264
Retained profit	35,285	13,264
Deposits	7,391,461	8,061,471
Total assets	11,553,261	12,104,939

No of Employees: 4,128

SOCIETÀ ITALIANA PER CONDOTTE D'ACQUA SPA

See CONDOTTE D'ACQUA SPA, SOCIETÀ ITALIANA PER,

SOCIETÀ ITALIANA PER IMPRESE E GESTIONI

SIGE SpA

Piazza San Fedele 2, 20121 Milano
Tel: (02) 85321
Telex: 311062 SIGEFI I
Telefax: 8532353

Chairman of the Board: Dr Luigi Arcuti
Directors: Dott Giorgio Mariotti (General Manager)

PRINCIPAL ACTIVITIES: Investment and merchant banking
Parent Company: IMI - Instituto Mobiliare Italiano
Subsidiary Companies: SIGECO; SIGE Capital Markets; SIGE
   FIDUCIARIA; TRADITAL; DUOMO

Principal Shareholders: IMI
No of Employees: 245

TORO ASSICURAZIONI SPA
Via Arcivescovado n 16, 10121 Torino
Tel: (011) 57331
Cable: Premiofisso Torino
Telex: 221567 toroas I
Telefax: (011) 543587

Chairman: Umberto Agnelli; Deputy Chairman: Benedetto
   Salaroli
Directors: Francesco Torri (Managing Director and General
   Manager), Jean d'Arenberg, Fausto Bacchetti, Sigismund von
   Braun, Davide Croff, Gabriele Galateri di Genola, Paolo
   Geisser, Franzo Grande Stevens, Pietro Carlo Marsani,
   Francesco Paolo Mattioli, Alberto Midana, Sergio Pirinfarina,
   Enrico Recchi
Senior Executives: Francesco Coggiola (Secretary)

PRINCIPAL ACTIVITIES: Insurance and reinsurance of all
   classes in Italy and abroad
Subsidiary Companies: Allsecures-Preservatrice SpA; Allsecures
   Vita SpA; La Vittoria Riassicurazioni SpA; DAS-Difesa
   Automobilistica Sinistri SpA; Giano SpA; Le Continent IARD
   SA; Le Continent Vie SA (France); L'Union Générale du Nord
   SA (France)
Financial Information:

	1986	1987
	L'000,000	L'000,000
Premium income	631,848	730,604
Investment income	1,464,090	1,579,185
Profit before tax	73,336	64,800
Profit after tax	52,550	58,800
Retained profit	124,504	148,521
Dividends	25,850	27,006
Dividends per ord share	230L	240L
Dividends per savings share	250L	260L
Earnings per ord share	478L	535L
Ordinary share capital	57,750	57,750
Pref share capital	24,750	24,750
Savings share capital	27,500	27,500
Shareholders funds	582,528	615,422

Principal Shareholders: IFI/IFIL (53.79%)
No of Employees: 1,440

Major Companies of LIECHTENSTEIN

BANK IN LIECHTENSTEIN AG

PO Box 85, FL-9490 Vaduz
Tel: 075/5 11 22
Cable: Bank Vaduz
Telex: 88 92 22 bil fl (General); 88 92 44 bil fl (Eurobonds); 88 92 11 (Forex)

Supervisory Board: Christian Norgren (Chairman), H.S.H. Prince Philipp von und zu Liechtenstein (Vice Chairman)

Management: Dr Egmond Frommelt (Chairman of Board of Management). Managers: Fritz Bühler, Remo Hediger, Jost Pilgrim, Alfons Wanger, Hermann Wille, Hans Rudolf Aebi, Josef Roos, Heinz Nipp (Deputy Manager)

PRINCIPAL ACTIVITIES: On a regional basis the bank offers all the services of a modern universal banking institution whilst internationally the emphasis is on investment advice and portfolio management services for private and commercial customers. The bank also participates in international bond market consortia

Group Companies: Bank in Liechtenstein (Frankfurt) GmbH, Frankfurt, Germany; Bank in Liechtenstein (UK) Ltd, London, UK; BIL, Trainer Wortham Inc, New York, USA; Bank in Lichtenstein AG, Representative Office, Hong Kong; Bilfinanz und Verwaltung AG, Zürich, Switzerland

Financial Information:

	31.12.86 SF'000	31.12.87 SF'000
Profit before tax	39,433	44,441
Profit after tax	35,466	40,130
Dividends	25,470	31,153
Dividends per share	12%	12%
Share capital	247,250	259,612
Shareholders funds	404,422	580,800
Deposits	3,083,129	3,322,674
Total assets	4,054,115	4,676,185

Principal Shareholders: The Prince of Liechtenstein Foundation
No of Employees: 392

VERWALTUNGS-UND PRIVAT-BANK AG

Banque Privée de Gérance SA

Private Trust Bank Corporation

Auelestrasse 6, Postfach 885, FL-9490 Vaduz
Tel: 075/5 66 55
Cable: Privatbank
Telex: 889200; Telefax: 075/2 66 97
Swift: VPBV LI22

Board of Directors: Dr Erich Seeger (President), Dir Olaf Walser, Prof Martin Hilti, Dir Emil Kuster, Dir Ernst Wanger, Dr Bruno B Güggi, Dr Rudolf Staub, Dr Gerard Batliner

Management Committee: Dr E H Batliner (General Manager), Dr J Hadermann (Manager), W Kaufmann (Manager), A Schurte (Manager), W Seger (Manager), G Erne (Manager)

PRINCIPAL ACTIVITIES: Money market dealing; Investment advice and portfolio management; FL numismatics; credits; savings; correspondents and brokers to all large financial centres in the world

Financial Information:

	31.12.86 SF'000	31.12.87 SF'000
Profit before tax	17,023	19,379
Profit after tax	15,018	17,721
Dividends	12%	12%
Share capital*	75,000	90,000
Shareholders funds	237,862	357,500
Deposits	1,710,258	2,030,439
Total assets	2,047,081	2,506,155

*plus non voting share capital

No of Employees: 265

Major Companies of
LUXEMBOURG

BAII
Banca d'America e d'Italia
Rue Notre Dame 37-39, 2240 Luxembourg
Tel: 2 90 81
Telex: 3746 Iubai lu

Management: Giancarlo Caine

PRINCIPAL ACTIVITIES: Banking
Financial Information:

	31.12.87 FLux'000
Total assets	1,851,600

No of Employees: 10

BANK OF CREDIT & COMMERCE INTERNATIONAL SA
BCCI
39 Boulevard Royal, L-2449 Luxembourg
Tel: 470391
Cable: BANCRECOM
Telex: 1240 LU
Telefax: 352 47 5185

President: Agha Hasan Abedi
Directors: Swaleh Naqvi, Yves C Lamarche, J D van Oenen, Agha Hasan Abedi, Ghanim Faris Al-Mazrui, Dr Alfred Hartmann, P C Twitchin, Khalid Salem Bin Mahfouz
Executives in Charge: Saleem Siddiqi (Central Audit), Imtiaz Ahmed (Central Credit), John Hillbery (International/EMP), A Hafeez (Corporate Affairs Central Office), Ameer H Siddiki (Corporate Identity and Special Projects), Masihur Rahman (Group Accounts), Shahid Jamil (Central Treasury), Nadir R Rahim (Human Resources), W S Al-Kaylani (Central Marketing International Business Development), A A Gillani (Investment), A M Haider-Mota (Legal Affairs), Pesi Sorab (Management Services), M A Faruqui (Planning and Strategy), P F Gutta (Systems and Operations)

PRINCIPAL ACTIVITIES: Commercial banking
Parent Company: Bank of Credit & Commerce International Holdings (Luxembourg) SA
Subsidiary/Affiliated Companies: Bank of Credit and Commerce International (Overseas) Ltd (Cayman Islands); Bank of Credit and Commerce (Botswana) Ltd (Botswana); Bank of Credit and Commerce Cameroon SA (Cameroon); Bank of Credit and Commerce Canada (Canada); Bank of Credit and Commerce (Emirates) (UAE); Bank of Credit and Commerce Ghana Ltd (Ghana); Bank of Credit and Commerce Gibraltar Ltd (Gibraltar); Bank of Credit and Commerce Hong Kong Ltd (Hong Kong); Bank of Credit and Commerce International (Lebanon) SAL (Lebanon); Bank of Credit and Commerce (Trinidad and Tobago Merchant Bankers) Ltd (West Indies); Bank of Credit and Commerce (Misr) SAE (Egypt); Bank of Credit and Commerce International (Nigeria) Ltd (Nigeria); Bank of Credit and Commerce Niger (Niger); Bank of Credit and Commerce SAE (Spain); Bank of Credit and Commerce International (Swaziland) Ltd (Swaziland); Bank of Credit and Commerce (Zambia) Ltd (Zambia); Bank of Credit and Commerce Zimbabwe Ltd (Zimbabwe); Banco de Credito e Comercio de Investimento SA (Brazil); Banco de Credito y Comercio de Colombia (Colombia); Bank of Credit & Commerce SA (Argentina); Banque de Commerce et de Placements SA (Switzerland); National Bank of Oman Ltd (SAO) (Sultanate of Oman); BCC Australia Ltd (Australia); BCC Credit and Finance (Uruguay) SA (Uruguay); BCC Finance and Securities (Thailand); BCCI Finance International Ltd (Hong Kong); BCCI Finance (Kenya) Ltd (Kenya); BCCI Leasing (Malaysia) SDN BHD (Malaysia); Credit and Finance Corporation Ltd (Cayman Islands); Kuwait International Finance Company SAK (Kuwait); Italfinance International SpA (Italy); PT BCC Pratama Leasing Indonesia (Indonesia); BCCI Finance NV (Netherlands Antilles)

Principal Bankers: National Westminster Bank (UK); all major banks throughout the world
Financial Information:

	31.12.86 US$'000	31.12.87 US$'000
Profit before tax*	22,298	8,618
Profit after tax	11,393	1,608
Retained profit	21,477	42,102
Dividend	-	7,000
Share capital	255,000	295,000
Shareholders funds	283,604	344,429
Deposits	5,942,805	4,878,841
Total assets	6,329,241	7,830,820

*after LLR

No of Employees: 13,727

BANQUE DE LUXEMBOURG SA
80 Place de la Gare, PO Box 2221, L-1022 Luxembourg
Tel: 49924-1
Cable: Luxbank, Luxembourg
Telex: 3757 Bluxh Lu; 2249 Bllux Lu
Telefax: 49-48-20; 49-26-36

Management Board: Gaston Zerr (President), R Reckinger (Vice-President), E Storck (Vice President)
Senior Executives: Robert Reckinger

PRINCIPAL ACTIVITIES: Bank
Parent Company: Crédit Industriel d'Alsace et de Lorraine, Strasbourg (71%); Deutsche Bank Cie Fin. Luxembourg (25%); Deutsche Bank Saar (3.9%)
Principal Bankers: Crédit Industriel d'Alsace et de Lorraine, Strasbourg; CIC Paris; CIC New York; CIC London; Générale de Banque, Brussels; Deutsche Bank, Frankfurt
Financial Information:

	31.12.86 FLux'000	31.12.87 FLux'000
Deposits	47,975,400	69,091,470
Total assets	54,766,750	74,125,200

No of Employees: 333

BANQUE GÉNÉRALE DU LUXEMBOURG SA
27 avenue Monterey, 14 Rue Aldringen, L-2951 Luxembourg
Tel: 47991
Cable: Generalbank-Luxemburg
Telex: 3401 bgl lu; 2742 bglex lu; 2471 bglbo lu; 3601 bgltit lu
Telefax: 47994388

Board of Directors: Georges Arendt (President), Georges Schwall (Honorary President), Xavier Malou (Vice President), Alain Georges (Managing Director), Remy Kremer (Managing Director), Paul Meyers (Managing Director), *André Bintz, *Raymond Bremer, *Roland De Cillia, Léon Derwa, *Armand Drews, Henri Fayt, Lucien Jung, Mme Georges Kipgen-Klensch, *Marcel Krier, Mme Lydie Polfer, *Aloyse Schiltz, Emmanuel Tesch, Michel Wurth, *Employee Representative
Board of Managing Directors: Alain Georges (President), Remy Kremer, Paul Meyers, Fernand Simon, Michel Waringo
Senior Executives: J Meyer (Manager), V Wolff (Manager), M Thinnes (Agency Manager), J Werner (General Counsel), M Reiter (Deputy Manager), E Cravatte (Deputy Manager), R Frising (Deputy Manager), P Hayot (Deputy Manager), P Massard (Deputy Manager), G Rommes (Deputy Manager), P Wolff (Deputy Manager)

PRINCIPAL ACTIVITIES: All banking services
Subsidiary Companies: Banque Générale du Luxembourg (Suisse) SA, Zürich, Switzerland
Principal Bankers: Générale de Banque (Brussels); European American Bank (New York); Banque Nationale de Paris (Paris); Banque Belge Ltd (London); The Royal Bank of Canada (Montreal)

Financial Information:

	31.12.86	31.12.87
	FLux'000	FLux'000
Profit before tax	1,578,863	1,844,329
Profit after tax	645,478	750,713
Retained profit	353,275	393,861
Dividends	265,305	329,707
Dividends per share gross	205.88FLux*	441.18FLux
Earnings per share	952FLux	1,004.5FLux
Share capital	2,440,202	2,690,410
Shareholders funds	6,358,263	7,656,626
Deposits	262,666,506	302,683,835
Total assets	297,151,004	339,018,632

*new shares. Old shares
  411.76 FLux

Principal Shareholders: Générale de Banque (44.6%); Majority of the capital held by Luxembourg shareholders
No of Employees: 1,647

BANQUE INTERNATIONALE À LUXEMBOURG SA

2 Blvd Royal, L-2953 Luxembourg
Tel: 4791/1
Cable: Internationbank
Telex: 3626 BIL LU; SWIFT: Luxembourg BILL LU LL
Telefax: 4791 2625

Supervisory Board: Gaston Thorn (Chairman), Marc Lambert (Deputy Chairman), Gérard Eskénazi (Deputy Chairman and Managing Director of the Group), Albert Dondelinger (Managing Director), Marc Assa (Director), Pierre Back (Director), Guy Berscheid (Director), Anne Brasseur (Director), Carlo Clasen (Director), Jean Duschène (Director), Maurits E Edersheim (Director), André Elvinger (Director), Georges Faber (Director), Gust Graas (Director), Pierre Gretz (Director), Régnier Haegelsteen (Director), Edmond Israel (Director), Henri Juda (Director), André Kieffer (Director), Baron Antoine de Schorlemer (Director), Gaston Schwertzer (Director), Pierre Schohier (Director), Eugène Storck (Director), Claude Vercambre (Director), Denise Weber-Ludwig (Director)
Executive Board: Albert Dondelinger (President), Jean-Donat Calmes (Senior Executive Vice-President), Jean Krier (Senior Executive Vice-President), Olivier Michon (Senior Executive Vice-President, Personal Advisor to the President of the Executive Board), François Steil (Senior Executive Vice-President)
Operational/Sectorial Responsibilities: President: Albert Dondelinger (Managing Director and President of the Executive Board), Jacques Kauffman (Senior Vice-President, General Counsel, Secretary General, Legal and Fiscal Matters and Participations), Charles Georges (Senior Vice-President, General Inspector, Internal Audit), Gaston Reinhard (Senior Vice-President, Human Resources), Jean Seywert (Senior Vice-President, Marketing). General Directorate Administration and Operations: François Steil (Senior Executive Vice-President and Executive Board Member, Organization and EDP), Jean-Pierre Werdel (Senior Vice-President, General Administration and Accounting), Raymond Loewen (First Vice-President, Securities Administration and Payments and Transfers). General Directorate Institutional Clientele and International Network: Jean-Donat Calmes (Senior Executive Vice-President and Executive Board Member, Management and Control of International Loans), François Moes (Senior Vice-President, Corporate Banking), Henri Wagener (First Vice-President, International Network), Fred Wagner (First Vice-President, Investment Funds). General Directorate Private Clientele and Domestic Network: Jean Krier (Senior Executive Vice-President and Executive Board Member, Management and Control of National Loans), Jean Seywert (Senior Vice-President, Domestic Network Middle Market companies), Claude Meiers (First Vice-President, Portfolio Managment Private Clientele). General Directorate Capital Markets and Financial Engineering: Olivier Michon (Senior Executive Vice-President and Executive Board Member),

Aloyse Reiff (Senior Vice-President, Treasury and Foreign Exchange), Fernand de Jamblinne de Meux (Senior Vice-President, Capital Markets Financial Engineering), Edmond Israel (Senior Executive Vice-President "Special Assignments", Marcel Schleder (Senior Executive Vice-President "Special Assignments")

PRINCIPAL ACTIVITIES: All banking and financial activities
Subsidiary Companies: (100% owned unless stated): Caisse Hypothecaire de Luxembourg (97.53%); PREFILUX SA (50%); BIL (Asia) Ltd; BIL Banque Internationale à Luxembourg Suisse SA; IB Finance SA (90%); BIL (France) (99.99%)
Financial Information:

	31.12.86	31.12.87
	FLux'000	FLux'000
Profit after tax	827,000	1,094,000
Dividends	472,000	607,000
Dividends per share net	380FLux	430FLux
Share capital	2,901,363	3,298,878
Customer deposits	235,350,000	288,010,000
Bank deposits	49,573,000	43,689,000
Total assets	320,210,000	369,887,000

No of Employees: 2,049

BAYERISCHE VEREINSBANK INTERNATIONAL SA

38-40 Avenue Monterey, PO Box 481, L-2014 Luxembourg
Tel: 45101-1; 45191
Cable: Vereinsbank
Telex: 2654 BVI Lu

Board of Directors: P Reimpell (President), A Puhlmann (Vice President)
Management Board: Dr K K Esswein (Managing Director) (Managing Director), Dr K Werhahn-Mees

PRINCIPAL ACTIVITIES: International bank
Parent Company: Bayerische Vereinsbank AG, Germany
Principal Bankers: Bank of Boston International, New York; Midland Bank PLC, London

No of Employees: 45

CAISSE D'EPARGNE DE L'ETAT DU GRANDE-DUCHÉ DE LUXEMBOURG

Banque de L'Etat

1 Place de Metz, L-2954 Luxembourg
Tel: 29851; 470401
Cable: Caissedepargne
Telex: 3417; 2287; 1535; 2839; 3456
Telefax: 27687; 491297; 491322; 492276

Board of Directors: Corneille Bruck (Président), Paul Koster (Vice Président), Armand Berchem, Robert Mangen, Yves Mersch, Antoine Weiss, Fernand Zurn
Management: Corneille Bruck (President-Directeur), Paul Koster (Sous-Directeur)

PRINCIPAL ACTIVITIES: All banking and financial activities
Financial Information:

	31.12.86	31.12.87
	FLux'000	FLux'000
Retained profit	646,100	726,124
Total assets	273,367,700	301,922,835

No of Employees: 1,384

COMMERZBANK INTERNATIONAL SA

11 Rue Notre Dame, L-2013 Luxembourg
Tel: 477911-1; 25411 (Foreign Exchange); 25934 (Foreign Exchange)
Cable: Commerzbank Luxembourg
Telex: 1292; 3462(Foreign Exchange); 2705 (Foreign Exchange); 1293 (Loans)
Telefax: 477911-270; Swift Address: cobadefxlux

Board of Directors: Dr W Seipp (President), Jacques Loesch (Vice-President), Martin Kohlhaussen, Jürgen Reimnitz, Dr Kurt Richolt, Alfred Knör (General Manager), Wolfgang Möller (Director-Manager), Klaus Tjaden (Director-Manager)
Management: Wolfgang Möller, Klaus Tjaden

PRINCIPAL ACTIVITIES: Banking services
Trade Names: Euromarket Transactions
Parent Company: Commerzbank AG, Germany
Principal Bankers: Commerzbank, New York; Commerzbank, London; Swiss Bank Corporation, Basel
Financial Information:

	31.12.87	31.12.88
	DM'000	DM'000
Profit for the year	45,000	45,000
Share capital	125,000	125,000
Deposits	15,500,106	15,958,815
Total assets	17,083,784	17,637,510

Principal Shareholders: Commerzbank AG (100%)
No of Employees: 124

DRESDNER BANK LUXEMBOURG SA (FORMERLY COMPAGNIE LUXEMBOURGEOISE DE LA DRESDNER BANK AG)

26 rue du Marché-aux-Herbes, PO Box 355, L-2013 Luxembourg
Tel: 47601 (General); 46 12 12 (Euro-Devises); 46 16 16 (Foreign Exchange)
Cable: Bankcompanie
Telex: 2558 drint lu (General); 2302 drifx lu (Euro-Devises); 2308 drifx lu (Foreign Exchange)
Swift-Code: CLBLLU LL

Supervisory Board: Jürgen Sarrazin (Chairman)
Management Board: Volker Burghagen (Managing Director), Wolfgang A Baertz (Managing Director)
Senior Executives: Norbert Hagedorn (Senior Manager, Loans and Syndications), Ernst Krause (Senior Manager, Trading and Chief Dealer), Bernd Ehinger (Senior Manager, Corporate Banking and Private Investor's Service), Dr H Peter Kaul (Assistant Manager Banking Relations), Rainer Bühler (Senior Legal Adviser)

PRINCIPAL ACTIVITIES: International banking (commercial and private)
Parent Company: Dresdner Bank AG, West Germany (100%)
Financial Information:

	31.12.86	31.12.87
	DM'000	DM'000
Share capital	160,000	160,000
Shareholders funds	1,752,500	1,869,600
Deposits	17,147,000	16,622,895
Total assets	19,103,300	18,839,883

No of Employees: 155

KREDIETBANK SA LUXEMBOURGEOISE

43 Bvd Royal, L-2955 Luxembourg
Tel: 47.97.1
Cable: Creditbank
Telex: 3418 kblux lu
Telefax: 47.26.67/282.67

Chairman: Jean L Blondeel
Executive Committee: Constant Franssens (President and Managing Director), Jean Adant (Executive Director), Damien Wigny (Executive Director), Lucien Pfeiffer (General Manager), Etienne Verwilghen (General Manager)
Management: Charles Ruppert (Financial Accounting and Data Processing Management), Charles Kerschenmeyer (Human Resources Management, Premises and Equipment), Jean-Marie Barthel (Operational Management), Marc Logelin (Money and Capital Markets), Wim Van Mulders (Corporate Finance and Syndication), Théo Kass (General Secretary), Raymond Pastoret (Credit Management), Léon Magnus (Financial Analysis and Information Department), Michel Sotil (Domestic Branches), Mathy Engelhardt (International Relations), M H Henry (Companies, Investment Funds, Financial Serivces, Private Banking)

PRINCIPAL ACTIVITIES: Banking, investment and financing
Parent Company: Associated bank: Kredietbank NV, Brussels, Belgium
Subsidiary Banks: KB International (Hong Kong) Ltd (Hong Kong); Kredietbank (Suisse) SA (Switzerland). Associated Bank: Brown Shipley Holdings Plc (UK)
Principal Bankers: All major banks
Financial Information:

	31.3.87	31.3.88
	FLux'000	FLux'000
Profit before tax	1,364,000	1,722,000
Profit after tax	829,000	1,122,000
Retained profit	386,000	535,000
Dividends	409,000	545,000
Dividends per share	608FLux	706FLux
Earnings per share net	1,233FLux	1,451FLux
Share capital	3,042,000	3,091,000
Shareholders funds	3,042,000	3,091,000
Deposits	219,873,000	245,571,000
Total assets	255,866,000	285,941,000

Principal Shareholders: A member of the Almanij-Kredietbank Group
No of Employees: 998

LUXEMBOURG

Major Companies of the NETHERLANDS

AEGON NV

Mariahoeveplein 50, PO Box 202, 2501 CE 's-Gravenhage
Tel: (31) 70-443210
Telex: 31657 aegn nl
Telefax: (31) 70-475238

Supervisory Board: O Vogelenzang (Chairman), J Bartels (Vice Chairman), J A Bakker, A W Dek, G Gerritse, J A P van Hoof, J A Houtzager, H J C ter Kuile, G van Schaik, W Scholten, Mrs S C Tóth, K Wetherell

Executive Board: J F M Peters (Chairman), E Oosterhoff (Vice Chairman), L J M Berndsen, F A A Stapels, K J Storm, H B van Wijk

Actuarial Adviser: A I M Kool. Secretaries to the Executive Board: R H Bolman

PRINCIPAL ACTIVITIES: Life insurance, accident and health insurance, other general insurance; holiday centres; property finance and construction

Subsidiary Companies: (100% owned unless stated): The Netherlands: AEGON Administratie BV; AEGON Bedrijfsfinanciering BV; AEGON Effectenbelegging BV; AEGON Hermes Beleggingen BV; AEGON International BV; AEGON Kantoren Holding BV; AEGON Levensverzekering NV; AEGON Nederland NV; AEGON Participatiemaatschappij BV; AEGON Praktijkfinanciering BV; AEGON Recreatie BV; AEGON Reinsurance NV; AEGON Schadeverzekering NV; AEGON Vastgoed Holding BV; AEGON Woning Holding BV; AEGON Woningfinanciering BV; Friesch-Groningsche Hypotheekbank NV (99.7%); Gemeenschappelijk Eigendom BV; Van Haften & Co NV; Merrem & la Porte Assurantiën BV; Nederlandse Verzekerings Groep NV; NV Maatschappij voor sparen, beleggen en verzekeren "Spaarbeleg"; BV "Tilburgia" Ontwikkelings-en Financieringsmaatshappij (99.5%). Other Countries: AEGON Allerlei Risico's NV (99.9%) (Belgium); AEGON Leven NV (Belgium); AEGON Ferienzentren GmbH (Germany); AEGON Insurance Company (UK) Ltd (UK); AEGON Union Levantina SA de Seguros y Reaseguros (99.7%) (Spain); Union Previsora SA (Spain); ENNIA Caribe NV (Curaçao); FGH Realty Credit Corporation (90%) (USA); N.E.N. Schadeverzekering NV (Surinam); Life Investors Inc (94.5%) (USA); Bankers United Life Assurance Company (USA); LeaseAmerica Corporation (USA); Life Investors Insurance Company of America (USA); NN Investors Life Insurance Company Inc (USA); Pacific Fidelity Life Insurance Company (USA); Monumental Corporation (USA); Monumental General Insurance Group Inc (USA); Monumental Life Insurance Group Inc (USA); National Old Line Insurance Company (USA); American Old Line Insurance Company (USA); Equity National Life Insurance Company (USA); AEGON Reinsurance Company of America (USA); AEGON US Holding Corporation (USA)

Financial Information:

	31.12.86 DFI'000	31.12.87 DFI'000
Premium income gross	4,941,881	5,522,434
Investment income	2,589,680	2,751,663
Gross income	401,721	412,380
Net income	327,127	337,505
Common dividends	45,110	135,575
Dividends per share	1.30DFI	3.75DFI
Net income per share	8.40DFI	8.67DFI
Share capital	219,295	223,605
Shareholders funds	2,708,050	2,081,808

No of Employees: 9,806

ALGEMENE-BANK NEDERLAND NV

ABN Bank

Vijzelstraat 32, PO Box 669, 1000 EG Amsterdam
Tel: (020) 299111
Cable: Genbank
Telex: 11417 abna nl
Wire SWIFT: ABNA NL 2A

Supervisory Board: D de Bruyne (Chairman), J D Hooglandt (Deputy Chairman), E K den Bakker (Deputy Chairman/Secretary), A H Heineken, E K den Bakker, S Orlandini, Prof. Dr. A C R Dreesmann, C T van der Molen, A G van den Bos, Prof. Dr. W Albeda, J J Kaptein, F A Maljers, Dr. A Batenburg, Mrs M Epema-Brugman, Prof. Dr. D J Wolfson, J Zantman, H B van Liemt, Dr G Bresser

Management Board: R Hazelhoff (Chairman), H Langman, P J Kalff, Prof. Dr. C J Oort, L D de Bièvre, P Ribourdouille, J H Geertsema, J M de Jong

Chief General Managers: G J Kalff, M Cram, W ten Berg, R P A Meijer, J J Oyevaar, Dr J J Kamp, W Brounts, R A Kleyn, Dr Th Beels, W van Driel

PRINCIPAL ACTIVITIES: Banking activities

Trade Names: ABN Bank

Subsidiary Companies: (100% owned unless stated): HOLLANDSCHE BANK-UNIE NV; BANK MEES & HOPE NV and its subsidiaries:- Theodoor Gilissen NV; Bank voor Zeeland NV; H Wesselius & Co BV (90%); Trustmaatschappij Mees & Hope BV. ABN ASSURANTIE HOLDING BV Holding company of: R Mees & Zoonen Assurantiën BV; ABV Assurantiën BV. OTHER AFFILIATED INSTITUTIONS:- ABN Auto Lease BV; ABN Hypotheken BV; ABN Onroerend Goed BV; ABN Onroerend Goed Lease BV; ABN Onroerend Goed Financieringen BV; ABN Trustcompany (Nederland) BV; Albank BV; Algemeen Fonds Midden-en Kleinbedrijf BV; Algemene Lease Maatschappij BV; Algemene Participatiemaatschappij BV; Algemene Particuliere Participatie Maatschappij BV; Algemene Particuliere Participatie Maatschappij II BV; Alleasehold BV; F Berger & Co BV; Bogamij Beheer BV (60%); Bogamij Management BV; D W Brand NV; BV Centrale voor Termijnfinancieringen CTF; NV Centrale Trust Compagnie; "CMV" Crediet Maatschappij Venlo NV, Steegh Asurantiën BV; BV De Factorij Factorbedrijf van de Algemene Bank Nederland; Gres Holding BV; HMR Holding BV; BV Hollandsche Belegging en Beheer Maatschappij; IDM Bank NV; Incasso Bureau Amersfoort BV; International Option Investment BV; Maatschappij voor Kapitaalgoederen BV; Nachenius & Dudok van Heel NV; De Wed Tjeenk & Co NV; De Twentsche Trust-Maatschappij NV (90%). Major subsidiaries and other investments outside the Netherlands, as well as Foreign branches of ABN, HBU and BMH: EUROPE: Austria: Algemene Bank Nederland (Österreich) AG. Belgium:- Algemene Bank Nederland (België) NV; ABN International Diamond Division; ABN Belgian Branch, Brussels; NV Kempische Leasing. Channel Islands, Jersey:- ABN Trustcompany (Jersey) Ltd. Denmark:- Algemene Bank Nederland; Alicon Børsmaeglerselskab A/S. Federal Republic of Germany:-Algemene Bank Nederland (Deutschland) AG. France:- de Neuflize, Schlumberger, Mallet SA (80%); Banque de Neuflize, Schlumberger, Mallet (71%); Caisse Industrielle de Crédit Automobile SA-CICA-; Société de Location Automobile et Matériel SA-SOLOCAM-(97.5%); Jacques François-Dufour, Jean-Louis Kervern SA (17%). Gibraltar:- Algemene Bank Gibraltar Ltd (68.9%). UK:- Mees & Hope Securities Ltd; Mees & Hope Corporate Finance Ltd; Mees & Hope Investment Management Ltd; ABN Securities (UK) Ltd. Greece:-Algemene Bank Nederland. Eire:- Algemene Bank Nederland (Ireland) Ltd; ABN Ireland Financial Services Group; Riada & Co (51%). Italy:- ABN Leasing Italia SpA; ABN Finanziaria SpA. Liechtenstein: ABN Management Services AG. Luxembourg:- ABN (Luxembourg) SA; ABN Trustcompany (Luxembourg) SA. Spain:-Algemene Bank Nederland; ABN Leasing España SA. Sweden:- ABN (Sverige) AB. Switzerland:- Algemene Bank Nederland (Schweiz) AG; Monaval Finanz AG; ABN Trustcompany (Suisse) SA; Mees & Hope Finanzgesellschaft AG; Mees & Hope Management AG. Turkey:- Hollandsche Bank Unie. AFRICA: Kenya:- Algemene Bank Nederland. Morocco:- Algemene Bank Marokko SA (50%). MIDDLE EAST: Saudi Arabia:- Albank Alsaudi Alhollandi (40%). Bahrain:- Algemene Bank Nederland. Lebanon:- Algemene Bank Nederland. United Arab Emirates:- Algemene Bank Nederland. ASIA: Hong Kong:- ABN Capital

Markets Far East Ltd; Hong Kong (BMH) Representative Office; ABN Finance Ltd; ABN Management Services (Hong Kong) Ltd; General Software Ltd (GSL). India:- Algemene Bank Nederland. Indonesia:- Algemene Bank Nederland; P.T. ABN Tungal Leasing Indonesia; P.T.Peranas Agung. Japan:- Algemene Bank Nederland. Korea:- Algemene Bank Nederland. Malaysia:- Algemene Bank Nederland. Pakistan:- Algemene Bank Nederland. People's Republic of China:- Representative Office Beijing; Singapore:- Algemene Bank Nederland; R Mees & Zoonen Assurianten BV. Sri Lanka:- Algemene Bank Nederland. Taiwan:- HBU. AUSTRALIA: ABN Australia Ltd; ABN Representative Office for Australia and New Zealand, Sydney. NORTH AMERICA: Canada:- ABN Bank Canada; ABN Canada Leasing Ltd; Canadian Representative Office; Algemene Bank Nederland NV; ABN Capital Markets Corporation. USA:- R Mees & Zoonen Assurantiën BV; ABN/La Salle North America Inc; ABN Bank International USA Inc; ABN Capital Markets Corporation; ABN Representative Office for Colombia, Mexico and Venezuela, Miami; LaSalle National Corporation; ABN Representative Office International Trust Affairs, New York; LaSalle Community Bancorporation Inc, Northbrook; LaSalle Northwest National Bank, Chicago; LaSalle Bank Lake View, Chicago; LaSalle Bank Northbrook; LaSalle Bank Westmont; LaSalle National Bank; LaSalle Bank of Lisle. CARIBBEAN: Aruba: ABN Trustcompany (Aruba) NV. Cayman Islands:- LaSalle National Bank. Virgin Islands:- ABN Trustcompany (Virgin Islands) Inc. NETHERLANDS ANTILLES: Curaçao: ABN Trustcompany (Curaçao) NV; Algemene Hypotheek Bank NV; Nederlandsche Handel-Maatschappij Trustkantoor Curaçao NV; Fides (Curaçao) NV. Sint Maarten:- ABN Trustcompany (St Maarten) Inc; Fides (St Maarten) NV. CENTRAL AND SOUTH AMERICA: Argentina:- Banco Hollando Unido; ABN Representative Office for Argentina, Paraguay, Uruguay and Chile, Buenos Aires; R Mees & Zoonen Assurantiën BV. Brazil:- Banco Hollando Unido; ABN Bank Representação Sociedade Civil Ltda; ABN Brazil Investment Fund SA; Aymoré-group consisting of: Companhia de Intercambio e Participaçoes "Interpar" SA; Banco Aymoré de Investimento SA; Companhia Aymoré de Crédito, Investimentos e Financiamentos; Aymoré Distribuidora de Títulos e Valôres Mobiliários SA; Aymoré Empreendimentos e Promoções SA; Credicenter Empreendimentos e Promoções Ltda; Cruzeiro Factoring Sociedade de Fomento Comercial Ltda; Aymoré Metais Ltda. Ecuador:- Banco Hollando Unido; Facturinsa, Facturas Internacionales SA (40%). Panama:- Algemene Bank Nederland. Paraguay:- Banco Holandis Unido. Surinam:- De Surinaamsche Bank NV (49%); Financieringsmaatschappij Paramaribo NV (49%); Surinaamse Trustmaatschappij NV; Surinaamse Administratie Maatschappij "Suramij" NV; Surinaamse Financierings Maatschappij "Surfimij" NV; Surinaamse Computer Maatschappij NV. Uruguay:- Banco Holandis Unido

Financial Information:

	31.12.86 DFl'000	31.12.87 DFl'000
Profit before tax	785,000	735,000
Profit after tax	527,000	517,000
Dividends	249,000	271,000
Dividends per ordinary share	2.7DFl	2.7DFl(1)
Dividends per preference share	-	12.00DFl
Earnings per share	56.74DFl	5.01DFl(2)
Share capital	977,000	1,101,000
Shareholders equity	5,143,000	5,845,000
Deposits	33,192,000	32,844,000
Total assets	146,663,000	150,856,000

(1) per share of DFl10 nominal value

(2) adjusted after share split up to DFl10 nominal value

No of Employees: 30,580

AMEV, NV

Archimedeslaan 10, PO Box 2072, 3500 HB Utrecht

Tel: (30) 579111

Telex: 47763

Telefax: (30) 522394

Supervisory Board: C C Nengermann (Chairman), A W Overwater (Vice Chairman), H J E van Beunigen, Ch O Eggink, J Kraaijeveld van Hemert, C J Lammers, F Roos, G H Veringa

Executive Board: J L M Bartelds (Chairman), jhr P J F van der Does de Willebois (Vice Chairman), F A J Smit, W J Nanninga, P M L Rosenberg

PRINCIPAL ACTIVITIES: Life assurance, general insurance and other financial services

Subsidiary Companies: (100% owned unless stated): Life Assurance: AMEV Levensverzekering NV; AMEV Vastgoed Maatschappij NV; AMEV Effectenbeleggingsmaatschappij NV; NV Levensverzekering Maatschappij HAV-Bank; NEFO NV (99.93%); AMEV-Ardanta NV; AMEV Beleggingsconsortium, maatschappij voor beleggen en verzekeren NV; Belgische Levensverzekering Maatschappij "Utrecht" NV (Belgium); l'Etoile 1905 SA (Belgium); Livsforsikrings-Aktieselskabet Utrecht (Denmark); Compagnie Française d'Assurances Vie UTRECHT ROYALE BELGE SA (50%) (France); Gresham Life Assurance Society Ltd (UK); Gresham Unit Assurance Ltd (UK); BILBAO Compañía Anónima de Seguros (76.51%) (Spain); Time Insurance Company (USA); Western Life/St Paul Life Insurance Company (USA); Union Security Life Insurance Company (USA); United Family Life Insurance Company (USA); Amev Life Assurance Company Ltd (Australia); Ka Wah AMEV Insurance Ltd (90%) (Hong Kong); MMGU Insurance Pte Ltd (40%) (Singapore). Non-life Insurance: AMEV Schadeverzekering NV; Belgische Verzekering-Maatschappij "Utrecht" Allerlei Risico's NV (Belgium); l'Etoile 1905 SA (Belgium); Compagnie Française d'Assurances IARD ROYALE BELGE UTRECHT SA (50%) (France); Bishopsgate Insurance Ltd (UK); Leadenhall Insurance Ltd (UK); AMEV General Insurance Company Ltd (Eire); BILBAO Compañía Anónima de Seguros (76.51%) (Spain); Time Insurance Company (USA); Western Life/St Paul Life Insurance Company (USA); American Security Insurance Company (USA); Union Security Life Insurance Company (USA); Standard Guaranty Insurance Company (USA); Superior Insurance Company/The Standard Plan Inc (USA); AMEV Insurance Holdings Ltd (Australia); Ka Wah AMEV Insurance Ltd (90%) (Hong Kong); MMGU Insurance Pte Ltd (40%) (Singapore). Other Activities: Finance: AMEV Praktijkvoorziening NV; AMEV Financieringen NV; AMEV Hypotheekbedrijf NV; Friesch-Hollandsche Hypotheekbank NV (99.87%); Kredietkas "Utrecht" NV (Belgium); Gresham Mortgage Ltd (UK); AMEV Finance Ltd (Australia). Property Development: AMEV Ontwikkeling Maatschappij NV; NV AMEV-IMO (Belgium); AMEV-CLAMART SA (France); Amev Investments Pty Ltd (Australia); Longtown Investments Pty Ltd (Australia). Other: NV AMEV Financiën; NV Brabant, maatschappij tot beheer en onwikkeling van deelnemingen; NV Beleggingsmaatschappij "ERK"; NV Domstad; V.O.F. AMEV PPM; De Grenswisselkantoren NV (40%); Kuiken NV (21.84%); Singelgracht BV (80%); Gronilux Holding BV (40%); Gebroeders Bakker BV (45.8%); Gresham Unit Trust Managers Ltd (UK); AMEV Suisse SA (Switzerland); AMEV Venture Management Inc (USA); AMEV Capital Corporation (USA); General Furniture Leasing Company (USA); AMEV Advisers Inc (USA); AMEV Asset Management Inc (USA); Amev Equity Pty Ltd (Australia). Holding Companies: AMEV Nederland NV; AMEV International NV; AMEV België NV; AMEV Far East NV; Société d'Investissements et de Participations SA (50%) (France); AMEV (UK) Ltd (UK); AMEV Holdings Inc (USA); Amev Australia Ltd (Australia); AMEV South East Asia Ltd (Hong Kong)

Financial Information:

	31.12.86	31.12.87
	DFI'000	DFI'000
Turnover	7,909,687	7,639,352
Premium income	5,756,277	5,449,497
Investment income(1)	1,790,699	1,695,267
Income-other activities	526,168	576,165
Profit before tax	433,054	339,597
Profit after tax(2)	322,523	291,695
Retained profit	185,070	151,998
Dividends	137,453	139,697
Dividends per share(3)	2.55DFI	2.55DFI
Earnings per share(3)	6.00DFI	5.35DFI
Share capital	150,049	152,276
Shareholders funds	2,400,430	2,217,468

(1) Insurance sector

(2) Including extraordinary result

(3) Figures for previous years have been adjusted to take account of bonus and rights issues

No of Employees: 11,147

AMSTERDAM ROTTERDAM BANK NV
AMRO Bank

Foppingadreef 22, PO Box 283, NL-1000 EA Amsterdam

Tel: (020) 289393

Cable: Amrobank

Telex: 11006 amsterdam nl

Supervisory Board: H F van den Hoven (Chairman), F H Fentener van Vlissingen (Vice-Chairman), W H Brouwer, W Dekker, J J J van Dijck, J J Endtz, A Heijn, Mrs M J 't Hooft, X Malou, A G Maris, B E Ruys, G van Schaik, B Scheepmaker, E de Villegas de Clercamp, O Vogelenzang, E P Wellenstein, E G G Werner

Board of Managing Directors: R J Nelissen (Chairman), F Hoogendijk, H H Foppe, M J Drabbe, Th A J Meys, R van Ommeren, R W J Groenink, R W F van Tets. Adviser to the Board of Managing Directors: H J Witteveen. Secretary to the Board of Managing Directors: P G M van Lierop

Senior Executive Vice Presidents: A J Deknatel, G D Dirks, H Heemskerk, Z O H M Baron van Hövell tot Westerflier, W G Jiskoot, J L Rijnja, W P M van der Schoot

PRINCIPAL ACTIVITIES: A wide range of banking activities

Subsidiary Companies: Equity participations consolidated: (100% owned unless stated): Amro Australia Ltd (Australia); Amro Bank (Asia) Ltd (Singapore); Amro Bank und Finanz AG (Switzerland); Amro Bank (Luxembourg) SA (87%) (Luxembourg); Amro Bank Overseas NV (Netherlands Antilles); Amro Effectenbewaarbedrijf NV; Amro (Finance and Securities) Ltd (Hong Kong); Amro Handelsbank AG (99.6%) (Germany); Amro International (Asia) Ltd (Japan); Amro Lease NV; Amstel Lease NV; Auto Lease Holland BV; Beheer- en Administratiemaatschappij Lafico BV; Bicker Caarten & Obreen NV; NV Consultants & Assistants "Consultass"; EBC Amro Bank Ltd (UK); NV Export-Financiering Maatschappij; Finata Bank NV; Firenta BV; Hondius en Zoon Assurantiën BV; International Factors Nederland BV; Maatschappij tot Financiering van Bedrijfspanden NV; Nationale Bank voor Middellang Krediet NV including: Nationale Maatschappij voor Vliegtuigfinanciering BV; Nationale Trust Maatschappij NV; Particuliere Participatie Maatschappij Amro BV; Pierson, Heldring & Pierson NV including: Pierson, Heldring & Pierson (Curaçao) NV (Netherlands Antilles), as well as the equity participations mentioned in that company's annual report; BV Projektontwikkelingsmaatschappij Amro; Rijnlandse Disconto Bank CV; Top Lease BV. Equity participations not consolidated: (The percentage held is not given where it is less than 10%): Allgemeine Deutsche Investment Gesellschaft mbH (Germany); Amsterdam Depositary Company NV (50%); Banco Bradesco de Investimento SA (Brazil); Banque Européenne pour l'Amérique Latine (B.E.A.L.) SA (45.4%) (Belgium); Banque Européenne de Credit SA (14.3%) (Belgium); Edesa SA (Luxembourg); European American Bancorp (20%) (USA) holding company of: European American Bank & Trust Co (New York) and European American Banking Corporation (New York). European Arab Holding SA (Luxembourg); Finanzierungsgesellschaft Viking (12%) (Switzerland); Holland Venture Holdings CV (40%); Hudig-Langeveldt Groep BV (20%); The International Investment Corporation for Yugoslavia SA (Luxembourg); Maatschappij voor Industriële Projecten NV; Massonaud Fontenay SA (30%); De Nationale Investeringsbank NV; Nederlandsche Credietverzekering Maatschappij NV (29%); Nederlandse Financierings-Maatschappij voor Ontwikkelingslanden NV; Nederlandse Participatie Maatschappij NV; Société Internationale Financière pour les Investissements et le Développement en Afrique (Switzerland); Swiss American Securities Inc (12.5%) (USA); Tuba Holding International BV (50%)

Financial Information:

	31.12.87	31.12.88
	DFI'000	DFI'000
Profit before tax	668,000	813,000
Profit after tax	479,000	587,000
Retained profit	251,066	300,055
Dividends	227,731	286,723
Dividends per share	4.60DFI	5.20DFI
Earnings per share	9.67DFI	10.64DFI
Share capital	1,001,000	1,163,000
Shareholders equity	5,107,000	5,902,000
Deposits	132,460,000	151,064,000
Total assets	143,930,000	168,102,000

No of Employees: 23,198

ASSURANTIECONCERN STAD ROTTERDAM ANNO 1720 NV

Blaak 333, NL-3011 GB Rotterdam

Tel: (010) 4017200

Telex: 23266

Telefax: (010) 4125490

Supervisory Board: J de Groot, P J van der Giessen, Chr Karelse, C J A van Lede, D G Postma

Management Board: Prof Dr L M van Leeuwen, Mr C J de Swart

PRINCIPAL ACTIVITIES: Non-life insurance, life insurance and mortgage business

Subsidiary Companies: NON-LIFE INSURANCE: NV Maatschappij van Assurantie, Discontering en Beleening der Stad Rotterdam anno 1720; NV Verzekeringmaatschappij Woudsend anno 1816; NV Verzekeringmaatschappij Koning & Boeke; Assurantie Maatschappij "de Nieuwe Prudentia" NV. LIFE INSURANCE: Levensverzekeringmaatschappij stad Rotterdam NV; Levensverzekeringmaatschappij de Waerdye NV; De Verzekerings Unie Levensverzekeringmaatschappij NV; Levensverzekeringmaatschappij Woudsend NV; Belstar Verzekeringen NV. OTHER OPERATIONS: NV Bouwkas Rohyp; Hoofdpoort Hypotheken BV; Beleggingsmaatschappij de Hoofdpoort NV; NV Beleggingsmaatschappij van Koning & Boeke; Woningbouw- en Woningexploitatiemaatschappij de Hoofdpoort NV; Woningbouw- en Woningexploitatiemaatschappij stad Rotterdam NV

Principal Bankers: Algemene Bank Nederland NV; Amsterdam-Rotterdam Bank NV; Nederlandsche Middenstandsbank NV

NETHERLANDS

Financial Information:

	1986	1987
	DFI'000	DFI'000
Premium income	1,167,585	1,283,530
Investment income	372,802	391,857
Profit before tax	75,657	82,956
Profit after tax	56,424	60,877
Retained profit	25,213	27,903
Dividend	26,933	28,206
Dividend per share	5.90DFI	6.05DFI
Earnings per share	11.52DFI	12.08DFI
Share capital	49,796	45,865
Shareholders funds	417,595	432,515

No of Employees: 1,100

BANK MEES & HOPE NV

Herengracht 548, Postbus 293, NL-1000 AG Amsterdam
Tel: (020) 527 91 11
Cable: meesbank amsterdam
Telex: 11 424 bmha nl

Board of Managing Directors: Drs J Schrok

PRINCIPAL ACTIVITIES: Banking activities
Parent Company: Algemene Bank Nederland NV
Associated Companies: Theodor Gilissen NV; H Wesselius & Co
  BV; Bank voor Zeeland NV; Mees & Hope Management AG
  (Switzerland); Mees & Hope Finanzgesellschaft AG
  (Switzerland)

No of Employees: 2,100

BARY, H ALBERT DE, & CO NV

448-456 Herengracht, 1017 CA Amsterdam
Tel: (020) 5554911
Cable: Barybank
Telex: 12029 bary nl
Telefax: (020) 5554 428

Supervisory Board: H Kopper (Chairman), F Hoogendijk (Deputy
  Chairman), M J Drabbe, P K R von Goerschen, R Habicht, B
  Henny, D G Postma, J Puhl, P-R Wagner
Managing Directors: B A Wilken (Senior Managing Director), W
  O Koenigs (Senior Managing Director), A F Baron van
  Lijnden, E Stein
Managers: P Dubbeld, W Holst, K de Vries. Divisional
  Managers: A W den Hartog, P R Jaeger, H J Jansen, R F
  Reinders Folmer, N W Sickler, R P Visser, F Y M van Zinnicq
  Bergmann

PRINCIPAL ACTIVITIES: Banking activities
Principal Bankers: Montagu & Co, London; Citibank NA, New
  York; Deutsche Bank AG, Frankfurt/Main
Financial Information:

	30.9.86	30.9.87
	DFI'000	DFI'000
Profit before tax	15,031	11,164
Profit after tax	11,431	10,164
Retained profit	11,431	10,164
Dividends	3,750	4,500
Dividends per share	15%	15%
Earnings per share	15%	15%
Share capital	30,000	30,000
Shareholders funds	109,779	117,661
Deposits	3,309,901	3,163,582
Total assets	3,581,765	3,497,858

Principal Shareholders: Amsterdam-Rotterdam Bank NV,
  Amsterdam (50%); Deutsche Bank AG; Frankfurt/Main (50%)
No of Employees: 390

BOUWFONDS NEDERLANDSE GEMEENTEN NV

Westerdorpstraat 66-68, 3871 AZ Hoevelaken
Tel: (03495) 39111
Telex: 79244 bouwfonds
Telefax: (03495) 39555

Supervisory Board: C M L Roozemond (Chairman), W Griffioen
  (Deputy Chairman), J Visser, Mevr Drs R M ten Cate-Dhont, B
  W Cazemier, Drs J P A Gruijters, J W Masman, Drs P H M de
  Roij, H L M Savelsbergh, A H M Timmermans, Drs M J H van
  de Ven, P J R L Verheggen, A H M van den Wildenberg, H J
  L Wouters, Drs J van der Zwaag
Board of Executives: Drs H J Viersen (Chairman), Drs W
  Dijksterhuis, B P L Stal, Drs J W M Simons
Senior Executives: A P Schilham (Secretary)

PRINCIPAL ACTIVITIES: Financial interests including mortgages
Subsidiary Companies: (100% owned unless stated):
  Bouwfonds Woningbouw NV; NV Bouwfonds Adviesgroep;
  NV Grondbedrijf van de NV Bouwfonds Nederlandse
  Gemeenten; NV Ontwikkelingsmaatschappij Monnickendam-
  Poelwijk; NV Hypothecair Krediet- en Bouwrisicoverzekering
  Maatschappij; NV Bemiddeland Orgaan van de NV
  Bouwfonds Nederlandse Gemeenten; NV Woningesploitatie
  Maatschappij Interim; Boufonds Woningbeheer NV; Combi-
  hypotheken NV; NV Bouwfonds Informatica; Data Process
  Informatica NV; Automatiseringsfonds Nederlandse
  Gemeenten NV; L&T Informatica BV; Instituut voor
  Professionele Computer Opleidingen (IPCO) BV; NV
  Levensverzekering Maatschappij Ubo (66.5%); 't Hooge Huys
  Verzekeringen NV (66%); BV Park Stokhorst (50%); Alkmare
  BV(50%); Parkstede BV (34%); Parkstede CV (34%)
Principal Bankers: NV Bank voor Nederlandsche Gemeenten
  te's-Gravenhage; Amsterdam-Rotterdam Bank NV;
  Nederlandsche Middenstandsbank NV; Algemene Bank
  Nederland NV; Coöperative Centrale Raiffeissen-
  Boerenleenbank BA; Nederlandse Credietbank NV; F van
  Lanschot Bankiers NV; Postbank NV
Financial Information:

	31.12.86	31.12.87
	DF1'000	DFI'000
Turnover	326,936	341,497
Profit before tax	64,976	71,865
Profit after tax	47,822	48,935
Dividend	9%	9%
Share capital	6,847	7,987
Shareholders funds	392,310	433,677
Total assets	8,816,363	9,240,836

Principal Shareholders: Exclusively Dutch Municipalities (86% of
  all Municipalities in the Netherlands are shareholders in the
  company)
No of Employees: 668

CAR FINANCE INTERNATIONAL BV

Steenovenweg 1, PO Box 329, 5700 AH Helmond
Tel: 0031-4920-62421
Cable: CFHM
Telex: 51101 CFHM
Telefax: 0031-4920-36995

Supervisory Board: Ir A H C Deleye (Chairman), Drs H Th M
  Bevers, Ir F W L R J M Sevenstern, Drs J H Goris, L
  Schouten, M L J M Maassen (Secretary to the Board)
Management: F H Monster (General Manager), L J J Rijkers
Senior Executives: P J G M Litjens (Accounts Manager), T M
  Rutten (Controller), Jhr Drs E van Rijckevorsel (Commercial
  Manager), H A B Smits (Manager, Carned Lease BV)

PRINCIPAL ACTIVITIES: Credits; foreign exchange;
  participations
Parent Company: Volvo Car BV
Subsidiary Companies: Car Industry Insurance BV; Carned
  Lease BV; Car Finance Belgium NV

Principal Bankers: F van Lanschot Bankiers; Banque de Suez; Rabobank; Crédit Lyonnais; Trinkhaus & Burkhardt

Financial Information:

	31.12.86	31.12.87
	DFl'000	DFl'000
Sales turnover	2,345,466	2,304,194
Profir before tax	4,116	7,374
Profit after tax	4,116	7,374
Retained profit	4,116	7,374
Earnings per share	412DFl	737DFl
Share capital	10,000	10,000
Shareholders funds	25,188	32,547
Total assets	368,760	385,618

No of Employees: 33

CENTRAAL BEHEER U.A.

Prins Willem-Alexanderlaan 651, Postbus 700, NL-7300 HC Apeldoorn

Tel: (055) 79 91 11

Telex: 36366 cbapd nl

Management Board: Jhr G F Boreel (Chairman)

PRINCIPAL ACTIVITIES: Insurance including life; reinsurance

Principal Bankers: ABN; Amro; NMB; Rabo

Financial Information:

	DFl'000
Premium income	1,150,000

No of Employees: 1,980

CREDIT LYONNAIS BANK NEDERLAND NV

PO Box 1045, 3000 BA Rotterdam

Tel: (010) 4695911

Cable: Credionais

Telex: 21366 clbn nl

Telefax: (010) 4148391

Board of Directors: G H van Driel (Chairman)

Executive Board: J J Brutschi (Chairman Executive Board), H S Gonggrijp, M L A Anfrie, A C J M Nollen, J M Griffault

Senior Executives: D Rackwitsz (Head Directorate Personnel), T W R Doorman (Head General Secretariat), C J Frohn (Head Internal Audit Department), H A R Röell (Head Legal Department), C J M R Roovers (Head Directorate Automation and Organisation), W C vd Weg (Head Directorate Payments), B H H Roelvink (Head Directorate Corporate Banking and Marketing), L L v d Woerd (Head Directorate Domestic Network), H J G Wyers (Head Directorate Corporate Affairs), B F Armstrong (Head Directorate General Accounting, Planning and Control), E L A Van Emden (Head Directorate Liquidities), R Straus (Head Staff Unit "Special Affairs"), M Rahusen (Head Directorate Foreign Affairs), J v d Laar (Head Directorate Credits)

PRINCIPAL ACTIVITIES: Bank activities

Parent Company: Crédit Lyonnais SA (94.67%)

Major Subsidiary Companies: Netherlands: Avista Bank NV; NV Maatschappij tot Exploitatie van Onroerende Goederen "Fortuna"; Lentjes & Drossaerts NV; CLN Oyens & Van Eeghen NV; CLN Oyens Trust BV; NV Sallandsche Bank; CLN Assurantiën BV. USA: Bankhaus Wölbern & Co; The Slavenburg Corporation, USA. Netherlands Antilles: CLN Informatica BV; Nederlandse Credietbank NV; CLBN Real Estate & Finance Co NV

Principal Bankers: All Principal Banks

Financial Information:

	31.12.86	31.12.87
	DFl'000	DFl'000
Gross result	81,217	100,146
Allocation to VAR	77,442	98,273
Profit before tax	3,775	1,873
Profit after tax	139	-
Share capital	349,427	599,427
Total risk capital	535,381	1,162,584
Deposits	14,735,094	23,085,148
Total assets	15,270,475	24,247,732

Principal Shareholders: Crédit Lyonnais SA (94.67%)

No of Employees: 3,953 plus 241

DE NEDERLANDSCHE BANK NV

See NEDERLANDSCHE, DE, BANK NV

DELTA-LLOYD VERZEKERINGSGROEP NV

Spaklerweg 4, NL-1096 BA Amsterdam

Tel: (020) 5949111

Telex: 18678 delta nl

Telefax: (020) 93 79 68

Management Board: Dr G Zoutendijk (Chairman), G Walstra (Deputy Chairman), A A Anbeek van der Meijden, G W Baron van der Feltz

PRINCIPAL ACTIVITIES: Insurance, life and non-life

Trade Names: Delta Lloyd

Parent Company: Commercial Union Assurance Company plc, London, UK

Subsidiary Companies: Delta Lloyd Levensverzekering N.V.; Delta Lloyd Schadeverzekering N.V.; N.V. Noord Braband; N.V. Nationaal Spaarfonds

Principal Bankers: Amsterdam-Rotterdam Bank NV; Algemene Bank Nederland NV; Nederlandsche Middenstandsbank NV

Financial Information:

	DFl'000
Revenues	2,702,000

No of Employees: 3,000

F. VAN LANSCHOT BANKIERS NV

See LANSCHOT, F. VAN, BANKIERS NV

FRIESLAND BANK

Zuiderstraat 1, 8911 BN Leeuwarden

Tel: (058) 99 44 99

Cable: fbank

Telex: 46120

Telefax: (058) 99 45 91

Board of Directors: H Westenbrink (Chairman), G van Veenen, G de Jong, H J Bierma, P Miedema

Supervisory Board: R Posthumus (Chairman), L J Hilarides, J D Bierma, H Politiek, H Kroes

Board of Management: J R Kuperus (President), H Nijman (General Manager), K Hornstra Moedt (Deputy General Manager), T H de Boer (Deputy General Manager)

PRINCIPAL ACTIVITIES: International bank

Subsidiary Companies: Fries Trustkantoor bv; Friesland Bank Assurantiën bv; BV Volmachtbedrijf FBA/JWM; Passage- en Reisebureau J W van der Meulen bv; Makelaardij Friesland bv; Makelaardij Groningen bv; Makelaardij Drenthe bv; FGD Beheer bv

Financial Information:

	31.12.86 DFl'000	31.12.87 DFl'000
Operating revenue	81,230	88,309
Profit before tax	16,632	20,190
Profit after tax	12,560	15,148
Retained profit	12,560	15,148
Share capital	40,525	40,536
Shareholders funds	127,691	146,813
Guarantee funds	151,598	169,287
Deposits	1,617,654	1,711,152
Total assets	2,639,017	2,894,452

No of Employees: 560

H ALBERT DE BARY & CO NV
See BARY, H ALBERT DE, & CO NV

HAGEMEYER N.V.
Rijksweg 69, PO Box 5111, 1410 AC Naarden
Tel: (02159) 57611
Cable: Dihagem Naarden
Telex: 43997 dihag nl
Telefax: 02159-47850

Supervisory Board: F O J Sickinghe (Chairman), E J van der Hagen (Vice Chairman), P A W Roef, A Batenburg, O Vogelenzang, R J Higgerson, A Salim, M V Pangilinan
Management Board: A H Land (Chairman), J J Hollenkamp (Vice Chairman), Ph A Betsch
Executive Staff: Directors: A A J van der Kleij, A A M Lauxtermann, J D Heyning, F L Dickt, J M Howgate. Division and Group Managing Directors:- H J J Birner, H W N A Jansen, M H R Corke, M P M Castenmiller, W J Imhoff, R P Proctor, A A B Teves (Corporate Secretary)

PRINCIPAL ACTIVITIES: General importing, exporting and trading; interests include housewares, audio/video, photo/repro/optics, clocks, watches, stationery, food, liquor, cosmetics, toiletries, leisure goods, do-it-yourself articles, non consumer technical and chemical products, luxury cars and trucks
Subsidiary Companies: (100% owned unless stated):- EUROPE: TRADING COMPANIES: Netherlands: BV Amsterdamse Rijtuigmaatschappij (A.R.M.) (99.72%); Comforta Nieuwkoop BV; Comforta Shoes BV; Haagtechno BV; Hadee BV; Kalorik Nederland BV; Nedtronics BV; Carel Teulings BV (49%). Belium: Artsel NV; Kalorik SA; Beltronics SA. England: Hagemeyer (UK) Ltd. Germany: TEC GmbH. RETAIL COMPANY: Netherlands: Allwave BV. INDUSTRIAL COMPANIES: Netherlands: Bobé BV; Koninklijke Verenigde Leder BV. MISCELLANEOUS: Netherlands: Worldport BV; Haagexport BV; Hagemeyer Services BV; Hagemeyer Executive Services BV; Hagemeyer Vast Goed BV; Hagemeyer Electronics Holding BV; Hagemeyer Assurantiën BV; Servicom BV. AMERICA: USA: Consumer Products Holdings Inc (81.5%); GPX Inc; Hagemeyer Foods Inc; Liberty Imports Inc; Passport Foods Outlet Inc. Canada: Hagemeyer Canada Inc. Food Divisions: Van's International; David Ashley; Koffman Foods. Electronics Division: Lloyd's Electronics. Netherlands Antilles: Hagemeyer Caribbean Holding NV; Hagemeyer (Curaçao) NV; Hagemeyer (Sint Maarten) NV. Aruba: Hagemeyer (Aruba) NV; Martijn Trading Company NV. Surinam: Hagemeyer (Suriname) NV. ASIA/PACIFIC: Singapore: Hagemeyer (Singapore) Pte Ltd; Hagemeyer Electronics (Singapore) Pte Ltd; Hagemeyer Marketing Services Pte Ltd; Hagemeyer Sales & Services (1986); Hagemeyer Marketing (1986); Océ-Hagemeyer (Singapore) Pte Ltd (49%). Malaysia: Hagemeyer (Holdings) Sdn Bhd; Hagemeyer (Malaysia) Sdn Bhd; Hagemeyer Marketing Services Sdn Bhd; Hagemeyer Sales & Services; Hagemeyer Equipment; Hagemeyer Marketing; Sirah Sdn Bhd (70%); Hagemeyer Industries (M) Sdn Bhd (70%). Thailand: Hagemeyer (Thailand) Ltd; Alpina (Thailand) Ltd (25%). Hong Kong: Hagemeyer-Morison (Hong Kong) Ltd; European

Integrated Fashion Ltd; French Perfumes (Asia) Ltd; Océ-Hagemeyer (Hong Kong) Ltd (49%). Japan: Hagemeyer Export & Import Co Ltd. Taiwan: Hagemeyer (Taiwan) Co Ltd. Indonesia: PT Intraport Raya. Australia: Hagemeyer (Australasia) BV; Haco Pty Ltd; Hagemeyer Investment Co (Australia) Pty Ltd. New Zealand: Briscoes (New Zealand) Ltd. AFRICA: Nigeria: Hagemeyer (Nigeria) Ltd (40%); Colourprint Laboratories Ltd (24%). Kenya: Haco Industries Kenya Ltd (49%)

Financial Information:

	31.12.86 DFl'000	31.12.87 DFl'000
Sales turnover	1,215,966	1,307,427
Profit before tax	35,902	38,750
Profit after tax	25,431	31,170
Retained profit	17,014	21,795
Dividends	10,775	11,930
Dividends per share	2.80DFl	3.10DFl
Earnings per share	7.22DFl	8.76DFl
Share capital	76,968	76,968
Shareholders funds	161,643	167,057

Principal Shareholders: First Pacific Group
No of Employees: 2,194

IBM NEDERLAND NV
Joh Huizingalaan 265, Postbus 9999, 1006 CE Amsterdam
Tel: (020) 5133111
Telex: 15692 IBMHNL
Telefax: (020) 170726

Supervisory Board: G Savy (Chairman), Dr G Zoutendrijk (Vice-Chairman), Drs W B Blaisse, D J Harris, H Schmidt, D Sorgdrager, Drs J G Wackwitz
Management Board: Drs G H Dik (Communications Director), R van Hoeken (General Manager), Drs J H Creemers (Personnel Director), R Frese (International Operations Director), F Kales (Marketing Director), J A I Mous (Financial Director), J J M van Son (Staff Support Director)

PRINCIPAL ACTIVITIES: Information processing, office equipment, telecommunication systems, robotic systems, information network services, software
Parent Company: International Business Machines Corporation, USA
Principal Bankers: ABN; Amro; NMB
Financial Information:

	31.12.86 DFl'000	31.12.87 DFl'000
Sales turnover	2,063,870	1,688,578
Profit before tax	389,004	328,767
Profit after tax	249,671	208,565
Share capital	150,000	150,000

No of Employees: 5,820

INKOOPCOMBINATIE SAMEN STERK (BV)
See SAMEN STERK, INKOOPCOMBINATIE, (BV)

INTERPOLIS NV
Conservatoriumlaan 15, NL-5037 DM Tilburg
Tel: (013) 62 91 11
Telex: 52 291 inpos nl
Telefax: (013) 62 26 04

Supervisory Board: Commissarissen: Drs J H H Holtackers (Voorzitter), Ir A J Latijnhouwers (Vice-Voorzitter), Drs P H Coenemans, Prof Dr J E A M van Dijck, M H M F Gardeniers-Berendsen, Drs H G J Hermsen, P L A van Horen, H J A E M Klarenbeek, Drs G E G van der Kruijs, A Oudijk, A J L Schopman, Drs J J Schouten, A J A M Vermeer, Drs N H M W van Wersch, H C Zootjes
Executive Board: Hoofddirectie: Drs D J Groninger (Voorzitter), H I J M Verhoeven (Vice-Voorzitter)
Senior Executives: Divisiedirecteuren: Ir A W M van Bijsterveldt (Divisie Leven), J G Carpaij (Divisie Schade), Drs S C

Hooimeijer (Divisie De Twaalf Gewesten), Ir K Geertsema (Divisie Hagelunie). Concerndienstdirecteuren: J C J Blom (Buitendienst IPV), D v.d. Bunt (Algemene Zaken), H A J Hannen (Organisatie en Automatisering), Drs C Th M Roorda (Beleggingen), F A K Snel Act AG (Inter-Ré), H J Vilain Act AG (Economie en Administratie), W P H Wijngaard (Sociale Zaken), R Ferwerda (Concernsecretriaat/PR)

PRINCIPAL ACTIVITIES: Insurance
Subsidiary Companies: NV Interpolis BTL; NV Interpolis Schade; NV Brand-en Variaverzekeringsmaatschappij De Twaalf Gewesten; NV Hagelunie; NV Interpolis Vast Goed; NV Interpolis Fonds Eèn; NV Interpolis Fonds Twéé; Interpand NV; Interpand Twee NV
Principal Bankers: Rabo Nederland
Financial Information:

	31.12.86	31.12.87
	DFI'000	DFI'000
Total revenues	1,572,000	1,716,000
Premium income life	424,000	479,000
Premium income*	796,000	863,000
Investment income net	341,000	362,000
Profit before taxes	100,000	111,000
Profit after tax	69,000	78,000
Retained profit	69,000	78,000
Dividend	90	90
Dividend per share	6DFI	6DFI
Earnings per share	4.625DFI	5.204DFI
Share capital	15,000	15,000
Shareholders funds	694,000	757,000

*non-life

No of Employees: 1,576

KONINKLIJKE BORSUMIJ WEHRY NV
1 Carnegielaan, 2517 KH 's Gravenhage
Tel: (070) 526400
Cable: Borsumij
Telex: 31055 BORS
Telefax: (070) 506798

Advisory Board: H H Nauta (Chairman), J H Schlimmer (Vice Chairman and Delegate Member), G H van Driel, C J vd Putte, J Steinz, P J Berger
Board of Management: M van der Klooster (Chairman), H de Haan, F P Polman, W Swart

PRINCIPAL ACTIVITIES: Holding company whose subsidiaries are engaged in international trading, photo, film and video equipment, sports and camping equipment, footwear and ready-made clothing for ladies, men and children, toys, computer systems and software, watches and costume jewellery, building materials, calculating and money-counting machines, microfilm equipment, forklift trucks, smokers' requisites and business gifts, insurance, TV, audio and video equipment, technical warehouse and office equipment
Trade Names: Freetime, Arti-Mode
Subsidiary Companies: (Over 50% owned unless stated):
Netherlands: Borsumij Wehry Nederland BV; Borsumij Sport BV; Nijdam Sport BV; Borsumij Textiles BV; Borsumij Lingerie Internationaal BV; Sharp Optonica Nederland BV; Borka BV; SIO BV; Borsu International BV; Borsumij Bouwspecialiteiten BV; Hesco BV; Holland Systema BV; Service Systema BV; Garny Systema BV (45%); Herman Hijmans & Co BV; Handelsonderneming Mageon BV; Handelsonderneming Merhottein BV; Pollyflame International BV; BV Handelsmaatschappij Rijnstroom; Sanyo Business Machines Nederland BV; Overtoom Magazijn- en Werkplaats-inrichting BV; Overtoom Kantoor-Kantie-inrichtung BV; Photofinishing Holding International BV (30%); De Drie Hoefijzers Beheer BV (50%). Fodor Group: Fodor Foto BV; Fodor Radio BV; Fujifim Nederland BV; Olympus Nederland BV; Amfo Electronics BV. Belgium: Borsumij Europe NV; Geo Wehry & Cie SA; Borsumij Belgium NV; NV "Merhottein"; Overtoom International Belgium NV. Germany: Himex GmbH; Borsumij Sport GmbH; Geo Wehry & Co Intern. GmbH; Wehry Sport & Leisure Wear

GmbH; Intran GmbH; Boftex Im- und Export Handels GmbH; Borsu Computerhandels GmbH; Borsumij Kommunikations Systeme Handels GmbH. France: Borsumij Wehry France SA; Borsu International Sarl; Borsumij France SA; Sharp France SA; Enzo Varini France SA. Spain. Pollyflame Española SA. UK: Borsumij Wehry (UK) Ltd; Borka UK Ltd; Overtoom International (UK) Ltd: Boftex Ltd; Borsu International Ltd; Beale Tools Ltd; Beale Powertools Ltd; Beale Distribution Ltd; Beale Gardening Ltd. USA: Borneo Sumatra Trading Co Inc; Borsumij Wehry (USA) Inc. Japan: Geo Wehry International Ltd; Pollyflame Japan Ltd; Yashima Sansho Co Ltd (50%). South Korea: Je Woo Trading Co Ltd (50%). Hong Kong: Borneo Sumatra Trading Co (Hong Kong) Ltd; Boftex Ltd; Borsumy Import & Export (HK) Ltd; Pollyflame Concept (HK) Ltd; Parry Pacific Ltd (48%). Taiwan: Borsumij Wehry Co Ltd. Thailand: Borneo Sumatra Trading Co (Thailand) Ltd. Singapore: Borneo Sumatra Trading Co (Singapore) Private Ltd; Duncan Baron Zenith (FE) Pte Ltd (50%). Malaysia: Borneosew Medical Trading (M) Sdn Bhd (50%); Intran Co
Principal Bankers: Algemene Bank Nederland NV; Amsterdam-Rotterdam Bank NV; Standard Chartered Bank Limited; Nederlandse Middenstandsbank
Financial Information:

	31.12.86	31.12.87
	DFI'000	DFI'000
Sales turnover	1,590,994	1,795,561
Profit before tax	64,434	68,893
Profit after tax	44,131	45,234
Retained profit	27,601	27,176
Dividends	90%	96%
Dividends per share	4.50DFI	4.80DFI
Earnings per share	12.01DFI	12,02DFI
Share capital	18,366	18,810
Shareholders funds	182,909	214,547

No of Employees: 2,670

LANSCHOT, F. VAN, BANKIERS NV
Hooge Steenweg 29, 5211 JN 's-Hertogenbosch
Tel: (0) 73-15 39 11
Cable: Lanschotbank; Swift address: FVLBNL 22
Telex: 50641 (general)
Telefax: (0) 73-15 30 66

Supervisory Board: C van Veen (Chairman), W K N Schmelzer (Deputy Chairman), J J M Kaulingfreks, O W A Baron van Verschuer, A A Anbeek van der Meijden, H J A E M Klarenbeek, D A Ingham, D F Goodman, W Meijer
Management Baord: J C van Lanschot (Chairman), C L M de Quay (Deputy Chairman), L J C M le Blanc (Deputy Chairman), M C A Klinkenbergh, M G van Lanschot, A C van Herpen, H J Baeten. Secretariat to the Management Board: A W Jonker (General Manager Secretary), Mrs M E G van Erp (Secretary), F R J Baron van Hövell tot Westerflier (Secretary)
General Managers: P J M Aalberse, M J Brok, P P M Dingenouts, P A G van der Doelen, H van Engelen, C M den Engelsen, J J Heijboer, B J de Jong, Ch H A Kuijpers, M B Levit, C L A M Meijers, J G J H Meijs, C C J Montens, M W M Petri, P J J von Schijndel, J W M van Straaten, H Tijink. Regional Managers: H A M van Dijk, L W J Mertens, A T V M Schols, P R Zwart. Internal Auditor: J Remijn. Deputy General Managers: J H Aarts, P A C M Asselbergs, D Bansema, G G F M Berben, W C P Blom, H J J Debrauwer, H A H A M Franken, J B W M van der Groen, F Hilhorst, B L M Jansen, A N Klomp, G Lenselink, J van Leijenhorst, R L Meezen, J van der Schans, Th J M Stalenhoef, Mrs J C van Tol-van Streepen, A Tonnaer, A G H Veeke, F J Versteijlen, A M M Willems

PRINCIPAL ACTIVITIES: Domestic and international banking activities
Trade Names: F. van Lanschot Bankiers NV
Parent Company: Van Lanschot's Beleggings-Compagnie BV
Subsidiary Companies: (100% owned unless stated): F. van Lanschot Overseas NV with subsidiaries in Curaçao and

Jersey; Woonfonds Holland BV (95%); F. van Lanschot International Ltd London; Financieringsbank Hufima NV; F van Lanschot PPM BV; F van Lanschot Participaties BV; Affiliation: Van Lanschot Assurantiën BV with subsidiary company Van Lanschot Chabot BV

Financial Information:

	31.12.86 DFI'000	31.12.87 DFI'000
Profit before tax	33,708	34,266
Profit after tax	24,092	24,280
Retained profit	14,455	15,175
Dividends	9,637	9,105
Dividends per share	11.3%	10.7%
Earnings per share	28.3%	28.6%
Share capital	85,000	85,000
Shareholders funds	204,296	219,417
Deposits	4,596,354	4,464,257
Total assets	6,228,019	6,210,359

Principal Shareholders of Parent Company: National Westminster Bank PLC (40%); Rabobank Nederland (40%); Delta Lloyd Verzekeringsgroep NV (5.4%); private shareholders (14.6%)
No of Employees: 780

LEASE PLAN HOLDING NV
Postbus 3001, 1300 EB Almere
Tel: (03240) 9 39 11
Telex: 70878
Telefax: (03240) 9 34 79

Board of Supervisory Directors: Dr K J H Bruins (Chairman), Drs H Haarbosch, Drs J J de Kort, N J Nouwen, J van Rijn
Board of Managing Directors: Dr A C Goudsmit, H C J M van der Meulen, F Kuiper
Senior Executives: Drs J Lindeman (Director of Administration), Dr S C A Peters (Director of Automation)

PRINCIPAL ACTIVITIES: Holding company whose subsidiaries are engaged in financial leasing, equipment leasing, operational leasing/fleet management, fuel cost management
Subsidiary Companies: Lease Plan Nederland NV; Lease Plan Belgium NV (Belgium); Lease Plan France SA (France); Lease Plan Beteiligungs- und Leasinggesellschaft mbH (Germany); Lease Plan UK Ltd (UK); Lease Plan Leasinggesellschaft mbH (Austria); Lease Plan España SA (Spain); Lease Plan USA Inc (USA); Welti-Furrer Lease Plan AG (Switzerland); Lease Plan Australia Ltd (Australia). Franchise-agreements: Lease Plan Norge A/S (Norway); Lease Plan AS (Sweden)
Principal Bankers: BMH; NCB; Rabo
Financial Information:

	31.12.86 DFI'000	31.12.87 DFI'000
Turnover	528,393	644,843
Operating result	21,927	31,707
Net profit	15,135	20,207
Dividend	6,078	8,100
Share capital	37,505	43,505
Shareholders funds	99,350	127,640

Principal Shareholders: Bank Mees & Hope NV (66.67%); Rabobank Nederland (33.33%)
No of Employees: 656

NATIONALE-NEDERLANDEN N.V.
Prinses Beatrixlaan 15, 2595 AK The Hague
Tel: (070) 418080
Cable: Natassurance Den Haag
Telex: 31585
Fax: 070-475962 CCITT gr 3/2; 070-415092 CCITT grl

Board of Directors: P E E Kleyn van Willigen (Chairman), E H Van der Beugel (Vice Chairman), S Orlandini (Vice Chairman), E K den Bakker, R E M van den Brink, D de Bruyne, V Halberstadt, J Lanser, J Reehorst, C van Veen, H J Witteveen, R Zijlstra

Executive Board: T C Braakman (Chairman), O Hattink (Deputy Chairman), J J van Rijn (Deputy Chairman), H Huizinga, G F Jonckheer, S J Jonker, L van Zwol

PRINCIPAL ACTIVITIES: Life and general insurance; professional reinsurance; mortgages; finance for industry; instalment credit; personal loans and real estate project development
Trade Names: Nationale-Nederlanden
Subsidiary Companies: (100% owned unless stated): Netherlands: Nationale Nederlanden Nederland BV; Nationale-Nederlanden Life Insurance Co. N.V.; Nationale-Nederlanden General Insurance Co. N.V.; ; RVS Life Insurance N.V.; RVS General Insurance N.V.; SWS Maatschappij voor Verzekerd Sparen N.V.; Tiel Utrecht Insurances; N.V. Life Insurance Company Victoria-Vesta; N.V. General Insurance Company Victoria-Vesta; Nationale-Nederlanden International B.V.; Nationale-Nederlanden International General Insurance N.V. Belgium: RVS Insurances N.V.; De Vaderlandsche N.V.; Victoria-Vesta N.V. Greece: Proodos General Insurances S.A., Eire: Hibernian Life Association Ltd (50%). UK: The Life Association of Scotland Ltd; The Orion Insurance Co. PLC. Canada: The Halifax Insurance Co.; The Halifax Life Insurance Company; Western Union Insurance Co. USA: Nationale-Nederlanden U.S. Holdings, Inc; Nationale-Nederlanden U.S. Corporation; Midwestern United Life Insurance Co; Security Life of Denver Insurance Co; Wisconsin National Life Insurance Co; Nationale-Nederlanden U.S. Life Insurance Co; GeorgiaUS Corporation; Life Insurance Company of Georgia; Associated Doctors Health & Life Insurance Co.; Nationale-Nederlanden U.S. Property & Casualty Holdings, Inc; Nationale-Nederlanden North America Property & Casualty Corporation; Peerless Insurance Co.; The Netherlands Insurance Co.; Excelsior Insurance Co.; Indiana Insurance Companies; First of Georgia Insurance Group. Surinam: FATUM General Insurance N.V. Indonesia: P.T. Maskapai Asuransi Nasuha (49%). Malaysia: The Netherlands Insurance (Malaysia) Sdn. Bhd. (49%). Australia: Nationale-Nederlanden (Aust.) Ltd; Mercantile Mutual Holdings Ltd; Mercantile Mutual Insurance (Australia) Ltd; Mercantile Mutual Insurance (N.S.W. Workers Compensation) Ltd; Mercantile Mutual Insurance (N.S.W. Workers Compensation) Ltd; Mercantile Mutual Life Insurance Company Ltd; Mercantile Mutual Casualty Insurance Ltd; Australian General Insurance Company Ltd. Reinsurance: Nationale-Nederlanden Reinsurance Company N.V.; Nederlandse Reassurantie Groep Holding N.V. (51%); Nederlandse Reassurantie Groep N.V. (51%); "Transatlantica" Reinsurance Company N.V.
Financial Information:

	31.12.86 DFI'000	31.12.87 DFI'000
Premium income	11,883,467	11,795,528
Investment income	5,470,740	5,411,251
Profit before tax	855,061	923,288
Profit after tax	635,517	703,083
Retained profit	361,024	396,998
Dividends	274,493	306,085
Dividends per share	2.49DFI	2.65DFI
Earnings per share	5.76DFI	6.09DFI
Share capital	277,102	291,495
Shareholders funds	7,280,163	6,192,886

No of Employees: 22,691

NEDERLANDSCHE, DE, BANK NV
Westeinde 1, PO Box 98, NL-1000 AB Amsterdam
Tel: (020) 5249111
Cable: Netherbank
Telex: 11355; 14354; 16639 dnbam nl
Telefax: (020) 5243429

Board of Directors: Dr W F Duisenberg (Chairman)

Executive Directors: A Szász, H J Muller, Dr A H E M Wellink, J A Sillem, T de Swaan (Secretary). Alternate Executive Director: B A J M van Hellenberg Hubar

PRINCIPAL ACTIVITIES: Central bank

NEDERLANDSCHE MIDDENSTANDSBANK NV

NMB Bank

PO Box 1800, 1000 BV Amsterdam
Tel: (020) 563 91 11
Cable: nmbbank
Telex: 11402 nmb nl
Telecopier: (020) 563 57 00; SWIFT: NMBANL2A

Supervisory Board: K Fibbe (Chairman), H C Wytzes (Deputy Chairman), A M P van Aelst, W de Boer, A Geurtsen, J Lanser, G van Muiden, J A Oosterhoff, A W Overwater, W M Perquin, Prof N Tiemstra, Prof G M J Veldkamp, S Veninga, Prof J J A Vollebergh, J W Berghuis, J H Choufoer, C Maas
Board of Managing Directors: W E Scherpenhuijsen Rom (Chairman), G J Tammes (Vice Chairman), I E G van der Boor, H H Faber, R S H Mees, A A Soetekouw
General Managers: J H M Lindbergh (International Division), J M A Yutema (International Division), S A Andringa (International Division), H Oei (International Division), H A N de Brauw (Assistant General Manager International Division), W A J Beelaerts van Emmichoven (Investment and Securities Division), C J van Helbergen (General Manager Investment and Securities Division), W F Dutilh (Investment and Securities Division), F Kamst (Investment and Securities Division), D G C van Vliet (Assistant General Manager International Division), C C Vos (Assistant General Manager International Division)

PRINCIPAL ACTIVITIES: General banking
Subsidiary Companies: Computer Uitwijk Centrum B.V. (33%); Cofiton B.V.; Crediet- en Effectenbank N.V.(CenE); Crediet Service Bank N.V. (50%); B.V. Maatschappij voor onroerende goederen Het Middenstandshuis; N.V. Nationale Volksbank (NVB); NMB Lease N.V.; NMB Volmachtbedrijf B.V.; NMB Heller Factoring N.V. (50%); Nederlandsche Middenstands Financieringsbank N.V. (NMFB); Maatschappij voor Bedrijfsobjecten N.V. (MBO); Nederlandsche Middenstands Trust Maatschappij B.V.; Finemy B.V.; NV Bronnenbad Nieuweschans-Bunde (25%); Welvaert Financieringen N.V.; Paronyme B.V.; Centraal Wagenpark Beheer BV; Foppingadreef Leasing BV; Leaseco Nederland BV; Nederlandsche Merchant Bank nv; NMB Participatie Startfonds bv; Oudhof Effecten NV (40%); Atlas Venture BV (33%); NMB Bank (Luxembourg) SA (Luxembourg); Nederlandsche Middenstandsbank (Schweiz) A.G. (Switzerland); NMB Bank (Singapore) Ltd (Singapore); NMB Banque (France) S.A. (France); NMB (Hong Kong) Finance Ltd (Hong Kong); NMB Empreendimentos e participacoes Ltd (Brazil); NMKB Finance Company (Curaçao) N.V. (50%) (Netherlands Antilles); NMKB Trust (Curaçao) N.V. (50%) (Netherlands Antilles); Banco NMB Sudamericano S.A. (Uruguay); Middenbank Curaçao N.V. (Netherlands Antilles); NMB Bank (Deutschland) A.G. (Germany); Foreign branches: Curaçao, Hong Kong, London, New York and Tokyo
Financial Information:

| | 31.12.87 |
	DFl'000
Profit before tax	296,000
Profit after tax	221,000
Retained profit	140,000
Dividends	81,000
Dividends per share	9.00DFl
Earnings per share	20.42DFl
Share capital	455,000
Shareholders funds	2,520
Deposits	76,495,000
Total assets	80,145,000

No of Employees: 11,844

NEDERLANDSE REASSURANTIE GROEP HOLDING NV

Postbus 7846, 1008 AA Amsterdam
Tel: (020) 43 45 45
Cable: Energy
Telex: 11093
Telefax: (020) 438318

Board of Directors: J Kleiterp (Chairman), O Hattink (Vice Chairman), J L M Bartelds, E Coppola di Canzano, F Fegitz, D J Groninger, H Huizinga, T M West, R Wijkstra
Corporate Board: J H Holsboer (Chairman), S Boerhout (Vice Chairman), J W Beumer, Ms H Kronheim, P Zandhuis

PRINCIPAL ACTIVITIES: All classes of reinsurance
Subsidiary Companies: (100% owned unless stated): Risk Carriers Reinsurance and Insurance: Nederlandse Reassurantie Groep NV; Algemene Levensherverzekering Maatschappij NV; NRG London Reinsurance Company Ltd (UK); NRG Fenchurch Insurance Company Ltd (UK); NRG Antillean Reinsurance Company NV (Curaçao); NRG America Life Reassurance Corporation (USA); Philadelphia Reinsurance Corporation (USA); NRG America Syndicate Inc (40.5%); Vereenigde Assurantiebedrijven "Nederland" NV. Investment Companies: NV Beleggingsmaatschappij NRG; Reassurantie Beleggingen NV; Curaçao Reinsurance Investment Company NV (Curaçao); NRG Woningbouw BV; "Fidelio" Onroerend Goed BV; "Traviata" Onroerend Goed BV
Financial Information:

| | 30.6.87 | 30.6.88 |
	DFl'000	DFl'000
Premium income	1,074,752	1,037,528
Investment income	-	211,926
Profit before tax	34,281	43,398
Profit after tax	17,781	20,898
Retained profit	3,257	15,000
Dividend	6,750	9,000
Share capital	90,000	90,000
Shareholders funds	313,306	325,000

Principal Shareholder: Nationale-Nederlanden NV
No of Employees: 250

NMB-HELLER NV

Zamenhofdreef 49A, NL-3562 JV Utrecht
Tel: (030) 61 18 33
Telex: 47 67 8 nmbh nl

PRINCIPAL ACTIVITIES: Holding company with interests in finance including factoring
Financial Information:

	DFl'000
Turnover	3,700,000

No of Employees: 115

PAKHOED HOLDING NV

60-68 Boompjes, PO Box 863, 3000 AW Rotterdam
Tel: (010) 4002911
Cable: pakh r
Telex: 22 112
Telefax: (010) 413 9829

Supervisory Board: J H Choufoer (Chairman), A W Overwater (Deputy Chairman), J Hoogendoorn, J J Kaptein, F Lagerwey, J F M Peters, A A van der Louw
Managing Board: H P H Crijns (Chairman), G Verhagen (Deputy Chairman), J E Jansen
Senior Executives: P S de Boorder (Control), C P van Veen (Control), H Goemans (Corporate Secretary/Public Affairs), S D Eikelboom (Economic Affairs and Corporate Planning Development), B A Seckel (Economic Affairs and Corporate Planning Development), F A Lauterbach (Organization and Automation), A Hoonaker (Employee Relations), W J A Avontur (Treasury)

PRINCIPAL ACTIVITIES: The Pakhoed Group of Companies provides services to business, trade and industry in the fields of tank storage, transport and distribution

Trade Names: Divisions: Paktank International, Pakhoed Corporation, Pakhoed Transport, Pakhoed Europa

Parent Company: PAKHOE

Subsidiary Companies and Participations: (100% owned unless stated): Paktank International BV:- Paktank Bonaire NV (Netherlands Antilles); Paktank Nederland BV; Nederlandse Opslag Maatschappij BV; Paktank GmbH (Germany); Paktank AB (Sweden); Paktank Industrial Distillation BV; Europak BV; EMPAK Europe BV; Paktank Laurenshaven BV; Paktank Maasvlakte BV; Paktank (Nigeria) Ltd (60%) (Nigeria); Bonaire Petroleum Corporation NV (50%) (Netherlands Antilles); Verenigde Tankrederij BV (50%); Paktank Méditerranée SA (47%) (Tunisia); Paktank Singapore Terminal (Pte) Ltd (42.5%) (Singapore); VTG-Paktank Hamburg GmbH (33.33%) (Germany); Maasvlakte Olie Terminal NV (14.29%); Maasvlakte Olie Terminal CV (6.1%). Pakhoed USA Inc:- Pakhoed Corporation; Paktank Corporation; Paktank Corporation-Deer Park Terminal; Paktank Corporation-Galena Park Terminal; Paktank Corporation-Westwego Terminal; Paktank Corporation-Richmond Terminal; Paktank Mid-States Terminals Inc; Paktank East Coast Inc; Paktank Atlantic Company (partnership); Paktank Philadelphia Inc; Paktank Philadelphia Company (partnership) (50%); EMPAK, Inc; Pakhoed Dry Bulk Terminals Inc; Manatee Terminals Inc; Paktank Bulk Services Inc; Paktank Florida Inc; Multi-Terminals Richmond Inc (all USA). Pakhoed Transport BV:- Container Terminals: Unitcentre BV. Multi-Purpose Port Terminals: Multi-Terminals Waalhaven BV; Multi-Terminals International BV. Shipbroking Agencies/Containers: NBK Maritime BV; Trend Container Services BV; Container Parts Rotterdam (CPR) BV; Dijkstra Container Service BV; Oceanfreight BV; Pakhoed Container Services BV; Sea Containers Nederland BV; Cross-Ocean CV (50%). Storage and Forwarding: Pakhoed Amsterdam BV; Belgian Pakhoed NV; Nederlandsch Bevrachtingskantoor BV; Pakhoed AG (Switzerland); Zaans Veem BV; Pakhoed Forwarding BV; Pakhoed CFS BV; Pakbox BV. Spezialized Transport: Brucargo Express (Rutges) NV (Belgium); Fritz Koch GmbH, Internationaler Güterfernverkehr und Spedition (Germany); Rutges SA (France); Rutges AG (Switzerland); Rutges BV; Rutges Ltd (UK). Pakhoed Europa BV:- Avia Presto BV. Pakhoed Vastgoed BV:- Pakhorst Gesellschaft zur Entwicklung von Immobililienprojekten mbH (Germany); Friehorst Gesellschaft zur Entwicklung von Immobilienprojekten mbH (51%) (Germany); Deltahorst Gesellschaft zur Entwicklung von Immobilienprojekten mbh (51%) (Germany); Blauwhoed BV (48%). Holding Companies:- Labda NV (Netherlands Antilles); Petrodutch BV; Pakantil Engineering NV (Netherlands Antilles); Pakhoed Energie BV; Pakhoed Investeringen BV; Paktantil Insurance NV (Netherlands Antilles); Pakhoed Assurantiën BV; Univar Corp (35%) (USA). On May 16th 1988 Pakhoed Holding NV took over the Broere Group of Companies:- Gebr. Broere BV; Tankvaart Dordrecht BV; Tank Terminal Rotterdam BV; Antwerp United Tanker Agencies NV (Belgium); Agence Maritime Eurotank NV (Belgium); Tees Storage Company Ltd (50%) (UK); Depositos del Norte SA (50%) (Spain); Terminales Quimicos SA (13%) (Spain)

Principal Bankers: Bank Mees & Hope NV; Amsterdam-Rotterdam Bank NV; Algemene Bank Nederland NV

Financial Information: Holding Company

	31.12.86	31.12.87
	DFI'000	DFI'000
Gross turnover	1,521,175	1,321,492
Net turnover	719,305	668,147
Profit before tax	63,239	67,035
Profit after tax	33,609	68,631
Retained profit	18,561	51,911
Dividends	15,048	16,720
Dividends per share	3.60DFI	4.00DFI*
Earnings per share	8.04DFI	16.42DFI
Share capital	300,000	300,000
Shareholders equity	362,617	375,840

*plus 2% share premium bonus

No of Employees: 3,342

PIERSON, HELDRING & PIERSON N.V.

Herengracht 214, 1016 BS Amsterdam
Tel: 020-211188
Cable: Piersonco
Telex: 16388
Telefax: 020-258164

Supervisory Board: A Schuitemaker (Chairman), H J E van Beuningen, A C Helfrich, P C van de Hoek, A Jiskoot, Th A J Meys, R J Nelissen, C C Nengerman, J C van Ouwerkerk, P J Vinken, F J de Wit

Board of Managing Directors: J Kleiterp (Chairman), F J Loudon (Deputy Chairman), L H Wurfbain (Vice Chairman), J W Ort, J P J Barth, G J Bergman, W van Veelen (Commercial Banking), Baron Ch W van Boetzelaer van Oosterhout (Securities and Investments), S E Beelaerts van Blokland (Merchant Banking), A van Marken (Trust Services), G A Greidanus, B van Marken, H B de Zeeuw (Treasury), E A J van de Merwe (Accounting and Controlling), B F A de Haas, P A Blumenthal

PRINCIPAL ACTIVITIES: Banking

Parent Company: Amsterdam-Rotterdam Bank N.V.

Subsidiary Companies: Bank Oppenheim Pierson International S.A., Luxembourg; Bank Oppenheim Pierson (Schweiz) A.G., Zürich; Pierson Capital Management Inc, Philadelphia, PA; Pierson Sal. Oppenheim Inc, New York, NY

Financial Information:

	31.12.86	31.12.87
	DFI'000	DFI'000
Profit before tax	116,694	87,695
Profit after tax	88,442	68,884
Retained profit	70,560	50,540
Dividends	14,200	14,200
Share capital	94,375	94,375
Shareholders funds	341,470	338,302
Deposits	5,709,833	6,058,998
Total assets	7,603,122	7,640,406

No of Employees: 1,329

POSTBANK NV

Haarlemmerweg 506-512, PO Box 21009, 1000 EX Amsterdam
Tel: +31.20.5849111
SWIFT Code: PSTBNL 2A
Telex: 10102 pstbnl
Telefax: (3120) 5842540

Management Board: G J A van der Lugt (Chairman), H W Unger, C J M Scholtes, M Minderhoud, H K Verkoren

Senior Executives: E Spilker (Administration and Automation Division), M A D H Schalekamp (Retail Banking and Payments Division), J H J Zegering Hadders (Corporate Division), P Lindeboom (Treasury Division), M Hendriksen (Mortgage Division)

PRINCIPAL ACTIVITIES: Postbank NV offers a wide range of financial products and services for both private and corporate clients. It has a dominant position in the domestic funds transfer and savings market in The Netherlands

Financial Information:

| | 31.12.87 |
	DFI'000
Profit before tax	318,000
Profit after tax	249,000
Retained profit	144,000
Dividend per share	4DFI
Share capital	300,000
Shareholders funds	1,942,000
Deposits	46,791,000
Total assets	55,198,000

Principal Shareholders: A limited liability company with the state as the sole shareholder
No of Employees: 10,429

RABOBANK NEDERLAND
Coöperatieve Centrale Raiffeisen-Boerenleenbank BA

Croeselaan 18, PO Box 17100, 3500 HG Utrecht
Tel: (030) 909111
Cable: Raboned
Telex: 40200 Rabo nl

Board of Directors: Baron O W A van Verschuer (Chairman), C G A Mertens (Deputy Chairman), J L M Niers, R Renkema (Secretary), Dr N H Douben (Deputy Secretary), D Luteijn, G van den Berg, Dr P W Moerland
Executive Board: H H F Wijffels (Chairman), H J A E M Klarenbeek (Deputy Chairman), R B J van Eldik, P M Burghouts, F H J Boons, H van den Broek, J van Rijn
General Managers: F H Brust (General Manager and Secretary), C J B Ebeling, Dr J Th Adolfse, H C G Bak, W Bakker, A J Bierings, C F Broekhuyse, W L van Dinten, W J H Draisma, H G Gentis, H M Geukers, Dr W M van der Goorbergh, F M Joosten, H E van de Kerk, K J Merk, D W Okker, J C J Oosterwaal, E M Rintel, J Ruiter, G P M van der Schrieck, M C Snoeys, H Steensma, A C Steketee, J J Verhaegen, P Vernède, H Visser, J M Vullings, M F L Zegers, W H I Zwart
Senior Executives:

PRINCIPAL ACTIVITIES: Banking activities
Subsidiary Companies: (100% owned unless stated): Domestic:- Rabohypotheekbank NV; NV Onroerend Goed Mij "Gebeka"; BV Onroerend Goed Maatschappij Valpron; Beleggingsmaatschappij "Boxtel" BV; De Lage Landen BV; De Lage Landen Factors BV; De Lage Landen Leasing BV; Translease De Lage Landen BV; De Lage Landen Financiering BV (95%); Credit Management Support BV; Translease BV; Ondernemend Vermogen Nederland CV; Rabo Nederland Biotech Venture Fund PPM BV; Rabo Nederland Biotech Venture Fund BV; Rabobank Nederland Participatiemij BV; Vastgoed Rabo Nederland BV; Beleggingsmaatschappij Basis Amsterdam BV; Beleggingsmaatschappij Camminghaburen BV; PPM Ondernemend Vermogen Nederland CV; Ondernemend Vermogen Nederland Beheer BV; Beheermaatschappij München BV; Rabobank Computerdiensten BV; Bodemgoed BV (Real estate company); Bruns ten Brink Groep BV (Insurance broker); BV Fondsenbeheer Rabobank; NV Handelmij "Het Zuiden" (Real estate company); Rabo Onroerend Goed Nederland BV; Schretlen en Co NV (Securities credit institution); BV Administratiekantoor Schretlen en Co; BV Trustmaatschappij Schretlen en co; HP Verbist BV (Private loan broker); Beheermaatschappij Bonn BV; Incassobureau Trajectum BV; Beheermaatschappij Frankfurt BV; BV Bewaring Huisfondsen Rabobank; Nederlandse Scheepshypotheekbank NV (99%); Nieuwe bedrijvencentrum "Basis" BV. Abroad:- NV Vastgoedmaatschappij "Croeselaan" (Belgium); Rabobank Curaçao NV (Netherlands Antilles); Rabobank Trustmaatschappij Curaçao NV (Netherlands Antilles); Rabobank Nominees (Singapore) Pte Ltd (Singapore);

Allgemeine Deutsche Credit-Anstalt AG (84%) (Germany); MAL Anlagenleasing GmbH (75%) (Germany)

Financial Information:

| | 31.12.86 | 31.12.87 |
	DFI'000	DFI'000
Profit before tax	913,000	1,010,000
Profit after tax	687,000	694,000
Net profit	685,000	692,000
Group equity	7,771,000	8,431,000
Total assets	139,689,000	145,535,000

No of Employees: 32,646

RIVA NV

Wieger Bruinlaan 104, 2132 AX Hoofddorp
Tel: 02503-15105
Telex: 11141 riva nl
Telefax: 02503-14748

Board of Directors: J W Verouden, J Visseher
Management: J E Demper (Secretary)
Senior Executives: A L van Dalen (Controller), J H Kuyper (Treasurer), L J Hartog (Head Personnel Department), P Mikmak (Head Property Services and Maintenance)

PRINCIPAL ACTIVITIES: Holding company with interests in the distribution of motor vehicles; automobile leasing; finance, gas trade and distribution
Subsidiary Companies: (100% owned unless stated): Riva Amsterdam BV; Riva Amstelveen BV; Riva Amsterdam-Noord BV; Riva's-Gravenhage BV; Riva Voorburg BV; Lisser Automobiel Bedrijf BV; EGAC BV; Auto Lichtstad BV; Interleasing Nederland BV; Interleasing Nederland Truck BV; Interleasing Belgium NV (75%) (Belgium); Unilease BV (50%); BV Techn. Handelsbedrijf De Groot (THG); Kors Automobielbedrijf BV (50%); Kistemaker's Handelmij BV; Rivafoam BV; Dealer Cash Plan BV; MAR-GAS Internationaal BV; MAR-GAS Nationaal BV; V.O.F.Gasdepot De Kellen; Rigros Automaterialen BV
Principal Bankers: Bank Mees & Hope NV; Amro Bank NV

Financial Information:

| | 31.12.86 | 31.12.87 |
	DFI'000	DFI'000
Sales turnover	500,000	578,000
Profit before tax	19,591	22,683
Profit after tax	15,292	15,608
Retained profit	11,529	11,666
Dividends	3,763	3,942
Dividends per share	2.10DFI	2.20DFI
Earnings per share	8.53DFI	8.71DFI
Share capital	4,480	4,480
Shareholders funds	88,758	99,001

No of Employees: 843

SAMEN STERK, INKOOPCOMBINATIE, (BV)

Archimedesstraat 17, 4816 BA Breda
Tel: (076) 785911
Telex: 54093 samst nl
Telefax: (076) 874666

Supervisory Board: J A van Buul (Chairman), Prof A P van Gent, A C Middelman, E Th A C Albada Jelgersma, J Elzas
Management Board: A H M van Roosmalen, O van Leeuwen, Drs F P J M Van Wegen, Drs R v Oers, Drs F F M Kuijer

PRINCIPAL ACTIVITIES: Importers, wholesalers, buying-groups, full-service organisation for retailers in textiles, clothing, furniture, household goods, luxury goods, sports goods, shoes
Trade Names: Topman, Dundee, Nouvelle, Raffine, Mode Mondiale, La Ligne Exclusive, Bébèl, Dunkey, Corline, Bumper, Buck, Topstof, Artilaine, Woonvisie, Sfeervol Huis, Esfera, Mondiluxe, Sportpoint, Club Fizzick Surfsquad, Spun, Geschenkengilde, Goed Idee, Inversa, Topdream
Subsidiary Companies: Assurantie- en Bedrijsadviesbureau Archimedes bv; Indaco-Topstof bv; Save BV (50%);

WoonAdviesCentrum BV; Projectontwikkelingsmij Doornkade BV; Detamar BV; Sambel NV (Belgium); Sambuy Ltd (Hong Kong); Samen Fung Ltd (50%) (Hong Kong)

Principal Bankers: Algemene Bank Nederland; Nederlandsche Middenstands Bank; Amsterdam-Rotterdam Bank

Financial Information:

	DFl'000
Sales turnover	783,000

Principal Shareholders: Members of the Company
No of Employees: 251 (full time equivalents)

WESTLAND/UTRECHT HYPOTHEEKBANK NV

MR Treublaan 7, NL-1097 DP Amsterdam
Tel: (120) 56 04 911
Cable: 16129
Telex: 16129 wuh nl
Telefax: 5604220

Chairman: J F Visser
General Managers: M B Bolle, W Davidzon. Staff: J R Voûte (External Affairs), F E van Haastert (Secretary), W van Pelt (Legal and Fiscal Affairs)
Senior Executives: F D Korfker, (Lending Activities), S E Amgwerd (Foreign Activities), J H L A M Sopers (Group Funding and Treasury Department)

PRINCIPAL ACTIVITIES: Banking services
Parent Company: Nationale-Nederlanden NV
Subsidiary Companies: Effectenmakelaardij Amstgeld NV; Frankfurter Boden Kreditbank AG (Germany); W U H Finanz AG (Switzerland); Westland/Utrecht (Deutschland) GmbH (Germany); LAS Investment Funding Ltd (UK)

Financial Information:

	31.12.86 DFl'000	31.12.87 DFl'000
Profit before tax	17,369	30,194
Profit after tax	15,869	28,694
Retained profit	-	4,801
Dividend	-	8,887
Share capital	139,000	139,000
Shareholders funds	224,207	225,873
Total assets	7,966,527	7,897,858

No of Employees: 531 (worldwide)

Major Companies of
NORWAY

BERGEN BANK A/S

PO Box 826, Torgalmenning 2, 5001 Bergen
Tel: (05) 21 76 00
Cable: Bergenbank
Telex: 42018
Telefax: (05) 21 19 11

Board of Directors: O Lund (Chairman), E G Greve (Managing Director and Chief Executive), P Munthe, T Andreassen, S Haugland, Else Bugge Fougner, O Rønnestad, S Thomassen. Deputies for the Managing Director and CEO: J Faye, T Brekke, R Lien, O Vevle, I Børsheim, T Grøndahl, S Eidem, Ø Fjeldstad

Group Management: E Gade Greve (Managing Director & CEO); J Faye (First Deputy Managing Director). Special Counselling Economic Policy: I Børsheim (Deputy Managing Director). Bergen Division: R Lien (Deputy Managing Director). Capital Markets Division: Ø Fjeldstad (Deputy Managing Director). Regional Banking Division: O Vevle (Deputy Managing Director). International Division: T Grøndahl (Deputy Managing Director). Foreign Exchange, Treasury: S Eidem (Deputy Managing Director). Financial Planning and Accounting: O Johannessen (General Manager). Information Systems: E Lindmo (General Manager). Personnel: K Natvik (General Manager)

PRINCIPAL ACTIVITIES: Commercial bank

Subsidiary Companies: (over 50% interest): Bergen Bank International SA (Luxembourg); Bergen Bank Asia Ltd (Singapore); Bergen Bank Corporation (USA); BB Mortgage Trust Ltd (UK); Bergen Bank Invest as; Fabin A/S; A/S Kreditt-Finans; A/S Viul Tresliperi; Jan Jensen as; Telefinans A/S; Alf Løkas as; F Beyer A/S; F Beyer Bok-og Papirhandel A/S; A/S Emma EDB; Bodd Investment A/S; ND Securities; Nord Og Syd Kredittforsikring A/S; Nord Og Syd Finans A/S; Økonomi Partner A/S; A/S Wallendahl & Søn; A/S Arna Gårdselskap; A/S Havnegata 2, Åndalsnes; A/S Salbukollen; A/S Vestlandske Kalkbrud og Eiendomskompagni; Heavy Invest A/S

Financial Information: Group figures

	31.12.86	31.12.87
	NKr'000	NKr'000
Profit before tax	687,000	338,500
Profit after tax	539,800	210,900
Retained profit	370,900	6,200
Dividends	168,900	204,700
Dividends per share	15.50NKr	14.00NKr
Earnings per share	80.30NKr	31.40NKr
Share capital	1,021,800	1,356,000
Shareholders funds	2,701,700	3,444,800
Deposits*	28,184,700	32,421,400
Total assets	80,617,100	99,317,700

*other than from banks

Principal Shareholders: Skandinaviska Enskilda Banken; Bolig og Næringskreditt; UNI Forsikring; K-Holding; Gjensidige Liusforsikring
No of Employees: 4,206

CHRISTIANIA BANK OG KREDITKASSE

7 Stortorvet, N-0179 Oslo 1
Tel: (47 2) 48 50 00
Cable: Xaniabank
Telex: 71043 (General Business); 71320 (Foreign Exchange Dept)
Telefax: (47 2) 48 47 49 (General Business)

Board of Directors: V Hveding (Chairman, Non-Executive), J A Holter (Deputy Chairman, Non-Executive), L Aune, T Bergem, L Smith, K K Husby, J A Krog, D Thorkildsen, T Moursund (President and Chief Executive Officer)
Management: T Moursund (President and Chief Executive Officer)

Senior Executives: S W Rostoft (Executive Vice President); G ter Jung (Executive Vice President), J Lange Paulsen (Executive Vice President), M Haga (Executive Vice President), J Whist (Executive Vice President)

PRINCIPAL ACTIVITIES: Banking and issuing house
Subsidiary Banks: Christiania Bank Luxembourg SA; Christiania Bank Aktiebolag, Stockholm; Christiania Banking (US) Corporation; Christiania Bank, London Branch; Christiania Bank, Singapore Branch; Christiania Bank, New York Branch
Financial Information: Group figures

	31.12.86	31.12.87
	NKr'000	NKr'000
Profit (loss) before tax	763,900	(316,468)
Profit (loss) after tax	511,200	(176,650)
Retained profit (loss)	511,200	(176,650)
Dividends	197,971	-
Dividends per share	16.20NKr	-
Earnings (loss) per share	60.69NKr	(16.12NKr)
Share capital	1,315,900	1,402,437
Shareholders funds	2,808,800	2,916,331
Deposits	86,147,000	90,246,172
Total assets	103,851,000	117,073,200

No of Employees: 4,893

DEN NORSKE CREDITBANK
DnC

Kirkegaten 21, 0153 Oslo 1
Tel: +47 2 481050
Cable: Creditbank
Telex: 78175 DnC N; 78295 DnC N (Foreign Exchange)
Telefax: +47 2 481870

Board of Directors: Jon R Gundersen (Chairman), Svein Erik Amundsen (Vice Chairman), Atle Bergshaven, Liv Hatland, Truls Holthe, Finn A Hvistendahl, Evlyn Raknerud, Kristian Rambjør
Managing Director: Kristian Rambjør. Deputy Managing Directors: Harald Arnkværn, Lars U Thulin, Ivar J Saunes, Harald Moen, Erling Naper, Stein Wessel-Aas, Odd Trender, John Giverholt

PRINCIPAL ACTIVITIES: International bank
Wholly Owned Subsidiaries: Den norske Creditbank (Luxembourg) SA (Luxembourg); Den norske Creditbank Plc (UK); DnC Finance Corporation (USA); DnC America Banking Corporation (USA); DnC Ship Mortgage International Bank NV (Netherlands); DnC Ltd (Hong Kong); Den norske Creditbank Sverige (Sweden). Branches Abroad: Singapore, Hong Kong, New York. Representative Offices Abroad: Beijing, Frankfurt, Hamburg, Houston, Cairo, Osaka, Tokyo, Rio de Janeiro
Financial Information: Consolidated figures

	31.12.86	31.12.87
	NKr'000	NKr'000
Profit (loss) before tax	515,700	(1,447,800)
Profit (loss) after tax	437,900	(1,470,100)
Retained profit	437,900	-
Earnings (loss) per share	24NKr	(77NKr)
Share capital	1,741,600	1,899,900
Subordinated loan capital (1)	1,301,500	2,265,900
Subordinated loan capital (2)	2,072,000	1,433,000
Conditional tax-free reserves	1,312,400	970,800
Funds	989,600	289,300
Shareholders funds	5,675,500	4,959,000
Customer deposits	42,871,000	47,383,000
Total deposits	93,117,500	93,771,900
Total assets	126,425,200	127,544,700

(1) ordinary
(2) primary perpetual

Principal Shareholders: Private individuals
No of Employees: 6,058

GJENSIDIGE FORSIKRING

PO Box 6738, St Olavs pl, N-0130 Oslo 1
Tel: (02) 48 08 90
Telex: 19036 reass n
Telefax: (02) 48 03 83

Chairmen: Harald Schjoldager, Ola T Ruud
Corporate Management: Helge Kvamme (Managing Director),
Leif R Lae (Deputy Managing Director, Erik Garaas (Deputy
Managing Director), Bjørn Elvestad (Finance Director), Lars
Austin (Deputy Managing Director), Svein Aandahl (Deputy
Managing Director)

PRINCIPAL ACTIVITIES: Gjensidige Forsikring is the common
designation for Livsforsikringsselskapet Gjensidige,
Gjensidige Norsk Skadeforsikring and 47 mutual fire
insurance companies. Gjensidige Forsikring has about 200
sales and service offices throughout Norway. Covering all
forms of insurance, the company is staffed by specialists in
life and non-life insurance
Subsidiary/Associated Companies: (100% owned unless stated):
Pensjonstjenester A/S; Agder Assuranceselskab A/S; Data-
Drift A/S (50%); Elcon Finans A/S (96.1%); Saminvest A/S.
Reinsurance companies: ARP (France); Scan Re (50%) (UK);
Hudson Life Reassurance (8%) (USA)
Financial Information:

	31.12.87	31.12.88
	NKr'000	NKr'000
Premium income	4,006,000	4,599,000
Financial income	1,143,000	1,513,000
Total resources	12,476,000	14,295,000

No of Employees: 1,946

HAFSLUND NYCOMED A/S

Slemdalsveien 37, PO Box 5010, Majorstuen, Oslo 3
Tel: (02) 46 38 30
Cable: Elpower Oslo
Telex: 76837 hafsl n
Telefax: (02) 46 03 28

Board of Directors: T Mikalsen (Chairman), A Aakvaag, H
Schjoldager, T Brøvig, B Myrvold, O H Henriksen, A
Theodorsen, K Tjønneland
Executive Management: S Aaser (President and Chief Executive
Officer), Gert W Munthe (Executive Vice President, Energy
Operations), F C Schreuder (Executive Vice President,
Investor Relations/Community Relations/Corporate
Development), Ø A Brøymer (Executive Vice President and
Chief Financial Officer), Berit Wenaas (Executive Vice
President, Organisational Development/Personnel), S H
Annexstad (Deputy Chief Executive Officer, President
Nycomed and Executive Vice President, Pharma and Imaging)
Group Management: J Gørrissen (Metals Division), G W Munthe
(Energy Operations), K-E Lindberg (Agricultural Division), S H
Annexstad (Pharmaceutical Operations)

PRINCIPAL ACTIVITIES: Electricity generation and distribution;
ferro-silicon smelting; contracting and consultancy services;
agriculture and forestry; financial activity; pharmaceuticals
and biotechnology; welding equipment
Subsidiary Companies: Energy: Power Division; A/S Kykkelsrud;
A/S Vamma Fossekompagni; Hafslund Power Int AS;
Hafslund Industries Inc (USA); Hafslund Power Inc (USA);
Swift River/Hafslund Co (60%) (USA); Swift River Operations
Inc; Greenville Steam Company; Thermal Power Inc; Hafslund
Consultants Inc (USA); Hafslund Engineering AS; AS
Norelektro; Norpower AS; Hafslund Hydropower AB (90%)
(Sweden). Pharmaceutical Industry: Nycomed AS; Nycomed
AB (Sweden); Nycomed Aps (Denmark); Jean Mette AS
(99.8%); Trombo AS; AS Nycotron; Nycomed Ltd (UK);
Nycomed BV (Netherlands); Nycomed SA/NV (Belgium);
Nytech Ltd (51%) (UK); Seragen Inc (71.4%) (USA). Other
Areas of Operation: Actinor Olje A/S; Hafslund Metall AS;
AlMet as (70%); K/S AlMet as (63%); Hafslund Ltd (90%)

(UK); Logan & Allan (UK); Tennant Nordic Alloys (UK);
Tennant Metallurgical Ltd (UK); Tennant Consultancy Services
(UK); Tennant Smith Recycling (76%) (UK)
Principal Bankers: Christiania Bank
Financial Information: Group figures

	31.12.86	31.12.87
	NKr'000	NKr'000
Sales turnover	2,526,000	2,114,000
Profit (loss) before tax	(109,000)	159,000
Profit (loss) after tax	(134,000)	111,000
Retained profit (loss)	(161,000)	82,000
Dividends	28,000	29,400
Dividends per share	3.00NKr	3.00NKr
Earnings per share	2.90NKr	17.80NKr
Share capital	193,000	196,000
Shareholders funds	422,000	183,000

Principal Shareholders: Mosvold Farsund and associated
companies (15.32%); Orkla/Borregaard (12.94%); Tharald
Brøvig (6.25%); Aktivum Invest (5.62%)
No of Employees: 2,692

JONAS ØGLÆND A/S

See ØGLÆND, JONAS, A/S

MØLLER-GRUPPEN

Frysjaveien 31, Sagene, Oslo 4
Tel: (02) 23 38 85
Cable: Stromcar
Telex: 71129
Telefax: 231503

Board of Directors: Jan H Møller (Chairman), Peter P Stabell,
Ellen Schage, Knut Schage, Harald Møller Jr, Per Gundhus,
Helge Sandvig
Management: Jan H Møller (Group Managing Director), H
Sandvig, E Wettre, O Nord, J Sannem, K Kopstad

PRINCIPAL ACTIVITIES: Motor trading; agents for Volkswagen,
Audi, MAN and Dodge; production of toys, sport and leisure
goods; plastic products; container and transport services; car
leasing and rental
Parent Company: Møller-Gruppen a.s.
Subsidiary Companies: Harald A Møller A/S; MAN VW Last og
Buss A/S; Møller US Import A/S (Chrysler); Harald A Møller
Investment A/S; Hamax-VEFI A/S; Hamax A/S;
Systemtransport A/S; Norca Ind. Ltd (Canada) and 13
Volkswagen/Audi/MAN-VW Dealer Companies out of a total
of 76
Principal Bankers: Kreditkassen
Financial Information: The Møller Group

	31.12.86	31.12.87
	NKr'000	NKr'000
Sales turnover	3,441,000	3,095,000
Result*	81,722	63,678
Net result	3,625	28,539
Net group result	3,311	28,226
Share capital	38,500	38,500
Shareholders equity	403,100	445,100

*before year-end
appropriations

Principal Shareholders: Møller and Schage Family
No of Employees: 1,360

NORSK KOLLEKTIV PENSJONSKASSE AS

Postboks 4300-5013 Nygårdstangen, N-5000 Bergen
Tel: (05) 21 10 20
Telefax: (05) 32 54 31

Board of Directors: John Lybak
Management: Jo Aabakken

PRINCIPAL ACTIVITIES: Provision of life and pension insurance

Financial Information:

	NKr'000
Premium income	1,000,000

No of Employees: 400

ØGLÆND, JONAS, A/S
PO Box 115, 4301 Sandnes
Tel: (04) 605000
Cable: Øglænd
Telex: 30343 jonas n
Telefax: (04) 605509

Board of Directors: Tore Lærdal (Chairman), Kai Engedal, Carl Graff-Wang, Steinar Manengen, Bjørn Ravnås, Alf Hjetland, Harald Rinaldo

Management: Arne Norheim (Chief Executive Officer), Ivar Rusdal (Vice President), John T Rønneberg, Olaf Myklatun, Wenche Meldahl, Arne Sanne, Gunnstein Austigard, Bjørn Thuen, Per Gribbestad

PRINCIPAL ACTIVITIES: Wholesale and retail sale of clothing for men, women and children; manufacture and marketing of bicycles, workwear, development and motor cycle accessories and exercise equipment; real estate development

Trade Names: CUBUS, PIONER, DBS, TEMPO, BODYGUARD

Subsidiary Companies: Cubus a.s; Øglænd DBS A.S; Øglæend Pioner A.S; Øglæend Systems A.S; Øglæend Teknologi A.S; Total Forsikring & Finans A.S.

Principal Bankers: Den Norske Creditbank; Christiania Bank og Kreditkasse

Financial Information: Group data

	31.12.86 NKr'000	31.12.87 NKr'000
Sales turnover	1,179,983	1,187,340
Profit before tax	23,166	35,668
Profit after tax	17,583	25,450
Retained profit (loss)	165	(1,149)
Dividends	4,800	-
Dividends per share	4.0NKr	-
Earnings per share	9.33NKr	12.66NKr
Share capital	72,000	79,260
Shareholders funds	21,594	6,811

Principal Shareholders: Dressman Group, Oslo; Aritmos AB, Helsingborg; Midelfart & Co A/S, Drammen; Made in as, Oslo; Harald A Møller Investment A/S, Oslo; NSV Invest A/S, Oslo; Aasmund Lærdal A/S, Stavanger

No of Employees: 1,775

ORKLA BORREGAARD A.S (FOLLOWING MERGER OF ORKLA INDUSTRIER A.S & BORREGAARD A.S IN OCT 1986) PARENT COMPANY OF THE ORKLA BORREGAARD GROUP
Box 308, N-1324 Lysaker
Tel: +472 50 10 80
Telex: 77529
Telefax: +472 50 16 91

Board of Directors: The Members of the Board elected by the shareholders are: Nils J Astrup (Chairman), Westye Egeberg, Truls Holthe, Gunnar Thommessen, Torvild Aakvaag. Elected by the Employees: Kjell Henning Johansen, Roger Johansen, Kjell Kjönigsen, Egil Friberg (Observer)

Group Management: Jens P Heyerdahl d.y.(Group President and Chief Executive Officer), Odd Vigmostad (Deputy Group President and Chief Operating Officer Consumer Products), Finn Jebsen (Executive Vice President and Chief Operating Officer Investment Activities)

Group Staff: Lars P Fasting (Senior Vice President Group Development), Trygve B Jörgensen (Senior Vice President Personnel/Organisation), Ludvik Sandnes (Senior Vice President Economy/Finance), Odvar Thorsen (Senior Vice President Legal), Jon Trøften (Vice President Information)

PRINCIPAL ACTIVITIES: Industry: Production of edible oil, fat, detergents, toiletries and cosmetics; manufacture of biscuits, processed food including jam, frozen and canned vegetables, frozen and fresh meat and pizza products; snacks; magazine publishing; regional newspapers; process industries based on natural resources including production of chemicals based on wood-fibre; specialty pulp and paper; hydro electric power production; forestry; land-based and offshore engineering; fine chemicals; trading; stationery wholesale; diary publishing; book production. Investments: Portfolio management, property and financial services

Subsidiary Companies: (100% owned unless stated): Consumer Goods: A/S De-No-Fa og Lilleborg Fabriker; Elico A/S; Ello A/S; Fredrikstad Blikk- & Metallverkfabrikk A/S; A/S Sætre Kjeksfabrik; Stabburet A/S; Viking Askim A.S; Polly A.S; Smak's Salater A/S; Emil Moestue as; Emo as; A/S Høvellast; Viking Fottøy A.S; Borregaard Trading AB (Sweden); Borregaard SARL (France). Media: Orkla Media A.S; Orkla Communication A.S; Ernst G Mortensens Forlag A/S; Direct Mail A/S; Forenede Trykkerier A/S; Norsk Presse Service A/S; Moss Avis A/S; Fredriksstad Blad and Sarpen (85%); Tönsbergs Aktietrykkeri (81%); A/S Sandefjords Blad og Trykkeri; A/S Gjengangeren (58.9%). Process Industry: Borregaard Industries Ltd; Borregaard Fine Chemicals A.S; Trafford Chemicals (UK); Borregaard GmbH (Germany); Borregaard Marketing Company (UK); Norsk Elektrokemisk Aktieselskab (89%); A/S Holmen-Hellefos; Vafos Brug A/S; Orkla-Exolon A/S & Co (50%); Salvesen & Thams's Comm. A/S; Borregaard Engineering A/S; Borregaard og Kiær skoger. Investments: 1. Securities Section. 2. The Orkla Eiendom Group. 3. The Orkla Finance Group

Principal Bankers: Christiania Bank og Kreditkasse; Den Norske Creditbank

Financial Information:

	31.12.86 NKr'000	31.12.87 NKr'000
Sales turnover gross	6,864,000	7,028,000
Profit before tax*	412,000	420,000
Profit for the year	61,600	123,000
Profit for the group	60,800	121,400
Dividends per share	12.50NKr	12.50NKr
Earnings per share	51.60NKr	72,40NKr
Share capital	654,836	721,699
Equity capital	1,859,900	2,050,100

*and appropriations

Principal Shareholders: At December 31st 1987: Orkla Borregaard A.S; K-Holding A/S; Bergen Bank; Den norske Creditbank; Fearnley & Eger; Dyno Industrier; UNI Forsikring Gjensidig Liv; DnC-Invest A/S; The Mitsubishi Bank; Storebrand Skadeforsikring; Storebrand Livsforsikring; Det Gjensidige Forsikringsselskap; Stormgård A/S; Royal Bank of Scotland; Norfond-Finans

No of Employees: 6,795

STOREBRAND A/S
Haakon VII's gate 10, PO Box 1380 Vika, N-0114 Oslo 1
Tel: (02) 31 10 20
Cable: Storebrand, Oslo
Telex: 71057 STORO N
Telefax: (02) 31 11 01

Board of Directors: Øystein Eskeland (Chairman), Harald Bastiansen, Finn Jebsen, Kasper K Kielland, Benedicte Berg Schilbred, Torvild Aakvaag, *Knut Eben, *Anne-Lise Løfsgaard, *Bjørn Volden, *Elected by the employees

Management: Jan Erik Langangen (Managing Director and Chief Executive Storebrand A/S), Bjørn Kristoffersen (Managing Director, Storebrand Forsikring A/S), Odd Espolin Johnson (Managing Director, Storebrand Livsforsickring A/S), Lloyd Briggs (Managing Director, Storebrand International A/S), Leif Landsverk (Managing Director, Storebrand Finans A/S)

NORWAY

PRINCIPAL ACTIVITIES: Holding company with interests in general, fire, marine and life insurance; international reinsurance; interests in finance and investment

Subsidiary Companies: Storebrand Forsikring A/S:- Storebrand Øst A/S; Storebrand Arendal A/S; Storebrand Haugesund A/S; Storebrand Vest A/S; Storebrand Trygd A/S; Storebrand Nord A/S; Europeiske Reiseforsikring A/S; Det norske Garantiselskab A/S; Forsikrings-Aksjeselskapet Vega; Det Norske Brandforsikringsselskap Fram A/S; A/S Risk Management. Storebrand International A/S:- Norsk Pensjonservice A/S. Storebrand International A/S:-Storebrand International Insurance A/S; Christiania Holding Company Inc (USA), with Subsidiary Companies, Storebrand Insurance Company (UK) Ltd, Norden Insurance Company (UK) Ltd, Aksjeselskabet DUO. Storebrand Finans A/S:- Custos Finans Øst; Custos Fians Nord A/S; Custos Finans Sør A/S; Custos Finans Vest A/S; Carl Kierulf & Co A/S; Forsikrings-Aksjeselskapet Kreditt-Atlas; Finansformidling A/S; Custos Card A/S; Storebrand Venture A/S; Storebrand Finans AB; Storebrand Finans A/S Danmark; Storebrand Finans Ltd

Principal Bankers: Den norske Creditbank

Financial Information:

	31.12.86	31.12.87
	NKr'000	NKr'000
Premium income	4,137,700	5,230,000
Investment revenues-		
insurance	830,700	797,100
Profit (loss)*	327,000	(37,800)
Net profit (loss)	149,600	(96,500)
Dividends	68,200	63,600
Dividends per share	11NKr	2NKr
Share capital	636,000	636,000

*before allocations

Principal Shareholders: Kommunal Landspensjonskasse (10.1%); Platou Investment (8.1%); Orkla-Borregard (4.9%); Realkredit (2.7%)

No of Employees: 3,857

UNI FORSIKRING

Postboks 1155-Sentrum, N-0107 Oslo 1
Tel: (02) 31 50 50
Cable: Brannkassen
Telex: 19716 nobra n (reinsurance); 17525 nomar n (marine); 72891 noil (offshore)
Telefax: (02) 48 96 30; (02) 31 57 90

Board of Directors: Ole Hope, Knut Martinsen, Astrid Lørdal Frøseth, Jon Hals, Trygve Horgen, Kåre Moe, Victor Pedersen, Sverre Pettersen, Helen Stockinger, Charles Jensen

Central Administration: Thorleif Borge (Managing Director), Olav Smedal, Thorstein Øverland, Johan W Amundsen, Svein Flugstad, Harald Taule, Jan Ørmen

PRINCIPAL ACTIVITIES: All classes of life and non-life insurance, direct as well as indirect

Trade Names: UNI Forsikring

Parent Company: Norges Brannkasse and Norske Folk

Principal Bankers: Christiania Bank og Kreditkasse

Financial Information:

	NKr'000
Premium income	5,900,000

No of Employees: 2,050

UNION BANK OF NORWAY
ABC Bank

14-18 Kirkegaten, Oslo 1
Tel: 472 31 90 50
Telex: 19470 ubn bk; 19550 (foreign exchange department)

Board of Directors: Ole Sjetne (Chairman)
Management: Olav Eikeland, Ove Fløtaker
Senior Executives: Geir Bergvoll (General Manager International Division)

PRINCIPAL ACTIVITIES: Banking

Subsidiary Companies: Union Bank of Norway International SA (Luxembourg)

Financial Information:

	31.12.87
	NKr'000
Total assets	4,392,620

VESTA GROUP PLC

N-5020 Bergen
Tel: (05) 17 10 00
Cable: Vesta
Telex: 42030 vesta n
Telefax: (05) 17 18 99

Board of Directors: Einar Falck (Chairman), Turid Emhjellen, Hans Peter Fuglesang, Per Hagelien, Gaute Henriksen, Jacob Neumann, Chr Rieber, Knut Vartdal, Finn Aagren

Group Management: Jöhan Fr Odfjell (President, Vesta Group plc), Ove Steinar Larsen (President, Skadeforsikringsselskapet Vesta A/S), Aksel Irgens (President, Livsforiskringsselskapet Hygea), Gudmund Rønningen (President, A/S Nevi), Frithjof H Frederiksen (President, Polaris Assuranse A/S), Ola Strøm (Senior Vice President and Chief Financial Officer, Vesta Group plc), Erik Brekke (Senior Vice President, Public Relations)

PRINCIPAL ACTIVITIES: Insurance and financing

Subsidiary Companies: Skadeforsikringsselskapet Vesta A/S: Kreditt Garanti A/S; Vesta Reinsurance Co Ltd A/S; Stavanger Vesta Forsikring A/S; Vesta UK Insurance Co Ltd; Vesta Re Australia; Vestkap A/S. Livsforsikringsselskapet Hygea; A/S Nevi: A/S Nevi Finans Norge; Finax A/S; Nevi Corporate A/S; Nevi Finans AB (Sweden); Nevi Finans A/S (Denmark); Multifinans A/S (Denmark); E Møllerseftf A/S (Denmark); Nevi Baltic plc; Polaris Assuranse A/S

Financial Information:

	NKr'000
Gross operating income	8,000,000

Principal Shareholders: Investa (24.8%); Norske Folk (9.9%); Bergen Bank (7.4%); Industriforsikring (2.8%); Norsk Hydros Pensjonkasse (2.7%); Jacobson & Ponsbach (2.6%); K-Avkastning (1.6%); Real-Invest (1.2%); Finans-Invest (1.2%); DnC, Own holdings (1.2%)

No of Employees: 2,880

Major Companies of
PORTUGAL

BANCO BORGES & IRMÃO

Praca do Municipio 31, PO Box 2153, 1104 Lisbon Codex
Tel: 321011
Telex: 14367 Borsex p

Executive Board: Antonio Carlos de Magalhães Fernandes Tato (Chairman), José Manuel Júdice da Costa Nunes da Glória (Vice Chairman), Cláudio da Silva Alves, Daniel Jorge Coimbra Palhares, Ricardo Alves de Castro Tavares
Management Board: Adriano António Teixeira, Afonso António Cantuárias Costa, Álvaro de Jesus Dias Duarte, António José Faria de Barros, António Manuel Castanheiros Santos, Armando de Araújo Camps, Augusto Pinto Martins, Carlos Armando Gonçalves Gomes, Carlos Mendes, Ernesto João da Costa Figueiredo Teixeira, Fernando de Almeida Abrantes, Francisco da Silva Monteiro Júnior, José Alves de Castro Sousa Rio, José Luís Figueira Vaz de Carvalho, Laurindo Ernesto Alves Ferreira, Marcel Vasconcelos Mascarenhas Galvão Mexia, Rui Duarte Ortigão Almeida Osório

PRINCIPAL ACTIVITIES: Bank

Principal Shareholders: Nationalised Institution
No of Employees: 4,600

BANCO DE PORTUGAL

Rua do Comércio 148, 1100 Lisbon
Tel: 362931
Cable: BANGAL
Telex: 16554; 16771
Telefax: 364843

Governor: José Alberto Tavares Moreira
Vice Governors: António Carlos F Palmeiro Ribeiro, Walter Waldemar Pego Marques
Other members of the Board: Alberto José dos Santos Ramalheira, Abel António Pinto dos Reis, José de Matos Torres, Rui Manuel P Chancerelle Machete, José Luis Nogueira de Brito, Luis Miguel Pizarro Beleza

PRINCIPAL ACTIVITIES: Central bank
Financial Information:

	31.12.86	31.12.87
	Esc'000	Esc,000
Total income and profits	277,836,200	249,326,300
Profits for the year	17,803,800	16,335,900
Capital and reserves	64,186,000	66,557,000
Total assets	2,439,976,000	2,528,466,000

No of Employees: 3,008

BANCO ESPIRITO SANTO E COMMERCIAL DE LISBOA

PO Box 2105, Avenida de Liberdade 195, 1200 Lisbon
Tel: 578005
Cable: Arieb-Lisbon
Telex: 12191 arieb p
Telefax: 574924

Board of Directors: Alexandre Vaz Pinto (Chairman), Antonio José Sacadura Vieira Monteiro (International), Carlos Mascarenhas de Almeida, Mário Martins Adegas, Mário de Sousa
Senior Executives: José de Castro (Marketing Department). International Division: Luis Neto (General Manager), Miguel Parreira (Senior Manager and Head of Correspondent Banking), Luis Spencer Martins (Senior Manager, Loans and Credit), José Luis Peixoto (Senior Manager, Foreign Operations) João F Matos (Manager, Exchange Control), Jerónimo de Castro (Manager, Planning and Control). Correspondent Banking: António de Mendia (Assistant Manager, EEC Countries, North America, Japan, Australia and New Zealand), Pedro Cardoso (Assistant Manager, other European Countries), Diniz Ferreira (Assistant Manager, Africa, Central and South America and Asia)

PRINCIPAL ACTIVITIES: All banking activities including domestic clearing facilities, management of international currency transactions and transfers, import and export funding
Foreign Branches: London, New York, Nassau. Representative Office: Madrid
Principal Bankers: Correspondents: folha anexa
Financial Information:

	31.12.86	31.12.87
	Esc'000	Esc'000
Profit before tax	1,997,000	2,920,000
Profit after tax	1,950,000	2,742,000
Capital and reserves	18,500,000	26,825,000
Deposits	533,793,000	612,427,000
Total assets	688,856,000	811,707,000

Principal Shareholders: Nationalised Institution
No of Employees: 6,433 (worldwide)

BANCO NACIONAL ULTRAMARINO

24 Rue Augusta, PO Box 2069, 1102 Lisbon Codex
Tel: 369981; 369991
Cable: Ultranacio
Telex: 13305 bnul p; 13586 bnul p

Executive Board: Maria Manuela Morgado (Chairman), Dr Abel Pinto Repolho Correia, José Maria da Cunha Rego de Amorim (International), Abel Pinto Repolho Correia, Vitor Manuel Carvelho Madureira, Carlos Prieto Traguelho
Senior Executives: Margarida Barros (Correspondent Banks and Syndicated Loans), Margarida Santos Martins (Operations Division), Luis de Almeida Braga (Foreign Exchange Division), Edmundo Rocha (International Division)

PRINCIPAL ACTIVITIES: Banking

No of Employees: 5,569

BANCO PINTO & SOTTO MAYOR

28 Rua do Ouro, 11003 Lisbon
Tel: 37 02 61
Cable: Ottos
Telex: 12516; 18307 ottos
Telefax: 52 42 67

President: A J N Loureiro Borges
Executive Directors: Miguel Krupenski, M M Marinho Babosa, R C C Fortes da Gama, José V S Cardoso da Silva

PRINCIPAL ACTIVITIES: Bank
Financial Information:

	23.3.87
	Esc'000
Profit after tax	1,212,192
Share capital	11,500,000
Participation certificates	8,000,000
Total assets	726,105,654

No of Employees: 6,791

BANCO PORTUGUÊS DO ATLÂNTICO

Praça D.João 1, 28 Porto
Tel: 27971
Cable: TIANDA
Telex: 22720; 22203 TIANDA P

Board of Directors: Dr João dos Santos Oliveira (Chairman), Dr Eduardo Manuel da Silva Rocha, Dr João José Mendonça Esteves da Silva, Dr Licínio Alberto Almeida Cunha, Dr Abílio Xavier Frederico Giraldo do Rosário
Central Managers: Dr António Campos, Eng. António Carvalho Amaro, Dr António Inácio Oliveira Carvalho, Dr António Luís P Drummond de Sousa, Dr Armando Augusto D Morais e Castro, Dr Augusto Joaquim Rodrigues Lopes, Dr Francisco Dias Rosa Jr, Dr João Mendes da Silva, José Manuel Filipe Rocha, Dr José Palla e Carmo, Dr José de Sousa Mendes, Dr Mário Macias Fernandes, Dr Miguel Ribeiro Cadilhe*, Dr Pedro Manuel Libano Monteiro, René Durval Freitas Souto, Dr Vasco Soares da Veiga, Victor Hugo Martins, *appointed by the Ministry of Finance

PORTUGAL

PRINCIPAL ACTIVITIES: Banking
Affiliated Banks: Banco Comercial de Macau (Macau); Banque
Portugaise à Luxembourg SA (Luxembourg)
Financial Information: Consolidated figures

	31.12.86	31.12.87
	Esc'000	Esc'000
Share capital	9,000,000	12,500,000
Deposits	717,586,000	1,074,372,082
Total assets*	1,597,701,000	2,171,153,027

*including contra accounts

No of Employees: 6,361

BANCO TOTTA & AÇORES
Rua Aurea 88, Lisbon 1100
Tel: 369421; 361301
Cable: Tottaçores
Telex: 12266; 16711; 18401; 12144

Board of Directors: Dr Raúl de Almeida Capela (Chairman)
Executive Board: Dr Alipio Pereira Dias, Dr Artur Manuel Pires
Chambel, Dr Augusto Leite Ferreira dos Santos, Dr Fernando
Mário Teixeira de Almeida, Dr José Dias Dinis

PRINCIPAL ACTIVITIES: Bank
Subsidiary Bank: Banco do Oriente SARL
Financial Information:

	1986	1987
	Esc'000	Esc'000
Deposits	470,959,040	562,943,808
Total assets	531,734,010	848,299,445

No of Employees: 4,740

CAIXA GENERAL DE DEPOSITOS
Largo do Calhariz, 1109 Lisboa
Tel: 361987
Telex: 12621

Chairman: Dr A Oliveira Pinto (and President); Vice Chairman:
Dr José Pires Lourenço
Directors: Dr A Morais Cruz, António Gonsalves, António
Redondo
Senior Executives: Dra Marília Baptista (Correspondent Banking
Division), Dr Sobral Torres (Investment Banking Division),
António Redondo (Documentary Credits Division)

PRINCIPAL ACTIVITIES: Banking
Financial Information:

	31.12.86	31.12.87
	Esc'000	Esc'000
Deposits	1,302,044,567	1,484,089,384
Total assets	1,561,711,721	1,845,791,211

No of Employees: 9,650

COMPANHIA DE SEGUROS IMPÉRIO EP
See IMPÉRIO, COMPANHIA DE SEGUROS, EP

COMPANHIA DE SEGUROS MUNDIAL
CONFIANÇA
See MUNDIAL CONFIANÇA, COMPANHIA DE SEGUROS,

FIDELIADADE GRUPO SEGURADOR EP
Largo do Corpo Santo 13, Lisboa
Tel: 360321
Telex: 15407 fidvid p
Telefax: 372208

Chairman: Dr Manuel Raminhos Alves Melo

PRINCIPAL ACTIVITIES: Insurance

Financial Information:

	1986	1987
	Esc'000	Esc'000
Premium income	13,182,722	15,393,753
Investment income	1,337,755	1,795,748
Profit before tax	76,608	130,727
Profit after tax	76,608	130,727
Retained profit	66,608	-
Share capital	1,500,000	1,500,000

No of Employees: 1,869

GUÉRIN-RENT-A-CAR
156 Av Repúb, 8000 Faro
Telex: 56738 Guerin p

PRINCIPAL ACTIVITIES: Car rental services
Financial Information:

	Esc'000
Sales turnover	6,000,000

No of Employees: 1,230

IMPÉRIO, COMPANHIA DE SEGUROS, EP
62 Rua Garrett, Lisboa 2
Tel: (019) 36 23 21, 32 82 81
Cable: Segurimpe
Telex: 12385; 16644 sagimp p

Board of Directors: P Moreira (Chairman), C Borges, M Correia,
A Esteves, T Gil
Senior Executives: R Pedrosa, L Torres (Secretary)

PRINCIPAL ACTIVITIES: Insurance company; general, aviation,
motor, marine life, reinsurance
Financial Information:

	Esc'000
Premium income	9,000,000

Principal Shareholders: State Owned
No of Employees: 1,200

MUNDIAL CONFIANÇA, COMPANHIA DE
SEGUROS,
Largo do Chiado 8, 1200 Lisbon
Tel: 326131; 325884
Telex: 15065 MUNDAL P
Telefax: 364194

Chairman: Pedro Rogério Seixas Vale

PRINCIPAL ACTIVITIES: Insurance
Principal Bankers: Banco Pinto & Sotto Mayor; Banco Fonsecas
& Burnay
Financial Information:

	31.12.86	31.12.87
	Esc'000	Esc'000
Premium income	15,000,000	17,383,000
Investment income	2,409,768	2,670,966
Profit before tax	700,183	807,417
Profit after tax	700,183	807,417
Share capital	1,500,000	1,500,000

No of Employees: 1,612

RAR-SOCIEDADE DE CONTROLE (HOLDING)
SA

RAR Group
Rua do Passeio Alegre 612/624, 4100 Porto
Tel: 672849; 676839
Telex: 27086
Telefax: 676819

Chairman: Jõao Macedo Silva
Directors: Vergílio Folhadela Moreira, José Frederico Jordão,
Rui Macedo Silva, João Nuno Macedo Silva (Sociedade de
Importação SA)

PRINCIPAL ACTIVITIES: Sugar refining and distribution, chocolate, coffee, ice cream, wholesalers, banks, finance and investment, venture capital, real estate and property, general trading and commodity trading, freight forwarding, road haulage, fish processing and wines

Trade Names: RAR, JUBILEU, BUONDI, "GELLATTI MOTTA" (exclusive representatives in Portugal)

Parent Company: Sociedade de Importação SA

Subsidiary Companies: RAR-Refinarias de Acúcar Reunidas; Imperial; Montarroio; Bagão Félix; Uniave; GAIAL; Acembex; Portugel; Vimompor; Maestro; RAR Imobiliária; RAR Soc Investimentos; RAR Soc Capital de Risco; Rartransportes; BPI; BCI; Sefis; Global; Global-Vida; Copam; Soc Lusitana de Destilação; Sucral; Medicalis; Luís & Loureiro etc

Principal Bankers: Banco Português do Atlântico; Banco de Comércio e Indústria

Financial Information: Group figures

	31.12.87	31.12.88
	Esc'000	Esc'000
Group sales turnover	20,831,000	25,429,000
Profit before tax	2,472,000	-
Profit after tax	2,388,000	-
Share capital*	2,000,000	7,500,000

*RAR Holding

Principal Shareholders: Sociedade de Importacão SARL

No of Employees: 1,200 (RAR Group)

RENAULT GEST-SOC. COMÉRCIO DE AUTOMÓVEIS SA

Av Infante D.Henrique 349A, 1800 Lisboa
Tel: 319311
Telex: 42658 rngest p
Telefax: 333237

Chairman: H de Malézieux
Directors: J Silva Pais, V Pacheco

PRINCIPAL ACTIVITIES: Trade in motor vehicles

Trade Names: Renault, Renault Gest

Parent Company: Renault Portuguesa SA

Principal Bankers: Crédit Franco-Portugais; Crédito Predial Português; Banco Totta & Acores; Banco Pinto & Sotto Mayor; Banco Espirito Santo & Comercial de Lisboa

Financial Information:

	31.12.87	31.12.88
	Esc'000	Esc'000
Sales turnover	44,060,980	54,657,426
Profit before tax	1,054,922	1,308,581
Profit after tax	663,241	902,537
Retained profit	474,482	*
Dividends	188,759	*
Dividends per share	274Esc	-
Earnings per share	96Esc	798Esc
Share capital	688,900	1,130,000
Shareholders funds	688,900	1,130,000

*not decided yet

Principal Shareholders: Renault Crédit International; IPE-Investimentos e Participações do Estado, SA; Confranpor

No of Employees: 94

SOCIEDADE FINANCEIRA PORTUGUESA SARL

Rua Castilho 39-14 e 15/B, 1296 Lisboa Codex
Tel: 533 113; 536824; 577220
Cable: SOFINANCIERA-LISBOA
Telex: 18317 sofin p

PRINCIPAL ACTIVITIES: Investment banking services

Financial Information:

	Esc'000
Turnover	6,000,000

No of Employees: 100

UNIÃO DE BANCOS PORTUGUESES

80 Praça de João 1, 4000 Oporto Codex
Tel: 20961; 28981
Cable: UNIPOR
Telex: 26061 UNIBA P

Senior Executives: Guedes da Silva, Fernando Bravo, Sacadura Botte, Jorge Paúl, Antonio Seixas, Marcelino Soares, Armando Esteves, Bana e Costa, Armando Nogueira

PRINCIPAL ACTIVITIES: Bank

Subsidiary Companies: Walter E Heller Factoring Portuguesa SARL; Lusoleading-Sociedade de Locação Financeira Mobiliára, SARL; UNICRE; Cartão International de Crédito, SARL; The Bank of Lisbon and South Africa (South Africa); Banque Portugaise à Luxembourg (Luxembourg)

Financial Information:

	1986	1987
	Esc'000	Esc'000
Deposits	323,275,161	358,870,021
Total assets	549,305,955	672,237,701

No of Employees: 4,318

PORTUGAL

Major Companies of SPAIN

ASEPEYO, MUTUA PATRONAL DE ACCIDENTES DE TRABAJO NO 151

Vía Augusta 36, 08006 Barcelona
Tel: (93) 217.41.44; 217.45.17
Telefax: (93) 237.35.79

President: Alvaro Alvarez Lipkau; Vice Presidents: Antonio Serra Santamans (Vicepresidente 1), Nemesion Fernandez-Cuesta Illana (Vicepresidente 2)
Directors: Jorge Serra (Director General), Ramón Armengou Foix (Director General Adjunto), Pedro Echeverria Cornet (Gerente Secretario General)
Senior Executives: Jesús Serra Bayona (Director de Inversiones), Antonio Mur Restoy (Director Administrativo), Andrés Henriquez Lluch (Director Sanitario), Clemente Bronte Conde (Director de Personal)

PRINCIPAL ACTIVITIES: Workmen's compensation insurance services
Financial Information:

	31.12.86	31.12.87
	Pta'000	Pta'000
Revenues	26,633,000	28,070,000

No of Employees: 1,074

ASISTENCIA SANITARIA INTERPROVINCIAL SA
ASISA

Caracas 12, 28010 Madrid
Tel: (91) 419 01 91
Telex: 44454 asi e

President: Francisco Carreño Castilla
Directors: Fernando Martin Aparicio (Managing Director), Andrés Rodríguez Villa (Managing Director)
Senior Executives: Tomás Diéz Paso (Finance and Administration)

PRINCIPAL ACTIVITIES: Insurance inclusing health insurance
Financial Information:

	Pta'000
Revenues	13,800,000

No of Employees: 315

AURORA-POLAR, SA DE SEGUROS Y REASEGUROS

Plaza Federico Moyúa 4, 48009 Bilbao (Vizcaya)
Tel: (94) 415 92 00
Telex: 31491 APBI E

President: Eduardo de Aguirre Alonso-Allende
Directors: Luis Herrando Prat de la Riba (Managing Director), Roberto Arana (General Director), José Luis Picaza (Administration and Financial Director), Juan José Lecanda Cayero (Management Director), Carlos Javier Biurrun (Risk Management Director), Antonio Hormaeche Ortiz (Life Assurance Director), Jesús Quilez Manrique (Human Resources Director), Pedro Mier (Publicity and Marketing Director), Francisco Javier Otalora (Secreatary)

PRINCIPAL ACTIVITIES: Insurance
Financial Information:

	Pta'000
Revenues	13,600,000

No of Employees: 426

BANCA CATALANA SA

Paseo de Gracia 84, 08008 Barcelona
Tel: (93) 215 71 00
Cable: Bancacatalana
Telex: 54661

Chairman: Alfredo Sáenz Abad; Vice Chairman: Jordi Sagnier Haussmann

Directors: José Angel Merodio Zubiarrain, Francisco Gómez Roldán
Senior Executives: Ramon Buisan (International Division)

PRINCIPAL ACTIVITIES: Bank
Subsidiary Companies: Banco Industriel de Cataluña; Banco Industrial del Mediterraneo; Banco de Barcelona
Financial Information:

	31.12.86	31.12.87
	Pta'000	Pta'000
Share capital	15,350,000	15,349,538
Deposits	433,450,000	619,095,594
Total assets	897,620,000	1,121,476,808

Principal Shareholders: Banco de Vizcaya (88.7%)
No of Employees: 3,263

BANCA MARCH SA

Avenida Alejandro Rosselló 8, 07002 Palma de Mallorca
Tel: (971) 46 15 00
Cable: MARCH
Telex: 68661/68611

Chairman: José Carlos March Delgado; Executive Vice Chairman: Alfredo Lafita Pardo
Directors: Joaquin Fernandez Puig, José María Navarro Ouva, Antonio Rodriguez Robles, Enrique Piñez López, Carmen Delgado Roses, Juan Gimeno Tuset, Simón J Galmes Cerdó, Cristobal Rodríguez Pizá
Senior Executives: Gabriel Cortès Cortés (International Division), Pedro Valdivia (Leasing)

PRINCIPAL ACTIVITIES: Banking
Subsidiary Banks: Gestemar SA (77.2%); Finuma SA (79.12%); Banco NatWest-March (50%); Banco de Asturias (79.15%); Trade Services International Inc; Balboa Finance Ltd (UK); Balboa Finance SA (Switzerland)

No of Employees: 855

BANCO ARABE ESPAÑOL
ARESBANK

Paseo de la Castellana 35, 28046 Madrid
Tel: (91) 419 10 11
Cable: ARESBANK
Telex: 43806 AREB E

Chairman: Abdulla A Saudi; Vice Chairmen: Abdulmuhsen Yousef Al-Hunaif, José Sartorius Alvarez de Bohorquez
Directors: Salem A Zenaty (Managing Director)
Senior Executives: Luis Vañó Martinez (General Manager), F Javier Abad Hernando (Assistant General Manager)

PRINCIPAL ACTIVITIES: Banking services
Subsidiary Banks: Banco de la Exportacion SA (BANEX) (89%)

Principal Shareholders: Libyan Arab Foreign Bank (30%); Kuwait Foreign Trading Contracting and Investment Company (30%); Private sector (16.65%)
No of Employees: 172

BANCO ATLÁNTICO SA

Avda Diagonal 407 bis, 08008 Barcelona
Tel: (93) 237 12 40
Cable: Banticoex
Telex: 52042

Chairman: Abdula A Saudi; Vice Chairman: Rafael Martinez Cortina
Directors: Antonio Sánchez Pedreño
Senior Executives: José Sartorius (International and Financial Division)

PRINCIPAL ACTIVITIES: Bank
Parent Company: Arab Banking Corporation (ABC); Banco Exterior de España; Banco Arabe Español (ARESBANK)
Subsidiary Banks: Banco de Iberoamerica

SPAIN

Financial Information:

	31.12.86	31.12.87
	Pta'000	Pta'000
Share capital	14,625,243	18,281,554
Deposits	399,886,862	490,044,729
Total assets	943,960,000	1,214,570,803

No of Employees: 3,476

BANCO CENTRAL SA
BC

Calle de Alcalá 49, 28014 Madrid
Tel: 532.88.20; 532.88.10
Cable: Centroban
Telex: 27732; 22722

Board of Directors: Alfonso Escámez López (Chairman), Luis Coronel de Palma (Vice Chairman), Epifanio Ridruejo Brieva (Director General Manager), Enrique Marsáns Comas, Luis de Ussía y Gavaldá, Alfonso Güell y Martos, Manuel Garí de Arana, Francisco Daurella Franco, Javier Gil de Biedma y Vega de Seoane, Gabriel Lodares Fontecha, José Vilá Marsáns, María de los Desamparados Villalonga de Jáudenes, Tomás Parejo Camacho, Antonio Escámez López, Pedro Ballvé Lantero, José Ferrer Sala, Fernando Abril Martorell, Antonio Beteré Cabeza, Romualdo García Ambrosio. Secretary General: Juan Bule Hombre

Management: General Managers: Feliso Martínez Picazo, Juan Manuel Echevarría Hernández, Luis Blázquez Torres, Felipe Navalpotro Moreno, Antonio Escámez Torres, Antonio Campos Campos, José Castelló Martínez. General Controller: Eugenio Posada Puntero. Joint General Managers: José María Cucurull Gil, Fernando Bustamante y García de Arboleya, Pedro Carnicero Valdivia. Deputy General Managers: Javier Rios Mingarro, Manuel Godino Muñoz, Francisco Doménech López, José Martínez Ros, José Ramírez Labrador, Jesús Gómez Rodríguez, José Antonio Muñoz López, Senén Marqués Pascual, Ricardo Gutiérrez Acero, Tomás Martín-Consuegra Pozuelo, Luis Ortiz Sánchez, Jesús Manuel Basanta Teijeiro, Lisardo Peláez Acero, Alberto Cadena Escuer, Félix Martín Gómez, Francisco Javier Compo López, José Luis del Valle Doblado, Juan Ignacio Muñiz Entrialgo. General Controller, International: Angel Fernández López. General Inspector: Mariano Ordóñez Piñero

PRINCIPAL ACTIVITIES: Commercial banking

Trade Names: Banco Central, SA

Consolidated Banks: Banco Internacional de Comercio SA; Banco Gallego SA; Banco de Granada SA; Banco de Fomento SA; Banco de Valencia SA; Centrobanco Panamá (Panama); Banco Central of Canada (Canada); Banco Central Corp (Puerto Rico); Banco Central of New York (USA); Banco de Asunción (Paraguay); Banco Popular Argentino (Argentina)

Financial Information:

	31.12.86	31.12.87
	Pta,000	Pta'000
Profit before tax	26,620,000	32,182,000
Profit after tax	20,620,000	22,182,000
Retained profit	10,101,000	10,000,000
Dividends	10,519,000	12,272,000
Dividends per share	150.00Pta	175.00Pta
Earnings per share	379.61Pta	458.92Pta
Share capital	35,063,000	35,063,000*
Shareholders funds	153,563,000	184,601,000
Deposits	1,940,520,000	2,061,494,000
Total assets	2,614,803,000	2,903,629,000

*plus Pta 7,013 million

No of Employees: 18,051 (BC only)

BANCO DE ANDALUCIA

José Antonio Primo de Rivera 18, Jerez de la Frontera 11400
Tel: (956) 34 76 00
Cable: Andalubank
Telex: 75014
President: Fernando de Solís Atienza
Directors: Miguel Gros Obradors (General Manager)
Senior Executives: Lorenzo Martinez Garcia (International Division)
PRINCIPAL ACTIVITIES: Banking
Financial Information:

	31.12.86	31.12.87
	Pta'000	Pta'000
Share capital	2,716,155	2,716,000
Deposits	186,512,000	190,793,000
Total assets	419,626,000	468,379,000

No of Employees: 1,653

BANCO DE BILBAO-VIZCAYA

Ibañez de Bilbao 28, 48009 Bilbao
Tel: 4202000
Cable: Bancobao Admon
Telex: 32055 bbac
Telefax: 34-4-427 3641

Chairmen: José Angel Sánchez Asiaín, Pedro Toledo Ugarte; Vice Chairman: Emilio de Ybarra y Churruca

Board of Directors: Angel Galindez Celayeta, Federico Lipperheide Wicke, Enrique de Sendagorta y Arramburo, Eduardo de Aguirre Alonso-Allende, Gervasio Collar Zabaleta, Juan Manuel de Zubiria y Uhagon, Javier Prado Urquijo, José Manuel Delclaux Barrenechea, Ramon Icaza y Zabalburu, Faustino García Monco y Fernandez, Ignacio Alzola y de la Sota, José Raimundo de Basabe y Manso de Zuñiga, José María Illera Maiz, Ignacio Zubiria Mac-Mahon, Vicente Eulate Mac-Mahon, Fernando de Ybarra y Lopez-Doriga, Pedro Zubiaga y Aldecoa, Carlos Gonzalez Rivero, Carlos Delclaux Oraa, Francisco Javier Aresti y Victoria de Lecea, Guillermo Barandiaran y Alday, Francisco Javier Aresti y Victoria de Lecea, Guillermo Barandiaran y Alday, Francisco Hurtado de Saracho y Epalza, Joaquin Nebreda de Miguel, José Domingo Ampuero y Osma, Luis Lezama Leguizamon Dolagaray, Andres Vilariño Maura, Alfredo Saenz Abad, José Antonio Saenz-Azcunaga Usandizaga, Marcial Echenique Talavera, Luis María de Ybarra y Zubiria, José Javier Gurpide Huarte, Enrique Mas Montañes, Alvaro Rengifo Calderon, Ricardo Muguruza Garteizgogeascoa, Angel Corcostegui Guraya, José Aureliano Recio Arias. Secretary to the Board of Directors: José María Concejo Alvarez

PRINCIPAL ACTIVITIES: Commercial and merchant bank offering a wide range of services through a national network of branches (3,208 Group; 2,436 BBV); issuing house; industrial finance and hire purchase; leasing; insurance; unit trusts; property development; over 80 outlets abroad

Subsidiary Companies: Banca Cataláña; Banco de Extremadura; Banco del Comercio SA; Banco del Oeste SA; Banco Industrial de Bilbao SA; Bilbao Merchant Bank SA; Banc International SA; Banca Mora SA; Banco Bilbao Vizcaya (Deutschland) AG; Banco Bilbao Vizcaya (Panama) SA; Banco Comercial de Mayaguez SA; Banco de Bilbao (Gibraltar) Ltd; Banco de Bilbao (Suisse) SA; Banque de Gestion Financiere (Belgium) SA; Bilbao Vizcaya Bank (Jersey) Ltd and 41 portfolio companies, 23 Financial service companies, 11 real estate companies, 13 other companies

Financial Information:

	31.12.87	31.12.88
	Pta,000	Pta,000
Group profit before tax	97,807,000	122,180,000
Profit after tax	71,055,000	85,086,000
Retained profit	49,705,000	54,361,000
Dividends	21,350,000	28,000,000*
Dividends per share	188Pta	250Pta
Earnings per share	638Pta	760Pta
Share capital	44,705,000	78,400,000
Shareholders funds	273,469,000	469,425,000
Deposits	5,559,370,000	6,257,814,000
Total assets	6,200,344,000	7,082,248,000

*in addition to the dividend a share premium of 2.725 million pesetas was paid this year-non recurrent

No of Employees: 32,027 Group

BANCO DE MADRID

Carrera de Sán Jerónimo 13, Aptd 554, Madrid 28014
Tel: 429 24 43
Cable: Bandri
Telex: 27687 (Madrid Head Office); 42801, 48626 Badri E, (Foreign Dept)
Telefax: 4296605

Board of Directors: Pablo de Garnica Gutierrez (Chairman), Félix Muelas Medrano, José Luis Fominaya Cisneros, Jose Gimenez Almenara, José Luis Sela Alverez
General Managers: José Gimenez Almenara, Liberat Hosta Romagosa
Senior Executives: Vicente Fernandez López (Assistant General Manager, Foreign), Carlos Lluberas Ramirez (Head, International Relations), E Ibarz Soria (Treasury Department)

PRINCIPAL ACTIVITIES: Bank
Parent Company: Banco Español de Crédito
Principal Bankers: Amsterdam: Algemene Bank Nederland; Amsterdam Rotterdam Bank; Bank Mees & Hope; Postbank NV; Nederlandsche Middenstansbank NV. Beijing: Bank of China. Brussels: Banque Bruxelles Lambert; Banque Commerciale de Bruxelles; B Nagelmackers; Genérale de Banque; Kredietbank NV. Copenhagen: Andelsbanken AS; Danebank; Copenhagen Handelsbank AS; Den Danske Bank; Jyskebank; Privatbanken AS. Frankfurt: Bank für Gemeinwirtschaft; Commerzbank AG; Deutsche Bank AG; Dresdner Bank AG. Geneva: Société de Banque Suisse. Helsinki: Kansallis Osake Pankki; Okobank; Union Bank of Finland. Hong Kong: Kwong On Bank Ltd; Security Pacific Asian Bank Ltd. London: Standard Chartered Bank; Barclays Bank; Midland Bank; National Westminster Bank; The Royal Bank of Scotland. Luxembourg: Banque Génerale de Luxemburg; Banque de l'Union des Coopérateurs Luxembourgeois SA. Melbourne: Australia and New Zealand Banking Group Ltd. Milan: Banca Commerciale Italiana; Credito Italiano; Banca Popolare di Milano; Nuovo Banco Ambrosiano. Montreal: The Royal Bank of Canada; Bank of Montreal. New York: Security Pacific International Bank; American Express; Banesto Banking Corporation; Chemical Bank; Citibank; Continental Bank International. Oslo: Bergen Bank; Den Norske Creditbank. Paris: Banque de Baecque Beau; Banque Française du Commerce Extérieur; Banque Nationale de Paris; Banque Paribas. Crédit Commercial de France. San Francisco: Bank of America. Singapore: Algemene Bank Nederland. Sydney: Westpac Banking Corporation. Tokyo: The Mitsui Bank Ltd; The Sumitomo Bank Ltd; The Fuji Bank Ltd; The Dia Ichi Kangyo Bank Ltd; Bank of Tokyo; The Sanwa Bank Ltd. Vienna: Creditanstalt Bankverein; Genossenschaftliche Zentralbank AG. Zürich: Crédit Suisse; Union des Banques Suisses

Financial Information:

	31.12.86	31.12.87
	Pta'000	Pta'000
Operating income before tax	2,409,000	3,695,300
Income for the year	7,819,000	9,719,000
Retained profit	655,000	500,000
Dividends	-	254,000
Dividends per share	-	20Pta
Earnings per share	25Pta	59Pta
Share capital	12,718,125	6,359,062
Shareholders funds	13,369,000	14,123,000
Deposits	134,321,229	226,648,313
Total assets	183,919,433	321,623,569

Principal Shareholders: Banco Español de Crédito (98%)
No of Employees: 2,337

BANCO DE SABADELL

Plaza de Catalunya 1, 08201 Sabadell (Barcelona)
Tel: (3) 726 21 00
Cable: Sabadellex
Telex: 59817 BANSA E
Telefax: (3) 726 97 33

President: Juan Corominas Vila
Directors: Félix Crusafón Griera (Vicepresidente), Tomás Casañas Guri, Buenaventura Garriga Brutau, Antonio Ferrer Sabater, Domingo Fatjó Sanmiquel
Senior Executives: Juan Oliu Pich (Director-General Manager), Esteve Ma Faus Mompart (General Secretary), José Montañés Also (Deputy General Manager), Llibert Barcons Freixas (Deputy General Manager), Joan M Desvalls Maristany (Deputy General Manager), José Oliu Creus (General Secretary, Technical)

PRINCIPAL ACTIVITIES: All banking activities
Financial Information:

	31.12.86	31.12.87
	Pta'000	Pta'000
Share capital	5,250,000	6,300,000
Deposits	398,105,000	505,104,000
Total assets	491,496,000	981,529,000

No of Employees: 3,221

BANCO DE SANTANDER

Paseo de Pereda 9-12, Santander
Tel: 22 11 00; 22 12 00
Cable: Bansander
Telex: 35833

Chairman: Emilio Botin Rios; Deputy Chairman: Pablo Tarrero Rivero; Deputy Chairman: Jaime Botín-Sanz de Sautuola
Directors: Santiago Corral y Pérez, Ignacio Soler de la Riva, Juan José Martínez Vázquez, César Martínez Beascoechea, Luis Bergé Abarca, Ramón Quijano Secades, Joaquin Chapaprieta Ornstein, Angel Jado Becerro de Bengoa, Juan Secades y González-Camino, Alberto Folch Rusiñol, Dimas Blanco Valdivieso, Rafael Alonso Botín, José Luis Diaz Fernández, John J Creedon, José Miguel Serrano Goyria
Vice Presidents: Rafael Alonso Botin, José Luis Diaz Fernandez, Rodrigo Echenique Gordillo, José Ma. Fernandez-Castañera, Carlos Gerardo Garcia Rodriguez, José Luis M. Marauri Bujanda, Julio Novo Polidura, Matias Rodriguez Inciarte, Enrique Soroa de las Cuevas

PRINCIPAL ACTIVITIES: All banking services
Consolidated Santander Group: Spanish Banks: Banco Santander; B. Santander de Negocios; Banco de Murcia; Banco Comercial Español; Banca Jover. Spanish Companies: Bansaleasing; Santander de Factoring; Hipotebansa; Gesbansander; Gesbansa; Bansander Noble Lowndes; Bansafina and 28 instrumentality companies. Foreign Banks: CC-Bank (Germany); Banco Español (Chile); Banco Santander International (USA); Banco Santander (Uruguay); Banco Santander (Puerto Rico); Banco Santander (Panama); Banco Santander (Argentina); B. Santander Trust & Banking Corp.

(Bahamas). Foreign Companies: Santander Finanz A.G. (Switzerland); Santander International Fund Management Company; Bansander Leasing (Puerto Rico) and 13 instrumentality companies

Financial Information:

	31.12.86	31.12.87
	Pta'000	Pta'000
Profit before tax	32,298,000	40,027,000
Profit after tax	22,156,000	26,701,000
Retained profit	81,450,000	124,314,000
Dividends	8,351,000	11,021,000
Dividends per share	90Pta	110Pta
Earnings per share	238.36Pta	256.59Pta
Share capital	43,687,000	52,030,000
Shareholders funds	126,535,000	154,290,000
Deposits	1,768,765,000	2,040,209,000
Total assets	2,773,248,000	2,928,839,000

No of Employees: 9,700

BANCO EXTERIOR DE ESPAÑA

Carrera de San Jerónimo 36, 28014 Madrid
Tel: 429.44.77
Cable: Extebank
Telex: 22033; 23698
Telefax: 429.83.42

President: Miguel Boyer Salvador; General Secretary: Miguel Vizcaíno Calderón

Directors: Rafael Martínez Cortiña (Vice Presidente y Consejero Delegado), José Luis Mora Mallo (Director General), Ramón Fernández Rubies (Director General), Joaquín Casuso Posada (Director General), Calixto Rios Pérez (Director General)

Assistant General Managers: Nemesio Fernández-Cuesta Illana, José Tejón Martínez, Mariano Casuso Posada, Luis Trujillano Puya, José A Morales Arranz, Francisco Stuyck Collado, Juan Zapata Gallego, Eugenio Dominguez García, César Rodríguez Alonso, Fernando Vivar Mira, Antonio Ruiz Ramos

PRINCIPAL ACTIVITIES: A commercial bank specialising in export credit and general foreign transactions

Subsidiary Banks: Banco Exterior-France SA; Banco Exterior-UK; Banco Exterior-Deutschland; Banco Exterior-Bélgica SA; Banco Exterior-Nederland NV; Banco Exterior-(Suiza) SA; Extebank-NY; Banco Exterior SA-Argentina; Banco-Exterior SA-Chile; Banco Exterior SA-Panamá; Banco Exterior SA-Paraguay; Banco Exterior SA-Uruguay; Banco Cantábrico; Banco de Alicante SA; Banco Simeón SA; Associated Banks: Banco Exterior de los Andes y de España SA; Banco Atlántico SA; Banco Exterior de Guinea Ecuatorial y de España; Banco Exterior Misr SA; Comercial Companies: Financeira do Comercio Exterior SARL; Panamá Investment Holding SA; Exterior International Ltd; Financiera Exterior SA

Financial Information:

	31.12.86	31.12.87
	Pta'000	Pta'000
Profit before tax	7,456,000	8,766,000
Profit after tax	5,956,000	6,574,000
Retained profit	2,104,000	2,400,000
Dividends	3,852,000	4,165,000
Dividends per share	92.50Pta	100.0Pta
Earnings per share	18.5%	-
Share capital	20,823,000	20,823,000
Shareholders funds	33,202,000	-
Deposits	685,451,000	756,000,000
Total assets	2,534,317,000	2,578,761,000

Principal Shareholders: The Spanish State (51.41%); Spain's Industrial Official Holding (10%); Bank of Spain (7.74%); Private Shareholders (30.85%)

No of Employees: 12,779

BANCO HISPANO AMERICANO

Plaza de Canalejas 1, 28014 Madrid
Tel: (91) 222 46 60
Cable: Hispamer
Telex: 44.404

Board of Directors: Luis de Usera y López-González (Honorary Chairman), Eugenio Rodriguez Pascual (Honorary Deputy Chairman), Claudio Boada Vilallonga (Chairman and Chief Executive Officer), José Ma Amusátegui de la Cierva (Deputy Chairman and Managing Director), José Manuel Arburúa Aspiunza, Fernando de Asúa Alvarez, Antonio Barrera de Irimo, Antonio Basagoiti Amézaga, Javier Benjumea Puigcerver, Leopoldo Calvo Sotelo y Bustelo, Pedro Durán Farell, Juan Lladró Dolz, Albert Oliart Saussol, Rafael del Pino y Moreno, Enrique Prada Rodriguez-Viforcos, Jürgen Reimnitz

Management Team: Chairman and Chief Executive Officer: *Claudio Boada Villalonga. Deputy Chairman and Managing Director: *José María Amusategui de la Cierva. General Managers: *Antonio Basagoiti García-Tuñón, *Benito Tamayo Hernáez, Emilio Novela Berlin, *Baldomero Falcones Jaquotot, *José Capón Rey, Miguel Geijo Baucells, *Antonio José Zoido Martinez, *Ramón Colao Caicoya. Secretary: Antonio de Huyos González. Deputy General Managers: *Alvaro Fernández Villaverde y de Silva, *Jorge Hay Guajardo-Fajardo, *Member of Management Committee

PRINCIPAL ACTIVITIES: Banking and financial services

Trade Names: Banco Hispano Americano

Subsidiary/Associated Banks: Banco Hispano Americano Benelux; Hispano Americano Sociedade Investimentos; Banco Mercantil de Tarragona SA; Banco Hispano Americano Ltd; Hispano Hipotecario SA; Corporación Financiera Hispamer; Corporación Inmobiliaria Hispamer; La Estrella; Banco del Norte; Banco de Jerez; Banif

Financial Information: Financial Group Consolidated Figures

	31.12.87	31.12.88
	Pta'000	Pta'000
Turnover	327,055,000	365,742,000
Profit before tax	6,705,000	51,459,000
Profit after tax	4,270,000	35,768,000
Retained profit	1,545,000	18,692,000
Dividends	8,175,000	13,915,000
Dividends per share	100Pta	170Pta
Earnings per share	19Pta	398Pta
Share capital*	40,877,000	41,483,000
Shareholders funds	142,979,000	164,581,000
Deposits and debt secur	2,587,086,000	2,186,506,000
Total assets	2,859,856,000	3,150,788,000

*500 Ptas per share

No of Employees: 17,908

BANCO PASTOR SA

Cantón Pequeño 1, 15003 La Coruña
Tel: (981) 22 41 00
Cable: Pastor
Telex: 82178

Chairman: Carmela Arias y Díaz de Rábago (Countess of Fenosa); Vice Chairmen: Francisco Gómez Fernández, Josaquín Arias y Díaz de Rábago

General Managers: Ramón Linares Martín de Rosales (Director and Executive General Manager), Alfonso Porras del Corral (Director and General Manager), Francisco José Mañas López (General Manager), Vicente Arces Mosquera

Senior Executives: Ignacio de Pablo Burgoa (Assistant General Manager-International), Jaime Crespo (Deputy General Manager-Treasury Department), José Maria Delgado (Manager-Public Relations)

PRINCIPAL ACTIVITIES: Commercial and international banking

Principal Bankers: Morgan; Chase; Citibank; Manufacturers; Deutsche Bank; BHF; Midland; BNP; CCF

Principal Shareholders: Fundación Pedro Barrié de la Maza

No of Employees: 3,423

BANCO POPULAR ESPAÑOL

Velazquez 34, Madrid 1

Tel: + 34 91 435 36 20

Cable: Popularbanex

Telex: 22511 BPE M

Board of Directors: Luis Valls (Chairman), Javier Valls (Deputy Chairman), Gabriel Gancedo (Secretary). Executive Directors: Bank's Managers Association (A professional body of more than 2500 managers and officers of the Bank) (Management and Control), Francisco Donate (Organization and Human Rescources), Juan Manuel Fanjul (Legal Advisor), Manuel Laffón (Andalucia), Enrique Pérez-Sala (Catalonia), Giorgio Stecher (International), Pedro Valls (Presidency). Directors: José María Casadellá, José María Catá, Gonzalo Fernández de la Mora, José Manuel Hernández-Suárez, Luis Miralles, Casimiro Molins, Luis Montuenga, Miguel Nigorra, José Ramón Rodriguez, Javier Serra, Fernando Solís, Rafael Termes (Managing Director, under suspension of functions while Chairman of the private Spanish banking association)

General Management: Ildefonso Ayala (Chief Executive). Secretaries: Benjamín Galiana (Managing Committee), Manuel Martín (Economics and Miscellaneous), Rodolfo Velarde (Chief Financial Officer). Managers: Angel Gómez (Consumer Banking), Ricardo Lacasa (Banking Subsidiaries), Fermín Peláez (Auditing), Mariano Peláez (Credit Committee), David Vives (Merchant Banking), José Luis Sanz (Risk and Productivity)

Principal International Officers: José Fernández (Foreign Treasury), José Antonio Fuentes (Foreign Trade Promotion), Julio Garcia (Equity and Capital Markets), Jaime Gutiérrez de Terán (Foreign Transactions Accounting), Roberto Higuera (International Financing), Francisco Manso (Foreign Risk Control), Luis Pintado (Correspondent Banking), Rafael Ruiz de Luna (Foreign Exchange Trading), Fernando Soto (Multinationals)

PRINCIPAL ACTIVITIES: International bank

Subsidiary Companies: (100% owned unless otherwise stated): Banking: Banco de Andalucia (63.02%); Banco de Castilla (84.68%); Banco de Crédito Balear (52.82%); Banco de Galicia (78.57%); Banco de Vasconia (88.99%); Banco Popular Industrial (Eurobanco) (98.4%). Finance Companies: Financiera de Expansión de Ventas (Fiventas) (97.52%); Heller Factoring (50%); Iberleasing (95.79%). Investment Companies: Análisis de Inversiones Mobiliarias (Andinsa) (73.16%); Cartera Balear (52.82%); Cartera de Valores de Galicia (Cartegal) (98.4%); Centro de Inversiones; Compañía General de Inversiones Mobiliarias; Depósitos Comerciales; Estudios Financieros de Valores (73.14%); Gestora Europea de Inversiones (Eurogestión) (99.2%); Gestora Popular; Inversora Vasconia (71.16%); Mobiliaria Continental de Inversiones (98.66%); Mobiliaria General; Móbiliaria Nacional Inversora; Sogeval (99.17%); Unión Mobiliaria; Unión de Valores (Unival) (73.16%); Universal de Inversiones (Unisa) (95.32%); Valores Mediterráneos (95.24%); Valtisa (95.24%). Real Estate Companies: Cedalca; Finespa; Inmobiliaria Estasa (94.48%); Inmobiliaria Lecasa (94.48%); Inmobilaria Mallorquina (52.82%); Inmobiliaria Nervasa (94.48%); Inmobiliaria San Nicolás de Bari (63.02%); Inmobiliaria Viagracia (99.89%); Inmobiliaria Villasa (94.48%); Inmobiliaria Vivesa (94.48%); Inmuebles Salamanca (84.68%); Promoción Goya-Velázquez (99.88%); Renta Balear (52.82%). Financial subsidiaries abroad: Banco Popular Español-Cayman (Grand Cayman); Bolsa (80%) (France); Popular Español International (Netherlands)

Financial Information: Consolidated figures

	31.12.86	31.12.87
	Pta'000	Pta'000
Income before tax	33,599,000	38,054,000
Income after tax	22,399,000	24,285,000
Retained earnings	6,954,000	8,591,000
Dividends	5,248,000	8,092,000
Dividends per share	243.00Pta	280.00Pta
Earnings per share	675.28Pta	746.23Pta
Common stock	10,799,000	14,450,000
Shareholders equity	66,603,000	86,999,000
Customer deposits	1,281,960,000	1,395,448,000
Total assets	1,667,202,000	1,811,226,000

No of Employees: 11,682

BANCO URQUIJO UNION SA

Paseo de la Castellana 46, Madrid 28046

Tel: 4353612; 4353025

Cable: Urquijo

Telex: 27602; 27256; 45651; 45654 BUKMD

Telefax: 275.81.83/276.93.75/276.24.45

Board of Directors: José Maria Amusátegui (Chairman), Francisco Novela (Managing Director), Antonio Basagoiti (Director), José Capón (Director), Ramón Colao (Director), Emilio Novela (Director), Miguel Geijo (Director), Antonio Zoido (Director), José María Ramírez (Secretary of the Board)

Senior Management: Javier Ruiz-Ogarrio (General Manager-Affiliated Companies), Mario Fernández Peña (General Manager-Corporate and International Banking), Jesüs Lladó (Assistant General Manager, Capital Markets), Ricardo Rullán (Assistant General Manager, International Banking), Manuel M Villanueva (Assistant General Manager, Treasury Financial Asst and Foreign Exchange), Gustavo Puceiro (Deputy General Manager Commercial Banking), Santiago Montesinos (Deputy General Manager Planification and Control), Leoncio Afonso (Deputy General Manager Organisation and Human Resources)

International Banking: Ricardo Rullán (Assistant General Manager), Manuel García Cano (Manager-Foreign Trade), Carlos Cañal (Manager-International Banking Relations), Juan Pedro Uscola (Manager-International Finance Dept), José Miguel Calleja (Manager Operations). Treasury, Financial Assets and Foreign Exchange: Manuel M Villanueva (Assistant General Manager), Francisco Javier Torres (Chief Dealer-Foreign Exchange and Money Market), Gonzalo de Cordoba (Chief Dealer-Treasury and Financial Assets). Capital Markets: Jesus Lladó (Assistant General Manager), Alfredo Alvarez Pickman (Manager Promotion), Pablo Prada (Manager Operations)

PRINCIPAL ACTIVITIES: All banking activites

Trade Names: Urquijo Union

Parent Company: Banco Hispano Americano (96%)

Subsidiary Banks: Through Urquijo International NV, The Hague, Netherlands:- Banco Hispano Americano Ltd (London) (50%)

Principal Bankers: Bankers Trust; Barclays Bank; Chase Manhattan; Morgan Guaranty Trust; Citibank; Commerzbank; Deutsche Bank; National Westminster Bank; Manufacturers Hanover Trust

Financial Information:

	31.12.86	31.12.87
	Pta'000	Pta'000
Profit before tax	1,044,000	1,500,000
Profit after tax	1,044,000	1,500,000
Retained profit	1,044,000	1,500,000
Share capital	23,342,000	23,342,000
Deposits	467,535,000	538,227,000
Total assets	519,659,000	596,401,000

No of Employees: 2,513

BANCO VITALICIO DE ESPAÑA, CÍA, ANÓNIMA DE SEGUROS Y REASEGUROS

Paseo de Gracia 11, 08007 Barcelona
Tel: (93) 318 56 00
Telex: 54447 vitae e
Telefax: 318 22 76

President: Alfonso Escámez López; Vice President: Manuel Gari de Arana
Directors: José Luis Pérez Torres (Director General), Joaquin Mas de Pablo (Financial Director), Juan José Loscos Sesen (Administration Director), Vicente Tardío Barutel (Technical Director), Ramón Baulies (Personnel Director)
Senior Executives: Enrique Benavente Mata (Secretary)

PRINCIPAL ACTIVITIES: Insurance
Financial Information:

	Pta'000
Revenues	52,800,000

No of Employees: 601

BANCO ZARAGOZANO SA

Coso 47, 50003 Zaragoza
Tel: 23 69 00
Cable: Banzano
Telex: Zaragoza 58044 BANZA E
Telefax: 976-21 56 78

Chairman: José Ramon Alvarez Rendueles; Deputy Chairman: Eduardo Solans Abadías
Directors: Juan Luis Barrenechea, Antonio García Bernal, Jorge Oriol Vilanova, Jaime Requeijo González, Juan José Calvo Pujol, Guillermo Visedo Navarro, Antonio Hernández-Gil Alvarez-Cienfuegos, Eugenio Domingo Solans, José Ma Mas Millet
Chief Executive: José Luis Arrietta Barrenechea

PRINCIPAL ACTIVITIES: Banking
Subsidiary Companies: (100% owned): Banco de Toledo SA; Leasing Banzano SA; Financiera Banzano SA; Gesbanzano SA; Banzano Hipotecario SA; Sureste SA
Financial Information:

	31.12.87
	Pta'000
Profit before tax	2,640,827
Profit after tax	1,832,427
Retained profit	971,357
Dividend	861,070
Dividend per share	75Pta
Earnings per share	230Pta
Share capital	5,740,467
Shareholders funds*	9,173,004
Deposits	220,487,958
Total assets	284,700,289

*After profit distribution

No of Employees: 2,125

BANKINTER

Banco Intercontinental Español SA

Paseo de la Castellana 29, 28046 Madrid
Tel: 419 95 00
Cable: Bankinter
Telex: 27285; 42760; 45629; 41286 BANKI E
Telefax: (341) 419 33 87

President: Jaime Botín
Directors: Juan Arena de la Mora (General Manager), Emilio Botín-Sanz de Sautuola y Garcia de los Rios, Emilio Botín-Sanz de Sautuola y Lopéz, Santiago Corral, Joaquin Chapaprieta, Juan Secades, Paul A Verburgt
PRINCIPAL ACTIVITIES: Banking
Subsidiary Companies: (100% owned unless stated):
Hispamarket SA; Intermobiliaria SA; Intergestora SA; Bankinter Financiera SA; Gesbankinter SA; Aircraft SA

(93.2%); Bankinter S.C.H. SA; Bankinter Leasing SA; Bankinter de Inversiones S.I.M. SA
Financial Information:

	31.12.86	31.12.87
	Pta'000	Pta'000
Profit before tax	8,140,000	12,217,000
Profit after tax	5,655,000	8,772,000
Dividend	1,720,000	2,422,000
Dividend per share	290Pta	285Pta
Earnings per share	952Pta	616Pta
Share capital	7,428,000	10,685,000
Deposits	134,972,000	322,528,000
Total assets	412,829,000	691,140,000

No of Employees: 1,494

BASANDER DE LEASING SA

Paseo de la Castellana 75, 28046 Madrid
Tel: (91) 581 30 00

Chairman: Matías Rodríguez Inciarte
Directors: José María Espi Martínez

PRINCIPAL ACTIVITIES: Leasing services
Financial Information:

	Pta'000
Sales turnover	38,000,000

Principal Shareholders: Banco de Santander
No of Employees: 11

BILBAO CIA. ANMA. DE SEGUROS Y REASEGUROS

Paseo del Puerto 20, E-48.990 Neguri-Getxo
Tel: (94) 443 64 00
Telex: 32224 BILBO E

President: Emilio Sorda Gómez
Directors: José Miguel Sorda Gallano (Managing Director)
Senior Executives: Juan Ramón Eganã Rodríguez (Publicity)

PRINCIPAL ACTIVITIES: Insurance services
Financial Information:

	1987	1988
	Pta'000	Pta'000
Premium income	12,950,700	15,467,300
Investment income	1,644,800	2,854,500
Profit (loss) before tax	65,800	(836,600)
Profit (loss) after tax	65,800	(836,600)
Retained profit	65,800	-
Earnings per share	8.16%	-
Share capital	806,400	998,400
Shareholders funds	2,302,700	2,464,500

Principal Shareholders: Amev NV (75%) (Netherlands)
No of Employees: 518

CAIXA DE BARCELONA

Avda Diagonal 530, Barcelona 08006
Tel: (3) 201.6666
Telex: 97214
Telefax: (3) 202.0204

Chairman: José-Juan Pintó Ruiz
Directors: Juan Torres Picamal (Managing Director)
Senior Executives: Josep Morist Puig (Manager International Division)

PRINCIPAL ACTIVITIES: Savings bank
Financial Information:

	31.12.86	31.12.87
	Pta'000	Pta'000
Profit before tax	-	8,001,000
Profit after tax	-	6,360,000
Retained profit	-	3,188,000
Share capital	29,869,000	31,954,000
Deposits	569,991,000	664,432,000
Total assets	670,291,000	813,242,000

CAJA DE AHORROS DE ZARAGOZA, ARAGON Y RIOJA

CAZAR

Plaza Paraíso 2, 50008 Zaragoza
Tel: (976) 22 10 81
Cable: Cajamonte Zaragoza
Telex: 58743/4 cazin e
Telefax: (76) 21 18 47

Chairman: Fernando Almarza Laguna de Rins
Directors: Amado Franco Lahoz (General Manager)
Senior Executives: Alfonson de Pedro Alfaro (Head of
International Division)

PRINCIPAL ACTIVITIES: Banking
Financial Information:

	31.12.87
	Pta'000
Total assets	503,386,000

CAJA DE AHORROS Y MONTE DE PIEDAD DE MADRID

Plaza Celenque 2, 28013 Madrid
Tel: (91) 445 32 00
Cable: cajamadrid
Telex: 23090

Chairman: Felipe Ruiz de Velasco y de Castro
Directors: Mateo Ruiz Oriol (Managing Director)
Senior Executives: Carlos Martínez Martínez (Head of
International Division)

PRINCIPAL ACTIVITIES: Banking
Financial Information:

	31.12.87
	Pta'000
Total assets	1,386,718,000

CAJA DE PENSIONES LA CAIXA

Caja de Pensiones para la Vejez de Ahorros de Cataluña y Baleares

Avenida Diagonal 621-629, 08028 Barcelona
Tel: 34-3-404 60 00
Cable: CAIXA
Telex: 50321
Telefax: 34-3-339 57 03

Chairman: Juan Antonio Samaranch; *President:* José Vilarasau
(and Chief Executive Officer)
Executive Vice-Presidents: Ricardo Fornesa, Abel del Ruste,
Isidro Fainé, Antonio Brufau. Assistant Vice-Presidents: Josep
M Samaranch, Juan J Toribio, Joaquín Clotet, Antonio
Massanell
General Managers: Manuel Ventura (Controller), Joan M Solá
(Planning), Eugenio Martínez (Investment), Jaume Lanaspa
(On leave. President, Banque pour la Construction et
l'Equipement), Tomás Muniesa (Customer Resources and
Capital Market). Deputy General Managers: Pere Turró
(Assistant to Executive Vice-President), Josep Massip
(Internal Audit), Josep M Miranda (General Assistant
Secretary, Legal), Josep Gener (Services Management),
Esteve Valero (Real Estate), Angel Monell (Area Manager,
City of Barcelona), Joan J Cuesta (Community Welfare
Projects), Climent Vilella (Area Manager, Barcelona Province-
East), Gerard Palacín (Operations Audit), Josep M Gasol
(General Secretary, GrupCaixa SA), Alfonso Maristany
(International Division Manager), Francisco Casals (Managing
Director, AgroCaixa SA), Josep F Vilardell (On leave.
President, Société de Crédit et de Banque de Monaco), Josep
M Sabaté (Area Manager, Lleida), Francisco Saldaña (Area
Manager, Barcelona Provice-West), Miquel Casellas (Area
Manager, Girona), Robert Leporace (Area Manager,
Tarragona), Angel Docampo (Area Manager, Balearic Islands),
Josep M Carrau (President's Office and Research
Department), Josep Forcada (Internal Administration),
Sebastián Sastre (Legal Services), Antonio Vila (Commercial).
Assistant General Managers: Jordi Campamá (Special
Branches), Javier J Paso (Taxation Services), José A Ruiz
(Area Manager, Northern Spain Area), Santiago Arnán
(Information Systems), Xavier Vidal (EDP Production), Angel
Pes (Assistant Area Manager, City of Barcelona).
Administrators: Jordi Hereter (Community Welfare Projects
Personnel), Joan Baldé (Financial Services), Albert Grau
(Investment), Joan Pagespetit (New Products and Sevices),
Alberto Defez (Decentralized EDP Services), Gloria Trías
(Personnel, GrupCaixa SA), Celestí de Manuel (Services
Administration), José A Campo (Assistant Area Manager,
Northern Area), F Xavier Rebull (Personnel), Jesús Escolano
(Risk Control and International). Departmental Managers:
Antoni García (Administration), Juan J Estarellas (International
Division), Ferran Matia (Building and Maintenance), José M
Arenas (Self-service Systems), Jordi Rodríguez (Head Office
Branch), Josep M Mòdol (Controller), Javier I Peyra
(Commercial Planning), Lluís Ramos (Payment Systems), José
Cerveró (Head Office Branch - "Diagonal"), Manuel Rosales
(Security), Anton Gasol (Cost Accounting), Juan Viadel
(Organization and Projects), Alfonso López de Murillas
(Personnel), Xavier Alfaras (Community Welfare Projects
Planning and Management), Francisco Romero (Treasury),
Fernando Cánovas (Controller GrupCaixa SA), Rosa Cullell
(Communication), Hipólito Fernández (General Secretary,
AgroCaixa SA), Abel Zuazua (Assistant Managing Director,
AgroCaixa SA), José Viladot (Loans)

PRINCIPAL ACTIVITIES: Savings bank
Financial Information: Figures not consolidated

	31.12.87	31.12.88
	Pta'000	Pta'000
Profit before tax	16,226,000	20,191,000
Profit after tax	14,500,000	16,662,000
Retained profit	7,250,000	8,331,000
Rserves	116,619,000	125,808,000
Deposits	2,028,363,000	2,516,916,000
Total assets	2,431,335,000	3,008,648,000

No of Employees: 6,412

CAJA DE SEGUROS REUNIDOS SA

CASER

Plaza Lealtad 4, 28014 Madrid
Tel: (91) 446 30 08

President: José Joaquín Sancho Dronda; *Vice President:*
Valentín Arroyo Ruipérez
Directors: Eloy Estévez Ramos (Director Generai), Joaquín Rollo
Brillo (Managing Director), Antonio Reina Gallego (Director of
Finance and Administration), Felipe Ruíz de Velasco (Legal
Director and Secretary)

PRINCIPAL ACTIVITIES: Insurance services
Subsidiary Companies: Minerva; Unión Condal de Seguros e
Intercaser
Financial Information:

	Pta'000
Revenues	35,000,000

No of Employees: 446

CATALANA DE COBROS Y FACTORING SA

CACYF

Rosellón 192, 08008 Barcelona
Tel: (93) 254 42 00
Telex: 52456 cacyfe

President: José María Sanclimens Genescá (and Director
General)
Directors: Carles Bellmonte Vila (Assistant Director General),
Antonio Grandía Rosell (Financial Director), Victor Torio Yarza
(Factoring Director), José María Jobani Riba (Debt Director)
Senior Executives: Luís Solana (Secretary)

SPAIN

PRINCIPAL ACTIVITIES: Financial services including debt collection and factoring facilities
Financial Information:

	Pta'000
Revenues	11,900,000

No of Employees: 107

CATALANA/OCCIDENTE, GRUPO ASEGURADOR
Av Alcalde barnils s/n, Apto 29, 08190 St Cugat d Vallés (Barcelona)
Tel: (93) 674 02 08
Telex: 52908 CATOC

President: José María Juncadella (La Catalaña); Joaquin de Cárdeñas (Occidente)
Directors: Jésus Serra Santamans (Managing Director), José María Serra Farré (Director General), Javier Villavecchia (Director General)
Senior Executives: Vicente Closa, Francisco Carné, Jésus Serra Farré, Valentin Oliveras

PRINCIPAL ACTIVITIES: Insurance
Financial Information:

	Pta'000
Revenues	27,930,000

No of Employees: 1,014

CENTRAL DE LEASING SA
LICO
Miguel Angel 23, 28010 Madrid
Tel: (91) 410 21 54
Telex: 27310 LICO E

Chairman: Jesús Muzas Rubio
Directors: Tomás Pérez Ruiz (Director General and Manager), Manuel P Bernáldez (Secretary and Personnel Director), Miguel Angel Espadas Giménez (Financial Director), Antonio Campillo del Campo (Commercial Director), Jaime Pastorín Tapia-Ruano (Technical Director), Francisco de Lorenzo Fontela (Publicity Director)

PRINCIPAL ACTIVITIES: Leasing services
Financial Information:

	Pta'000
Sales turnover	15,000,000

No of Employees: 80

COFIC
Compañia de Financiacion SA
Magellenes 3, 28015 Madrid
Tel: (91) 447 30 00
Telex: 22900

Chairman: Ignacio Gómez-Acebo y Duque de Estrada
Directors: Michel Bricont (Managing Director), José María Jerez (Commercial Director), Francisco Piñeiro Martes (Commercial Director), Guillermo Moreno Rodríguez (Finance Director)

PRINCIPAL ACTIVITIES: Financial services
Financial Information:

	Pta'000
Turnover	74,000,000

No of Employees: 16

CORPORACION MAPFRE COMPAÑIA INTERNACIONAL DE REASEGUROS SA
See MAPFRE, CORPORACION, COMPAÑIA INTERNACIONAL DE REASEGUROS SA

ESTRELLA, LA,
Gran Via 7, 28012 Madrid
Tel: (91) 232 65 14
Telex: 45197

President: Benito Tamayo Hernáez; Vice President: Enrique Prada y Rodríguez de Viforcos
Directors: Felipe Pujol Plans
Senior Executives: Antonio de Hoyos (Secretary)

PRINCIPAL ACTIVITIES: Insurance services including life, accident and general risk
Financial Information:

	Pta'000
Revenues	54,250,000

Principal Shareholders: Aetna Life and Casualty Company (43.84%)
No of Employees: 694

FERNANDO ROQUÉ TRANSPORTES INTERNACIONALES SA
See ROQUÉ, FERNANDO, TRANSPORTES INTERNACIONALES SA

FINAMERSA, ENTIDAD FINANCIACION
Lagasca 88, 28001 Madrid
Tel: (91) 431 48 78
Telex: 43860 FINAM E

Chairman: Benito Tamayo Hernáez
Directors: Teodoro Bergés Madrid, Angel Rodríguez de Gracia, Alvaro Domecq Bertrán de Lis (Commercial Director), José Antonio Corrales (Secretary)

PRINCIPAL ACTIVITIES: Finance and investment
Parent Company: Banco Hispano Americano
Financial Information:

	Pta'000
Turnover	63,000,000

No of Employees: 637

FIRST CHICAGO SPAIN SA, ENTIDAD DE FINANCIACIÓN
Paseo de la Castellana 18, 28046 Madrid
Tel: (91) 435 33 44
Telefax: (91) 435 33 44

Senior Executives: Rolf K Zion (Manager)

PRINCIPAL ACTIVITIES: Finance and credit services
Financial Information:

	Pta'000
Revenues	20,500,000

Principal Shareholders: First Chicago International Finance Corporation (50%)
No of Employees: 10

FISEAT
Financiera Seat SA
Velázquez 36, Aptd 14002, 28001 Madrid
Tel: (91) 431.13.00
Cable: Fiseat
Telex: 23736 fseat E
Telefax: 2.75.30.46

Board of Directors: Eberhard Müller (Presidente), Pablo Vidal Rius (Vice Presidente), Rafael Cabello de Alba y Gracia, Emilio del Castillo López, Juan José Cobo Medina, José Luis Gómez Tristán, Tomás Lamamie de Clairac Alonso, Johannes Maniura, Pedro Moriyón Costales, Francisco Seco Sánchez-Seco, Luis Villar Seco, José Martínez Rodríguez (Secretario y Letrado Asesor)
Management Board: Nicolás Andino Ruiz (Presidente-Gerente), Manuel García Moreno (Director-General-Gerente), José María Alonso Conde (Director-Gerente)

PRINCIPAL ACTIVITIES: Finance company for motor credit
Trade Names: FISEAT
Parent Company: SEAT

Financial Information:

	31.12.87 Pta'000
Profit after tax	60,789
Share capital	6,207,390

Principal Shareholders: SEAT
No of Employees: 304

FORD CREDIT SA

Paseo de la Castellana 135, E-Madrid
Tel: 571 13 86
Telefax: 571 21 99

President: Ingvar M Sviggum
Directors: Michael H Higgins (Managing Director)

PRINCIPAL ACTIVITIES: Distribution of motor vehicles; car finance services
Parent Company: Ford Motor Company, USA
Financial Information:

	1986 Pta'000	1987 Pta'000
Sales turnover	147,170,202	240,633,655
Profit before tax	1,604,319	2,751,121
Profit after tax	1,046,560	1,800,999
Retained profit	941,904	1,620,899
Dividend	419,900	-

No of Employees: 139

GDS LEASINTER SA

Tuset 5-11, 2a planta, 08006 Barcelona
Tel: (93) 201 71 33

Directors: Jorge Torra Marti (Director General), Antonio Sala Marti (Financial Director), Francisco Javier Aleix Fernández (Commercial Director), José Manuel Patiño Baro (Operations Director)

PRINCIPAL ACTIVITIES: Leasing of equipment and motor vehicles
Parent Company: Caja de Ahorros de Barcelona
Financial Information:

	Pta'000
Sales turnover	20,600,000

No of Employees: 46

HERCULES HISPANO DE SEGUROS Y REASEGUROS

C/Alcalá 17, Madrid
Tel: (91) 532-66-10
Cable: Herculesa
Telex: 46352
Telefax: (91) 521.85.89; (91) 532.93.12

President: Rafael Estevez Bartolome
Directors: Justo Martinez Martin, Jose Luis Guerrero Guerrero
Senior Executives: Jose Joaquin Manjon Simon (Director Pensiones), Ricardo Alonso-Guevillas Soley (Asesoria Juridico-Fiscal), Juan Antonio Baron Gomez (Dpto Automoviles), J E Moran y Gutierrez de Teran (Expansion), Carmen Gonzalez-Bueno Lillo (Dr Vida), Luis Miguel Fernandez-Solorzano (Dpto Diversos y Reaseguros), Concepcion Dotor Lapesinos (Administracion)

PRINCIPAL ACTIVITIES: Insurance and reinsurance operations; pension funds and related activities
Subsidiary Companies: Hercules Dos SA; Hercules Pensiones
Principal Bankers: Exterior de España

Financial Information:

	31.12.86 Pta'000	31.12.87 Pta'000
Premium income	21,118,593	7,144,827
Investment income	2,132,095	1,969,611
Profit before tax	230,167	412,708
Profit after tax	230,167	412,708
Retained profit	185,367	316,708
Dividend	44,800	96,000
Dividend per share	6.66Pta	-
Dividends per share	40Pta	50Pta
Earnings per share	119.87Pta	214.95Pta
Share capital	960,000	960,000

No of Employees: 204

LA ESTRELLA
See ESTRELLA, LA,

LA UNIÓN Y EL FÉNIX ESPAÑOL
See UNIÓN Y EL FÉNIX ESPAÑOL, LA,

LEASING BANCAYA SA
LISCAYA

Paseo de la Castellana 110, 28046 Madrid
Tel: (91) 262 13 05
Telex: 43241

President: José María de Arana y Aizpurúa
Directors: Antonio Maria Echavarren Villate (Director General)
Senior Executives: Fernando Quintana (Secretary)

PRINCIPAL ACTIVITIES: Specialists in medium and long-term finance for industrial and construction projects
Financial Information:

	Pta'000
Revenues	30,500,000

No of Employees: 70

LISBÁN SA, CIA DE LEASING

Paseo de la Castellana 79, Edificio U.A.P., 28046 Madrid
Tel: (91) 455 30 50
Telex: 23804

President: Tomás Lamamié de Clairac
Directors: Enrique Pellevaz Avello (Director General), Marcelo Llagostera Ferré (Financial Director), Roberto Vilches Navarro (Commercial Director), Luis Donaire García-Aranda (Risk Management Director)

PRINCIPAL ACTIVITIES: Leasing services
Financial Information:

	Pta'000
Sales turnover	11,500,000

Principal Shareholders: Bank of America (20%) (USA)
No of Employees: 45

MAPFRE, CORPORACION, COMPAÑIA INTERNACIONAL DE REASEGUROS SA

Paseo Recoletos 25, 28004 Madrid
Tel: (341) 581 11 00
Telex: 42544 MAFR-E/43225
Telefax: (341) 581 11 34

Board of Directors: Ignacio Hernando de Larramendi y Montiano (Chairman), Rafael Galarraga Solores (Vice Chairman), Filomeno Mira Candel (Managing Director), Carlos Alvarez Jiménez, Francisco Cailá Mestre, Julio Castelo Matrán, Dewey P Clark, Julio Caubin Hernández, Juan Fernández-Layos Rubio, Carlos Guilherme Ivens Ferraz de Mesquita, Manuel J Lagares Calvo, Luis Leguina Cebreiros, Gregorio Martin Rodriguez, Manuel Ocón Terrasa, Domingo Sugranyes Bickel, Luis Villar Seco, José Manuel Martinez Martinez (Director and General Manager), Alberto Manzano Martos (Director and Secretary of the Board)

SPAIN

Executive Commission: Ignacio Hernando de Larramendi y
Montiano (Chairman), Rafael Galarraga Solores, Manuel J
Lagares Calvo, José Manuel Martinez Martinez, Filomeno
Mira Candel, Domingo Sugranyes Bickel

PRINCIPAL ACTIVITIES: Reinsurance, and holding company of
direct insurance companies and financial entities

Parent Company: Mapfre Mutualidad De Seguros y Reaseguros
a Prima Fija

Subsidiary/Associated Companies: (100% owned unless stated):
Insurance Companies: Mapfre Vida (97.6%); Mapfre Industrial
(97.2%); Mapfre Caución y Crédito (96.5%); Mapfre Andalucia
(91.3%); Mapfre Cataluña y Baleares (94.4%); Mapfre Levante
(93.7%); Mapfre Noroeste (93.3%); Mapfre Norte (92.9%);
Mapfre XL. Financing Companies: Mapfre Finanzas (93.3%);
Mapfre Finanzas de Andalucia (96.6%); Mapfre Finanzas de
Atlántico (95.7%); Mapfre Finanzas de Levante (94.9%);
Mapfre Finanzas de Navarra, Aragón y Rioja (84.8%); Mapfre
Finanzas del Sur (93.7%); Mapfre Hipotecaria (93.4%); Mapfre
Leasing (93.9%). Others with Majority Interest: Mapfre
Inversiones (97.6%); Invermapfre (70%); Mapfre Internacional
(99.5%); Servicios Financieros Mapfre (97.6%); Gesmapfre
(97.6%); Mapfre Consultores de Seguros y Reaseguros
(97.6%); Mapfre Finanzas Servicios; Inmobiliaria Calvo
Bárbara. Foreign Companies: Mapfre Corporation of Florida
(USA); Mapfre do Brasil (98.4%) (Brazil); Caja Reaseguradora
De Chile (50.01%) (Chile); Seguros Universales (40%)
(Guatemala); Aconcagua Seguros (49%) (Argentina); Inversora
Austral (Argentina); Euroamérica Seguros (45%) (Chile);
Seguros Caribe (37.2%) (Colombia); Promotora Carima (70%)
(Colombia); Universo Assicurazioni (5.8%) (Italy); Varese
Assicurazioni (5%) (Italy); Companhia de Seguros Macau (4%)
(Macao); Reaseguradora Nuevo Mundo (2.5%) (Panama);
Ibero Assistencia (Portugal); Maghrebia (5%) (Tunisia).
Minority Interests: Mapfre Inmuebles (43.8%); Central de
Leasing (Lico) (8.3%); Dinver (9.8%); Dinversit (6.8%);
Gesprofon (29.3%)

Financial Information: Consolidated figures

	31.12.86	31.12.87
	Pta'000	Pta'000
Gross premium income	31,094,000	32,369,000
Net premium income	25,002,300	26,160,000
Net investment income	5,062,000	7,211,000
Profit before tax	1,846,000	2,885,000
Profit after tax	1,581,000	2,445,000
Net income(1)	1,498,000	2,354,000
Dividend(2)	20%	25%
Dividend per share	84Pta	125Pta
Earnings per share	164Pta	218Pta
Share capital	4,809,000	5,610,600
Shareholders funds	15,216,000	21,511,600

(1) less minority
shareholders' participation
(2) equals 100 pta

Principal Shareholders: Mapfre Mutualidad De Seguros (Majority
Shareholder)

No of Employees: 1,526

MAPFRE, MUTUA PATRONAL DE ACCIDENTES DE TRABAJO

Carretera de Pozuelo a Majadahonda, km 3.500, 28220
Majadahonda (Madrid)
Tel: (1) 638.10.11
Telex: 48.125
Telefax: (1) 638.05.15

President: José Antonio Rebuelta Garcia
Directors: José Márquez y Alvarez de Toledo (Vice President)
Senior Executives: Carlos Alvarez Jiménez (General Manager),
José Manuel Garcia Pablos (Controller), Miguel Angel
Montalvo De Lamo (Administration Manager), Enrique Hevia
Campomanes (Technical Manager), José Angel Fernando
Pablo (Safety Manager), Alfonso Iglesias Arrieta (Legal
Manager)

PRINCIPAL ACTIVITIES: Employers' mutual insurance company
specialising in work related accidents

Financial Information:

	31.12.86	31.12.87
	Pta'000	Pta'000
Premium income	24,030,507	30,080,691
Investment income	2,004,072	1,684,766
Profit before tax	3,357,499	3,447,000
Mutual funds	11,705,115	13,792,133

No of Employees: 1,319

MAPFRE MUTUALIDAD DE SEGUROS

Carretera Pozuelo a Majadahonda, km 3, 800, 28220
Majadahonda (Madrid)
Tel: (91) 637 66 11
Telex: 23898 MAPFRE E
Telefax: (91) 638 10 11

President: José Antonio Rebuelta García; Vice Presidents: José
Joaquín Márquez, Alvarez de Toledo
Directors: Julio Castelo Matrán (Managing Director), Manuel
Ocón Terresa (Director General), Arturo Platas (Personnel
Director), Ignacio Hernando de Larramendi y Montiano
(President of the Control Board), José Antonio Navez Muñiz
(Technical Director)
Senior Executives: Alberto Manzano (Secretary)

PRINCIPAL ACTIVITIES: Insurance
Financial Information:

	Pta'000
Revenues	23,000,000

No of Employees: 1,017

MAPFRE VIDA SA

Avda General Perón 40, 28020 Madrid
Tel: (91) 581.14.01
Telex: 46468 MPVI E
Telefax: (91) 581.15.92

President: Juan Ferñandez-Layos Rubio (also Managing
Director)
Directors: Sebastián Homet (General Manager), Arturo
Fernández Pacheco (Deputy Commercial General Manager),
Ma Jesús Ferñandez-Antón (Deputy Actuary-General
Manager) Antonio Romero (Administration), Miguel Marbán
(Human Resources), Rafael San Martin (Data Processing)

PRINCIPAL ACTIVITIES: Insurance services including medical
cover

Financial Information:

	1987
	Pta'000
Revenues	22,832,040
Premium income	16,616,211
Investment income	7,339,387
Profit before tax	1,049,553
Profit after tax	1,031,051
Retained profit	454,551
Dividend	550,000
Dividend per share	220Pta
Earnings per share	412Pta
Share capital	2,500,000

Principal Shareholders: Corporacion Mapfre
No of Employees: 420

MUSINI

Sociedad Mutua de Seguros y Reaseguros a Prima Fija

Padilla 46, 28006 Madrid
Tel: (91) 468 10 00
Telex: 48799 mini e
Telefax: (91) 276 90 47

President: Alvaro Muñoz López

Directors: Fernando Núñez Núñez (Financial Director), Jose A Carro del Castillo (Technical Director)
Senior Executives: J J Pardo (Secretary)

PRINCIPAL ACTIVITIES: Mutual insurance society
Trade Names: Musini
Subsidiary Companies: Interbolsa, SA; Servifinanzas SA
Financial Information:

	1987	1988
	Pta'000	Pta'000
Premium income	23,233,000	25,345,000
Investment income	3,933,000	6,669,000
Profit before tax	3,036,000	3,567,000
Profit after tax	2,343,000	2,640,000
Retained profit	1,435,000	1,515,000
Dividend	908,000	1,125,000
Shareholders funds	10,605,000	13,264,000

No of Employees: 108

MUTUA MADRILEÑA AUTOMOVILISTA, SOCIEDAD DE SEGUROS A PRIMA FIJA

Almagro 9, 28010 Madrid
Tel: (91) 410 77 66

Chairman: Juan Martín Martín; Vice Chairman: Alberto Comenge y Gerpe

PRINCIPAL ACTIVITIES: Automobile insurance
Financial Information:

	Pta'000
Revenues	22,000,000

No of Employees: 280

OCASO SA, COMPAÑÍA DE SEGUROS Y REASEGUROS

Princesa 23, 28008 Madrid
Tel: (91) 247 21 00
Telex: 44181/47671 SEGUROCASO
Telefax: (91) 247 45 83

President: Isabel Castelo
Directors: Fernando Sadornil (Director General), Emilio Orts (Financial Director)
Senior Executives: Concepción Ortega (Secretary), José Luis González Carazo (Secretary General)

PRINCIPAL ACTIVITIES: Insurance
Financial Information:

	1988
	Pta'000
Revenues	20,856,000

No of Employees: 690

PLUS ULTRA CÍA ANONIMA DE SEGUROS Y REASEGUROS

Plaza de las Cortes 8, 28014 Madrid
Tel: (91) 429 90 68
Telex: 22079 plus

President: Eustaquio Rodríguez Alvarez; Vice President: Gerardo Arostegui Gómez
Directors: Julio Gilsanz Arrola (Director General), José Luis Segimon (Secretary)

PRINCIPAL ACTIVITIES: Insurance services
Financial Information:

	Pta'000
Revenues	20,000,000

Principal Shareholders: Münchener Ruckversicherungs Gesellschaft (Germany) (10%)
No of Employees: 541

ROQUÉ, FERNANDO, TRANSPORTES INTERNACIONALES SA

Vía Layetana 26, 08003 Barcelona
Tel: (93) 310 00 00
Cable: Fernanroqué
Telex: 54747; 52931
Telefax: 93-3103815

Board of Directors: Fernando Roqué Rivero (Chairman), José Angel Martínez, Jorge Roqué Vilá, *Herman Plantenga, *Antonio de Fortuny, Jose Arus Codina, *On behalf of Van Gend & Loos NV (Netherlands)
Management: José Angel Martínez González (General Manager), Pablo Masclans Solana (Subdirector General), Jose Sans Camardón (Financial Manager), Pablo Masclans Solana (Commercial/Production Manager)

PRINCIPAL ACTIVITIES: International transport, customs brokerage, warehousing and insurance
Trade Names: Fernando Roqué
Principal Bankers: Banco Exterior de España; Banco de Bilbao; Banco Central; ABN Bank; Banco de Londres y América del Sur
Financial Information:

	Pta'000
Sales turnover	33,600,000

Principal Shareholders: Fernando Roqué Rivero; Antonio Roqué Rivero; Van Gend & Loos NV
No of Employees: 500

SANITAS SA

Francisco de Rojas 8, 28010 Madrid
Tel: (91) 445 90 16
Telex: 45503

President: Marcial Gómez Segueira; First Vice President: Eliseo Pestaña Núñez (and Secretary); Second Vice President: Pedro Patricio de Augustín Jiménez
Directors: Alfonso Martínez Gastey (Director General)

PRINCIPAL ACTIVITIES: Insurance
Financial Information:

	Pta'000
Revenues	13,800,000

No of Employees: 340

SANTA LUCÍA SA

Plaza de España 15, 28008 Madrid
Tel: (91) 241 93 87

President: Manuel Alvarez López
Directors: Teresa Otero Carreira (Managing Director and Secretary), José Luís Diáz Martínez (Director General), Rafael López Herrero

PRINCIPAL ACTIVITIES: Insurance
Financial Information:

	Pta'000
Revenues	17,500,000

No of Employees: 141

TORRAS HOSTENCH, SA

Gran Via Corts Catalanes 678, Barcelona 8010
Tel: (93) 318 31 00
Telex: 52558 THSA
Telefax: (93) 317-1836

Board of Directors: President: Sheik Fahad Mohammed Al-Sabah. Vice Presidents: Fouad Khaled Jaffar, Bruce Amager Dawson, Francisco Javier de la Rosa Marti. Chief Executive: Jorge Nuñez y Lasso de la Vega. Directors: Khaled Al-Sabah, Muneer Al-Sabah, Trevor D Ball, Narciso de Mir Faura, Alfonso Ferrari Herrero, Manuel Guasch Molins, Pedro Misol Herrador, Richard Robinson, Jose Pedro Perez-Llorca Rodrigo, Jose Ma Sot Casas. Secretary: Juan Pique Vidal

Management: Jorge Nuñez y Lasso de la Vega (Chief Executive)
Senior Executives: Miguel Soler (Chief Financial Officer)

PRINCIPAL ACTIVITIES: Pulp, printing and writing, cardboard, self-adhesives, paper conversions, industrial stationery; school and office stationery; chemicals, food processing, services, distribution
Trade Names: Torras Hostench, SA, Enri, Luchador, Folk, Adestor, Sils-Foc, Sils-Interfoc, Sils-Intemper, Sils-Mobel
Subsidiary Companies: Papelera del Mediterráneo SA; Cartón y Papel Recuperado SA; SA CROS; EBRO-Compañia de Azucares y Alcoholes SA; Beta Capital; ERPO; Amaya, Compañia de Seguros y Reaseguros; Cellulose des Ardennes
Financial Information: Group figures

	31.12.86	31.12.87
	Pta'000	Pta'000
Sales turnover*	30,439,000	46,530,000
Profit before tax	9,544,000	14,161,000
Earnings for year	6,943,000	15,086,000
Retained profit	15,997,000	62,459,000
Dividends	2,348,622	5,197,000
Dividends per share	960Pta	950Pta
Earnings per share	3,903Pta	3,210Pta
Share capital	2,445,000	7,336,000
Shareholders equity	18,442,000	69,795,000

*paper only

Principal Shareholders: Kuwait Investment Office
No of Employees: 3,282 (paper only)
90,1010

UNINTER LEASING SA (UNILEASING)
Prim 12, 28004 Madrid
Tel: (91) 522 91 81
Telex: 42648 UNEF E

Chairman: Benito Tamayo Hernáez
Directors: Fernando Fontana Hervada (Managing Director), Arturo Clavero Roda (Director General)

PRINCIPAL ACTIVITIES: Leasing services
Parent Company: Banco Hispano Americano
Financial Information:

	Pta'000
Sales turnover	27,800,000

Principal Shareholders: Campagnie Bancaire SA (14.8%) (France)
No of Employees: 138

UNIÓN Y EL FÉNIX ESPAÑOL, LA,
Compañía de Seguros Reunidos SA
Paseo de la Castellana 33, 28046 Madrid
Tel: (91) 419 85 50
Cable: UNION FFNIX
Telex: 27759 MDUF E

President: Jaime Argüelles Armada; Vice Presidents: Enrique Sáinz Ortueta, M François Pereire
Directors: Eduardo Carriles Galarraga, Felix Mansilla García, Francisco Arán López, Fernando Sáinz Quintana, Enrique Marco Fuentes, Domingo Solsona Ejalde, Miguel Angel de Gregorio Colmenares, Antonio Rodríguez Medina

PRINCIPAL ACTIVITIES: Insurance
Financial Information:

	Pta'000
Turnover	140,000,000

WINTERTHUR SOCIEDAD SUIZA DE SEGUROS
Plaza Francesc Maciá 10, 08036 Barcelona
Tel: (93) 321 78 00 - 321 79 00
Telex: 52236 WTHUR E

Directors: José Cercos Martínez (Director General and Spanish Representative), Juan Dionisio Martínez López (Financial Director), Antonio Bofarull Roset (Commercial Director), Pedro Cabrafiga Cristina (Technical Director), Enrique Jané Calleja (Services Director), Pedro Luna Candela (Accidents Claim Director)
Senior Executives: Federico Boix Serra (Secretary)

PRINCIPAL ACTIVITIES: Insurance services
Parent Company: Winterthur Schweizerische Versicherungsgesellschaft, Switzerland
Financial Information:

	Pta'000
Revenues	20,600,000

No of Employees: 737

Major Companies of SWEDEN

AMF
Arbetsmarknadsförsäkringar
Pensionsförsäkring
Regeringsgatan 103-105, S-11388 Stockholm
Tel: 08/7280500

PRINCIPAL ACTIVITIES: Insurance services
Financial Information:

	SKr'000
Premium income	4,000,000

ANDERSON, PAUL, INDUSTRIER AB
Box 730, 721 20 Västerås
Tel: 021/15 50 00
Telex: 40609 aros s
Telefax: 021/115125

Board of Directors: B Frid (Chairman), R Bergstrand, Bengt
Magnusson (Managing Director), J Andberger, B Mattson, B
Frid, Fritz Staffas, C Stiigh, L Strömberg, L Svanbom, B Tagg
Management: Bengt Magnusson (Managing Director)

PRINCIPAL ACTIVITIES: Building contracting; foundry; plant
hire; real estate management; finance company
Subsidiary Companies: Andersons Fastigheter; Byggpaul AB;
Ekebro AB; Kramo AB; Metodbyggen AB
Principal Bankers: Handelsbanken
Financial Information:

	1986	1987
	SKr'000	SKr'000
Sales turnover	1,980,600	2,225,000
Profit before tax	38,200	48,900
Profit after tax	32,000	43,400
Dividends	5,968	6,638
Dividends per share	5.50SKr	6.60SKr
Earnings per share	19.20SKr	20.90SKr
Share capital	54,200	55,200
Shareholders funds	117,400	157,100

No of Employees: 2,392

AXEL JOHNSON AB
See JOHNSON, AXEL, AB

BOFAB
Svensk Bostadsfinansiering AB
Box 16166, 10324 Stockholm
Tel: 08/7965000

PRINCIPAL ACTIVITIES: Financial services including credit
operations
Financial Information:

	SKr'000
Turnover	650,000

CATENA (AB)
PO Box 500, 40127 Göteborg
Tel: 031-80 16 00
Telex: 28239 CATENA S
Telefax: 031-15 47 22

Board of Directors: Lars Bylund (Chairman), Claes Beyer, Olle
Blomqvist, Bo Nyhlén, Gunnar Carleson, Gunnar Dahlsten,
Lennart Jeansson, Carl-Gösta Attling, Josef Schilcher
Management: Bo Nyhlén (Managing Director), Olle Thelander
(Deputy Managing Director), Anders Leijon (Automobile
Retailing), Hans Olbing (Group Controlling), Håkan Larsson
(Finance)

PRINCIPAL ACTIVITIES: The Catena Group comprises the
following four business areas: Automobile retailing: the
companies involved primarily sell Volvo passenger cars,
trucks and buses and conduct substantial repair shop
operations and spare parts sales. Trading and Industry:
conducts international trading in forestry products, sales of
engines for marine and industrial applications, agency
operations and manufacture of components for the
automotive industry. Finance: offers a broad range of
services, including leasing, installment financing, capital
management, real estate financing and financing of
international trade. Real Estate: manages real estate for
automobile dealerships, industrial and residential real estate
etc. Operations also include management of a share portfolio
Trade Names: AB Catena
Subsidiary Companies: AB Bilia; AB Bil & Truck; Ernst Nilson
AB; AB Verdexa; Lastbilsbolaget i Stockholm AB;
Lastbilsbolaget i Göteborg AB; Catena Norge A/S; Ekman
Liebig AB; Gustaf Fagerberg AB; AB Probo; Fastighets AB
Bohus; Probo Pacific Pte Ltd (Singapore); Bilvex AB;
Frösunda Motor AB; Bilia Motor A/S (Denmark); AB Bilorama;
AB Umeå; Bilkompani; Motoria Bil AB; Vexia AB; Autopart
Sweden AB; Rattia AB; Fastighets AB Nattskiftet; Imatech
AB; Bil AB Nils; Arfwidson Umeå Bilkompanis; Fastighets AB;
Fastighetsbolaget Etnografen AB; Bilia Motor A/S (Denmark);
Cimbrer Bil A/S (Denmark); AB Visby; Motorcentral; AB
Catusafe; Dalbo Bil AB; Bildepan i Säffle AB; Catena
Deutschland GmbH
Financial Information:

	31.12.86	31.12.87
	SKr'000	SKr'000
Sales turnover	8,535,000	11,209,700
Profit before tax	724,000	450,000
Profit after tax	254,500	83,200
Retained profit	35,600	116,900
Dividend	33,000	42,000
Dividend per share	2.00SKr	2.00SKr
Earnings per share	12.01SKr	7.73SKr
Share capital	300,000	420,000
Shareholders funds	629,800	680,000

Principal Shareholders: AB Volvo; Skanska AB; Tredje
Löntagarfonden; Skandia
No of Employees: 5,871

CUSTOS (AB)
Norrlandsgaten 16, Box 1738, 111 87 Stockholm
Tel: 08/24 47 35
Telex: 10412 custos
Telefax: 08-119074

Board of Directors: B Haak (Chairman)
Management: L Öberg (Managing Director), O Leksell (Deputy
Managing Director), Ingemar Sanders (Director of Finance)

PRINCIPAL ACTIVITIES: Industrial holding company and
investment trust with interests in a wide range of Swedish
companies including real estate, wood and pulp, automobile
and engineering companies
Subsidiary/Associated Companies: Finans AB Vestos;
Fastighetsaktiebolaget Hufvudstaden; AB Värdefinans;
International Invest AB; AB Arcas Finans; AB Crestos
Principal Bankers: S E Banken
Financial Information:

	31.12.86	31.12.87
	SKr'000	SKr'000
Sales turnover	959,000	947,000
Profit before tax	575,000	510,000
Profit after tax	491,000	489,000
Retained profit	1,230,000	1,209,000
Dividends	179,000	198,000
Dividends per share	3.94SKr	4.25SKr
Earnings per share	15.00SKr	11.00SKr
Share capital	844,000	1,165,000
Shareholders funds	2,373,000	2,747,000

Principal Shareholders: Skanska AB; AB Volvo; Investment AB
Cardo; Skandinaviska Bankens Pensionsstiftelse;
Investmentbolaget Øresund; Trygg-Hansa
No of Employees: 426

SWEDEN

FERMENTA AB
Apelbergsgatan 57, S-11137 Stockholm
Tel: +46-8-23 83 50
Telex: 12792 Ferm S
Telefax: +46-8-113124

Board of Directors: Bert Sjolin (Chairman), Bertil Holmberg
(President)
Senior Executives: Lars Landstrom (Executive Vice President
and Chief Financial Officer)

PRINCIPAL ACTIVITIES: Holding company for chemical and
pharmaceutical companies and companies offering financial
services
Subsidiary Companies: SDS Biotech (USA); Pierrel (Italy);
Fermenta Products; Independent Finans
Principal Bankers: Gotabanken, Handelsbanken
Financial Information:

	31.12.88
	S'000
Sales turnover	4,920,000
Profit before tax	189,000
Profit after tax	162,000
Earnings per share	0.59SKr
Share capital	469,500
Shareholders funds	753,000

No of Employees: 2,500

FOLKSAMGRUPPEN
Bohusg 14, 10660 Stockholm
Tel: 08/772 6000
Cable: Folksam
Telex: 19084 folksam s
Telefax: 08/714 7611

Supervisory Board: Stig Malm (Chairman), Hans Dahlberg
(Managing Director), Åke Lundqvist (Deputy Chairman), Björn
Rosengren (Deputy Chairman) Lage Andreasson, Uno Ekberg,
Inge Granqvist, Rolf Holmgren, Ralf Hultberg, Ulla Hultén, Leif
I Lewin, Henry Lindmark, Sigvard Marjasin, Sture Nordh, Curt
Persson, Lars Svensson, Rolf Trodin, Bertil Whinberg, Ulf
Åberg
Chief Executives of the Folksam Group: Hans Dahlberg, Tage
Gardelin, Per-Olof Granstedt, Gunnar Grosse, Rut
Hammarström, Erland Strömbäck, Arnold Thunström

PRINCIPAL ACTIVITIES: Insurance and reinsurance
Trade Names: The Folksam Group
Group Companies: Folksam Mutual Life Insurance Society;
Folksam General Mutual Insurance Society; Folksam
International Insurance Co Ltd
Subsidiary Companies: Folksam International Insurance
Company (UK) Ltd (UK); Folksamerica Reinsurance Company
(USA)
Financial Information:

	31.12.87
	SKr'000
Premium income	6,773,700
Investment income	2,142,500
Operating result	1,714,600
Total assets	23,625,000

No of Employees: 3,991

FÖRENINGSBANKERNAS BANK
Grev Turegatan 30, PO Box 5844, S-102 48 Stockholm
Tel: +46 8 782 30 00
Cable: Jordbank Stockholm
Telex: 10627 Foba S
Telefax: +46 8 60 78 65

Board of Directors: Nils G Åsling (Chairman)
Management: Ove Israelsson (Managing Director)
Senior Executives: P G Hemborg (International Division),
Thomas Ekman (Trading Division)

PRINCIPAL ACTIVITIES: Bank

Financial Information: Group Figures

	31.12.87
	SKr'000
Deposits	12,186,000
Total assets	13,048,000

No of Employees: 150

FÖRSTA SPARBANKEN
Hamngatan 31, S-103 31 Stockholm
Tel: 46-8-700 70 00
Cable: Fospbank Stockholm
Telex: 17588 fospsg s
Telefax: 46-8-11 86 77

Board of Directors: Lars Lindmark (Chairman)
Management: Bertil Sjöstrand (Group President), Karl-Henrik
Pettersson (President)
Senior Executives: Nils-Olof Malmström (Executive VP
Stockholm), Owe Aronsson (Executive VP Göteborg), Bo
Lönn (Executive VP), Kurt Jansson (Executive VP Göteborg),
Gunnar Schonberg (General Manager International), Carl-
Henrik Falkenberg (Head of Correspondent Banking Division),
Per Smedberg (Head of Foreign Exchange Division), Tommy
Österberg (Head of Eurodeposits and Funding), Lars-Olof
Malmström (Head of Bond Dealing and New Issues), Björn
Bergström (Head of Syndicated Loans Division), Stefan
Alvolin (Head of Documentary Credits Division), Karl Axel
Jonsson (Regional Foreign Manager, Stockholm), Anders
Vage (Regional Foreign Manager, Göteborg)

PRINCIPAL ACTIVITIES: Banking
Subsidiary Companies: Mälarstadens Förvaltnings AB;
Värdetjänst Väst AB; Första Sparbankens Värdeppersfonder
AB; AB Sydsidan; Västifinans AB
Financial Information:

	31.12.86	31.12.87
	SKr'000	SKr'000
Profit before tax	230,000	122,000
Profit after tax	85,000	100,000
Untaxed reserves	1,593,000	1,657,000
Shareholders equity	433,000	533,000
Deposits	20,683,000	22,772,000
Total assets	29,417,000	34,196,000

No of Employees: 2,500

FÖRVALTNINGS RATOS AB
See RATOS, FÖRVALTNINGS, AB

GÖTABANKEN
Sveavägen 14, S-103 77 Stockholm
Tel: +46-8-790 40 00
Cable: Gotabank
Telex: 19420 Gotban s
Telefax: +46-8-790 41 45

Board of Directors: Sven Erik Ragnar (Chairman), Kristian von
Sydow (First Vice Chairman), Gabriel Urwitz (Second Vice
Chairman), Staffan Crona, Lil Karhola-Wettergren, Christopher
Rappe, Rune Hallåker, Lennart Henrikson, Lil Karhola-
Wettergren, Gunnar Malmenström (President and Chief
Executive), Sven Olving, Curt Stjernlöf, Bert Sjölin, Roland
Svensson. Deputies: Kai Holmbeck, Mikael Kamras, Ulf
Lignell (Executive Vice President). Appointed by the
Government: Sören Mannheimer, Lars F Tobisson. Appointed
by the Employees: Gösta Bankell, Rolf Schönning, Holger
Larsson, Bernt Pettersson
Central Management: Gunnar Malmenström (President and
Chief Executive Officer), Björn Salomonson (Deputy to the
President, Executive Vice President), Ulf Holmström
(International and Corporate Division, Executive Vice
President), Ulf Lundahl (Private Banking, Executive Vice
President), Henning Burlin (Eastern Region, Executive Vice
President), Bengt Hammarlind (Stockholm Region, Executive
Vice President), Kenneth Odéus (Western Region, Executive

Vice President), Carl-Johan Persson (Southern Region, Executive Vice President), Stefan Petersson (Central Sweden Region, Executive Vice President)

Senior Executives: P O Johansson (Senior Vice President, Foreign Exchange), V Jangeberg (Senior Vice President, Administration), L Elmer (Senior Vice President, International Credits), B Sevelius (Senior Vice President, International Corporate Finance)

PRINCIPAL ACTIVITIES: Bank
Parent Company: Gota-Gruppen AB
Subsidiary Companies: (100% owned): Götabankens Fastighets AB; Merkantil Leasing Deutschland GmbH (Germany); Fastighets AB Filen; Gotabanken Inc (USA); Hägglöf & Ponsbach Gota Securities Ltd (UK); Gotabanken (Luxembourg) SA (Luxembourg). Affiliated Companies: FondSpar Förvaltnings AB; Banque Transatlantique SA (France); Sunnmörsbanken A/S (Norway) Bankdata AB; Sunnmörsbanken A/S (Norway) and representative offices in Hong Kong, Singapore, China, Thailand and Norway
Principal Bankers: All principal banks
Financial Information: Group figures

	31.12.86 SKr'000	31.12.87 SKr'000
Profit before tax(1)	473,000	941,000
Profit after tax(1)	84,000	737,000
Dividends	71,000	84,000
Dividends per share(2)	9.50SKr	9.50SKr
Dividends per share (3)	1.25SKr	9.50SKr
Earnings per share	28SKr	9SKr
Shareholders equity	625,000	1,278,000
Deposits(4)	19,946,000	21,626,000
Total assets	47,822,000	48,888,000

(1) and appropriations
(2) regular
(3) extra
(4) Götabanken

No of Employees: 2,298 (Götabanken)

HANSA ÖMSEDIG SAKFÖRSÄKRING

Hansa Mutual General Insurance Co
Flemminggatan 18, S-106 26 Stockholm
Tel: (08) 785 10 00
Cable: Trygghansa
Telex: 19887 Hansa s

Board of Directors: A Åkerman (Chairman), S-E Johansson (Deputy Chairman)
Management: Björn Sprängare, Lars Lönnborg, Lars-Göran Nilsson, Lars Pihlgren, Per Welin

PRINCIPAL ACTIVITIES: Insurance
Subsidiary Companies: Trygg-Hansa
Financial Information:

	31.12.87 SKr'000
Premium income	5,027,000
Investment income	10,723,000

No of Employees: 3,500

INCENTIVE AB
Arsenalsgaten 2, PO Box 1723, S-111 87 Stockholm
Tel: 08/23 45 00
Cable: Incentive
Telex: 173 71 incent s
Telefax: 08-108161

Board of Directors: Sten Gustafsson (Chairman), Lars V Kylberg (President), Hans Cavalli-Björkman, Karl-Axel Linderoth, Sture Eckeskog, Fredrik Lundberg, Curt Nicolin, Bengt-Ola Nygren, Hans Stahle, Assar Svensson, Peter Wallenberg, Tore Burström (Deputy), Inge Mauritzson (Deputy)
Executive Management: Lars V Kylberg (President), Erik Hagström (Finance), Lennart Lindqvist (Chairman of Carl

Munters and AKA Industriprodukter), Assar Svensson (Chairman of DIAB-BARRACUDA, Orrefors Glasbruk and Skega)

PRINCIPAL ACTIVITIES: Holding company with interests in building and industrial components; industrial control equipment and instruments; earth moving plant; mechanical handling equipment; automotive components
Subsidiary Companies: (100% owned unless stated): AKA Industriprodukter AB; AB Almex; DIAB-BARRACUDA AB; Engström & Nilson Maskin AB; Garphyttan Industrier AB (55.7%); Victor Hasselblad AB (58.1%); AB HNJ Intressenter; Incentive Group Inc; Incentive Research & Development AB; AB Lorentzen & Wettre; AB Carl Munters; AB Orrefors Glasbruk (63.4%); Skega AB; Tour & Andersson AB; Incentive Kredit AB; Incentive Elektronik AB
Principal Bankers: Skandinaviska Enskilda Banken
Financial Information:

	31.12.86 SKr'000	31.12.87 SKr'000
Sales turnover	6,153,000	7,077,000
Profit before tax	802,000	408,000
Profit after tax	621,000	198,000
Retained profit	179,000	587,000
Dividends	65,300	78,900
Dividends per share	3.65SKr	4.50SKr
Earnings per share	9.45SKr	12.90SKr
Share capital	346,000	481,000
Shareholders funds	1,021,000	994,000

No of Employees: 10,201

INDUSTRIVÄRDEN-GROUP
Hamngatan 6, 111 47 Stockholm
Tel: 08/237830
Telefax: 08/208033

Board of Directors: B Rydin (Chairman)
Management: Kjell Brändström (Managing Director)
Senior Executives: L Bertmar, O Wesström

PRINCIPAL ACTIVITIES: Portfolio Investment Company with holdings in major Swedish companies and subsidiaries in real estate and industry
Parent Company: AB Industrivärden
Subsidiary/Associated Companies: Fastighets AB Fundament; Bilrom; PLM
Principal Bankers: Svenska Handelsbanken
Financial Information:

	31.12.86 SKr'000	31.12.87 SKr'000
Group sales turnover	672,000	2,420,000
Profit before tax	461,000	513,000
Profit after tax	542,000	453,000
Retained (loss)	(317,000)	(369,000)
Dividends	144,000	144,000
Dividends per share	5.25SKr*	5.25SKr
Earnings per share	12.50SKr	14.00SKr
Share capital	546,000	549,000
Shareholders funds	1,973,000	2,055,000

*of which 0.50SKr was bonus dividend

Principal Shareholders: Förvaltnings AB Aeolus (SCA); Ratos (Förvaltnings AB Ratos); Investment AB Beijer; Trygg-Hansa; Pension Funds of Handelsbanken
No of Employees: 2,615

JOHNSON, AXEL, AB
S-103 75 Stockholm
Tel: +46 8 788 50 00
Telex: 17153 axmanag s

Board of Directors: Antonia Ax:son Johnson (Chairman)
Management: Göran Ennerfelt (President and Chief Executive Officer)

SWEDEN

PRINCIPAL ACTIVITIES: Raw materials trading including oil, coal, ores and metals; specialty oil products; industrial trading, particularly in stainless steel; fluid handling equipment, marine equipment; industrial plant and machinery; east-west trade in engineering products and lubricants; trade with Nordic countries in special steels, marine products for ships and the offshore sector, sales of electrical, electronic and chemical products; medical and dental equipment; financial services; since April 1988 retail and wholesale trade in Sweden

Subsidiary Companies: (100% owned unless stated): A Johnson & Co HAB; Bjurströmspålen AB. Within subsidiary group AXINTER: Axel Johnson International AB; A Johnson & Co Eastern AB; Cellma Contracting AB; KAB Kjellbergs AB; Handels AB Mercator; J Åkerman AB; Mercatsia Inc (USA); Nordisk Elektra AB; Svenk Teknisk Byrå AB; A Johnson & Co Wien GesmbH (Austria); A Johnson & Co Oy (Finland); Axel Johnson Marine Oy (Finland); A Johnson & Cie SA (France); KaMeWa SA (France); A Johnson & Co (Hellas) Ltd (Greece); A Johnson & Co Ltd (UK); George Meller Ltd (UK); A Johnson & Co (Holding) BV (Netherlands); Axel Johnson Steel International BV (Netherlands); Axel Johnson Industrial BV (Netherlands); Johnson-Saunders SA (Belgium); Axel Johnson & Co NV (Belgium); A Johnson & Co (Portugal) Lda (90%) (Portugal); A Johnson & Co SA (Spain); A Johnson & Co International Ltd (Switzerland); A Johnson & Co GmbH (Germany); A Johnson & Co Anlagentechnik GmbH (Germany); ISA Internationale Schiffahrts-Agentur GmbH (60%) (Germany); AKO GmbH (Germany); Axel Johnson & Co Africa (Pty) Ltd (South Africa); A Johnson & Co (SEA) Pte Ltd (Singapore); Wemex Engineering Asia (Pte) Ltd (Singapore); A Johnson & Co (Asia-Pacific) Ltd (Hong Kong); A Johnson & Co (Asia-Pacific) Ltd (South Korea); Axel Johnson Corporation (HK) Ltd (Hong Kong); Axel Johnson Corporation (Australia) AB (Australia); Axel Johnson do Brasil Ind e Com Ltda (Brazil); Axel Johnson de Mexico SA de CV (Mexico). Within subsidiary group AXOIL: Axel Johnson Petroleum AB; Axoil Ltd (UK); Axel Johnson Petroleum International Ltd (Bermuda); Axel Johnson Petroleum (Asia) Pte Ltd (Singapore). Within subsidiary group AXORE: Axel Johnson Ore & Metals AB (Sweden); Axel Johnson Minerais et Métaux SARL (France); Axel Johnson Ore & Metals AG (Switzerland); Axel Johnson Ore & Metals GmbH (Germany); Axel Johnson Ore & Metals Inc (USA). Within subsidiary group AXCHEM: Axel Johnson Chemicals AB; AB Axel Christiernsson; Olson-Dinol Engineering AB. Within subsidiary group DINOL: Dinol International AB; Dinol Svenska AB; Dinol Norge A/S (Norway); Dinol Oy (Finland); Dinol GmbH (Germany); Dinol France SARL (France); Dinol Italia SRL (Italy); Dinol Properties Ltd (UK); Dinol Korea Company Ltd (70%) (Korea); Dinol Chemicals (SEA) Pte Ltd (Singapore); Guard Insurance Company Ltd (Barbados). Within subsidiary group AXMED: Axel Johnson Instrument AB; STI-Gruppen AB; DAB Dental AB; Nordenta AB; Svenska Dental Instrument AB; Axel Johnson Radiologi AB; Interdental AB; Interdental Väst AB; AB Axel E Madsen; A Johnson & Co A/S (Denmark); Kjellbergs Successors A/S (Denmark); Kjellbergs Successors BV (Netherlands); A Johnson & Co A/S (Norway). Within subsidiary group NORTEC: Nortec Electronics AB: AB Distex; Nordisk Arrayteknik AB; Nordisk Elektronik AB; Nortelco AB; Traco AB; Nortec Electronics A/S (Denmark); A/S Nordisk Elektronik (Denmark); Nordisk Mikrosystem ApS (Denmark); Distributören Interelko A/S (Denmark); Nortec Electronics A/S (Norway); Oy Nortec Electronics Finland AB (Finland); Oy Fintronic AB (Finland); Oy Comdax AB (Finland); Oy Ribbert AB (75%) (Finland). Miscellaneous: A Johnson & Co Credit AB; Axel Johnson Insurance Co Ltd (Bermuda); Axel Johnson Petroleum Trading AB; AB Kratos; HB Nydo Energi

Financial Information:

	31.12.86	31.12.87
	SKr'000	SKr'000
Consolidated turnover	5,558,800	4,316,800
Profit before tax	31,200	800
Profit after tax	15,500	31,000
Net profit	15,500	31,000
Share capital	50,000	8,200
Shareholders equity	126,900	239,800

No of Employees: 17,500 Group

KULLENBERG, YNGVE, BYGGNADS AB
Gyllenkrooksgatan 10B, S-403 32 Göteborg
Tel: 031-81 5 00
Telex: 21760 Kbygg S

Board of Directors: Governor Erik Huss (Chairman)
Management: Rolf Ericsson (Managing Director)

PRINCIPAL ACTIVITIES: Building and civil engineering
Parent Company: Fenix Developement AB
Subsidiary Companies: Lindahl Byggnads AB; Kullenberg Markbostäder AB; Kullenberg Ombyggnader AB; Järphag Unenge AB; Yngve Kullenberg Maskin AB
Principal Bankers: Skandinaviska Enskilda Banken
Financial Information:

	SKr'000
Sales turnover	980,000

No of Employees: 1,170

L E LUNDBERGFÖRETAGEN AB WITH SUBSIDIARIES
See LUNDBERGFÖRETAGEN, L E, AB WITH SUBSIDIARIES

LUNDBERGFÖRETAGEN, L E, AB WITH SUBSIDIARIES
Lundbergkoncernen
Box 2157, 600 02 Norrköping
Tel: 011/21 65 00

Board of Directors: Lars Erik Lundberg (Chairman), Fredrik Lundberg (Chief Executive Officer), Kurt Pålsson (President), Ulf Gorän Calmfors, Bo Berggren, Sven Andersson, Hans Malmberg

PRINCIPAL ACTIVITIES: Construction industry, real estate property, process engineering, finance, investment
Trade Names: Lundbergs
Subsidiary Companies: (100% owned unless stated): L E Lundberg Ekonomiservice AB; Lundberg International AB; Förvaltnings AB LE Lundberg; Byggnads AB L E Lundberg; Industri AB Lunden; AB Benzlers; Broddway AB; Förenade Ljud AB; Hymo AB; Moramast AB; Primpac AB; L E Lundberg Finans AB; L E Lundberg Byggen KB; AB Katrineholms Husbyggen; Jonsons Rör KB; Bankeryds Rör AB; SIAB (43.6%); Incentive (30.4%); Alfa-Laval (13.7%); Östgöta Enskilda Bank (49.9%)
Financial Information:

	31.12.86	31.12.87
	SKr'000	SKr'000
Sales turnover	1,860,000	2,245,000
Profit before tax	424,000	425,600
Profit*	1,079,200	920,900
Profit after tax	314,100	298,700
Retained profit	834,800	1,072,400
Dividends	59,800	73,600
Dividends per share	1.30SKr	1.60SKr
Earnings per share	11.70SKr	10.00SKr
Share capital	460,000	460,000
Shareholders funds	1,397,500	1,637,200

*after extraordinary items

Principal Shareholders: Lundberg family (59.3%); Trygg Hansa Group (4.8%); SPP and AMF pension funds (3.6%); S-E-

Banken's pension fund (2.4%); AB Investor (2.4%); Förvaltnings AB Providentia (2.4%); Skandia (1.9%)

No of Employees: 2,031

NORDBANKEN

PO Box 7133, S-10387 Stockholm

Tel: Int.+ 468 796 3000 (General); Int. + 468 237 360 (Foreign Exchange)

Cable: NORDBANKEN; S.W.I.F.T.: Norrsess

Telex: 17616 Nordsts (General); 11640 Nordexs (Foreign Exchange)

Telefax: Int.+468 21 34 82

Board of Directors: Tomas Matsson (Chairman), Sigvard Ando (Deputy Chairman), Rune Barnéus (Chief Executive), Mårten Carlsson, Erik Ehn, Bertel Finskas, Ms Anita Hård, Samme Lindmark, Gunnar Lundh (Deputy Chief Executive), Leif Marklund, Ms Görel Thurdin, Olle Måberg, Per Olsson, Wiking Sjöstrand

Management: Rune Barnéus (President and Chief Executive), Gunnar Lundh (Deputy Chief Executive), Carl-Johan Bouveng (International Banking Division), Bengt-Åke Eriksson (Marketing), Jakob Grinbaum (Money Market), Lars Hjortengren (Southern Region), Claes Humbla (Personnel), Thorbjörn Högberg (Administration), Arne Liljedahl (Economy/Controller), Tomas Lindstrand (Management Secretariat), Kjell Pettersson (Data/Organisation), John Roos (Corporate Banking Division), Lennart Söderberg (Corporate Finance), Sven Söderberg (Auditing Department, deputy member), Bernt Öhrn (Stockbroking Division), Claes Östberg (Northern Region)

PRINCIPAL ACTIVITIES: A full range of domestic and international banking and financial services to customers and correspondents around the world

Subsidiary Companies: (100% owned unless stated): Nordbanken Fastighets AB; Nordbanken Finans AB; Nordbankens Fastighetsbyrå AB; Nordbankens Fondförvaltning AB; Sundsvallsbankens Aktiesparfondförvaltning AB; Bostadsföreningen SEB u.p.a.; Stockholms Tomträttskassa AB (51%); Arbuthnot Latham Bank Ltd, (London, UK); Nordbanken North American Inc (USA)

Principal Correspondents: Deutsche Bank, Dresdner Bank (Frankfurt); Hambros Bank, Midland Bank (London); Irving Trust, Security Pacific International Bank (New York)

Financial Information: Nordbanken only

	31.12.86	31.12.87
	SKr'000	SKr'000
Profit before tax (1)	260,000	144,700
Profit before tax (2)	383,800	335,700
Profit after tax	91,000	75,000
Retained profit	2,300	2,600
Dividends	62,300	74,700
Dividends per share	8.00SKr	4.00SKr
Earnings per share	24.20SKr(3)	6.70SKr
Share capital	128,200	467,300
Shareholders equity	499,400	753,600
Deposits	15,612,600	15,277,200
Total assets	27,188,800	31,420,000

(1) excluding non-taxed allocations

(2) including non-taxed allocations

(3) 11.21Skr (adjusted)

Principal Shareholders: Custos (14.1%); Oobab Trading (10.4%); Skandia (5%); Trygg Hansa (5%); SPP (4.3%); LKAB (4.1%)

No of Employees: 1,494 (Nordbanken); 1,863 (Group)

ÖSTGÖTA ENSKILDA BANK

Box 7523 Smålandsgatan 2, 103 92 Stockholm

Tel: 08-7960400

Cable: ostgotabanken

Telex: 130 80, 13081, 13586, 16554 Oebank S

Telefax: 46-8-21 30 83; 21 36 45

Board of Directors: Torbjörn Ek (Chairman)

Senior Management: Thord Söderlund (President and Chief Executive Officer), Gunnar Almesåker (Executive Vice President, Head of Administrative Services), Lennart Söderberg (Executive Vice President Head of Branch Office Operations), Staffan Lindblom (Senior Vice President, Head of International Division), Lennart Löfstrand (Senior Vice President, Secretary of the Board, Head of Legal Services), Tommy Glanz (Senior Vice President, Head of Östgöta Enskilda Fondkommission, Trustee and Investment Division), Lars Erik Westin (Senior Vice President, Head of Investment Management Division)

PRINCIPAL ACTIVITIES: Bank

Subsidiary Companies: Östgöta Enskilda Finans AB; Östgöta Enskilda Fastighets AB; Östgöta Enskilda Fondförvaltning AB

Financial Information:

	31.12.86	31.12.87
	SKr'000	SKr'000
Profit before tax	56,400	43,200
Profit after tax	17,700	26,500
Retained profit	12,800	12,400
Dividends	12,392	12,674
Dividends per share	1.47SKr	1.50SKr
Earnings per share	8.39SKr	4.19SKr
Share capital*	56,300	84,500
Deposits	4,781,500	4,626,100
Total assets	11,844,000	12,692,200

*1987 stock split: each share with a par value of 50SKr was exchanged for five shares with a par value of 10SKr each. The bonus issue was carried out at the same time, increasing the share capital by SKr28.2 million through a transfer from unappropriated earnings. The bonus issue gave shareholders one new share for every two old shares

Principal Shareholders: L E Lundbergföretagen AB

No of Employees: 569

PATRICIA AB

Box 16174, 10324 Stockholm

Tel: 08/743 05 70

PRINCIPAL ACTIVITIES: Financial services

Parent Company: AB Thisbe

Financial Information:

	S'000
Turnover	6,600,000

PAUL ANDERSON INDUSTRIER AB

See ANDERSON, PAUL, INDUSTRIER AB

PKBANKEN

Hamngatan 12, S-103 71 Stockholm

Tel: +468 781 80 00

Cable: pkbank, Stockholm

Telex: 19310; 11471 pkbank s

Telefax: +468 20 08 46

Board of Directors: Regular Members: Tony Hagström (Chairman Non-Executive), Nils Landqvist (Vice Chairman

Non-Executive), Olof Lund (Vice Chairman, Non-Executive), Bengt Berg, Johan Björkman, Olof Hakelius, Göran Hållén, Uno Janson, Birgitta Johansson-Hedberg, Ebbe Krook, Gösta Källén, Bertil Lövgren (Employee Representative), Sigvard Marjasin, Christer Ragnar, Birgit Ragnarsson (Employee Representative), Gösta Renell, Björn Rosengren, Peter Sahlin, Bertil Zachrisson, Christer Zetterberg. Deputy Members: Nils-Gunnar Anderby (Employee Representative), Peter Enström, Lars-Olof Hjalmar, Rune Jansson, Bo Klevborn, Inger Rudholm (Employee Representative), Åke Smids

Executive Management: Christer Zetterberg (President and Chief Executive Officer), Christer Ragnar (President Corporate Banking), Göran Hållén (President Consumer Banking)

PRINCIPAL ACTIVITIES: Commercial bank

Subsidiaries: (100% owned unless stated): PKfinans AB; PKpartner AB; PKfonder Administrations AB; PKinkasso AB (75%); PKkredit AB; Pundet AB; PKfastigheter AB; Kommunalåneinstitut AB; PKbanken International (Luxembourg) SA (Luxembourg); PK English Trust Group PLC (UK); PK English Trust Isle of Man Ltd (Isle of Man); PKbanken International (Finland) Oy (Finland); PKbanken International (US) Banking Corp (USA); PKbanken International (Asia) Ltd (Singapore); PK International (Hong Kong) Ltd (Hong Kong). Affiliated Bank: Banque OBC-Odier Bungener Courvoisier SA (France)

Financial Information:

	31.12.86	31.12.87
	SKr'000	SKr'000
Profit before tax	1,062,000	1,400,000
Profit after tax	255,000	734,000
Dividends	255,000	277,500
Dividends per share	8.50SKr	9.25SKr
Earnings per share	34.75SKr	35.00SKr
Share capital	750,000	750,000
Reserves	837,000	800,000
Shareholders funds	623,000	1,102,000
Deposits	143,197,000	133,764,000
Total assets	177,726,000	181,448,000

Principal Shareholders: The Kingdom of Sweden
No of Employees: 4,653

PROGRAMATOR (AB)

Box 20072, 161 02 Bromma
Tel: (08) 799 3500
Telefax: (08) 98 74 75

Board of Directors: Lars Irstad (Chairman and Chief Executive Officer), Leif Pilhage, Anders Rydin, Anders Skarin, Fritz Staffas. Deputies: Tore M Danielsen, Ulf Risberg, Zerny Sandh

Management: Lars Irstad (Chairman and Chief Executive Officer)

PRINCIPAL ACTIVITIES: Data processing operations; technical consulting operations in the construction and civil engineering areas and in contracting in certain specialised areas; financial services in the areas of leasing and asset based financing; trading operations in which the industrial trading group functions as an intermediary of know-how and logistics in the area of international trade

Subsidiary Companies: (100% owned unless stated): Programator Service AB; Programator Invest AB; Finansor AB (99.9%); Programatic Oy (Finland); Programatic Ven Man Oy (Finland); Krammdata AB (91%); AVD i Sverige AB; Argos AB; Programator Mellansverige AB; AU Network AB (51%); AB Jacobson & Widmark AB (59%); Programatic A/S (Denmark); Systemex AB; Grapple AB; Depea AB; Mimer Software AB; Industriell utrustning AB; Aureus AB; Programatic International AB; Alpro Information Services AB (51%); Programator Gåvleborg AB; IHG Holding AB (26%); Systemtech AB

Financial Information:

	31.12.86	31.12.87
	(8 mths)	(12 mths)
	SKr'000	SKr'000
Sales turnover	1,007,400	2,067,000
Income before tax(1)	76,400	156,500
Income before tax	34,800	78,200
Net income	22,400	41,900
Dividend per share	2.75SKr	5.00SKr
Earnings per share(2)	14.41SKr	27.89SKr
Share capital	44,400	44,800
Shareholders equity	97,400	131,400

(1) and appropriations
(2) after taxes actually paid

Principal Shareholders: Lars Irstad; Leif Pilhage; AGA AB; WASA; Natur & Kultur; Sparbankernas Aktiefonder
No of Employees: 4,342 (Group 3,701; Affiliated Companies 641)

RATOS, FÖRVALTNINGS, AB

Drottninggatan 2, S-111 51 Stockholm
Tel: 08/700 17 00
Cable: Ratos
Telex: 19355
Telefax: 08/10 25 59

Board of Directors: E Söderberg (Chairman), S Söderberg, Ole Damgaard-Nielsen, Harry Faulkner, Ulf Laurin, Harry Norin, Johan Söderberg
Management: Sven Söderberg (Managing Director), Thomas Mossberg (Finance Director)

PRINCIPAL ACTIVITIES: A mixed investment company with a portfolio of Swedish stocks and bonds, real estate companies, companies in the hotel business, transport, heating and sanitation

Subsidiary Companies: Dahl Invest International A/S (Denmark); Scandic Hotel AB; Nordisk Transport & Spedition AB; Förvaltnings AB Stancia. Associated Company: Ratos MA
Principal Bankers: Skandinaviska Enskilda Banken
Financial Information:

	31.12.86	31.12.87
	SKr'000	SKr'000
Sales turnover	6,722,200	7,093,900
Profit before tax	367,500	1,726,400
Profit after tax	230,500	1,562,100
Retained profit	152,000	1,471,100
Dividends	78,500	91,000
Dividends per share	6.25SKr	7.25SKr
Earnings per share	20.80SKr	124.70SKr
Share capital	631,100	631,100
Shareholders funds	1,631,700	3,131,100

Principal Shareholders: The family Söderberg; Skandia Trygg-Hansa; Pk-banken funds
No of Employees: 7,099

SIEMENS-GRUPPEN I SVERIGE (THE)
"Siemens Group of Sweden (The)"

S-171 95 Solna
Tel: (0) 8-730 70 00
Telex: 19958
Fax: (0) 8-98 60 17

Supervisory Board: Dr B Rudewald (Managing Director of Siemens-Elema AB, the Parent Company), J Bottheim (Chairman of the Board of Siemens-Elema AB)
Management Board: L-O Hjalmar (Managing Director of Siemens AB), D Obenaus (Financial Director Siemens AB), C G Myrin (Director of Finance Siemens-gruppen and Financial Director Siemens-Elema AB)

PRINCIPAL ACTIVITIES: Siemens-Elema AB, parent company of the Siemens-gruppen, industrial plant (medical engineering); Siemens AB, 100% subsidiary of Siemens-Elema AB, electrical and electronic engineering

Trade Names: Siemens-Elema

Parent Company: of Siemens-Elema AB (and the "Siemens-gruppen i Sverige") is Siemens Beteiligungen AG in Zürich, which is a 100% subsidiary of Siemens AG, Berlin/Munich

Subsidiary Companies: Siemens AB, Stockholm

Principal Bankers: Svenska Handelsbanken; Goetabanken

Financial Information:

| | 30.9.86 | 30.9.87 |
	SKr'000	SKr'000
Group sales turnover	2,809,900	3,120,900
Profit before tax	242,600	132,600
Profit after tax	84,200	70,600
Retained profit	55,600	105,100
Dividends	36,000	36,000
Dividends per share	15.0SKr	15.0SKr
Earnings per share	35.0SKr	-
Share capital	120,000	120,000
Shareholders funds	44,100	43,800

No of Employees: 3,261

SKANDIA INSURANCE COMPANY LTD

Sveavägen 44, S-103 50 Stockholm

Tel: +46-8-788 11 00

Cable: SKANDIA STOCKHOLM

Telex: 19347 SKANDIA S

Telefax: +46-8-10 31 74

Board of Directors: Sven Söderberg (Chairman), Clas Reuterskiöld (Vice-Chairman), Peder Bonde, Hans Delborg, Gunnar Ericsson, Sten Gustafsson, Tom Hedelius, Sture Lindquist, Curt G Olsson, Rolf Olténg, Bo Rydin, Pentti Talonen, Anders Wall, Björn Wolrath (Managing Director). Appointed by the Swedish Government: Karin Söder, Bo Södersten, Lars Tobissen. Elected by the employees: Bengt Green, Christina Torsslow. Deputy Members of the Board: Karl-Eric Halén (Executive Vice President, Skandia), Björn Hall (Executive Vice President, Skandia), Leif Victorin (Executive Vice President, Skandia). Appointed by the Swedish Government: Bengt Kronbled, Göran Borggård, Filip Fridolfsson. Elected by the employees: Ethel Almbjär, Bertil Holmström

Group Management: Björn Wolrath (President and Chief Executive Officer), Björn Hall (Executive Vice President, Finance), Leif Victorin (Executive Vice President, Insurance Operations). Departmental Heads: Tor Bystedt (Legal Affairs), Karl-Eric Halén (Executive Vice President, Administrative Director of Departments and Service Units), Bengt Johansson (Group Accounting). Insurance Operations: Lars Arfwidson (Group Insurance Division), Göran Ericsson (Consumer Division), Göthe Gustafsson (Marketing Division), Bo Wahllöf (Commercial Division). Financial Operations: Christer Dahlström (Skandia Investment), Torsten Johansson (Stocks), Christer Käck (Bonds), Bernt Johansson (Real Estate). Service Units: Valborg Anderzon (Personnel), Olli Aronsson (Skandia Data), Kenneth Borensjö (Office Services)

PRINCIPAL ACTIVITIES: Parent company of a group engaged in insurance in all classes

Subsidiary Companies: (100% owned unless stated): Skandia Life Insurance Company Ltd; Skandia Investment AB; Skandinavisk Försäkringsservice AB (95%); Husdjursförsäkrings AB Sleipner; Internationell Assurans AB; Utredningskliniken AB; ProSkandia AB; Amphion Holdings Inc; Anticimex AB (50%); Bokredit AB (34%); Sickla Industrifastighet AB (33%); Skandia International Holding AB (45%); Svenska Kreditförsäkringsaktiebolaget (31%); Trans Secure SA (50%) (Luxembourg)

Financial Information: Group figures

| | 31.12.86 | 31.12.87 |
	SKr'000	SKr'000
Group premium income	9,231,000	9,931,000
Group investment income	6,816,000	6,952,000
Profit before tax*	104,000	139,000
Profit after tax*	120,000	144,000
Dividends	120,000	144,000
Dividends per share	2.00SKr	2.40SKr
Earnings per share	7.00SKr	9.00SKr
Share capital	300,000	300,000
Shareholders equity*	522,000	546,000

*figures of Parent Company only

Principal Shareholders: Tidnings AB Marieberg (10.5%); Investment AB Beijer (7.5%); Opab Förvaltnings AB (7.4%); AB Investor (5.3%); Skandia Group Pension Foundation (3.8%); Förvaltnings AB Providentia (3.7%); AB Industrivärden (2.7%); Investment AB Öresund (2.3%); Assurance Générales de France Vie (2%); Trygg Ömsesidig Livförsäkring (1.8%)

No of Employees: 4,563 Group

SKANDINAVISKA ENSKILDA BANKEN

Kungsträdgårdsgatan 8, S-106 40 Stockholm

Tel: +46 8 763 50 00

Cable: Essebank

Telex: 11000 essebi s

Telefax: +46 8 763 83 89

Board of Directors: Directors elected by the Annual General Meeting: Curt G Olsson (Chairman), Peter Wallenberg (First Deputy Chairman), Lennart Johansson (Deputy Chairman), Nils Linander (Deputy Chairman), Rolf Cavallius, Sten Gustafsson, Pehr G Gyllenhammar, Bengt Haak, Ernst Herslow, Curt Nicolin, Carl Henrik Nordmark, Hans Stahle, Sven Söderberg, Knut Wachtmeister, Mrs Karin Westerberg, Karl Erik Önnesjö, Hans Cavalli-Björkman (Managing Director and Chairman of the Executive Management), Jacob Palmstierna (Managing Director and Deputy Chairman of the Executive Management). Directors appointed by the Government: Leif A Andersson, Mrs Ingegärd J Oskarsson, Lennart Pettersson. Directors appointed by the employees: Lars Ahlborn, Rolf Blom. Deputy Directors elected by the Annual General Meeting: Göran Ahlström, Rune Andersson, Bo Berggren, Bertil Bertilsson, Olle Blomqvist, Sven Borelius, Gösta Bystedt, Tore Daun, Mrs Antonia Ax:son Johnson, Rune F Persson, Stefan Persson, Karl-Erik Sahlberg, Mrs Ingegaard Troedsson, Tom Wachtmeister, Sven Wallgren, Ulf Wikman, Rutger Barnekow (Managing Director), Bo C E Ramfors (Managing Director), Per Åke Harrison (First Deputy Managing Director). Deputy Directors appointed by the employees: Gunnar Boberg, P-O Magnusson

Executive Management: Hans Cavalli-Björkman (Chief Executive of the Malmö Group), Jacob Palmstierna (Chief Executive of SEB International), Bo C E Ramfors (Chief Executive of the Göteborg Group), Rutger Barnekow (Chief Executive of the Stockholm Group), Per Åke Harrison (Deputy Chief Executive of SEB International)

PRINCIPAL ACTIVITIES: Banking

Subsidiary Companies: Aktiv Placering AB; AB Arsenalen; FinansSkandic AB and its subsidiaries Eurocard AB; FS FinansInkasso AB; AB Vendax; FinansSkandic (Luxembourg) SA (Luxembourg); Deutsche Skandic Leasing GmbH (Germany); FinansSkandic Corporation (USA); FinansSkandic (UK) Ltd; FinansSkandic (South East Asia) Pte Ltd (Singapore); AB Garnisonen; SEB-Fastigheter AB; Svensk Fastighetskredit AB. Subsidiary Banks: Deutsch-Skandinavische Bank AG (Germany) with a branch in Hamburg and its subsidiary Deutsch-Skandinavische Bank (Luxembourg) SA (Luxembourg); Enskilda Securities, Skandinaviska Enskilda Ltd (UK); Skandinaviska Enskilda Banken Corporation (USA) and its subsidiary SEBC Funding Incorporated (USA); Skandinaviska Enskilda Banken

(Luxembourg) SA (Luxembourg); Skandinaviska Enskilda Banken (South East Asia) Ltd (Singapore). In addition Skandinaviska Enskilda Banken has three branch offices (in London, Hong Kong and New York) and eleven representative offices abroad

Financial Information:

	31.12.86 SKr'000	31.12.87 SKr'000
Profit before tax(1)	3,949,000	4,165,000
Profit after tax	2,667,000	3,429,000
Dividends	474,000	557,000
Dividends per A share	3.60SKr	4.25SKr
Earnings per share(2)	14.87SKr	15.64SKr
Share capital	1,204,000	1,284,000
Shareholders funds(3)	13,416,000	16,793,000
Deposits	139,022,000	162,686,000
Total assets	191,053,000	233,688,000

(1) and appropriations
(2) after 50% standard tax
(3) including untaxed reserves

Principal Shareholders: AB Custos; Knut and Alice Wallenberg Foundation; Svenska Personal-Pensionskassan
No of Employees: 9,149 Group

SKÅNSKA BANKEN

S-205 40 Malmö
Tel: +46 40 35 10 00
Cable: Skanskabank
Telex: 33209 SYDBANK S (General Traffic)
Telefax: +46 40 35 14 57

Board of Directors: Carl-S. Roos (Chairman)
Executive Management: Axel Roos (Managing Director), Sven Vernerson (Deputy Managing Director), Arne Arstam (Legal Department), Bo Axelsson (Administration Department), Kjell Gunnarsson (Småland Region), Lennart Hansson (Treasury/Finance), Sten Johansson (Audit Department), Lars Lindgren (Malmö Region), Reino Lindgren (Stockholm Region), Kaj Norberg (International Division), Axel Roos jr (Skåne Region), Christen Roos (Stockbroking and Trustee Department)

PRINCIPAL ACTIVITIES: Banking business, domestic and international (Regional Bank)
Representative Office: In London. Affiliated Bank: Inter Nordisk Bank AS, Oslo, Norway
Financial Information:

	31.12.86 SKr'000	31.12.87 SKr'000
Profit before tax	102,800	120,000
Profit after tax	23,500	59,300
Retained profit	64,403	68,523
Dividends	19,380	21,420
Dividends per share	9.50SKr	10.50SKr
Earnings per share	73.00SKr	71.00SKr
Share capital	102,000	102,000
Shareholders funds	672,008	791,556
Deposits	6,262,149	6,497,835
Total assets	16,399,492	15,651,564

No of Employees: 660

SKARABORGSBANKEN

Fack Radhusgatan 5, Box 612, 541 29 Skövde
Tel: 0500/777 00
Cable: Skaraborgsbank
Telex: 67034
Telefax: 0500/86302

Board of Directors: Bengt Nyström (Chairman), Kurt Pålsson (Deputy Chairman), Gösta Karlsson, Paul Jansson (Appointed by the Government), Anders Rapp, Håkan Davidsson, Ivar Virgin, Tore Johansson, Gunnar Dahlsten, Erik-V Ramberg, Rigmor Dahl (Representative of the Employees), Alf Arvidsson

(Representative of the Employees), Bengt Skoglund (Managing Director)
Management: Bengt Skoglund (Managing Director), Sven Wåhlgren (Deputy Managing Director, Head Trust Division), Lennart Lundberg (Deputy Managing Director, Head Stockholm Central Office), Björn Rosén (Deputy Managing Director, Head Branch Network except Stockholm), Lennart Nilsson (Legal Advisor), Kristina Ottosson (Head Finance), Sten-Börje Nävarp (Head Administration)

PRINCIPAL ACTIVITIES: Bank
Parent Company: Gotagruppen AB
Subsidiary Companies: Skaraborgsmäklaren AB; Skaraborgsbankens Finans AB; Fastighets AB Guller; Banco Fastighetsfinansiering AB; Banco Fonder
Principal Bankers: Midland Bank; Chase Manhattan Bank; Deutsche Bank
Financial Information:

	31.12.86 SKr'000	31.12.87 SKr'000
Profit before tax	129,500	95,600
Profit after tax	17,172	42,015
Retained profit	16,799	17,771
Dividends	16,200	18,900
Dividends per share	1.80SKr	2.90SKr
Earnings per share	7.40SKr	5.30SKr
Share capital	90,000	90,000
Reserves	183,971	209,786
Deposits	4,030,372	4,621,685
Total assets	7,327,723	9,094,339

Principal Shareholders: GOTA-Gruppen AB, Employee foundations
No of Employees: 533

SPAFI

Sparbankernas Fastighets Finansierings AB

Regeringsgatan 19, 10534 Stockholm
Tel: 08/7902100

Management: Leif Sjöqvist

PRINCIPAL ACTIVITIES: Financial services including credit operations
Financial Information:

	SKr'000
Turnover	14,600,000

SPARBANKERNAS INTECKNINGS AB
SPINTAB

Regeringsgatan 19, 10534 Stockholm
Tel: 08/7902100

Management: Leif Sjöqvist

PRINCIPAL ACTIVITIES: Financial services including credit facilities
Parent Company: Sparbankernas Bank
Financial Information:

	Skr'000
Turnover	20,000,000

No of Employees: 200

SVENSKA HANDELSBANKEN

Blasieholmstorg 11, 10328 Stockholm
Tel: (468) 7691000
Cable: Handelsbank
Telex: 11090

Board of Directors: J Wallander (Chairman), K Brändström, L Claesson, P-O Edin, J Ekman (Deputy Chairman), G Ennerfelt, P-O Eriksson, C-E Feinsilber, T Fälldin, T Hedelius (President), N Holgerson, L-O Håkansson, G L Johansson, U Laurin, P Lindberg, B Lindegren, G Lundblad, B Löf, B J Persson, B Rydin, G Sundblad, A Wall, S Agrup
Management: J Hedelius (President)

Executive Vice Presidents: Regional Bank Divisions:- J Setterberg (Stockholm City), H Lundholm (Central Sweden), A Mårtensson (Western Sweden), G Rundström (Southern Sweden), C Källströmer (Northern Norrland), R Larson (Greater Stockholm), L Spetz (Southern Norrland), B Strage (Eastern Sweden). Central Head Office:- M Abrahamsson (Central Administration Division), B Damberg (Central Credit Division), G Björling (Central Division of Investment Banking), K Wahlström (Central Controller's Division), S Hallström (Control Computer Department). Central International Divisions:-A Ljungh (Division Head), S Wikstrom (Foreign Exchange)

PRINCIPAL ACTIVITIES: Bank

Subsidiary Companies: (100% owned unless stated): Fastighets AB Filia; Sigab (95%); Svenska Finans AB (95%); Fastighets AB Blasieholmen; SHB Kapitalplacering AB; Svenska International PLC; Svenska Handelsbanken SA (Luxembourg); Svenska Handelsbanken Inc (USA); Svenska Handelsbanken Asia Ltd (Singapore); Nordic Monitoring Services Ltd (UK); Oy Rahameklarit AB (50%) (Finland)

Financial Information: Consolidated Figures

	31.12.86	31.12.87
	SKr'000	SKr'000
Profit before tax(1)	3,427,100	2,605,100
Retained profit	231,200	272,200
Dividends	320,700	383,300
Dividends per share	3.00SKr(2)	3.60SKr
Earnings per share	16.10SKr(2)	11.90SKr
Share capital	1,211,400	1,211,400
Shareholders funds	979,200	1,347,700
Deposits	66,853,000	84,198,000
Total assets	156,553,000	202,561,000

(1) and allocations
(2) after bonus issue and split

Principal Shareholders: The Oktogonan Foundation (5.3%); The Trygg-Hansa Group (5.2%); The Skandia Group (5.1%); SPP (Pension Insurance) (4.7%); Sparbankernas aktiefonder (5.1%); Industrivärden (3.3%); The Folksam Group (2.2%); Trefond (1.8%); Handelsbankens aktiefonder (1.2%); PK bankens aktiefonder (1.6%); Wasa-koncernen (Insurance) (1.6%); Skandia International (1.1%)

No of Employees: 6,447

SWEDBANK

Sparbankernas Bank

Brunkebergstorg 8, S-10534 Stockholm
Tel: +46 8-222320
Cable: SWEDBANK
Telex: 12826 swedbnk s
Telefax: 08-7968092

Board of Directors: Ingemund Bengtsson (Chairman), Åke Janson (First Vice Chairman), Per-Göran Nyberg (Second Vice Chairman), Erik Hovhammar, Leif Lewin, Carl Eric Stålberg, Kjell Andersson, Bernt Gröön, Kurt Lodenius, Göran Collert (President and Chief Executive), Roland Duvebäck, Erland Nygren, Karl-Erik Andersson, Gunnar Berg, Jan Rydh. Appointed by the Government: Gunnar Hofring. Deputy Members: Anders Larsson, Johan Engström, Ingrid Jansson, Christer Malm, Jerker Pohl, Ove Rydin. Appointed by the Employees' Trade Unions: Ragnar Berglind, Robert Svanberg, Gunilla Hohmann, Lars Thuresson

Executive Management: Göran Collert (President), Ove Rydin (Executive Vice President and Deputy to the President), Sören Andersson (Executive Vice President), Johan Engström (Executive Vice President)

Senior Executives: Staff Functions: Johan Stiernstedt (Management Secretariat), Hubert Fromlet (Economic Research Department), Tord Beite (Legal Counsel), Anna Persson (Personnel), Lars Bergstig (Marketing and Information), Bengt Sand (Auditing)

PRINCIPAL ACTIVITIES: Banking

Subsidiary Companies: (100% owned unless otherwise stated): Sparbankernas Bostadfinansiering AB SPABO; Kommunkredit AB (94%); Sparbankernas Jordbrukscredit AB; Sparbankernas Aktiefonder Administrations AB; Sparbankernas Inteckningsaktiebolag SPINTAB; Sparbankskort AB; Industrifinans AB IFAB; Sparbankernas Fastighetsfinansierings AB SPAFI; Bankkorts Administrativ Butiksservice AB BABS (50%); Industrifinans Service AB (51%); Industrifinans International SA; Industrifinans Datainvest AB; S & I Datainvest AB (50%); FennoScandia Bank Ltd (50%) (UK); Banque Nordeurope SA (50%) (Luxembourg). Representative offices in New York, London, Moscow, Helsinki, Oslo and Copenhagen and representation in Beijing

Financial Information: Consolidated Figures

	31.12.86	31.12.87
	SKr'000	SKr'000
Profit before tax	690,000	757,000
Profit after tax	518,000	611,000
Retained profit	498,000	591,000
Share capital	360,000	360,000
Shareholders funds(1)	3,749,000	4,341,000
Deposits(2)	160,897,000	175,192,000
Total assets	171,998,000	187,781,000

(1) Including untaxed reserves for bonds
(2) Including long term borrowing and deposits from financial institutions

Principal Shareholders: The Swedish Savings Banks
No of Employees: 1,200

TRYGG-HANSA

Fleminggatan 18, 106 26 Stockholm
Tel: 08/785 10 00
Telex: 19887 hansa s
Telefax: 8-51 87 88

Board of Directors: Alf Åkerman (Chairman), C-S Giesecke
Management: Björn Sprängare (Managing Director)

PRINCIPAL ACTIVITIES: Life insurance and non-life insurance
Parent Company: Hansa ömsesidig sakförksäkring (Hansa Mutual General Insurance Co)
Subsidiary Companies: Hansa Företag Försäkrings AB (Hansa Commercial & Industrial Insurance Co); *Hansa Trafik ömsesidig sakförsakring (Hansa Commercial Motor Insurance Co); *Hansa SJÖ Försäkringsaktiebolag (Hansa Marine Insurance Co); *Hansa International Försäkringsaktiebolag (Hansa International Insurance Co); *These three form Hansa Trade
Financial Information: Group figures

	31.12.86	31.12.87
	SKr'000	SKr'000
Premium income(1)	3,651,000	3,742,000
Premium income(2)	4,663,000	5,027,000
Investment income(1)	2,726,000	2,959,000
Investment income(2)	1,056,000	1,057,000
Profit before tax	2,907,252	865,479
Profit after tax	457,019	783,439

(1) Life insurance
(2) Non-life insurance

No of Employees: 2,823 Office Staff, 624 Field Staff

WASA INSURANCE

S-10376 Stockholm
Tel: 08/796 4 000

Board of Directors: Sigvard Ando (Chairman)
Management: Carl Erik Hedlund (President and Chief Executive Officer)

PRINCIPAL ACTIVITIES: Insurance services

SWEDEN

Subsidiary Companies: WASA International; WASA LIV; WASA Industrial; WASA NORD; WASA Marine; WASA GÖTA; WASA European; WASA SkÅNE; WASA Credit and Guarantee

Financial Information:

	1988
	SKr'000
Premium income	5,400,000

No of Employees: 1,900

WERMLANDSBANKEN

Box 1008, S-651 15 Karlstad
Tel: (+46-54) 17 40 00
Cable: WERMLANDSBANK
Telex: 660 33 WEBANKK S
Telefax: (+46-54) 18 14 90

Supervisory Board: Sven Erik Ragnar (Chairman of the Board), Carl-Johan Wettergren (Deputy Chairman of the Board), Lennart Hemmingson, Lars Ander, Sune Lundh, Bengt Magnusson, Gunnar Schotte, Ulla-Britt Green, Göran Bergholtz, Rune Ström, Erik Willers, Ingemar Essén, Ingvar Larsson, Karin Söder, Valter Hedlund, Torbjörn Öberg (Staff Representative)

Management Board: Sven Erik Ragnar (Chairman of the Board), Gunnar Schotte (Managing Director and Chief Executive), Tord Olofsson (Deputy Managing Director), Anders Mattsson (Deputy Managing Director), Sten Lindblad (Deputy Managing Director), Aake Lindell, Kjell Mellander, Aake Lönnberg, Ulla-Britt Green (Staff Representative), Anita Carlsson (Staff Representative)

PRINCIPAL ACTIVITIES: Banking business
Trade Names: Wermlandsbanken
Subsidiary Companies: Fastighets AB Merkurius; Wemlandsbankens Finans AB
Principal Bankers: Philadelphia International Bank; Midland Bank PLC; Crédit Lyonnais; Deutsche Bank AG
Financial Information:

	31.12.86	31.12.87
	SKr'000	SKr'000
Profit before tax	100,000	112,000
Profit after tax	31,000	63,000
Retained profit	14,000	25,000
Dividends	20,000	25,000
Dividends per ordinary share	13.00SKr	16.00SKr*
Earnings per ordinary share	68.80SKr	58.00SKr
Dividends per preference share	2.50SKr	2.50SKr
Ordinary share capital	78,000	78,000
Preference share capital	4,000	4,000
Shareholders funds	735,000	926,000
Deposits	5,485,000	6,148,000
Total assets	12,961,000	14,298,000

*of which 2SKr was a special bonus dividend

Principal Shareholders: GotaGruppen AB; Stiftelsen Wevas; Nya Wermlands-Tidningens AB
No of Employees: 738

YNGVE KULLENBERG BYGGNADS AB

See KULLENBERG, YNGVE, BYGGNADS AB

Major Companies of SWITZERLAND

A&E LEASING AG

Stampfenbachstrasse 117, CH-8035 Zürich

Tel: (01) 363 10 00

Telex: 54 481 ael ch

Supervisory Board: B Burkhardt (President), Dr W Lüem, A S Schneiter

Management Board: A Studer, H Dick

Senior Executives: C Dogan, B Kessler, R Hämmerli, B Herzog,

PRINCIPAL ACTIVITIES: Leasing of industrial machinery

Financial Information:

	1987
	SF'000
Sales turnover	229,000

No of Employees: 51

AARGAUISCHE KANTONALBANK

Bahnhofstrasse 58, CH-5001 Aarau

Tel: (064) 21 77 21

Cable: Kantonalbank

Telex: 981 200

President: H Rauber

Senior Executives: Dr W Bolleter, Dr H Bolliger, H-R Gloor

PRINCIPAL ACTIVITIES: Bank

Financial Information:

	31.12.86	31.12.87
	SF'000	SF'000
Profit after tax	6,762	6,729
Share capital	160,000	180,000
Reserves	61,620	64,920
Deposits	2,393,973	2,592,992
Total assets	4,811,990	5,074,183

ABB ASEA BROWN BOVERI LTD

Affolterstrasse 52, CH-8050 Zurich

Tel: (41) 01 317 71 11

Telex: AB 823 790 32

Telefax: (41) 01 317 73 21

Board of Directors: Dr Fritz Leutwiler (Co-Chairman, Chairman of the Board of BBC Brown Boveri Ltd), Dr Curt Nicolin (Co-Chairman, Chairman of the Board of Asea AB), Dr Bernd H Müller-Berghoff (Deputy Chairman and President of BBC Brown Boveri Ltd), Christian Norgren, Dr Stephan Schmidheiny, Dr Gaston Thorn, Heinrich Weiss

Management Board: Percy Barnevik (President and Chief Executive Officer), Arne Bennborn (Executive Vice President), Erwin Bielinski (Executive Vice President), Sune Carlsson (Executive Vice President), Thomas P Gasser (Deputy Chief Executive Officer), Dr Ebehard von Koerber (Executive Vice President), Göran Lindahl (Executive Vice President), Göran Lundberg (Executive Vice President), Bertold Rouracken (Executive Vice President), Bert-Olof Svanholm (Executive Vice President), Edwin Somm (Executive Vice President), Lars Thunell (Executive Vice President), Leonardo Vanotti (Executive Vice President)

PRINCIPAL ACTIVITIES: ABB develops, produces, sells, and services systems and products in a wide range of areas generally related to the production, distribution, and application of electricity. Its principal activities are in the fields of power generation plants for primary energy - coal, gas, oil, water, nuclear; high-voltage transmission of electricity with such products as switchgear, transformers, relays and cables; medium- and low-voltage distribution with substations, switchgear and installation. These three segments together contribute almost half of total sales. Further major areas of activity are public transportation such as high-speed trains, locomotives, urban transportation systems; electrical drives, process automation, and metallurgy in the industry sector; and environmental control technology, where the main focus is on air handling and treatment. ABB's activities also include robotics, instrumentation, power lines, general contracting, superchargers, and certain specialities within telecommunications, advance plastics, and local businesses in installation material, wholesale and service. The Financial Services Segment provides service in the areas of financing, leasing, treasury operations, insurance, trading, and portfolio management for companies within the ABB Group and for third parties

Parent Companies: ASEA AB, Sweden and Brown Boveri Ltd, Switzerland

Subsidiary Companies: ABB Asea Brown Boveri Ltd has some 800 subsidiary companies

Financial Information:

	1988
	US$'000
Sales turnover	17,832,000
Income before tax	536,000
Net income	386,000
Dividends	100,000
Share capital	1,250,000
Shareholders equity	3,122,000

Principal Shareholders: ASEA AB and BBC Brown Boveri Ltd

No of Employees: 170,000

ALLGEMEINE TREUHAND AG
ATAG

Aeschengraben 9, Postfach 2149, CH-4002 Basel

Tel: (061) 54 88 88

Telex: 965505 atag ch

Telefax: (061) 23 40 60

Supervisory Board: Dr Peider Mengiardi (Präsident), Karl Müller (Vizepräsident), Fritz Christen (Bern), Dr Markus Schär (Bern), Dr Heini Wiki (Zürich), Eugen Harter (Basel), Charles Jaquier (Lausanne), Theodor Fässler (Mörigen), Leonhard Gysin (Liestal), Ernst Höhn (Kronbühl), Riccardo Jagmeth (Zürich), Cuno Pümpin (Jona), Walter Ryser (Gümlingen), Charles Schaefer (St Sulpice)

Central Management: Dr P Mengiardi (Basel), M Attenhofer (Zürich), W Fischer (Bern), PD Dr A Frey (Zürich), E Harter (Basel), C Jaquier (Lausanne), K Müller (Zürich), Dr M Schär (Bern), Dr U Neuenschwander (Biel), Dr H Wiki (Zürich), Dr U Widmer (Zürich)

Branch Office Management: Dr J Engeli (Aarau), E Harter (Basel), Dr W Glaus (Bern), Dr U Neuenschwander (Biel), J Brosi (Chur), J Bardy (Freiburg), J Kettiger (Genf), G Schlaeppi (Lausanne), J Hertig (Neuenburg), Dr B Grossmann (St Gallen), M Leysinger (Solothurn), Dr H Wiki (Zürich), A Kämpfen (Brig), M Gisler (Kreuzlingen), C Riand (Sion), M Burri (Winterthur)

PRINCIPAL ACTIVITIES: Auditing and economic advisory services; economic information services

Trade Names: Allgemeine Treuhand AG (ATAG)

Subsidiary Companies: Afidium Treuhand AG; Arthur Young & Co AG; Fiduciaire de Gestion et d'Informatique; Hangartner Treuhand AG; IHA Institut für Marktanalysen AG; Implementa AG; Intercontrol AG; Interdata AG; KOREAG Kontroll & Revisions AG; Retrema Treuhand AG; Fiduciaire Lémano SA; SRC Revisiongesellschaft AG; Tutor Versicherungsberatung AG; Funk Treuhand AG/Gestion Financière SCHIBLER SA; Fiduciaire C Riand & Cie SA; BFO Treuhand AG; Bösch & Edthofer AG; Eldag Informatik AG; Opus Development AG; T & R Grischa AG; Datron AG

Financial Information:

	31.12.86	31.12.87
	SF'000	SF'000
Sales turnover	183,200	227,095
Profit before tax	15,880	16,427
Profit after tax	12,940	12,450
Retained profit	12,000	12,000
Dividend	540	540
Dividend per share	180SF	180SF
Share capital	3,000	3,000
Shareholders funds	81,700	88,950

Principal Shareholders: Dr Manfred Hoessly-Stiftung
No of Employees: 1,720 Group

ALPINA VERSICHERUNGS-AG

Alpina Cie d'Assurances SA

Alpina Insurance Co Ltd

Seefeldstrasse 123, CH-8034 Zürich
Tel: (01) 259 5151
Cable: Versalpina
Telex: 52078 Alpin CH
Telefax: (01) 259 5255

Board of Directors: F Gerber (President), Dr A Weber (Vice President), Dr R E Bitterli, H Dennler, Dr A Egli, Dr H Hereth, H Egenter, F Martinoli, R Cloos, P Meier, P Penggli, P Rüdisser, B Schiess, P Christen, E Knecht, T Thomann, K Ehrenbaum
Management Board: H Egenter (General Manager), F Martinoli (Manager), R Cloos (Deputy Manager), P Renggli (Deputy Manager), P Rüdisser (Deputy Manager), B Schiess (Deputy Manager), P Christen (Assistant Manager), E Knecht (Assistant Manager), P Meier (Assistant Manager), T Thomann (Assistant Manager), K Ehrenbaum (Assistant Manager)

PRINCIPAL ACTIVITIES: All classes of insurance except life
Parent Company: Zürich Versicherungs-Ges
Financial Information:

	31.12.86	31.12.87
	SF'000	SF'000
Gross premium income	381,356	411,199
Gross investment income	44,945	44,990
Profit before tax	5,866	7,639
Profit after tax	4,756	5,210
Retained profit	4,756	5,210
Share capital	18,000	18,000
Shareholders funds	57,186	62,396

Principal Shareholders: Zürich Insurance Co Ltd
No of Employees: 690

ANDRÉ & CIE SA

Ch. de Messidor 7, CH-1002 Lausanne
Tel: (021) 21 11 11
Telex: 454 101

PRINCIPAL ACTIVITIES: International trade in foodstuffs and commodities including cereals, animal feedstuffs, tea, coffee and rice; financial services
Financial Information:

	SF'000
Sales turnover	8,800,000

No of Employees: 450

BALOISE INSURANCE COMPANY LTD (THE)

Baloise Life Insurance Company Ltd (The)

Aeschengraben 21, Box, CH-4002 Basle
Tel: (061) 55 70 00
Cable: Baloise
Telex: 962182 bvg ch
Telefax: (61) 55 85 85

Board of Directors: Dr Rudolf Th Sarasin (Chairman), Guy Sarasin (Vice Chairman), Dr Alex Krauer, Dr Christoph J C Albrecht, Dietrich Forcart, Sergio Grandini, Dr Martin Junger, Dr Franz Lütolf, Me Jacques Matile, Dr Hans-Peter Sigg, Dr Paul Wyss, Josef Zumstein
Management: Dr Robert Bauman (President and General Manager, Switzerland, Marine), Gianfranco Balestra (General Manager Finance/Administration), Luzius Gloor (General Manager Foreign Operations/Reinsurance/Marine)
Senior Executives: Heinz Beer (Senior Manager Sales Organisation), Dr Hans Krenger (Senior Manager Casualty/Property Lines), Dr Jacques Röthlisberger (Senior Manager Personal Lines)

PRINCIPAL ACTIVITIES: Insurance
Parent Company: Baloise-Holding
Subsidiary Companies: La Baloise (France); Basler Versicherungs-AG (Austria); The Baloise of America (USA); Deutscher Ring Vers.-Gruppe (Germany); Compañia Vascongada de Seguros y Reaseguros SA (Spain); Baloise Insurance Company (IOM) Ltd (Isle of Man); Mercator (Belgium); Levante (Italy); Erpin (Spain)
Financial Information:

	31.12.86	31.12.87
	SF'000	SF'000
BALOISE INSURANCE CO		
Premium income	1,261,300	1,320,200
Investment income	153,126	163,900
Financial result	86,631	70,147
Net profit for the year	26,037	26,657
Retained profit	13,000	13,000
Dividend	12,350	13,000
Dividend per share	19SF	20SF
Share capital	65,000	65,000
BALOISE LIFE INSURANCE CO		
Premium income	1,020,695	1,020,700
Investment income	384,736	410,100
Overall profit	161,336	175,344
Net profit for the year	5,336	5,344
Retained profit	3,500	3,500
Dividend	1,750	1,750
Dividend per share	350SF	350SF
Share capital	25,000	25,000

No of Employees: 8,386

BANCA DEL GOTTARDO

Gotthard Bank

Casella Postale 2811, Viale S. Franscini 8, 6901 Lugano
Tel: (091) 21 41 14
Cable: Gottardbank
Telex: 841051 bdg ch
Telefax: (091) 23 23 17

Board of Directors: Dr F Garzoni (Chairman), E Tamai (Vice Chairman 1), C von Castelberg (Vice Chairman 2), J Baumgartner, S Grandini, A Hartmann, M Ishizaki, C Jelmini, A Kammerer, B Müller, Y Okabe, G B Pedrazzini, H Tanner
Executive Board: F Bolgiani (President), A Campana (Executive Vice President), W Canepa (Executive Vice President), L Cereghetti (Executive Vice President), O Husi (Executive Vice President), T Tsukuda (Executive Vice President), S Bassi (Deputy Executive Vice President), F Charrey (Deputy Executive Vice President), F Manzoni (Deputy Executive Vice President) and Deputy General Manager)

PRINCIPAL ACTIVITIES: Bank
Trade Names: Gottardo
Parent Company: The Sumitomo Bank Ltd, Japan
Subsidiary/Associated Companies: (100% owned unless stated): Gotthard Bank International Ltd (Bahamas); Gotthard Finance International Inc (Panama); Ultrafin AG; Dreieck Leasing SA; Dreieck Finanz AG; Dreieck Fiduciaria SA; Gestivalor-Gestione Fondi SA; Banque de Gestion Privée (75%); Banque

Pariente (45%); AG für Fondsverwaltung (20%); Gertrust SA (20%); Gesfid SA (40%); Cogesta-Compagnia Finanziaria SpA (10%) (Italy)

Financial Information:

	31.12.86	31.12.87
	SF'000	SF'000
Cash flow	76,000	77,000
Net profit	39,250	40,000
Dividends	24,600	23,985
Dividends per share	20.00SF	18.00SF
Share capital*	123,000	133,250
Shareholders funds*	431,046	476,935
Deposits	3,726,058	4,043,882
Total assets	4,504,946	4,874,758

*and participation capital

Principal Shareholders: Sumitomo Bank Ltd
No of Employees: 690

BANK LEU LTD

32 Bahnhofstrasse, 8022 Zürich
Tel: +411 219 11 11
Cable: Bankleu
Telex: 812 174 leu ch
Telefax: +411 219 3197

Board of Directors: Prof Dr K Schiltknecht (Chairman)

General Management: Dr W B Schick (President), A Brändle (Executive Vice President), Dr W Frey (Executive Vice President), Dr G Kramer (Executive Vice President)

Senior Executives: W Dönz (Senior Vice President), P Freimuller (Senior Vice President), E Gassner (Senior Vice President), Dr A Grütter (Senior Vice President), P Heiniger (Senior Vice President), A Iten (Senior Vice President), R Kälin (Senior Vice President), R Lyner (Senior Vice President), K Rüegg (Senior Vice President), Dr H P Schaad (Senior Vice President), W Schenkel (Senior Vice President), Ch Scherrer (Senior Vice President), E Vetsch (Senior VP), A Wiprächtiger (Senior Vice President)

PRINCIPAL ACTIVITIES: Bank

Subsidiary Companies: Bank Leu Geneva Ltd, Geneva; Bank Heusser & Co Ltd, Basel; Banque Leu (Luxembourg) SA, Luxembourg; Leu Trust and Banking (Bahamas) Ltd, Nassau (Bahamas); Leu Securities Ltd, London (UK); Swiss Investment Company, Zürich; Fundus Marty Ltd, Trust and Management Company, Zürich; Leu Fondsleitung AG, Zürich; Merkur Immobilien AG, Zürich; Leu Travel Ltd, Zürich; Baugesellschaft zum Brunnen, Zürich

Financial Information:

	31.12.86	31.12.87
	SF'000	SF'000
Profit before tax	102,981	105,401
Profit after tax	55,725	61,108
Retained profit	45,884	40,479
Dividends	46,149	51,875
Dividends per SF500 share	100SF	100SF
Dividends per SF100 share*	20SF	20SF
Earnings per SF500 share	172.77SF	146.45SF
Earnings per SF100 share*	34.55SF	29.29SF
Share capital	200,400	240,000
PC capital*	65,941	75,335
Reserves	687,578	779,733
Shareholders funds	953,919	1,095,068
Deposits	12,125,956	12,462,562
Total assets	14,177,886	14,710,515

*Participation Certificate

No of Employees: 1,495

BANQUE CANTONALE DE BÂLE

Basler Kantonalbank

2 Spiegelgasse, Basle
Tel: 29.21.21
Cable: Kantonalbank Basle
Telex: BKB CH 96.59.59
Telefax: 25.84.34

Board of Directors: Dr L Burckhardt (Chairman)
Executive Board: H J Gallusser (Speaker)
Senior Vice Presidents: Dr U Rohde, A Gfeller, M Walliser

PRINCIPAL ACTIVITIES: Bank
Subsidiary Companies: 12 branches; 1 agency
Financial Information:

	1987	1988
	SF'000	SF'000
Profit before tax	17,061	17,362
Profit after tax	17,061	17,362
Retained profit	17,406	17,907
Dividends per share	10%	10%
Share capital	150,000	150,000
Deposits	2,211,026	2,241,794
Total assets	5,814,019	5,967,189

Principal Shareholders: Capital furnished and all engagements guaranteed by the Canton
No of Employees: 523

BANQUE CANTONALE DE BERNE

Kantonalbank von Bern

Cantonal Bank of Berne

8 Place Federale, 3001 Berne
Tel: (031) 66 61 11
Cable: Kantonalbank
Telex: 911122
Telefax: 22 53 85

Board of Directors: Dr F Hofmann (Präsident), Dr P Legler (Vizepräsident), Dr U Augsburger, W Bircher, Dr Ing ETH Th Fässler, P Gerber, P O Joss, H Junker-Pfister, H W Koenig, F Kummer, W Marti, Dr K Meyer, F Reimann, H-P Richterich-Imhof

General Management: General direktoren: K Meier, R Kammer, A Schneiter. Direktoren: E Ruprecht, R Soltermann. Stv. Direktoren: P Farine (Personalchef), E Gyger, E Horisberger, P Hutzli, H Schärer. Vizedirektoren: W Bütikofer, Th Christen, G Laveissière, B Maibach (Geschäftsführer Swiss Bankers Travellers Cheque Centre), Dr H-U Zurflüh. Rechtskonsulentin: R Brazerol. Inspektorat: Chefinspektor: K Eggenschwiler

PRINCIPAL ACTIVITIES: Bank
Financial Information:

	31.12.86	31.12.87
	SF'000	SF'000
Profit before tax	37,601	44,251
Profit after tax	29,233	30,499
Dividends per share	9%	9%
Share capital	275,000	275,000
Deposits	9,474,191	10,338,521
Total assets	10,297,157	11,217,255

Principal Shareholders: Canton of Berne
No of Employees: 1,347

BANQUE CANTONALE DU VALAIS

Rue des Cèdres 8, 1950 Sion
Tel: (027) 21 11 11
Telex: 472518
Telefax: (027) 23.23.93

Supervisory Board: Félix Carruzzo (Président), Aloys Copt (Vice Président), Guy Cotter, Maurice Gillioz, Adrian Mathier, Bernard Mudry, Hermann Pellegrini, Pierre Veuthey, Armand Zenhäusern

Management Board: Raymond Duroux (Directeur), Charles-Marc Muller (Directeur-Adjoint), Roland Spiess (Sous-Directeur), Frédéric Gollut (Sous-Directeur)

Senior Executives: Moritz Amacker (Chef de la Division Administrative), Philippe Clerc (Secretaire Général), Gérard Duroux (Chef de la Division Hypothécaire et Commerciale), Firmin Sierro (Chef de la Division Financière)

PRINCIPAL ACTIVITIES: Banking services
Financial Information:

	1986 SF'000	1987 SF'000
Profit before tax	11,686	12,156
Profit after tax	11,686	12,156
Retained profit	3,500	3,800
Dividend	8,186	8,356
Share capital	100,000	105,000
Deposits	2,852,986	2,949,671
Total assets	4,015,864	4,208,583

No of Employees: 429

BANQUE CANTONALE VAUDOISE

14 Place St François, 1002 Lausanne
Tel: (021) 21 21 11
Cable: Cantonalbank
Telex: 454 304 BCVL CH

Board of Directors: R Junod (Chairman), Marc-Henri Chaudet (Vice Chairman), Pierre Aubert, Hubert Barde, *Jean-Pierre Berger, *Philippe Bertholet, *André Groux, Antoine Hoefliger, Bernard Ischy, Charles Keuffer, *Jean-François Leuba, *Philippe Luquiens, *René Marguet, *Jacques Martin, Charles Montferrini, *André Perrin, *Christine Petitpierre, *Jean-Pierre Schmidely, Jean-Claude Veillon. Secretary: Paolo Gallone, *appointed by the Conseil d'Etat of the Canton of Vaud; the other directors are elected by the General Meeting of Shareholders

Executive Board: F Pahud (President), J Treyvaud (Member), A Koller (Member), R Pidoux (Member)

Senior Executives: Georges Bovay (Senior Vice President), Claude Chapuis (Senior Vice President), Claude Courvoiser (Senior Vice President), Daniel Descombes (Senior Vice President), D Maret (Senior Vice President), Michel Péclat (Senior Vice President), F Randin (Senior Vice President), M Weber (Senior Vice President)

PRINCIPAL ACTIVITIES: Bank
Financial Information:

	31.12.86 SF'000	31.12.87 SF'000
Profit before tax	61,552	67,591
Profit after tax	49,798	51,435
Retained profit	24,726	24,862
Dividends	27,000	28,000
Dividends per share	30.00SF	30.00SF
Earnings per share	55.33SF	53.58SF
Share capital	225,000	240,000
Shareholders funds	517,227	584,862
Deposits*	8,575,452	9,149,629
Total assets	9,678,584	10,339,006

*includes bank deposits, customers' deposits, medium and long term borrowings

Principal Shareholders: State of Vaud
No of Employees: 1,672

BANQUE DE L'ETAT DE FRIBOURG

Case postale, 170 Fribourg
Tel: (037) 20 71 11
Cable: Banquetat
Telex: 942431 bef ch
Telefax: (037) 20 75 07

Management Board: L Rigolet (Directeur), F Betticher (Sous-Directeur), J Sautaux (Sous-Directeur)

PRINCIPAL ACTIVITIES: Banking
Financial Information:

	31.12.86 SF'000	31.12.87 SF'000
Profit after tax	13,860	15,048
Share capital	70,000	70,000
Customer deposits	2,571,281	2,757,253
Total assets	4,155,755	4,539,176

No of Employees: 377

BANQUE HYPOTHÉCAIRE DU CANTON DE GÉNÈVE

Case postale 428, 1211 Génève 3
Tel: (22) 28 00 11
Telex: 423325

Management Board: Jacques A Chapuis (Chairman), Pierre Raisin (Vice Chairman), Alan Borner, Claude Ketterer, Jules Mabut, René Barro

PRINCIPAL ACTIVITIES: Banking services
Subsidiary Companies: Gesplan Analyse et Réalisation SA; PME Finance SA
Financial Information:

	1987 SF'000
Profit after tax	14,000
Shareholders funds	250,000
Total assets	4,823,000

No of Employees: 426

BANQUE PARIBAS (SUISSE) SA

Case postale, CH-1204 Genève
Tel: (22) 87 71 11
Cable: Paribas
Telex: 422 165 PB CH

Supervisory Board: G Eskenazi (Chairman)
General Managers: J P Rambaud

PRINCIPAL ACTIVITIES: Bank
Parent Company: Compagnie Financière de Paribas, France
Financial Information:

	1987 SF'000
Share capital	78,600
Shareholders funds	639,600
Total assets	6,055,000

No of Employees: 898

BASELLANDSCHAFTLICHE KANTONALBANK

Banque Cantonale de Bâle-Campagne

7 Rheinstrasse, 4410 Liestal
Tel: (061) 925 91 11
Cable: Kantbank
Telex: 966 166
Telefax: (061) 925 93 17

Board of Directors: Dr St Herbster (Präsident)
Managers: H Frey, W Löw, A Raschle, A Salathe

PRINCIPAL ACTIVITIES: Bank
Financial Information:

	31.12.86 SF'000	31.12.87 SF'000
Profit after tax	21,600	22,440
Retained profit	7,000	7,000
Share capital	170,000	170,000
Deposits	4,753,337	4,920,157
Total assets	6,222,000	6,474,192

No of Employees: 518

BERNER ALLGEMEINE VERSICHERUNGSGELLSCHAFT

Bernese Insurance Company

Postfach 2617, 3001 Bern
Tel: (031) 63 51 11
Cable: Bernervers
Telex: 912265 beas ch
Telefax: 635 600

Supervisory Board: Dr H Flückiger (Präsident und Delegierter), H Moser (Vizepräsident), H Ammann, Dr K Meier, Dr H Munz, Dr J Stämpfli, Dr H G Lüchinger
Management Board: Dr H Flückiger, H-U Ernst, G Macquat, U Maeder, R Vogelbacher, M Willen, H Zingg
Senior Executives: Stellvertretende Direktoren: H Schaad, M Schori, Dr B Steinemann. Vizedirektoren: P Beck, J Belz, Chr Brönnimann, A Gerber, A Guggisberg, W Hebeisen, H-P Hofer, W Hoffmann, H Hofstetter, B Jöhr, Dr U Karlen, P Klopfstein, P Krättli, W Küpfer, P Marelli, F Schneider, P Schneider, P Stammbach, Th Zimmermann. Prokura: W Ackermann, F Affolter, A Augsburger, L Balerna, S Beyeler, P Bezzola, E Boss, J Bürge, H Bürki, B Bütikofer, B Chatton, P-H Dubois, J Frei, K Frei, H Habegger, K Hagen, R Heinzelmann, W Huber, W Küchler, J-P Leu, E Moser, G Prillmann, E Reinhardt, U Schmid, J Schneider, J Schwaller, R Spadarotto, C Stauffer, H Steffen. Handlungsvollmacht: R Aberegg, J-C Benoit, K Blunier, W Döbeli, H Eggen, H Fischer, T Füeg, R Gehrig, M Gerber, M Graber, Chr Griesser, N Guisolan, R Gurtner, M Hayoz, A Henzi, R Heussser, E Holzer, P Käsermann, M Kunz, H Lanz, L Le Calvez, K Lötscher, L Martelozzo, A Messerli, W Michel, P Pfeuti, W Rawyler, M Schenk, C Studer, W Thomet, W Tonacini, C Wyss

PRINCIPAL ACTIVITIES: Accident/sickness, liability, fire, theft, water, glass, part/full cover, valuables, health insurance, machinery damage, machinery cover, fittings, buildings management, building guarantees, transport insurance; all classes of reinsurance
Trade Names: Berner Versicherung, Berner Insurance
Subsidiary Companies: Berner-Lebensversicherungs-Gesellschaft in Bern; Alba Allgemeine Versicherungs-Gesellschaft in Basel; Bernese Assicurazioni-Compagnia Italo-Svizzera di Assicurazioni e Riassicurazioni SpA, Rom; Bernese Vita-Compagnia Italo-Svizzera di Assicurazioni e Riassicurazioni sulla vita SpA, Rom
Financial Information:

	1986	1987
	SF'000	SF'000
Premium income	391,638	410,983
Investment income	34,700	39,200
Profit before tax	10,400	12,200
Profit after tax	8,100	9,300
Retained profit	2,100	2,600
Dividend	5,610	6,600
Dividend per share	85SF*	90SF
Earnings per share	122SF	127SF
Share capital	33,000	36,500
Shareholders funds	90,320	120,710

*plus the profit of Berner Leben/Alba 12.50SF

No of Employees: 1,222

BSI-BANCA DELLA SVIZZERA ITALIANA

Via Magatti 2, CH-6901 Lugano
Tel: (091) 587111
Cable: Banca
Telex: 841020 bsi ch
Telefax: (091) 58 76 78

Board of Directors: G Antognini (Chairman)
General Management: Dr G Ghiringhelli, Dr A Capra, J-M Simond, G Strini, M Blaesi, L Butti, A Frezza, R Kessler

PRINCIPAL ACTIVITIES: Commercial Bank

Subsidiary Banks and Companies: Banque Romande, Geneva; Domus Bank, Zurich; Adler Change AG, Basle, Banca della Svizzera Italiana (Overseas) Ltd, Nassau (Bahamas)
Financial Information:

	31.12.86	31.12.87
	SF'000	SF'000
Profit before tax	55,344	59,901
Profit after tax	42,007	45,423
Dividends	12%(1)	12%
Dividends per bearer share	60SF(2)	60SF(2)
Dividends per reg share	12SF(3)	12SF(3)
Dividends per share(4)	60SF(5)	60SF(5)
Dividends per PC nom share(6)	12SF(7)	12SF(7)
Share capital	145,500	158,525
Participation cert capital	60,858	73,182
Shareholders funds	491,750	586,525
Deposits	5,762,646	5,749,321
Total assets	6,508,492	6,591,084

(1) plus 2% bonus
(2) plus 10SF
(3) plus 2SF
(4) PC nom Fr 500
(5) plus 10SF
(6) PC nom Fr 100
(7) plus 2SF

Principal Shareholders: Uni Tower Holding participation (51%)
No of Employees: 1,548

COOP SCHWEIZ (GENOSSENSCHAFTSVERBAND)

Postfach 2550 Thiersteinerallee 12, CH-4002 Basel
Tel: 061 20 61 11; 061 336 66 66 (from 6 October 1989)
Cable: coopsuisse
Telex: coop ch 962 133
Telefax: 061 20 60 40

Supervisory Board: Kurt Graf (Präsident)
Management Board: R Leuenberger (Präsident-Warenbeschaffung Non-food), Frau E Nobel-Rüefli (Vizepräsidentin Personal, Public Relations, Presse), F Widmer (Direktor Marketing), A Felder (Direktor, Warenbeschaffung Food), Dr G Metz (Direktor Finanzer Logistik, Organisation, Informatik)
Senior Executives: H P Rumpel (Personal), Dr H Teuscher (Ausbildung), Dr M Schelker (Public Relations und Presse), A Grossert (Graphische Betriebe), M Häner (Zentrales Finanz- und Rechnungswesen), P Fitz (Zentrale Verkaufsfrontplanung), P Dürrenberger (Warenverteilung und Technik), J Berner (Organisation und Informatik), A Hardegger (Allgemeine Waren), D Duc (Frischprodukte), R Maurer (Zentraleinkauf und Eigenproduktion Nonfood), Dr P Paul (Generalsekretär), H Peier (Zentrales Marketing)

PRINCIPAL ACTIVITIES: Food and consumer goods cooperative
Subsidiary Companies: Coop Bildungszentren; Coop Frauenbund Schweiz; Patenschaft Coop; Coop Personalversicherung; Coop AHV-Ausgleichskasse; Coop Familienausgleichskasse; Coop Anlage-Genossenschaft; Coop Hostellerie AG; Coop-Lebensversicherungs-Genossenschaft; Coop Rechtsschutz; Genossenschaftliche Zentralbank AG; Popularis Tours AG; "St Johann" Lagerhaus & Schiffahrts-Gesellschaft; Argo AG; Arni AG; Bebié AG; Bell AG; Bernische Grossmosterei Worb AG; Coop-Mühle; Coop-Baucenter AG; Pasta Gala SA; Coop Waschmittel + Cosmetic AG; Diana Schuhgeschäfte; Genossenschaftsapotheken; Genossenschaftsverband für Fleischproduktion; Graphische Betriebe Coop Schweiz; Halba AG; Liwo GmbH; Minoterie Coop; Nutrex AG; OK Coop AG; Panofina AG; Radio TV Steiner AG; Rivaliment SA; Schweiz Genossenschaft für Gemüsebau
Principal Bankers: Genossenschaftliche Zentralbank AG

Financial Information:

	31.12.86 SF'000	31.12.87 SF'000
Sales turnover	8,435,700	8,514,100
Share capital	50,799	51,219
Shareholders funds	135,982	146,847

Principal Shareholders: The Members (Co-operative societies) of Co-op Schweiz
No of Employees: 37,923 Coop Group

CREDIT FONCIER VAUDOIS

Case postale 77, CH-1000 Lausanne 9
Tel: 021/20 60 11
Cable: Foncier Lausanne
Telex: CFV 454 970
Telefax: 021/238 183

Management: Francis Sauvain (Directeur Général), Pierre-François Etter (Directeur), Jean-Claude Grangier (Directeur)
Head Office: Sous-direceurs, attachés à la Direction générale: Roger Morel, Pierre Muller, Jean-Robert Zahnd (Chef du personnel). Sous-directeurs: Alexandre Chatelain, Ingo Christel, Henri Hagmann, André Jordan, Jean-François Stegmann
Senior Executives: Fondés de pouvoir chefs de service: Antoine Bender, Robert Bonetti, Jean-F Calame, Alain Dénéréaz, Charles Desponds, Jean-Pierre Jaques, Louis Jaquet, Bernard Marendaz, Daniel Margot, René Perreten, Georges Radice, André Ruffieux, Roland Valet. Fondés de pouvoir: Jean-Claude Ansermet, Luigi Bertolini, Pierre-A Corboz, Pierre-A Cornaz, Jean-Pierre Favre, Antoine Gonin, Norbert Ledermann, Jean-Pierre Mathey, René Meyer, Etienne Nicole, Eric Pantet, Jean-P Petitmaître, François Rapp, Martin Rohr, Michel Sueur, Gérard Thonney, Jean-M Vallélian, Pierre-A Violi, Jacques Winkelmann

PRINCIPAL ACTIVITIES: Mortgage bank
Financial Information:

	31.12.86 SF'000	31.12.87 SF'000
Turnover	356,801	386,374
Profit before tax	41,610	37,654
Profit after tax	28,000	25,080
Retained profit	16,000	13,080
Dividends	12,000	12,000
Dividends per share	50SF	50SF
Earnings per share	116.7SF	104.5SF
Share capital	120,000	120,000
Shareholders funds	346,479	358,859
Deposits	3,037,304	3,278,862
Total assets	7,221,490	7,679,602

Principal Shareholders: Canton de Vaud
No of Employees: 350

CREDIT SUISSE

Schweizerische Kreditanstalt

Paradeplatz 8, Postfach, CH-8021 Zürich
Tel: (01) 215 11 11
Cable: Credit
Telex: 812 412 cs ch

President: R E Gut (Chairman of the Board of Directors)
Executive Board: R A Jeker (President), Dr H von der Crone, Dr W Wirth, E Schneider, Dr M Kopp, Dr H-U Doerig, Dr K Widmer, Dr K Jenny, Dr R W Hug, H-J Rudloff, Dr H Geiger, H P Sorg, Dr R Rasi, R Stalder
Senior Executives: V Erne (Executive Vice President, Marketing and Advertising), Ch Ammann (Senior Vice President, Accounting, Controlling), W Guyer (Executive Vice President, Corporate Planning), U P Hänni (Senior Vice President, Internal auditors), H-R Tischhauser (Senior Vice President, Secretariat to the Chairman), P Bosshard (Executive Vice President, Delegate for North America), H-J Heun (Senior Vice President, Commercial Banking in Europe; except France, Belgium, Italy, Spain, Portugal, Israel and English-Speaking Africa), J Müller (Executive Vice President, Commercial Banking in Latin America), B Häusermann (Senior Vice President, Export financing), P Fankhauser (Senior Vice President, Lending by Swiss branches), P Sailer (Senior Vice President, Planning and support; commercial banking), J-J Guinand (Senior Vice President, Commercial Banking in France, Belgium, Italy, Spain, Portugal, Israel and French-Speaking Africa), A-E Lieb (Senior Vice President, Commercial Banking in the Middle and Far East), E Lienhard (Senior Vice President, Commercial Banking in Zurich), M M Meier (Senior Vice President, Commerical Banking in Switzerland), F W Schweizer (First Vice President Commercial Banking in Austria, the North of Europe, Eastern Europe, Greece and Israel), E Neuenschwander (Senior Vice President, Branch planning and support, Zurich region), F-J Groth (Senior Vice President, Special financing in Switzerland), M Wetter (Senior Vice President, Financial holdings in Switzerland, consumer credit and leasing), A Gremli (Senior Vice President, Commercial Banking in the Middle East), A M Banz (Senior Vice President, Special assignments), N Eisenring (First Vice President, Branches in West Switzerland), C Vela (Executive Vice President, Branches in the Ticino), R Andermatt (Senior Vice President, Commerical Banking in North America and Bahamas), V Senn (Senior Vice President, Branches in Central Switzerland), H Clavadetscher (Executive Vice President, Branches in the Grisons), E Schlauri (Senior Vice President, Branches in Northeast Switzerland), P Kappeler (Senior Vice President, Capital markets and corporate finance in Switzerland), H Wegmann (First Vice President, Capital markets in the Pacific region), A Horisberger (Senior Vice President, Swiss securities trading), C Zwick (Senior Vice President, Foreign securities trading), A Syz (Senior Vice President, Management support: securities), F Mani (First Vice President, Administration of securities portfolio), H Hofmann (Senior Vice President, Planning and support: finance), A Hirs (Executive Vice President, Sfr capital markets international, except the Pacific region), H R Zehnder (Senior Vice President, Syndications, swaps, securities sales), C Willi (Executive Vice President, Foreign exchange, precious metals, banknotes; money market, liquidity), E Locher (Senior Vice President, Banks in Switzerland), G G Goetz (Senior Vice President, Central banks), J Ackermann (Senior Vice President, Multinational services in Switzerland), B M Fenner (Senior Vice President, Multinational serivces international), H C Maurer (Senior Vice President, Asset management), M Vogel (Senior Vice President, Investment funds), A Bischofberger (Senior Vice President, Economic research), P Dellsperger (First Vice President, Investment research), P Klauser (Senior Vice President, Planning and support: investment business and customer deposits, E Scheller (Executive Vice President, Investment counselling/investment management overseas), A A Konrad (Senior Vice President, Investment counselling/investment management Europe), Th J Zimmermann (Senior Vice President, Investment counselling for institutional investors), M Koller (Senior Vice President, Special investment counselling), H R Messerli (First Vice President, Investment business, customer deposits), O J Dufner (Senior Vice President, Business management, lombard facilities), D Marti (Executive Vice President, Organization and information systems), W de Capitani (First legal Adviser, Legal department), K Arnold (Tax Adviser, Tax and inheritance services), H Häsli (Executive Vice President, Personnel), K F Baumann (Executive Vice President, Operations and services Head Office), A B Frey (Senior Vice President, Buildings and Premises), J-P Huwyler (Senior Vice President, Security)

PRINCIPAL ACTIVITIES: Banking
Trade Names: Schweizerische Kreditanstalt; Crédit Suisse; Credito Svizzero; Credit Suisse
Subsidiary/Associated Companies: (100% owned unless stated): Affida Bank; Bank Hofmann Ltd; Bank in Luzern; Bank

Neumünster (92.3%); Bank for Commerce and Securities; Société anonyme de participations à des entreprises privées (88.8%); City Bank (81.8%); CS Leasing Ltd; Finanz AG Zürich; Krefina Bank AG; Swiss-Kuwaiti Bank (70%); Schweizerische Kreditanstalt (Deutschland) AG (Germany); Gibraltar Trust Bank Ltd (Gibraltar); Buckmaster & Moore Ltd (UK); Swiss American Securities Inc (80%) (USA); Credit Suisse (Guernsey) Ltd; Sovereign Group Ltd (Guernsey); Credit Suisse Canada; Credit Suisse (Luxembourg) SA; Crédit Suisse (France); Swiss American Securities Inc. (80%) (USA) (Bahamas) Ltd; Credit Suisse Finance (Panama) SA; Credit Suisse France Ltd (Hong Kong) and branches throughout the world

Financial Information:

	31.12.87	31.12.88
	SF'000	SF'000
Gross Profit	1,151,385	1,179,323
Profit after tax	550,189	591,702
Retained profit	215,000	245,000
Dividends	326,571	347,699
Dividends per share(1)	100SF	100SF
Dividends per share(2)	8SF	8SF
Total	108SF	108SF
Dividends per share(3)	20SF	20SF
Dividends per share(4)	1.60SF	1.60SF
Total	21.60SF	21.60SF
Share capital	1,850,000	1,850,000
Shareholders funds	6,612,808	7,196,939
Deposits	65,596,264	72,240,071
Total assets	107,239,875	113,383,038

(1) CS bearer share
(2) CS Holding bearer PC
(3) CS registered share
(4) CS Holding registered PC

No of Employees: 15,055

ELVIA SCHWEIZERISCHE VERSICHERUNGS-GESELLSCHAFT

Bleicherweg 19, 8022 Zürich
Tel: (01) 209 51 11
Telefax: (01) 209 51 20

Supervisory Board: Dr Heinz R Wuffli (Präsident), Jean-Daniel Cornaz (Vizepräsident), Dr Nicolas Bär, Dr Bruno Hunziker, Edith Nobel-Rüefli, Dr Rolf Reidhaar, Bernhard Schneider, Gianfranco Cotti, Prof Dr Paolo Gallone
Management Board: Paul Hug (Generaldirektor), Adolphe Jonsson (stv Generaldirektor), Dr Pius Hofer (Direktor), Dr Urs Oberholzer (Direktor), Peter Steffen (Direktor), Gian Schaad (Direktor), Dr Josef Schwere (Direktor), Dr Jakob Werlenmann (Direktor), Peter Graf (stellv. Direktor), Theodor Häberli (stellv. Direktor), Hans Ueli Hauri (stellv. Direktor), Jürg Rohner (stellv. Direktor), Alfons Rüegg (stellv. Direktor), Daniel Stäheli (stellv. Direktor), Dr Werner Ott (stellv. Direktor), Beat Bär (Vizedirektor), Franz Bürgin (Vizedirektor), Jürg Frey (Viredirektor), Dr Klaus Froesch (Vizedirektor), Volker Fuhlrott (Vizedirektor), Georges Gredig (Vizedirektor), Claude Guggenheim (Vizedirektor), Dr Erwin Kümmerli (Vizedirektor), Reinhard Kunz (Vizedirektor), Jürgen Lohr (Vizekirektor), Josef Lüthi (Vizedirektor), Charles Oppenheim (Vizedirektor), Dr Gaudenz Prader (Vizedirektor), Philippe Signer (Vizedirektor), Franz Schmid (Vizedirektor), Ernest Varonier (Vizedirektor), Walter Wittmer (Vizedirector), Paul Zeier (Vizedirektor), Urs Erni (Vizedirektor), Hans-Ruedi Roth (Vizedirektor), Peter Steiger (Vizedirektor), Dr Hans-Ulrich Vollenwerder (Vizedirektor), Bruno Ramseier (Vizedirektor)

PRINCIPAL ACTIVITIES: Accident insurance; motor insurance; health insurance, house contents insurance; travel and transport insurance; single and group life insurance
Subsidiary and Associated Companies: ELVIA Leben; ELVIA Reseversicherungen; CAP Rechtsschutz-Versicherungs-Gesellschaft; Deggo AG, Generalunternehmung; ELVIA SA (Belgium); Savoia SpA (Italy); Savoia Vita SpA (Italy); Gruppe

La Equitativa (Spain and Portugal); gedo mbH; Data Processing Company (Germany)
Principal Bankers: All major Swiss Banks
Financial Information:

	31.12.86	31.12.87
	SF'000	SF'000
Premium income	742,916	813,252
Retained profit	17,808	19,568
Share capital	20,000	20,000
Shareholders funds	219,100	271,000

FIDES TRUST COMPANY

Postfach 656, 8027 Zürich
Tel: (01) 249 21 21
Telex: 815 388 fid ch
Telefax: (01) 249 29 78

Supervisory Board: Dr H von der Crone (Präsident), K Baumann, F Corbat, Prof Dr P Forstmoser, Dr P Gloor, Prof Dr E Kilgus, Dr P Ramelet, E Scheller, Dr W Wirth
Management Board: Dr E Baudat, Hans-Jorg Furrer, Dr G Kramer, Dr H Witt

PRINCIPAL ACTIVITIES: Advertising and commecial consultancy services; financial services; data processing services
Financial Information:

	1987
	SF'000
Sales turnover	174,000
Profit after tax	5,600

No of Employees: 1,224

GALENICA HOLDING LTD

Untermattweg 8, CH-3027 Bern
Tel: (031) 55 22 22
Cable: galenica bern
Telex: 911 791 gado ch

Supervisory Board: F Milliet (Chairman), Dr R Alther, J-P Bonny, Dr J F Bonvin, Dr M Brentano, M Bryois, Dr A Engel, W Frick, B de Kalbermatten, Dr R Hofstetter, H Hotz, G Monnerat, H-R Moser, Dr P Nussbaumer, M Rey, Dr H Schaffhauser, Dr F Schnetzer, Dr Ch Studer, L Varini
Management Board: F Milliet (Chairman), G Botteron, Dr B Secrétan, H Keller, R Cavillier, F Hirsbrunner, René Jenny, J Lehmann

PRINCIPAL ACTIVITIES: Production and wholesale trade in pharmaceuticals
Subsidiary Companies: Galenica Ltd (Wholesale); Vifor Ltd; Hausmann Laboratories Ltd; Galenica Representation Ltd; Panpharma Ltd; Adima Ltd; Pharmacal Ltd; Lab Fandre SA (France)
Principal Bankers: SBG
Financial Information:

	1987
	SF'000
Sales turnover	931,000
Profit after tax	12,300

Principal Shareholders: Pharmacists (exclusively)
No of Employees: 2,288

GENEVA LIFE INSURANCE COMPANY

La Genevoise, Compagnie d'Assurances sur la Vie

16 Avenue Eugène-Pittard, Case postale 332, 1211 Genève 25
Tel: (022) 47 92 22
Telex: 422478 geas ch
Telefax: (022) 47 02 47

Administrative Board: François Peyrot (Président), Jacques Darier, Thierry Barbey, Paul E Christen, Jacques Hubscher, Martin H E Hürlimann, William de Rham, Hans U Ryser, Lucien Schatz

Management Board: Béat de Watteville, Walter Jordi, René Lecoultre, Marcel Rimann, Jean Winkler, Raymond Dessingy, Marcel Künzler, Michel Pfenninger, Michel Riat, Michel Clerc, Arnaldo Galezia, Erwin Kühne, Jean-Francois Müller, Jean-Pierre Rodieux, Willy Zumofen

PRINCIPAL ACTIVITIES: Life insurance

Subsidiary Companies: La Genevoise France (France); Alboran SA (Spain)

Financial Information:

	31.12.86 SF'000	31.12.87 SF'000
Premium income	339,000	387,000
Investment income	100,700	109,800

No of Employees: 660

GENOSSENSCHAFTLICHE ZENTRALBANK AG
Banque Centrale Cooperative SA
Banca Centrale Cooperative SA
Co-operative Central Bank Ltd

3 Aeschenplatz, 4002 Basle
Tel: (061) 56 21 21 (from October 1989 (061) 286 21 21)
Cable: Zentbank
Telex: 962290
Telefax: (061) 22 45 95

Supervisory Board: Hans Thuli (President), Urs Käser (Vice President), M Albert-Louis Dupont-Willemin, Armin Frei, Rolf Leuenberger, Dr Diego Malaguerra, Jürg Messerli, Dr Gerhard Metz, Hervé Pétroz, Edith Nobel-Rüefli, Dr Walter Renschler, Dr René Schneider, Gilbert Tschumi, Nationalrätin Dr Lilian Uchtenhagen

Management: Chairman: Hans Thuli. President of the Executive Board: Marcel Hurni. General Managers: Fritz Leuenberger, Heinz Voegeli. Head of Correspondent Banking: Konrad Bretscher. Head of Foreign Exchange: Otmar Thaler. Head of Trade Finance: Paul Huber. Head of Investment Banking: Gustav Gass. Head of PR-Marketing: Jürg Flückiger. Head of International Division: Hans Voegeli

PRINCIPAL ACTIVITIES: Bank

Trade Names: GZB, BCC, CCB

Subsidiary Companies: Genossenschaftliche Zentralbank AG has 42 subsidiaries

Financial Information:

	31.12.88 SF'000
Profit after tax	19,303
Dividends	8%
Share capital	210,000
Deposits	2,958,158
Total assets	5,709,948

No of Employees: 619

GOTTHARD BANK

Casella postale 6957, 6901 Lugano
Tel: (091) 214114
Telex: 841051 bdg ch

PRINCIPAL ACTIVITIES: Banking

Financial Information:

	1987 SF'000
Profit after tax	40,000
Shareholders funds	476,900
Total assets	4,875,000

No of Employees: 690

GRAUBÜENDNER KANTONALBANK
Banque Cantonale des Grisons

2 Poststrasse, 7002 Chur
Tel: (081) 21 91 11
Cable: Kantonalbank
Telex: 851122 kbch ch
Telefax: (081) 22 67 29

Bank Council: Dr. rer.pol.Christian Jost (Präsident), Dr. iur.Hans Degiacomi, Dr. rer.pol.Jon Armon Campell, Reto Togni, Dr. rer.pol.Aluis Maissen, Peter Kasper, Marco Itin, Luigi Lanfranchi, Remi Capeder, Richard Allemann

Managing Board: Dr.iur. Christian Schmid (Direktor), Hans Casanova (Direktor), Johann Luzi Dolf (Direktor), Alfred Hattich (Stellvertretender Direktor), Bruno Plozza (Vizedirektor), Johann Cavegn (Vizedirektor), Georg Simmen (Vizedirektor), Willi Giovanoli (Vizedirektor), Christian Durisch (Vizedirektor)

Senior Executives: Eduard Gasser (Chef der Organisations-abteilung), Florenzio Olgiati (Chef-Stallvertreter der Hauptabteilung Kapitalanlagen), Dr.Iur. Adriano Oswald (Personalchef), Dr.iur. Peter Salzgeber (Direktionssekretär und Rechtskonsulent), Urs Ackermann (Regionalleiter), Reto Balzarini (Chef der Abteilung Dienste), Georg Barandun (Chef der Hypothekarabteilung), Christian Bargetzi (Leiter des Rechenzentrums), Siegfried Brunner (Chef der Hauptbuchhaltung), Reto Buchli (Chef der Emissionsabteilung), Josef Bundi (Chef der Kreditabteilung), Jakob Casal (Regionalleiter), Hans Christoffel (Chef der Sparkassenabteilung), Jakob Clavadetscher (Chef des Rechtsbüros), Edgar Degiacomi (Chef der Abteilung Zahlungsverkehr Ausland), Clemens Demont (Chef des Steuerbüros), Hanspeter Dolf (Chef der Börsenabteilung), Hansjürg Dupont (Prokurist der Anlageberatung), Jürg Frei (Regionalleiter), Arno Giovanoli (Chef der Devisenabteilung), Robert Hendry (Chef der Anlageberatung), Hans Hitz (Chef der Couponsabteilung), Eduard Jäger (Chef-Stellvertreter der Hypothekarabteilung), Mario Kurmann (Chef der Zweigstellenbuchhaltung), Hansjörg Ladner (Prokurist der Hypothekarabt.), Hans Lang (Prokurist der Kreditabteilung), Urs Mayer (Ausbildungsleiter), Arthur Meier (Chef der Abteilung Zahlungsverkehr Inland), Christian Metz (Chef der Hauptkasse), Werner Moser (Chef des Wechselportefeuilles), Martin Quinter (Regionalleiter), Pietro Salis (Chef der Depotabteilung), Viktor Schär (Prokurist der Kreditabteilung), Karl Sigrist (Chef der Datenverarbeitungsabteilung), Rainer Stocker (Chef der Liegenschaftenverwaltung), Rudolf Stocker (Chef der Kassenabteilung), Paul Thoma (Prokurist der Kreditabteilung), Werner Würth (Chef-Stellvertreter der Datenverarbeitungsabteilung)

PRINCIPAL ACTIVITIES: Bank

Trade Names: Kantonalbank

Financial Information:

	31.12.86 SF'000	31.12.87 SF'000
Profit before tax	19,033	21,748
Profit after tax	19,033	21,748
Dividends*	8%	8%
Dividends per share*	8.00SF	8.00SF
Dotation capital	180,000	200,000
Participation capital	7,500	15,000
Shareholders funds	259,666	300,423
Deposits	4,672,671	4,980,097
Total assets	5,860,418	6,330,716

*Participation capital

Principal Shareholders: Government Bank

No of Employees: 652

HELVETIA SCHWEIZERISCHE FEUERVERSICHERUNGS-GESELLSCHAFT

Dufourstrasse 40, Postfach 972, 9001 St. Gallen
Tel: (071) 26 51 11
Cable: Helvassur St. Gallen
Telex: 77 215 helv ch
Telefax: (071) 25 63 76

Supervisory Board: Dr Paul Bürgi (Präsident), W Hess (Vizepräsident), Dr H Wuffli (Vizepräsident), J.-D. Cornaz, U Forster, B Schneider, Dr A Bodmer, Dr P Kamer, Prof Dr A Meier, Fürsprech C Nüscheler, Dr R Perret, Dr K Rohner, D Schmidheini, Dr J Schönenberger, Dr H-L Schreiber
Management Board: Dr F Th Hefti (Generaldirektor), Dr A Kern (Direktor), Dr H Lanz (Direktor), B Leippold (Direktor), Dr O Rhiner (Direktor), Dr H Gisun (Direktor), Dr H J Gafafer (Direktor), R Künzler (Direktor), D Angster (stellv. Direktor), E Günthart (stellv. Direktor), H P Haller (stellv. Direktor), Dr J Hensel (stellv. Direktor), G Nissille (stellv. Direktor), Dr H Schmid (stellv. Direktor), W Schuhmacher (stellv. Direktor), E Walser (stellv.Direktor), A Wassmer (stellv. Direktor), H U Hohl (Beratendes Mitglied der Direktion), P Bächtiger (Vizedirektor), F Bieri (Vizediriktor), J Boesch (Vizedirektor), R Edelmann (Vizedirektor), J Küng (Vizedirektor), E Marghitola (Vizedirektor), W Niederer (Vizedirektor), P Riegg (Vizedirektor), Ch Röllin (Vizedirektor), R Schmitter (Vizedirektor), N von Schumacher (Vizedirektor)

PRINCIPAL ACTIVITIES: Insurance for accidents, common law liability, third party motor liability, damage caused by fire or the elements, transport, vehicle body work, theft, glass, water damage, security, machines, jewellery, special insurance, special reinsurance branches; operations in Switzerland, Canada, Germany, France, Netherlands, Italy and Austria
Subsidiary and Associated Companies: In Austria, Greece, Netherlands, Canada, Switzerland and Spain
Principal Bankers: All major Swiss Banks
Financial Information:

	31.12.86	31.12.87
	SF'000	SF'000
Premium income	832,518	850,649
Investment income	69,194	72,119
Profit after tax	17,541	20,132
Dividends	22,200	24,900
Dividends per share	50SF	54SF
Share capital	50,000	56,200
Shareholders funds	221,600	313,300

No of Employees: 2,314

HELVETIA-ACCIDENT SWISS INSURANCE COMPANY LTD

Bleicherweg 19, Postfach 4889, 8022 Zürich
Tel: (01) 209 51 11
Cable: Helvetia-unfall Zürich 2
Telex: 815635 HEL CH
Telefax: (01) 209 51 20

Supervisory Board: Dr. Heinz R Wuffli (Zumikon, President), Jean-Daniel Cornaz (Bülach, Vice President), Dr. Paul Burgi (St. Gallen, Vice President), Roberto Antonini (Bellinzona), Dr. Nicholas Bär (Zürich), Pierre Chessex (Lausanne), Ueli Forster (St. Gallen), Walter Hess (Effretikon), Dr. Bruno Hunziker (Aaarau), Dr. Dr. h. c. Werner Kämpfen (Zürich), Edith Nobel-Rüefli (Muttenz), Dr. Rolf Reidhaar (Thalwil) Bernard Schneider (Givisiez)
Management Board: Paul Hug (Managing Director), Adolphe Jousson, Bruno Scacchi (Export Sales Director), Peter Steffen (Data Processing and Information), Dr. Pius Hofer (Finance Director), Dr. Urs Oberholzer

PRINCIPAL ACTIVITIES: Accident, motor and sickness insurance; reinsurance
Subsidiary Companies: Helvetia Leben Insurance Company Ltd; CAP-Rechtsschutz-Insurance Company Ltd; Elvia-Insurance

Company Ltd; Deggo Ltd; Helvetia SA (Belgium); Savoia SpA (Italy); Comitas SpA (Italy)
Financial Information:

	1986	1987
	SF'000	SF'000
Premium income	742,916	813,000
Investment income	94,443	98,100
Profit after tax	17,808	19,600

Principal Shareholders: Number of Shareholders 4,674 (30.6.87)

HYPO ARGAU

Aargauische Hypotheken- & Handelsbank

Hauptstrasse 1, 5200 Brugg
Tel: (056) 42 01 11
Cable: HYPOBANK
Telex: 825108; 825107; 825106
Telefax: (056) 422160

Management Board: R R Summermatter (General Manager), Dr J N Brändle, Dr U Heger

PRINCIPAL ACTIVITIES: Banking services
Financial Information:

	1987	1988
	SF'000	SF'000
Profit before tax	22,437	23,881
Profit after tax	15,600	17,029
Dividend	13%	13%
Dividend per share	65SF	65SF
Share capital*	221,100	225,100
Shareholders funds	96,000	96,000
Deposits	1,893,000	2,037,000
Total assets	4,504,000	4,800,000
*including reserves		

No of Employees: 593

HYPOTHEKARKASSE DES KANTONS BERN

Schwanengasse 2, 3001 Bern
Tel: (031) 22 72 31
Telex: 33304 hkb ch

Management Board: R Martz, Dr Ch Prétat, M Engelmann, P Gander

PRINCIPAL ACTIVITIES: Banking
Financial Information:

	1987
	SF'000
Profit after tax	14,000
Shareholders funds	193,000
Total assets	6,287,000

No of Employees: 150

IC INDUSTRIES (INTERNATIONAL) SA

55 Chemin Moïse-Duboule, Case postale 231, CH-1211 Geneve 19
Tel: (022) 98 23 11
Telex: 22 884 abex ch
Telefax: (022) 98 98 06

Supervisory Board: Hans B Erni (Président), André de Pfyffer d'Altishofen, Philippe Braunschweig, Dr W F L Place, Claude Lüthi, John P Fagan, Frank T Westover, Lewis Robertson
Management Board: Hans B Erni

PRINCIPAL ACTIVITIES: Financial services
Principal Bankers: UBS Genève
Financial Information:

	SF'000
Turnover	200,000

No of Employees: 3,000

SWITZERLAND

INDUSTRIE-LEASING GROUP

Badenerstrasse 329, CH-8040 Zürich
Tel: (01) 491 33 55
Telex: 822 553 ilag ch
Telecopier: (01) 491 1371

Supervisory Board: Fritz Köhli (Präsident), Dr Jörg Auckenthaler (Vizepräsident), Walter Fuhrimann, Georg Schnell, Bert Wietlisbach
Management Board: Fritz Peter (Vorsitzender), Hanspeter P Högger (stellv Direktor), Jürg Heeb (Vizedirektor), Werner Domeisen (Vizedirektor), Ewald Siegrist (Vizedirektor), Dr Alfred Fischer (Vizedirektor), Jacques Bonjour (Vizedirektor), Jean-Frédéric Henchoz (Vizedirektor)

PRINCIPAL ACTIVITIES: Leasing of industrial equipment, cars and real estate
Parent Company: Swiss Bank Corporation, Basle, Switzerland
Subsidiary Companies: Aucreda AG; IL Immobilien-Leasing AG; Industrie-Leasing AG
Principal Bankers: Swiss Bank Corporation
Financial Information:

	1987 SF'000
Sales turnover	500,000
Profit before tax	4,200
Profit after tax	3,600
Dividends	6%
Share capital	30,000

No of Employees: 70

INSPECTORATE INTERNATIONAL AG

Postfach 1354, 3001 Bern
Tel: (031) 26 11 15
Telefax: (031) 25 16 85

Board of Directors: Werner K Rey (London, Chairman). Members: Hans Herzig (Muri), Martin Imbach (Meilen), Yann Richter (Neuchâtel), Prof Dr M Armin Seiler (Greppen), Fritz Berger (Berne)
Executive Board: Werner K Rey (President). General Managers: Francis A Holmes, Dr Ulrich Guggisberg, Dr Matthias U Jermann
Group Management: Harold A Ahlberg (PSI Holdings USA Network Security Corporation), Harvey Kinzelberg (The Meridian Group, USA), Timothy Ozark (The Meridian Group, USA), Ian Orrock (Meridian Group Services Ltd), Patrice Courbey (Meridian Group Services Ltd), Ulrich Monsch (ACI Automation Center International AG), Günther E Hering (Harpener AG), Jürgen Schippenkühler (Harpener AG)

PRINCIPAL ACTIVITIES: Consultation and inspection services; computer leasing
Subsidiary Companies: (100% owned unless stated): Inspection/Services: Inspectorate UK Holdings Ltd (Loddon, UK):- Pycraft & Arnold Ltd (London); Inspectorate Technical Workforce (Middlesborough), Griffith Group (Witham). Inspectorate USA:- PSI-Inspectorate-Lombard; Chas Martin Corp (Coral Gables). Inspectorate Italia SpA (Milan, Italy); Inspectorate International Inspection & Consulting GmbH (Düsseldorf, West Germany). Information Technology: The Meridan Group USA:- Meridian Leasing Corp (Deerfield); Meridian Equipment Corp (Deerfield); Meridian Financial Corp (Deerfield); Unilease Inc (New York). Meridian Group Europe:- CRE Leasing GmbH (Düsseldorf, West Germany); Meridian France SA (Paris); Meridian Leasing Ltd (London); Compex SpA (Milan) (30%); ACI Meridian (International) AG (Wettingen, Switzerland); AC Automation Center AG (Wettingen), AC Service GmbH (Offenbach aM, West Germany); AC Service GmbH (Vienna, Austria); Automation Center SA (Brussels, Belgium). Security Systems: Network Security Corp (Irving, Texas, USA); Network Multi-Family Security Corp (Dallas, Texas, USA); Multiwatch Security Corp (Houston, Texas, USA); Securican Inc (Montreal, Canada). Other Interests: Inspectorate Finance SA (Geneva, Switzerland); Harpener AG

(75%) (Dortmund, West Germany); Universal Engineering Ltd (UIAG) (40%) (Basle, Switzerland); CUC (60%) (Amsterdam, Netherlands)
Financial Information:

	31.12.86 SF'000	31.12.87 SF'000
Sales turnover	612,168	1,347,400
Income before tax(1)	42,336	143,500
Income(1)	35,290	100,889
Net income	35,194	85,672
Retained earnings(2)	44,895	112,500
Dividend	13,800	23,900
Dividend per share	24SF	26SF
Dividend per PC(3)	2.4SF	5.20SF
Earnings per share	70.46SF	105.31SF
Share capital	60,000	60,000
PC Capital(3)	15,000	31,000
Shareholders funds	225,200	404,200

(1) before minority interest
(2) end of year
(3) Participation Certificate

Principal Shareholders: Omni Holding AG, Bern, Switzerland (58%)
No of Employees: over 7,698

KLEINERT UNTERNEHMUNGEN HOLDING AG

Bubenbergplatz 8, 3001 Bern
Tel: (031) 21 81 11
Telex: 912 008 ku ch
Telefax: (031) 22 94 07

Management Board: V Kleinert (Vorsitzender), Dr K Weder, B Elsasser, H Stücklin

PRINCIPAL ACTIVITIES: Commercial real estate; industry; project management and financing
Financial Information: Consolidated

	31.12.86 SF'000	31.12.87 SF'000
Sales turnover	250,000	312,000

No of Employees: 700

LISCA LEASING AG

Morgartenstrasse 6, CH-8036 Zürich
Tel: (01) 241 90 37
Telex: 812 847 lis ch

Supervisory Board: Emil Hinnen (President), Dr Jost Grob, Gérald Senglet, Dieter Bäbler, Dr Heinrich Bolliger, Leo Bühler, Raymond Duroux, Dr Joseph F Felder, Hans Frey, Hans-Jurg Gallusser, Dr Romano Mellini
Management Board: Hugo Konrad

PRINCIPAL ACTIVITIES: Leasing
Financial Information:

	1987 SF'000
Sales turnover	239,000
Profit after tax	1,500

No of Employees: 60

LUZERNER KANTONALBANK

12 Pilatusstrasse, 6002 Lucerne
Tel: (041) 24 11 11
Cable: Kantonalbank
Telex: 862860 (Regular); 862855 (Securities); 868239 (Forex)
Telefax: (041) 23 38 87

Supervisory Board: Dr Theo Fischer (Chairman)
Senior Management: Karl Schmid (President), Franz Grüter (Senior Vice President), Fritz Studer (Senior Vice President), Karl Reichmuth (Senior Vice President). Investment Counseling and Portfolio Management: Monjek Rosenheim (Vice President), Antonio Fernandez (Vice President), Werner Henzen (Vice President), Hans Lötscher (Assistant Vice

President), Erwin Hartmann (Assistant Vice President), Beatrice Ulrich (Assistant Vice President). Banking Relations/Deposits: Oskar Heini (First Vice President). Documentary Credits and International Payments: Reto Frei (Assistant Vice President), Kurt Hächler (Assistant Vice President). Foreign Loans: Peter Bieri (Vice President). Foreign Exchange: Marcel Birrfelder (Chief Dealer), Sergio Sigrist (Assistant Vice President), Barbara Lehnhard (Assistant Vice President). Money Market: Hans Stübi (Dealer)

PRINCIPAL ACTIVITIES: State bank
Financial Information:

	31.12.86 SF'000	31.12.87 SF'000
Profit after tax	26,858	29,491
Share capital	400,000	430,000
Deposits	6,064,212	6,800,360
Total assets	9,332,681	10,310,366

No of Employees: 991

MIGROS BANK

Seidengasse 12, 8023 Zürich
Tel: (01) 229 81 11
Telex: 813464 migba ch
Telefax: (01) 211 12 44

Supervisory Board: J Kyburz (Präsident), Dr D Weber, H Born, Dr B Kaufmann
Management Board: S Benz, E Hort, Dr A Schwerzmann, F Reich

PRINCIPAL ACTIVITIES: Banking
Parent Company: Migros Genossenschafts-Bund
Financial Information:

	31.12.86 SF'000	31.12.87 SF'000
Profit before tax	18,124	20,035
Profit after tax	13,812	14,943
Retained profit	13,952	15,145
Dividend	7,750	8,875
Dividend per share	5%	5%
Share capital	170,000	200,000
Shareholders funds	241,500	277,500
Deposits	3,983,878	4,516,154
Total assets	4,547,500	5,112,900

No of Employees: 459

MIGROS-GENOSSENSCHAFTS-BUND

Limmatstrasse 152, Postfach 266, CH-8031 Zürich
Tel: (01)277.21.11
Cable: Migrosbund
Telex: 82 33 00 migrosbund zch
Telefax: 277 25 25

Supervisory Board: Pierre Arnold (Chairman), Dr Mario Pozzi (Vice Chairman), Jürg F Rentsch (Vice Chairman)
Management Board: Jules Kyburz (President/Co-ordination, Cultural and Social Services), Erich Gugelmann (Vice President/Marketing), Dr Ulrich Geissmann (Manufacturing), Eugen Hunziker (Planning and Logistics), Dr Beat Kaufmann (Finance), Dr Darius Weber (Diversification)

PRINCIPAL ACTIVITIES: Retailing of foodstuffs and household products through stores; interests in publishing, banking, shipping and insurance
Subsidiary Companies: Micarna SA; Jowa AG; Conserves Estavayer SA; Bischofszell Canning Factory AG; Chocolat Frey AG; Production AG Meilen; Mifa AG; Mibelle AG; Optigal SA; Seba Aproz SA; Fruidor SA; Dietiker & Co AG; Hotelplan International Travel Organization AG; Ex Libris Publishing House AG; Migrol Cooperative Society; Mifroma SA; Migros Distribution Center Neuendorf AG; Limmatdruck AG; Migros Enterprises Birsfelden AG; Zurich Shipping Company AG; Storage and Transport Company GmbH; Riseria Taverne SA; Cold Storage Warehouse AG St Margrethen; Neptune

International Holding AG; M-Informatics AG; PROBAG Environmental Technology AG
Financial Information:

	1986 SF'000	1987 SF'000
Sales turnover	10,923,600	11,275,900
Profit before tax	210,400	162,700
Share capital	3,555,351	3,765,000

No of Employees: 61,704

OERLIKON-BÜHRLE HOLDING AG

Hofwiesenstrasse 135, 8021 Zürich
Tel: (01) 363 40 60
Cable: bührleholding zürich
Telex: 59274 obh ch
Telefax: (01) 363 72 60

Corporate Management: Michael Funk (Chief Executive Officer)
Board of Directors: Dr Dietrich Bührle (Zumikon, Chairman), Philippe de Weck (Matran, Vice Chairman), Hortense Anda-Bührle (Zurich), Rudolf Reichling (Stäfa), Professor Edwin Rühli (Auslikon)
Managers: Dr Emil Blättler, Hans Dietz, Dr Herbert Menzi, Alfred M Niederer, Dr Otto Sager, Nik Schliep, Dr Fritz Strohmeier, Dr Ernst Winkler, Dr G J Zinsmeister. Assistant Managers: Dr Beat Baumgartner, Erwin Bucklar, Dr Hans Bunschi, Josef Hackl, Dr Hans-Peter Landerer, Dr Willy Schenkel, Bruno Schmid, Bruno Stalder, Dr Hans-Peter Walker

PRINCIPAL ACTIVITIES: Holding company whose interests include military equipment including ground anti-aircraft systems; weapon systems for vehicles, aircraft and the navy; "Adats" anti-aircraft and anti-tank guided missiles; fire control systems for anti-aircraft and field antillery; satellite assemblies; electronic components; measuring and medical systems; welding equipment; machine tools; textiles including natural and synthetic fibres, yarns and fabrics; footwear and fashion accessories; adhesives, rubber, tannery and chemical products; high vacuum products; thin films; aircraft; real estate; hotels; insurance
Subsidiary Companies: (Majority holdings only): Military Products Division: Machine Tool Works Oerlikon-Bührle Ltd; Oerlikon Singapore Pte Ltd (Singapore); Olkon Handelsgesellschaft für Metallprodukte mbH (Germany); Oerlikon Italiana SIpA (Italy); Montaggi e Collaudi SpA (Italy); Oerlikon Aerospace Inc (Canada). Contraves Division: Contraves AG; Contraves Italiana SpA (Italy); Contraves GmbH (Germany); Contraves Goerz Corporation (USA); Contraves Antriebstechnik AG; Contraves Industrial Products Ltd (UK); Prodotti Industriali Contraves SpA (Italy); Codix KK (Japan); Contraves Advanced Devices (Malaysia); Ingegneria Sistemi Elaborazione (Italy); Contraves Medical Instruments Inc (USA). Machines Division: Press-& Stanzwerk AG; Société Mécanique de Florange (France); Universal Giunti SpA (Italy); Oerlikon Motch Corporation (USA); Advanced Technic Inc (USA). Welding Division: Welding Industries Oerlikon Bührle Ltd; Oerlikon Soudométal BV (Netherlands); Arc Machines Inc (USA); Oerlikon Schweisstechnik GmbH (Germany); Soudométal SA (Belgium); Soudometal UK Ltd (UK); Holding Intercito SA (Panama); Europäische Holding Intercito AG; Fontargen AG; Fontargen GmbH (Germany); Oerlikon Welding Skandinavien A/S (Denmark); Oerlikon-Soudure France Sàrl (France); Soudometal Schweisselektroden GmbH (Germany); Oerlikon Electrodes (South Africa) (Pty) Ltd (South Africa); Fontargen SpA (Italy). Balzers Division: Balzers Limited (Liechtenstein); Balzers (USA); Balzers Optical Corporation (USA); Balzers Hochvakuum GmbH (Germany); Balzers SA (France); Balzers SpA (Italy); Balzers High Vacuum Ltd (UK); Balzers Hochvakuum AG; Nordiska Balzers AB (Sweden); Balzers Hochvakuum GesmbH (Austria); Deutsche Balzers GmbH (Germany); Balzers Union Ltd (Liechtenstein); SA Balzers NV (Belgium); Nihon Balzers KK (Japan); Balzers Metal-Lux SpA (Italy); Arthur Pfeiffer Vakuumtechnik Wetzlar

GmbH (Germany). Bally Division: Bally International Ltd; Bally Gesellschaft mbH (Austria); Bally Schoenhandel BV (Netherlands); Commerce de Chaussures Jacques Biver SECS (Luxembourg); SA Cortume Carioca (Brazil); SA La Tourelle St-François; Bally Schuhfabriken AG; Bally Arola AG; Chaussures Bally-France SA (France); SA Hatté & Henry (France); Bally Group (UK) Limited (UK); Bally's Shoe Co Ltd (UK); Bally Deutschland GmbH (Germany); Bally Inc (USA); Bally Retail Inc (USA); E Braun & Co Gesellschaft mbH (Austria); SA Chaussures Bally Schoenen NV (Belgium); Bally Canada Inc (Canada); Bally do Brasil (Brazil); Société Immobilière Holgina Sàrl (France); Bally Les Accessoires SA; SI Grand Pont No 1; Calzaturificio Savoy SA; Au Chat Botté Sàrl (France); Bally's Shoe Factories (Norwich) Ltd (UK); Bally London Shoe Co Ltd (UK). Textiles Division: Kunz Spinning Mills Ltd; The Dietfurt Spinning and Weaving Co Ltd. Subdivision Aircraft: Pilatus Aircraft Ltd; Pilatus Britten-Norman Ltd (UK); CIBA-Pilatus Aerial Spraying Co Ltd. Subdivision Real Estate: Oerlikon-Bührle Immobilien AG. Subdivision Hotels: Hotel Zürich AG; Hotel Storchen AG; Hotel Airport AG. Subdivision Limmat: Limmat Insurance Company; Limmat Insurance Company for Industry Ltd (Germany); Libera Consulting Actuaries; Rhine Reinsurance Company Ltd. Other Subsidiaries: Oerlikon-Bührle USA Inc (USA); Olkon Corporation (USA); Oerlikon-Bührle Rechenzentrum AG

Principal Bankers: Union Bank of Switzerland; Swiss Credit Bank; Swiss Bank Corp; Industrie & Handelsbank Zürich AG

Financial Information:

	31.12.86 SF'000	31.12.87 SF'000
Sales turnover(1)	4,660,500	4,108,000
(Loss) after tax(1)	(89,800)	(115,200)
(Loss) per share(2)	(58.45SF)	(74.65SF)
(Loss) per share(3)	(23.40SF)	(29.85SF)
(Loss) per share(4)	(23.40SF)	(29.85SF)
Share capital	380,000	380,000
Share capital(5)	24,000	24,000
Shareholders equity	3,450,200	3,252,100

(1) Oerlikon-Bührle Group
(2) Bearer shares
(3) Registered shares
(4) PC shares
(5) Participation Certificate

Principal Shareholders: Bührle Family (over 40%)
No of Employees: 29,915 group

PATRIA SCHWEIZERISCHE LEBENSVERSICHERUNGS GES AG

Patria Sté Mutuelle Suisse d'Assurances sur la Vie

26 St Alban-Anlage, PO Box 3855, CH-4002 Basle
Tel: (061) 55 11 55
Telex: Patr 963 985
Telefax: (061) 22 48 51

Management Board: B Pérusset (Vorsitzender der Geschäftsleitung)
Senior Executives: R Geissmann (Leiter Finanzen), H Weibel (Leiter Einzelversicherung), H Strübin (Leiter Kollectivversicherung)

PRINCIPAL ACTIVITIES: Life and health insurance
Subsidiary Companies: Patria Allgemeine Versicherungs-Ges; Patria Consulta; Hamburger Leben
Financial Information:

	31.12.87 SF'000
Gross premium income	916,000
Investment income	317,000
Profit before tax	6,800
Profit after tax	3,700

PAX SCHWEIZERISCHE LEBENSVERSICHERUNGS GESELLSCHAFT

Aeschenplatz 13, 4002 Basel
Tel: (061) 222000
Telex: 62 922 pax ch

Management Board: Heinz Buser, Anton Fehr, Hansjörg Batz, Dr iur Albert Poldes

PRINCIPAL ACTIVITIES: Life insurance
Financial Information:

	1987 SF'000
Premium income	343,000
Investment income	103,800

RENTENANSTALT

Schweizerische Lebensversicherungs-und Rentenanstalt

Société Suisse d'Assurances Générales Sur La Vie Humaine

Swiss Life Insurance and Pension Company

General Guisan Quai 40, Postfach, 8022 Zürich
Tel: (01) 206 33 11
Cable: rentenanstalt zuerich
Telex: 815 620 ra ch
Telefax: 2012977

Board of Directors: Dr Fritz Honegger (Chairman), Prof Dr Hans Merz (Vice Chairman), Dr G Clottu, Dr Paul Hofmann, Prof Dr Ernst Jaggi, Dr Rudolf Ulrich
Executive Officers: Dr Walter Diener (General Manager), Kurt Rutz (General Manager), Dr W Frauenfelder (General Manager), E Wehrle (General Manager)

PRINCIPAL ACTIVITIES: Life insurance company; life, annuities, group
Subsidiary Companies: Branch Offices in Germany, France, Netherlands, Belgium, UK. Swiss Life Subsidiaries: Swiss Life (España) (Spain); Swiss Life (Luxembourg) SA (Luxembourg); Swiss Life North American Services Inc (SLS) (Ridgewood, USA); Swiss Life North American Services Inc (SLS) (Eden Prairie, MN 55344, USA); Swiss Life North American Services Inc (SLS) (Campbell CA 95008, USA)
Financial Information:

	31.12.86 SF'000	31.12.87 SF'000
Premium income	4,364,000	4,771,000
Investment income	1,530,000	1,624,000

No of Employees: 4,054

SCHWEIZ VERSICHERUNG

Schweiz Insurance

Schweiz Assurance

Gotthardstrasse 43, PO Box 4289, CH-8022 Zürich
Tel: (01) 207 66 66
Cable: navigation
Telex: 815 415 swi ch
Telefax: (01) 201 54 88

Supervisory Board: Dr R Gamper (Präsident), P E Christen (Vizepräsident), Dr H Schönenberger, O H Staubli, Dr M Gherzi, Dr E Luk Keller, Dr F W Meyer, E Schneider
Management Board: Dr Reto Arpagaus, Siegfried W Drews, Dr Robert Feer

PRINCIPAL ACTIVITIES: General insurance company; reinsurance in all classes, including life
Subsidiary Companies: Switzerland General Insurance Co (London) Ltd (UK)

Financial Information:

	1987
	SF'000
Gross premium income	513,000
Investment income	60,400
Profit after tax	8,000

SCHWEIZER REISEKASSE

Neuengasse 15, CH-3011 Bern

Tel: (031) 22 66 33

Directors: Dr H Teuscher, W Bernet (Vice Director), P Luginbühl (Vice Director)

Senior Executives: J Haussener, F Gebel, E Gurther, H R Dähler

PRINCIPAL ACTIVITIES: Promotion of domestic tourism by issuing special holiday money

Financial Information:

	1987
	SF'000
Sales turnover	215,000

No of Employees: 73

SCHWEIZER VERBAND DER RAIFFEISENKASEN

Zentralbank der Raiffeisenkassen

Vadianstrasse 17, 9001 St Gallen

Tel: (071) 21 91 11

Telex: 71 231 rksg ch

Supervisory Board: Dr Gion Clau Vincenz (President), Roger Hügli (Vice President)

Management Board: G Botti, Dr rer.pol.Felix Walker, lic-oec Friedrich C Byland

PRINCIPAL ACTIVITIES: Banking, investment and financial services

Financial Information:

	1987
	SF'000
Profit after tax	6,800
Shareholders funds	162,000
Total assets	6,573,000

No of Employees: 456

SCHWEIZERISCHE MOBILIAR VERSICHERUNGSGESELLSCHAFT

Swiss Mobiliar Insurance Society

Bundesgasse 35, CH-3001 Berne

Tel: (031) 63 61 11

Cable: mobiliar

Telex: 912 732

Telefax: (031) 63 68 52

Board of Directors: W Augsburger (Chairman) with 28 members

Management Board: O Saxer (Managing Director), W Bosshart (General Manager), H Erb, E Fankhauser, A Ferrari, G Schelling, Ch Stalder, K Reif, R Schaer

Senior Executives: B Häusler (Secretary)

PRINCIPAL ACTIVITIES: Direct insurance in Switzerland. Reinsurance world-wide

Trade Names: Swiss Mobilair

Subsidiary Companies: Providentia Swiss Life Insurance Company; Der Anker Insurance Company (Austria)

Principal Bankers: Schweiz Volksbank; Union Bank of Switzerland; Swiss Bank Corp; Credit Suisse and others

Financial Information:

	31.12.86	31.12.87
	SF'000	SF'000
Premium income gross	947,499	1,011,844
Investment income	74,405	84,063
Profit before tax	39,877	44,680
Profit after tax	23,643	25,020
Retained profit	21,743	23,120
Own free capital	415,667	436,182

No of Employees: 2,460

SCHWEIZERISCHE NATIONAL-VERSICHERUNGS-GESELLSCHAFT

Compagnie d'Assurances Nationale Suisse

Compagnia di Assicurazioni Nazionale Svizzera

Swiss National Insurance Company Ltd

Steinengraben 41, PO Box 648, CH-4003 Basle

Tel: (061) 21 01 11

Cable: Universum

Telex: 962 380 snvg ch

Telefax: (061) 21 06 56

Supervisory Board: Dr R Holzach (President), Dr R Baumann (Vice President), Dr R Theler (Managing Director), Dr P Lardy, Dr H P Schär

PRINCIPAL ACTIVITIES: All classes of insurance; including life insurance through subsidiaries

Trade Names: National Versicherung

Subsidiary Companies: (100% owned unless stated): "Schweizerische National" Lebensversicherungs-Gesellschaft; Europäische Reiseversicherungs AG; "Schweizer-National" Versicherungs-Aktiengesellschaft in Deutschland (Germany); Nationale Suisse (France); "Nationale Suisse" Compagnia Italiana di Assicurazioni SpA (Italy); NATIONAL SUISSE VITA Compagnia Italiana di Assicurazioni SpA (Italy); Forsikrings-Aktieselskabet Trekroner (99.4%) (Denmark); Forsikrings-Aktieselskabet Trekroner-Liv (Denmark); Compagnie Européenne d'Assurance des Marchandises et des Bagages (99.1%) (Belgium); La Garantie Nationale SA (99.8%) (Belgium); National Suiza-Orión, Compagñia Española de Seguros y Reaseguros SA (Spain)

Financial Information:

	31.12.86	31.12.87
	SF'000	SF'000
Gross premium income	483,089	512,606
Investment income	52,115	55,639
Profit before tax	13,397	14,186
Profit after tax	9,687	10,410
Retained profit	463	634
Dividend	5,490	5,795
Dividend per share*	180SF	190SF
Earnings per share*	318SF	341SF
Share capital	30,500	30,500
Shareholders funds	140,963	144,134

*1,000 SF nom

No of Employees: 1,922

SIMONIUS, VISCHER & CO

Lindenhofstr 7, 4002 Basel

Tel: (061) 94 43 45

Cable: lanosa

Telex: 966 192 svcw ch

Management Board: Dr R Vischer, W Vorster

PRINCIPAL ACTIVITIES: International wool trade and finance

Principal Bankers: SBV

SWITZERLAND

ST GALLISCHE KANTONALBANK
Banque Cantonale de Saint-Gall
Banca Cantonale di San Gallo
Cantonal Bank of Saint Gall
PO Box 92, St Leonhardstrasse 25, 9001 St Gallen
Tel: (071) 20 31 31
Cable: Kantonalbank
Telex: 881188
Telefax: (071) 20 32 32

Executive Board Committee: Dr J Schönenberger (Chairman),
Dr L Gehringer (Vice Chairman)
General Management: K Lenz (General Manager, Commercial
Division), Josef Baumann (General Manager, Mortgage
Division), Dr H Hurni (General Manager, Administrative
Division)
International Services: A Fey (Deputy General Manager, Bank
Relations), Dr M Müller (Manager International
Loans/Syndic.Loans), E Ammann (Deputy General Manager,
Securities), E Schmid (Assistant Manager, Foreign
Exchange/Money Market), G Keller (Assistant Manager,
Transfer Department)

PRINCIPAL ACTIVITIES: Bank
Financial Information:

| | 31.12.86 | 31.12.87 |
	SF'000	SF'000
Profit after tax	23,054	26,000
Share capital	335,000	452,300
Total assets	9,083,995	9,935,000

Principal Shareholders: Kanton St Gallen
No of Employees: 1,039

SUVA
Schweizerische Unfallversicherungsanstalt
Fluhmattstrasse 1, CH-6002 Luzern
Tel: (041) 21 51 11
Telefax: (041) 21 58 28

Supervisory Board: L Generali, F Leuthy, H Allenspach, A
Baudois, B Boller, A Brunner, C Brunner, J-L Brunner, Dr J
Bucher, T Cantieni, P Dreyer, M Engel, H R Enggist, H
Fischer, G Goumaz, E Grimm, H U Hug, Prof Dr R Husser, C
Jelmini, M Konrad, M Kowalski, H Krähenbühl, K Kugler, M
Kündig, H Lang, Prof D R Lanz, Dr P Leibundgut, M Muller, A
Poretti, Dr G Richterich, W Rindlisbacher, R Roost, Dr W
Rutz, A Salzmann, T Scartazzini, A Scheuber, Dr E Steiner, F
Steinegger, G Tschumi
Management Board: Dr D Galliker, Dr W Seiler, Dr P Wüthrich,
Dr R Zufferey

PRINCIPAL ACTIVITIES: Insurance
Financial Information:

| | 1986 | 1987 |
	SF'000	SF'000
Premium income	1,833,000	1,923,000
Investment income	422,000	432,000

SWISS BANK CORPORATION
Schweizerischer Bankverein
Société de Banque Suisse
Società di Banca Svizzera
Aeschenplatz 6, CH-4002 Basle
Tel: (061) 20 20 20
Cable: Schweizerbank
Telex: 962334 bvbs ch
Telefax: (061) 20 45 76

Chairman: F Galliker; President of the Executive Board: W G
Frehner
General Managers: M Kühne, G Streichenberg, G Blum, R
Blattmann, A Togni, H-C Kessler, F Köhli, U Leber, E
Gautschi, K Steuber, E Balsiger, A Large. Deputy General
Manager: K H Martin
Sector Manager: P G Rogge. Senior Vice Presidents: P
Aebischer, J Auckenthaler, R J Baerlocher, U Bebler, M
Boeglin, W Brupbacher, G Condrau, *H Dietzi, H Ender, U
Fünfschilling, H Gander, Th Gassmann, F Güdel, K Haegi, M
Heer, W Hirter, H Kaufmann, J Kaufmann, F Keiser, H Künzli,
H Kyburg, P A Meier, U P Meier, R S Merten, H Merzweiler,
W Nanzer, M L Ospel, P Poyet, Ch F Puhr, W O Sala, C
Sarasin, Ch Schaub, W D Schlechthaupt, R Schneider, R
Scholer, A Schwietert, H Spörndli, H Steiger, R Stöckli, R M
Villiger, E von Wangenheim. First Vice Presidents: H U
Berger, R Bischofsberger, A Buttschardt, O Eichenberger, J
Gabathuler, S Gilgen, L Gremper, W A Guldimann, E Heri, A E
Huber, F Humm, Ch Im Obersteg, G Kurath, H Lerbscher, U
Leupin, H A Malär, *O Mörikofer, K Müller, J M Nusbaumer, U
Oschsenbein, H Oeri, M Péclard, J L Peyrot, P Pfenniger, H G
Roth, A Schlumberger, H Schöb, M R Staehelin, B Steck, B
Steffen, B Stettler, P J Stillhart, A Stocker, M Strässle, F
Strauch, W Studer, E A Sulser, *Ch Thalmann, E Tschirren, A
Voegelin, D Weichelt, U W Wepf, P Wittmann, A Wyss, *Legal
Advisers

PRINCIPAL ACTIVITIES: Universal Bank
Subsidiary Companies: (100% owned unless stated):
Switzerland: Swiss Deposit and Creditbank (Basel); Banque
Procrédit SA (Fribourg); Bank Finalba ltd (Zürich); Bank
Ehinger & Co Ltd (Basel); Armand Von Ernst & Co Inc (Bern)
(over 50%); Caisse d'Epargne du Valais (Sion) (over 50%);
Spar & Leihkasse Schaffhausen (Schaffhausen) (over 50%).
Subsidiaries Abroad: Schweizerischer Bankverein
(Deutschland) AG (Frankfurt); Frankfurter Münzhandlung
GmbH (Frankfurt); Société de Banque Suisse (Luxembourg)
SA (Luxembourg); Banque de Placements et de Crédit (Monte
Carlo); Swiss Bank Corporation (Jersey) Ltd (Jersey); Swiss
Bank Corporation (Canada) (Toronto); SBC Finance
(Delaware) Inc (Delaware); Swiss Bank Corporation
(Overseas) Ltd (Nassau); Swiss Bank & Trust Corporation Ltd
(Grand Cayman); SBC Finance (Cayman Islands) Ltd (Grand
Cayman); Swiss Bank Corporation (Overseas) SA (Panama);
SBC Australia Ltd (Sydney); SBC Securities Ltd (London and
Zürich); SBCI Swiss Bank Corporation Investment Banking
NV (Amsterdam); SBCI Swissf Bank Corporation Investment
Banking Inc (New York); SBCI Futures Inc (New York); SBCI
Hong Kong Ltd (Hong Kong); Schweizerischer Bankverein
Kapitalanlagegesellschaft mbH (Frankfurt); SBC Portfolio
Management International Ltd (London); SBC Investment
Services Ltd (Dublin); SBC Portfolio Management International
Inc (New York); SBC Portfolio Management International K.K.
(Tokyo); Banque Stern (Paris) (over 50%); Ducatel-Duval
(Paris) (over 50%); SBC Finance (Asia) Ltd (Hong Kong) (over
50%); SBCI Securities (Asia) Ltd (Hong Kong and Tokyo)
(over 50%)
Financial Information:

| | 1987 | 1988 |
	SF'000	SF'000
Profit before tax	922,681	936,665
Profit after tax	652,255	675,120
Retained profit(1)	11,441	13,047
Dividends	442,590	448,514
Dividends per share(2)	13SF	13SF
Earnings per share	19SF	20SF
Share capital	3,599,038	3,645,209
Shareholders funds	8,760,803	9,107,279
Customers deposits(3)	91,925,862	96,919,275
Total assets	146,189,868	154,108,811

(1) balance carried forward
(2) and participation
    certificate
(3) exclusive Banks

No of Employees: 17,477

SWISS LIFE INSURANCE & PENSION COMPANY

Genearl-Guisan-Quai 40, 8022 Zürich

Tel: (01) 206 33 11

Telex: 815 620 ra ch

Management Board: Dr iur W Diener, Dr phil W Frauenfelder, K Rutz

PRINCIPAL ACTIVITIES: Life insurance

Financial Information:

	1987
	SF'000
Premium income	4,771,000
Investment income	1,624,000

SWISS RE
Swiss Reinsurance Company
Schweizerische Rueckversicherungs-Gesellschaft
Compagnie Suisse de Réassurances

Mythenquai 50-60, Postfach, 8022 Zürich

Tel: (01) 208 21 21

Cable: swissre

Telex: 815 722 sre ch

Telefax: (01) 208 29 99

Board of Directors: W Diehl (Chairman)

General Managers: P E Christen, P-F Niquille, H Odermatt, H Schoenenberger, R Gamper. Deputy General Managers: H R Kaufmann, T Obrist, P-C Perrenoud, A W Saxer, W G Seifert

PRINCIPAL ACTIVITIES: All classes of reinsurance

Subsidiary Companies: SwissRe Holding Ltd; European General Reinsurance Company of Zurich; Bavarian Reinsurance Company Ltd (Germany); Swiss Reinsurance Company (UK) Ltd (UK); SwissRe Holding (North America) Inc (USA); North American Reassurance Company (USA); North American Reinsurance Corporation (USA); Swiss-Am Reassurance Company (USA); SwissRe Life Insurance Company (USA); Canadian Reassurance Company (Canada); Canadian Reinsurance Company (Canada); Switzerland General Insurance Company Ltd; SWITZERLAND Insurance Company (London) Ltd (UK); Schweiz CA de Seguros y Reaseguros (Spain); Fidelity and Deposit Company of Maryland (USA); North American Specialty Insurance Company (USA); SR Beteiligungen AG (Germany); Magdeburger Feuerversicherungs-AG (Germany); Magdeburger Lebensversicherungs-AG (Germany); Magdeburger Hagelversicherungs AG (Germany); Magdeburger Rechtsschutzversicherungs-AG (Germany); Vereinte Aachen-Berlinische Versicherung AG (Germany); Vereinte Eos-Isar Lebensversicherung AG (Germany); Vereinte Krankenversicherung AG (Germany); Saar-Rhein Allgemeine Versicherungs-AG (Germany); General Surety Holdings Ltd (UK); General Surety & Guarantee Co Ltd (UK); Palatine Insurance Company Ltd (UK)

Financial Information:

	1987
	SF'000
Group net premium income	10,665,000
Investment income	1,093,000
Group net profit	179,000
Share capital	154,000

Principal Shareholders: Public company with over 15,000 registered shareholders-no single shareholder owns more than 2%

No of Employees: 1,075 (Company); 13,792 (Group)

SWISS VOLKSBANK
Banque Populaire Suisse
Schweizerische Volksbank
Banca Popolare Svizzera

Weltpoststrasse 5, 3015 Bern

Tel: 32 81 11

Cable: Populdege

Telex: 912 486 svb ch

Telefax: 32 79 14

Board of Directors: Dr Rainer Weibel (Chairman), B Schneider (Vice Chairman), G. Cotti, T D Berg, Dr M Bohren-Hoerni, B Boller, R A Cornaz, F Loeb, C de Mercurio, W von Moos, M A Naville, W Reist, Dr I Rezzinico, Dr A W Roth, F Ruesch, Dr O Saxer, G Schilplin, A Stefani, Ch Vogele, Dr P Wyss

Executive Board: W Rüegg (President and Chief Executive), F Pedrazzoli (Executive Vice-President), R Beeler (Executive Vice-President), R Lienert (Executive Vice-President), R Bosshard (Executive Vice-President), A Keller (Executive Vice-President)

Senior Executives: Dr M Höhener (Secretary General), Hans-Rudolf Renfer. Senior Vice-Presidents Head Office: R Achermann, G Barone, H Bieri, B Bohlhalter, H Christen, P Dubler, M Fisler, W Geering, H R Küchler, C Meier, B Merki, W Meyer, G Perego, E Sigrist, Dr H Stocker, H Widmer

PRINCIPAL ACTIVITIES: All major banking activities

Trade Names: Swiss Volksbank

Subsidiary Companies: ABZ-Finanz-und Beteiligungsgesellschaft; Columna-Service AG; AG für Immobilien-Anlagefonds Devo; Devo Verwaltungs AG; Immobilien-Gesellschaft Glarus; Invoca AG; KAFAG AG für die Verwaltung von Anlagefonds; Swiss Volksbank Finance (Cayman Islands) Ltd

Financial Information:

	31.12.86	31.12.87
	SF'000	SF'000
Profit before tax	156,885	151,555
Profit after tax	116,351	115,546
Retained profit	27,687	26,815
Dividends	88,664	88,731
Dividends per share	75SF	75SF
Earnings per share	98.40SF	97.65SF
Share capital	716,500	716,500
Shareholders funds	1,832,735	1,861,219
Deposits	26,423,268	29,630,132
Total assets	28,910,042	32,195,416

No of Employees: 5,691

TELEKURS AG

Neue Hard 11, 8005 Zürich

Tel: (01) 275 21 11

Telex: 56892 tele ch

Telefax: (01) 271 80 10

Supervisory Board: Dr H von der Crone (Präsident), R Beeler, J P Chapius, D Forcart, E R Gautschi, Dr M Gsell, D Marti, E Ruprecht, Dr J Treyvaud, G Urban, Dr O Kramer, E Balsiger, Dr H Huschke, Dr G P Rossetti, Dr C Vital, E Fritschi

Management Board: Dr M Rüegg, M Nicollier, K Santelli, Dr W Zingg, J Bohren, M Storz

PRINCIPAL ACTIVITIES: Financial services including administration of estates and dealing in securities, EDP-services

Trade Names: Investdata-System, INDES, Valordata-System, Forex Sport, Swiss Index, SIC, eurocheque, EUROCARD (Switzerland), RIS

Subsidiary Companies: Telekurs (Deutschland) GmbH; Telekurs (UK) Ltd; Telekurs (France) SARL; Telekurs (Nederland) BV; Telekurs (North America) Inc; II EUROCARD (Switzerland) SA

Financial Information:

	31.12.86 SF'000	31.12.87 SF'000
Turnover	201,000	264,000
Retained profit	600,000	807,000
Share capital	15,000	15,000
Shareholders funds	15,000	15,000

Principal Shareholders: Swiss banks and stock exchanges
No of Employees: 1,334

THURGAUER KANTONALBANK

Banque Kantonale de Thurgovie

Bankplatz 1, 8570 Weinfelden
Tel: (072) 21 61 11
Cable: Kantonalbank
Telex: 882 324 tkbw ch; 882 323 tkb ch (Foreign Exchange)
Telefax: (072) 21 63 63

President: H Moll
Managers: L Bühler, H Michel

PRINCIPAL ACTIVITIES: Bank
Financial Information:

	31.12.86 SF'000	31.12.87 SF'000
Profit before tax	20,038	21,115
Profit after tax	18,024	18,763
Retained profit	1,621	1,678
Dividends	16,403	17,085
Share capital	400,000	400,000
Reserves	48,216	49,563
Deposits	4,094,689	4,399,149
Total assets	7,302,000	7,608,000

No of Employees: 621

UNION BANK OF SWITZERLAND

Union de Banques Suisses

Schweizerische Bankgesellschaft

Unione di Banche Svizzere

45 Bahnhofstrasse, 8021 Zurich
Tel: (01) 234 11 11
Cable: Bankunion
Telex: 813 811; 822 802
Telefax: (1) 236 51 11

Supervisory Board: Dr Robert Holzach (Honorary Chairman), *Dr Nikolaus Senn (Chairman), *Hans Rüegg (Vice Chairman), *Dr Gustav Tobler (Vice Chairman), Ulrich Ammann, Henri André, Dr Christoph Blocher, *Jean Carbonnier, Marc C Cappis, Charles R Firmenich, Dr H Goetz, Dr Hans K Jucker, Dr Martin Junger, Bruno de Kalbermatten, *Markus Kündig, Dr Giovanni Lombardi, *François Milliet, Dr Bernd H Müller-Bergoff, A P Reinhart, *Dr Hans-Peter Schär, *Dr Stephan Schmidheiny, Dr Hans-Peter Sigg, Dr Peter Spälti, Peter G Sulzer, *Members of the Board's Committee
Executive Board: Robert Studer (President). Executive Vice Presidents: Mathis Cabiallavetta, Dr Ulrich Grete, Robert Favarger, Stephan Haeringer, Guido Hanselmann, Hans Heckmann, Dr Hubert Huschke, Karl Janjöri, Heinz Müller, Urs Rinderknecht, Heinrich Steinmann, Dr Robert Sutz

PRINCIPAL ACTIVITIES: International bank
Subsidiary Companies: (100% owned unless stated): In Switzerland: Banks:-AKO Bank; Orca Bank Ltd; Bank Cantrade Ltd (85%); Saudi-Swiss Bank (51%); Bank Aufina; Swiss Mortgage and Commercial Bank (99.5%); Bank Rohner Ltd (80.3%); Mortgage and Commercial Bank Winterthur (27.6%); Banco di Roma per la Svizzera, Lugano (75%). Investment Companies:-Aufina Leasing + Factoring AG; "Banque Fédérale" Société de Particip. et Financ. (99.9%); Noreco Finance Corp (58%); Motor-Columbus Ltd (41.7%); UBS Finanzholding AG; Crédit Industriel SA; "Thesaurus" Continental Securities Corporation (99.9%); AKAG Anlage-

und Kapital AG (99%); Cantrade Participation Ltd, Zürich (85%). Other Participations:- Intrag Ltd (75%); Argor-Heraeus SA (75%); Klinik Hirslanden AG (79%); Kur-Zentrum Rheinfelden AG (65%); REVISUISSE Swiss Auditing Company (49.4%); Premex AG (33.3%); IFA Institute for Automation Ltd (95%); Swiss National Insurance Company Ltd (25.1%); Union Reinsurance Company (60.7%); Commercial Services Corporation-Coseco (76%); Casino-Kursaal de Montreux SA (64.4%); Losinger Ltd (21.3%); Tourism Finance Ltd (25%). Abroad: Banks and Investment Companies:- Union Bank of Switzerland (Luxembourg) SA (Luxembourg); Schweiz Bankgesellschaft (Deutschland) AG (Germany); Union Bank of Switzerland (Canada); European Brazilian Bank Ltd (Eurobraz) (13.7%) (UK); UBS (UK) Ltd (UK); UBS Philips & Drew International Ltd (50%) (UK); UBS Australia Ltd (Australia); Union Bank of Switzerland (Trust and Banking) Ltd (Japan); Banque Sudameris (12%) (France); P T Aseam Indonesia (9.8%) (Indonesia); Union Bank of Switzerland (Panama) Inc (Panama); Banque Internat. pour l'Afrique Occidentale (BIAO) (15%) (France); Intermex Holding SA (12%) (Luxembourg); Société Financière Européenne (SFE) (11.1%) (Luxembourg); Saudi International Bank Ltd (5%) (London, New York and Nassau); BPI-Banco Portugês de Investimento S.A.R.L. (3.4%) (Portugal); Union Bank of Switzerland (Securities) Ltd (UK); Union Bank of Switzerland (Underwriters) Ltd; Euro-Latinamerican Bank Ltd (Eulabank) (4.8%) (UK); Aseambankers Malaysia Bhd (5%) (Malaysia); UBS Securities Inc (USA); UBS Finance (Delaware) Inc (USA); UBS Finance NV (Netherlands Antilles); BAII Holdings (3.5%) (Luxembourg); Philips & Drew Ltd (UK); Korea Long Term Credit Bank (KLB) (2.4%) (South Korea)
Financial Information:

	31.12.86 SF'000	31.12.87 SF'000
Profit before tax	1,135,760	1,087,489
Profit after tax	776,228	752,792
Dividends	466,080(1)	505,200
Dividends per bearer share	120SF(2)	120SF
Dividends per registered share	24SF(3)	24SF
Dividends per PC share(6)	4.80SF(4)	4.80SF
Earnings per share	24%(5)	24%
Share capital	1,900,000	2,175,000
Shareholders funds	8,688,642	9,746,271
Customer deposits	82,026,441	91,884,151
Total assets	152,167,447	160,415,886

(1) plus anniversary bonus SF155,360,000
(2) plus anniversary bonus 40SF
(3) plus anniversary bonus 8SF
(4) plus anniversary bonus 1.60SF
(5) plus anniversary bonus of 8%
(6) participation certificate

Principal Shareholders: UBS is a publicly owned company and the shares are widely distributed
No of Employees: 20,881

UNION REINSURANCE COMPANY

Genferstrasse 27, 8027 Zürich
Tel: (01) 209 91 11
Telex: 53 344 reun ch
Telefax: (01) 201 18 95

PRINCIPAL ACTIVITIES: All classes of insurance
Financial Information:

	1987 SF'000
Premium income	475,000
Investment income	57,000
Profit after tax	6,500

VERITAS REINSURANCE CORPORATION LTD

Zugerbergstrasse 4, 6301 Zug
Tel: (042) 23 11 71
Telex: 865341 veri ch
Telefax: (042) 21 96 88

Supervisory Board: Dr Hans-Joachim Scherzberg (Präsident), Gerhard Ohligschläger (Delegierter), Dr Burkhard Gantenbein, Dr Andreas Renggli, Hans Rosenberger, Dr Richard Wiedemann, Max Zaugg
Management Board: Herbert Künzler (Direktor), Hans Petermann (stellv Direktor), Beat Landtwing (Vizedirektor)

PRINCIPAL ACTIVITIES: Reinsurance services
Parent Company: Vereinigte Haftpflicht Versicherung V.a.G., Hannover, Germany
Subsidiary Companies: Veritas Holdings Ltd (USA); Pilatus Re Ltd (Bermuda)
Financial Information:

	1986	1987
	SF'000	SF'000
Premium income	383,775	440,655
Profit after tax	2,655	2,870
Share capital	237,202	275,296

VITA LEBENSVERSICHERUNGS-GESELLSCHAFT

Mythenquai 10, 8022 Zürich
Tel: 01/465 65 65
Cable: VITA ZURICH
Telex: 813596 VITA CH

Management Board: Dr A Wenger (Generaldirektor), M Bardola, J-D Lavanchy, P Schenker, H Bänziger, K Werlen, G Humair, R Peyer

PRINCIPAL ACTIVITIES: Life assurance
Parent Company: "Zürich" Versicherungs-Gesellschaft, Zürich
Subsidiary Companies: Branch offices in Germany, Netherlands, France, Belgium, Italy, Spain, Denmark, Luxembourg, Hong Kong
Principal Bankers: Schweiz. Kreditanstalt
Financial Information:

	1987
	SF'000
Premium income	2,213,000
Investment income	725,100

No of Employees: 1,800

"WINTERTHUR" LEBENSVERSICHERUNGS-GESELLSCHAFT

Roemerstrasse 17, 8401 Winterthur
Tel: (052) 85 11 11
Cable: Winterthurleben winterthur
Telex: 896120 wile ch
Telefax: (052) 85 35 85

Board of Directors: Dr H Braunschweiler (Chairman), Dr P Spälti (Managing Director), Dr H Kundert (Vice Chairman), P Borgeaud, Dr M-H Chaudet, Prof. Dr M Hilti, G Kaiser, Dr F Leutwiler, Prof. Dr Verena Meyer, Dr L von Planta, J Revaclier, Dr N Senn, R Staubli
Management: Dr H R Studer (General Manager), Dr M Köhler (Manager), Dr M Clerckx (Manager), Dr F Kuenzler (Manager), A Schneiter (Manager)

PRINCIPAL ACTIVITIES: Life insurance
Parent Company: "Winterthur" Schweizerische Versicherungsgesellschaft (100%)
Subsidiary Companies: Branches in Belgium, Germany, Denmark, France, Netherlands, Canada, Spain and Italy
Principal Bankers: Union Bank of Switzerland

Financial Information:

	1986	1987
	SF'000	SF'000
Premium income	2,420,265	2,864,388
Investment income	731,736	825,258
Profit before tax	26,430	31,714
Profit after tax	10,977	11,035
Retained profit	1,904	2,339
Dividends	5,000	5,000
Dividends per share	100SF	100SF
Earnings per share	220SF	221SF
Share capital	50,000	50,000
Shareholders funds	95,904	97,339

Principal Shareholders: "Winterthur" Swiss Insurance Company
No of Employees: 1,400

"WINTERTHUR" SCHWEIZERISCHE VERSICHERUNGS GESELLSCHAFT
Société Suisse d'Assurances
Swiss Insurance Company

General-Guisan-Strasse 40, 8401 Winterthur
Tel: (052) 85 11 11
Cable: Wintassur Winterthur
Telex: 896232 widi ch

Board of Directors: *Dr Hans Braunschweiler (Chairman), *Dr Heinz Kundert (Vice Chairman), Dr Peter Spälti (Managing Director), Pierre Borgeaud, Dr Marc-Henri Chaudet, Giuseppe Kaiser, Dr Fritz Leutwiler, Prof Dr Verena Meyer, Dr Louis von Planta, Jean Revaclier, *Dr Nikolaus Senn, Robert Staubli, *Members of the Executive Committee
Management: Dr Peter Spalti (Managing Director). General Managers: Gerhard Christen, Dr H J Vollenweider, Dr Peter Portmann, Dr Rudolf Staub, Dr Hans-Rudolf Studer, Jacques Schupp, Dr Hans-Rudolf Stucki
Senior Executives: Dr Jürg Spiller (Corporate Secretary)

PRINCIPAL ACTIVITIES: Insurance
Subsidiary Companies: Winterthur-Life; The Federal; Winterthur-Rechtsschutz; Vitodurum Versicherungs-Gesellschaft; Vitodurum Holding, Neuchâteloise Générale; Neuchâteloise Vie; Winterthur-Garantie (Germany); Winterthur (Austria); Europeia (Portugal); Union et Prévoyance (Belgium); Winterthur-Assicurazioni (Italy); Intercontinentale (Italy); Veneta Assicurazioni (Italy); Veneta Vita (Italy); SAPA (Italy); Winterthur (UK) Holdings (UK); Provident Life (UK); United Standard (UK); Winterthur Insurance (UK) (UK); Winterthur U.S. Holdings (USA); Republic Financial Services (USA); Southern Guaranty (USA); Winterthur-Canada Financial (Canada); Citadel General (Canada); Citadel Life (Canada); Commonwealth Insurance (Cayman Islands); Winterthur-International (Bermudas); Winterthur-Participaçoes (Brazil); Itau-Winterthur Seguradora (Brazil)
Financial Information: Group figures

	31.12.86	31.12.87
	SF'000	SF'000
Gross premium income	7,655,800	8,139,900
Investment income	1,365,900	1,483,100
Profit before tax	298,800	320,900
Profit after tax	172,200	191,400
Retained profit	8,400	9,800
Dividends	78,081	78,631
Dividends per share	60SF	60SF
Share capital	135,200	135,200
Shareholders equity	2,413,300	2,404,200

Principal Shareholders: Public
No of Employees: 16,000

SWITZERLAND

ZÜRCHER KANTONALBANK
Bahnhofstrasse 9, Postfach, CH-8022 Zürich
Tel: (01)220 11 11
Cable: kantonalbank
Telex: 812 140 ZKB CH; 813 688 ZKB CH (Foreign Exchange)
Telefax: (01) 211 65 39

Board of Directors: B Schürch (Chairman), E Spillmann (Vice-Chairman), Dr jur H Weigold (Vice-Chairman), K Domeisen, E Frauenfelder, Dr R Gerster, H Hauser, Dr R Hux, P Keller, A Killias, O Ronner, E Rüfenacht, K Schärer, I Plüss (Secretary)
General Management: W Lüthy (General Manager, Commercial Division), Dr R Müller (General Manager, Administrative Services Division), E Hinnen (General Manager, Mortgage Division)
Senior Executives: W Albrecht, W Fluck, E Fritschi, T Guler, P Hasenfratz, R Müller, U Naef, K Stäubli

PRINCIPAL ACTIVITIES: Bank
Financial Information:

	31.12.86 SF'000	31.12.87 SF'000
Profit before tax	72,680	76,308
Profit after tax	72,680	76,308
Equity capital	850,000	925,000
Reserves	270,225	288,225
Deposits	25,432,949	29,067,266
Total assets	30,357,770	33,499,599

No of Employees: 3,653

ZÜRICH INSURANCE COMPANY
Mythenquai 2, 8022 Zürich
Tel: (01) 205 21 21
Cable: Assurich zuerich
Telex: 815555 zur ch
Telefax: (01) 201 33 97

Board of Directors: F Gerber (Chairman), Peter Hefti (Vice Chairman), Henry C M Bodmer, Peter Böckli, Bruno de Kalbermatten, Markus Kündig, Yves Oltramare, Kaspar Cassani, Vreni Spoerry-Toneatti, Helmut Maucher
General Management: Rolf F Hüppi (Chief General Manager), Rolf Hänggi (Deputy Chief General Manager), Hans-Peter Karlen, Heinrich Wirth, Charles Wyniger, Rolf Schäuble, Don Menzies

PRINCIPAL ACTIVITIES: Insurance including reinsurance and life insurance
Subsidiary/Associated Companies: Switzerland: Vita; Alpina; Turegum; Altstadt. Germany: Agrippina Versicherung; Agrippina Rück; Agrippina Leben; Patria; Agrippina Rechtsschutz; Deutsche Allgemeine; Zürich Kaution; Zürich Rechtsschutz. Austria: Zürich Kosmos. UK: Zürich Life; Bedford General. Spain: Hispania; Caudal. Italy: Danubio; Minerva; Minerva Vita; Sicurtà 1879; SIAR. France: Abri. Portugal: Metrópole. USA: American Guarantee; American Zurich; Zurich American of Illinois; Zurich American Life; Empire Fire and Marine; Empire Indemnity; Universal Underwriters Ins; Universal Underwriters Life; Zurich Reinsurance; Fidelity and Deposit (50%). Bermuda: Zurich International. Canada: Zurich Life of Canada. Brazil: Zurich-Anglo Seguaradora SA (50%). Australia: Zurich Australian Insurance; Zurich Australian Life; National & General. Morocco: Garantie Générale Marocaine (50%). Kenya: African International Insurances (50%)

Financial Information:

	31.12.87 SF'000
Group premium income	12,207,000
Group investment income	2,042,400
Premium income (1)	5,549,900
Investment income (1)	718,800
Profit after tax (1)	156,400
Dividends (1)	104,095
Dividends per share (1)	56SF
Dividends per share (1)(2)	28SF
Share capital (1)	150,000
Non-voting share capital(1)	55,000
Shareholders equity (1)	2,660,300
(1) Parent company	
(2) non-voting shares	

No of Employees: 25,160 (Group's Permanent Staff)

Alphabetical Index

INDEX: Alphabetical

INDEX: Alphabetical

Index by Country

GERMANY

INDEX: By Country